Lecture Notes in Computer Science   8404

Commenced Publication in 1973
Founding and Former Series Editors:
Gerhard Goos, Juris Hartmanis, and Jan van Leeuwen

Alexander Gelbukh (Ed.)

Computational Linguistics and Intelligent Text Processing

15th International Conference, CICLing 2014
Kathmandu, Nepal, April 6-12, 2014
Proceedings, Part II

 Springer

Volume Editor

Alexander Gelbukh
National Polytechnic Institute
Center for Computing Research
Av. Juan Dios Bátiz, Col. Nueva Industrial Vallejo
07738 Mexico D.F., Mexico
E-mail: gelbukh@gelbukh.com

ISSN 0302-9743
ISBN 978-3-642-54902-1
DOI 10.1007/978-3-642-54903-8
Springer Heidelberg New York Dordrecht London

e-ISSN 1611-3349
e-ISBN 978-3-642-54903-8

Library of Congress Control Number: 2014934305

LNCS Sublibrary: SL 1 – Theoretical Computer Science and General Issues

Typesetting: Camera-ready by author, data conversion by Scientific Publishing Services, Chennai, India

Printed on acid-free paper

Springer is part of Springer Science+Business Media (www.springer.com)

Preface

CICLing 2014 was the 15th annual Conference on Intelligent Text Processing and Computational Linguistics. The CICLing conferences provide a wide-scope forum for discussion of the art and craft of natural language processing research as well as the best practices in its applications.

This set of two books contains four invited papers and a selection of regular papers accepted for presentation at the conference. Since 2001, the proceedings of the CICLing conferences have been published in Springer's *Lecture Notes in Computer Science* series as volume numbers 2004, 2276, 2588, 2945, 3406, 3878, 4394, 4919, 5449, 6008, 6608, 6609, 7181, 7182, 7816, and 7817.

The set has been structured into 17 sections, representative of the current trends in research and applications of natural language processing:

- Lexical Resources
- Document Representation
- Morphology, POS-tagging, and Named Entity Recognition
- Syntax and Parsing
- Anaphora Resolution
- Recognizing Textual Entailment
- Semantics and Discourse
- Natural Language Generation
- Sentiment Analysis and Emotion Recognition
- Opinion Mining and Social Networks
- Machine Translation and Multilingualism
- Information Retrieval
- Text Classification and Clustering
- Plagiarism Detection
- Style and Spelling Checking
- Speech Processing
- Applications

The 2014 event received submissions from 57 countries, a record high number in the 15-year history of the CICLing series. Exactly 300 papers (third highest number in the history of CICLing) by 639 authors were submitted for evaluation by the international Program Committee (see Figure 1 and Tables 1 and 2). This two-volume set contains revised versions of 85 regular papers selected for presentation; thus the acceptance rate for this set was 28.3%.

In addition to regular papers, the books feature invited papers by:

- Jerry Hobbs, ISI, USA
- Bing Liu, University of Illinois, USA
- Suresh Manandhar, University of York, UK
- Johanna D. Moore, University of Edinburgh, UK

Table 1. Number of submissions and accepted papers by topic[1]

Accepted	Submitted	% Accepted	Topic
19	45	42	Semantics, pragmatics, discourse
14	43	33	Lexical resources
12	31	39	Machine translation and multilingualism
12	33	36	Practical applications
12	35	34	Emotions, sentiment analysis, opinion mining
12	40	30	Clustering and categorization
12	56	21	Text mining
11	48	23	Information retrieval
10	29	34	Underresourced languages
8	26	31	Syntax and chunking
7	44	16	Information extraction
6	18	33	Social networks and microblogging
5	16	31	Natural language generation
4	11	36	Noisy text processing and cleaning
4	16	25	Summarization
3	4	75	Spelling and grammar checking
3	9	33	Plagiarism detection
3	12	25	Word sense disambiguation
3	16	19	POS tagging
2	5	40	Coreference resolution
2	7	29	Computational terminology
2	7	29	Other
2	9	22	Textual entailment
2	13	15	Formalisms and knowledge representation
2	17	12	Named entity recognition
2	20	10	Morphology
1	6	17	Speech processing
1	10	10	Natural language interfaces
1	11	9	Question answering
0	3	0	Computational humor

[1] As indicated by the authors. A paper may belong to several topics.

These speakers presented excellent keynote lectures at the conference. Publication of full-text invited papers in the proceedings is a distinctive feature of the CICLing conferences. Furthermore, in addition to presentation of their invited papers, the keynote speakers organized separate vivid informal events; this is also a distinctive feature of this conference series. In addition, Professor Jens Allwood of the University of Gothenburg was a special guest of the conference.

With this event we continued with our policy of giving preference to papers with verifiable and reproducible results. In addition to the verbal description of their findings given in the paper, we encouraged the authors to provide a proof of their claims in electronic form. If the paper claimed experimental results, we asked the authors to make available to the community all the input data necessary to verify and reproduce these results: if it claimed to introduce

Table 2. Number of submitted and accepted papers by country or region

Country or region	Authors Subm.	Papers[2] Subm.	Accp.	Country or region	Authors Subm.	Papers[2] Subm.	Accp.
Afghanistan	1	1	–	Japan	22	8.33	3
Algeria	2	0.67	–	Jordan	12	3.33	–
Australia	8	3	1	Kazakhstan	6	1.67	1.67
Bangladesh	9	3	–	Korea (South)	12	3.5	0.50
Belgium	3	2	–	Latvia	6	2	1
Brazil	18	6.17	2.17	Malaysia	4	1.67	–
Bulgaria	1	1	–	Mexico	19	12.42	2.67
Canada	13	7	4	Mongolia	1	0.5	0.5
China	57	21.1	7.35	Morocco	5	3	–
Christmas Isl.	1	0.2	0.2	Nepal	12	6	2
Colombia	3	1	1	Norway	1	0.2	–
Croatia	1	0.33	0.33	Pakistan	4	1.83	–
Czech Rep.	20	11.4	3	Poland	2	2	–
Denmark	3	0.38	–	Portugal	5	2.5	1
Egypt	12	7	1	Romania	10	5.67	–
Ethiopia	5	4	2	Russia	9	5.17	–
Finland	5	2	2	Singapore	9	2.78	1.78
France	29	12.42	9.67	Slovenia	2	0.67	0.67
Germany	19	7.33	4.33	Spain	13	3.7	0.67
Greece	1	0.33	0.33	Sweden	5	4	1
Hong Kong	4	2	1	Switzerland	6	5	2
Hungary	3	1	–	Taiwan	5	1	–
India	136	75.1	10.33	Thailand	2	1	–
Indonesia	3	1	–	Tunisia	20	8.83	1.83
Iran	4	2	–	Turkey	3	2.83	1.5
Iraq	0	1	–	UK	10	3.83	3.33
Ireland	0	0.5	–	USA	48	21.48	7.17
Israel	14	7	2	Vietnam	5	1.67	–
Italy	6	2.5	1	*Total:*	639	300	85

[2] By the number of authors: e.g., a paper by two authors from the USA and one from UK is counted as 0.67 for the USA and 0.33 for UK.

an algorithm, we encouraged the authors to make the algorithm itself, in a programming language, available to the public. This additional electronic material will be permanently stored on the CICLing's server, www.CICLing.org, and will be available to the readers of the corresponding paper for download under a license that permits its free use for research purposes.

In the long run, we expect that computational linguistics will have verifiability and clarity standards similar to those of mathematics: In mathematics, each claim is accompanied by a complete and verifiable proof (usually much longer than the claim itself); each theorem's complete and precise proof—and not just a vague description of its general idea—is made available to the reader. Electronic

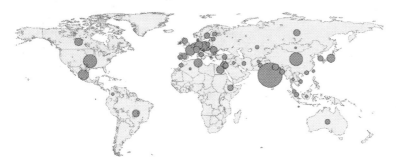

Fig. 1. Submissions by country or region. The area of a circle represents the number of submitted papers.

media allow computational linguists to provide material analogous to the proofs and formulas in mathematic in full length—which can amount to megabytes or gigabytes of data—separately from a 12-page description published in the book. More information can be found on www.CICLing.org/why_verify.htm.

To encourage providing algorithms and data along with the published papers, we selected a winner of our Verifiability, Reproducibility, and Working Description Award. The main factors in choosing the awarded submission were technical correctness and completeness, readability of the code and documentation, simplicity of installation and use, and exact correspondence to the claims of the paper. Unnecessary sophistication of the user interface was discouraged; novelty and usefulness of the results were not evaluated—instead, they were evaluated for the paper itself and not for the data.

The following papers received the Best Paper Awards, the Best Student Paper Award,[1] as well as the Verifiability, Reproducibility, and Working Description Award, respectively:

1st Place:	"A graph-based automatic plagiarism detection technique to handle artificial word reordering and paraphrasing", by Niraj Kumar, India
2nd Place:	"Dealing with function words in unsupervised dependency parsing," by David Mareček, Zdeněk Žabokrtský, Czech Republic
3rd Place:	"Extended CFG formalism for grammar checker and parser development," by Daiga Deksne, Inguna Skadiņa, Raivis Skadiņš, Latvia
and	"How preprocessing affects unsupervised keyphrase extraction," by Rui Wang, Wei Liu, Chris McDonald, Australia
Student:	"Iterative bilingual lexicon extraction from comparable corpora with topical and contextual knowledge," by Chenhui Chu, Toshiaki Nakazawa, Sadao Kurohashi, Japan

---

[1] The best student paper was selected among papers of which the first author was a full-time student, excluding the papers that received a Best Paper Award.

Verifiability: "How document properties affect document relatedness measures," by Jessica Perrie, Aminul Islam, Evangelos Milios, Canada

The authors of the awarded papers (except for the Verifiability award) were given extended time for their presentations. In addition, the Best Presentation Award and the Best Poster Award winners were selected by a ballot among the attendees of the conference.

Besides its high scientific level, one of the success factors of CICLing conferences is their excellent cultural program. The attendees of the conference had a chance to visit the wonderful historical and cultural attractions of the lesser-known country Nepal—the birthplace of the Buddha and the place where pagodas were invented before their spread to China and Japan to become an iconic image of East Asia. Of the world's ten highest mountains, eight are in Nepal, including the highest one Everest; the participants had a chance to see Everest during a tour of the Himalayas on a small airplane. They also attended the Seto MachindraNath Chariot festival and visited three historical Durbar squares of the Kathmandu valley, a UNESCO world cultural heritage site. But probably the best of Nepal, after the Himalayas, are its buddhist and hindu temples and monasteries, of which the participants visited quite a few. Even the Organizing Committee secretary and author of one of the best evaluated papers published in this set was the hereditary Supreme Priest of an ancient Buddhist temple!

I would like to thank all those involved in the organization of this conference. Firstly, the authors of the papers that constitute this book: it is the excellence of their research work that gives value to the book and sense to the work of all the rest. I thank all those who served on the Program Committee, Software Reviewing Committee, Award Selection Committee, as well as the additional reviewers, for their hard and very professional work. Special thanks go to Pushpak Bhattacharyya, Samhaa El-Beltagy, Aminul Islam, Cerstin Mahlow, Dunja Mladenic, Constantin Orasan, and Grigori Sidorov for their invaluable support in the reviewing process.

I would like to thank the conference staff, volunteers, and the members of the local Organizing Committee headed by Professor Madhav Prasad Pokharel and advised by Professor Jai Raj Awasthi. In particular, I am very grateful to Mr. Sagun Dhakhwa, the secretary of the Organizing Committee, for his great effort in planning all the aspects of the conference. I want to thank Ms. Sahara Mishra for administrative support and Mr. Sushan Shrestha for the website design and technical support. I am deeply grateful to the administration of the Centre for Communication and Development Studies (CECODES) for their helpful support, warm hospitality, and in general for providing this wonderful opportunity to hold CICLing in Nepal. I acknowledge support from the project CONACYT Mexico—DST India 122030 "Answer Validation through Textual Entailment" and SIP-IPN grant 20144534.

The entire submission and reviewing process was supported for free by the EasyChair system (www.EasyChair.org). Last but not least, I deeply appreciate the patience and help of Springer staff in editing these volumes and getting them printed in very short time—it is always a great pleasure to work with Springer.

February 2014   Alexander Gelbukh

Organization

CICLing 2014 was hosted by the Centre for Communication and Development Studies (CECODES), Nepal, and was organized by the CICLing 2014 Organizing Committee in conjunction with the CECODES, the Natural Language and Text Processing Laboratory of the CIC (Centro de Investigación en Computación) of the IPN (Instituto Politécnico Nacional), Mexico, and the Mexican Society of Artificial Intelligence (SMIA).

Organizing Chair

Madhav Prasad Pokharel

Organizing Committee

Madhav Prasad Pokharel (Chair)	Tribhuban University, Kathmandu
Jai Raj Awasthi (Advisor)	CECODES, Lalitpur
Sagun Dhakhwa (Secretary)	CECODES, Lalitpur
Bhim Narayan Regmi	CECODES, Lalitpur
Krishna Prasad Parajuli	CECODES, Lalitpur
Sandeep Khatri	CECODES, Lalitpur
Kamal Poudel	CECODES, Lalitpur
Bhim Lal Gautam	CECODES, Lalitpur
Krishna Prasad Chalise	CECODES, Lalitpur
Dipesh Joshi	CECODES, Lalitpur
Prajol Shrestha	NLP Engineer, Vision Objects

Program Chair

Alexander Gelbukh

Program Committee

Ajith Abraham
Rania Al-Sabbagh
Sophia Ananiadou
Marianna Apidianaki
Alexandra Balahur
Kalika Bali

Leslie Barrett
Roberto Basili
Pushpak Bhattacharyya
Nicoletta Calzolari
Nick Campbell
Sandra Carberry

Michael Carl
Hsin-Hsi Chen
Dan Cristea
Bruce Croft
Mike Dillinger
Samhaa El-Beltagy
Tomaž Erjavec
Anna Feldman
Alexander Gelbukh
Dafydd Gibbon
Gregory Grefenstette
Eva Hajicova
Sanda Harabagiu
Yasunari Harada
Karin Harbusch
Ales Horak
Veronique Hoste
Nancy Ide
Diana Inkpen
Hitoshi Isahara
Aminul Islam
Guillaume Jacquet
Sylvain Kahane
Alma Kharrat
Adam Kilgarriff
Valia Kordoni
Leila Kosseim
Mathieu Lafourcade
Krister Lindén
Bing Liu
Elena Lloret
Bernardo Magnini
Cerstin Mahlow
Suresh Manandhar
Diana Mccarthy
Alexander Mehler
Rada Mihalcea
Evangelos Milios
Dunja Mladenic
Marie-Francine Moens
Masaki Murata
Preslav Nakov
Costanza Navarretta
Roberto Navigli
Vincent Ng

Joakim Nivre
Attila Novák
Kjetil Nørvåg
Kemal Oflazer
Constantin Orasan
Ekaterina Ovchinnikova
Ivandre Paraboni
Saint-Dizier Patrick
Maria Teresa Pazienza
Ted Pedersen
Viktor Pekar
Anselmo Peñas
Octavian Popescu
Marta R. Costa-Jussà
German Rigau
Horacio Rodriguez
Paolo Rosso
Vasile Rus
Kepa Sarasola
Roser Sauri
Hassan Sawaf
Satoshi Sekine
Serge Sharoff
Grigori Sidorov
Kiril Simov
Vivek Kumar Singh
Vaclav Snasel
Thamar Solorio
Efstathios Stamatatos
Carlo Strapparava
Tomek Strzalkowski
Maosong Sun
Stan Szpakowicz
Mike Thelwall
Jörg Tiedemann
Christoph Tillmann
George Tsatsaronis
Dan Tufis
Olga Uryupina
Karin Verspoor
Manuel Vilares Ferro
Aline Villavicencio
Piotr W. Fuglewicz
Savas Yildirim

Software Reviewing Committee

Ted Pedersen
Florian Holz
Miloš Jakubíček

Sergio Jiménez Vargas
Miikka Silfverberg
Ronald Winnemöller

Award Committee

Alexander Gelbukh
Eduard Hovy
Rada Mihalcea

Ted Pedersen
Yorick Wiks

Additional Reviewers

Mahmoud Abunasser
Naveed Afzal
Iñaki Alegria
Hanna Bechara
Houda Bouamor
Janez Brank
Chen Chen
Víctor Darriba
Owen Davison
Ismaïl El Maarouf
Mahmoud El-Haj
Milagros Fernández-Gavilanes
Daniel Fernández-González
Corina Forascu
Kata Gabor
Mercedes García Martínez
Diman Ghazi
Rohit Gupta
Francisco Javier Guzman
Kazi Saidul Hasan
Radu Ion
Zahurul Islam
Milos Jakubicek
Antonio Jimeno
Olga Kolesnikova
Mohammed Korayem
Tobias Kuhn

Majid Laali
Yulia Ledeneva
Andy Lücking
Tokunbo Makanju
Raheleh Makki
Akshay Minocha
Abidalrahman Moh'D
Zuzana Neverilova
An Ngoc Vo
Mohamed Outahajala
Michael Piotrowski
Soujanya Poria
Francisco Rangel
Amir Hossein Razavi
Francisco Ribadas-Pena
Alvaro Rodrigo
Armin Sajadi
Paulo Schreiner
Djamé Seddah
Karan Singla
Vit Suchomel
Aniruddha Tammewar
Yasushi Tsubota
Francisco José Valverde Albacete
Kassius Vargas Prestes
Tim Vor der Brück
Tadej Štajner

Website and Contact

The webpage of the CICLing conference series is www.CICLing.org. It contains information about past CICLing conferences and their satellite events, including links to published papers (many of them in open access) or their abstracts, photos, and video recordings of keynote talks. In addition, it contains data, algorithms, and open-source software accompanying accepted papers, according to the CICLing verifiability, reproducibility, and working description policy. It also contains information about the forthcoming CICLing events, as well as contact options.

Table of Contents – Part II

Sentiment Analysis and Emotion Recognition

Opinion Mining and Social Networks

Machine Translation and Multilingualism

Information Retrieval

Text Classification and Clustering

Text Summarization

Plagiarism Detection

Style and Spelling Checking

Speech Processing

Applications

Table of Contents – Part I

Lexical Resources

Document Representation

Morphology, POS-tagging, and Named Entity Recognition

Syntax and Parsing

Anaphora Resolution

Recognizing Textual Entailment

Semantics and Discourse

Natural Language Generation

Sentence-Level Sentiment Analysis
in the Presence of Modalities

Yang Liu[1], Xiaohui Yu[1,3], Bing Liu[2], and Zhongshuai Chen[1]

[1] School of Computer Science & Technology, Shandong University, Jinan, China
[2] Department of Computer Science, University of Illinois at Chicago, Chicago, IL, US
[3] School of Information Technology, York University, Toronto, ON, Canada
{yliu,xyu}@sdu.edu.cn, liub@cs.uic.edu, czs046@163.com

Abstract. This paper studies sentiment analysis of sentences with modality. The aim is to determine whether opinions expressed in sentences with modality are positive, negative or neutral. Modality is commonly used in text. In a typical corpus, there are around 18% of sentences with modality. Due to modality's special characteristics, the sentiment it bears may be hard to determine. For example, in the sentence, *this cellphone would be perfect if it has a bigger screen*, the speaker is negative about this phone although there is a typically positive opinion word "perfect" in this sentence. This paper first presents a linguistic analysis of modality, and then identifies some features to train a support vector machine classifier to determine the sentiment orientation in such sentences. Experimental results on sentences with modality extracted from the reviews of four different products are given to illustrate the effectiveness of the proposed method.

Keywords: Opinion mining, Sentiment analysis.

1   Introduction

Almost all online retailers, such as *Amazon.com*, *eBay.com*, enable their customers to write reviews on products they have purchased. These reviews can help other customers to make informed purchase decisions. For product manufacturers, reviews also play an important role in keeping track of customers' opinions on their products and services. However, with the rapid growth in the number of reviews, it becomes very difficult for both customers and manufacturers to manually process the reviews and extract the sentiment information. It is thus highly desirable to perform automatic sentiment analysis for this purpose.

In the past few years, sentiment analysis (or, opinion mining) has received considerable attention. In particular, a large body of work is dedicated to the problem of sentiment classification at different granularities, i. e., document-level classification (determining whether opinions expressed in a whole opinion document are positive or negative)[20,22], sentence-level classification (classifying each sentence as positive, negative or neutral)[24,14,8], and aspect-based sentiment analysis (extracting opinion target and determining the speaker's sentiment orientation)[4]. Surveys of this area can be found in [19] and [9].

A. Gelbukh (Ed.): CICLing 2014, Part II, LNCS 8404, pp. 1–16, 2014.

We focus on sentiment classification at the sentence level. Although many general approaches have been developed for this problem, it is unlikely to have a one-technique-fits-all solution due to linguistic diversity. A divide-and-conquer approach is needed, which means that different kinds of sentences should be analyzed in different ways[15].

Table 1. Count of each modal verb

Modal verb	Can	Could	May	Might	Must	Need	Should	Will	Would	**Sum**
Count	534	214	63	43	43	2	84	453	409	1845

In this paper, we deal with an important type of sentences: sentences with modality. Such sentences have some special characteristics that make them difficult to handle. First, sentences with modality may express opinions opposite to those conveyed by the opinion words used. For example, the sentence *a removable battery would be nice* has a typically positive opinion word "nice", but expresses a negative sentiment. Second, in some cases, sentences with modality may express opinions even though they do not contain any apparent sentiment bearing word, e.g., *the vacuum should include a headlight*. The analysis above shows that sentiment analysis of sentences with modality is a challenging problem.

We find that sentences with modality take up a large share of customer reviews. The dataset we use is a customer reviews corpus that contains 9,152 sentences. We extract the sentences with modality using part-of-speech information, based on the fact that a modality sentence contains at least one modal verb whose tag is "MD". Table 1 shows the count of each modal verb appearing in our corpus. (Note that the total number of modal verbs is greater than the number of sentences with modality as many sentences contain more than one modal verb.) Table 2 shows the percentage of modal sentences and the total number of sentences in each corpus. The significant percentages of these sentences clearly demonstrate the need to develop techniques that can tackle them effectively.

Table 2. Percentage of sentences with modality

Corpus	% of mod. (total # of Sent.)
Cellphone	17.96(2037)
Vacuum	19.20(3124)
Matterss	16.11(1980)
Haircare	17.40(2011)

To the best of our knowledge, there is no existing study on sentiment classification of sentences with modality. This paper represents the first attempt along this direction. We start with a linguistic analysis to gain a good understanding of existing work on English modality. We then explore how each kind of modality influences sentiment. With the linguistic knowledge, we perform a computational study using machine learning techniques. In addition to some general features that might be related to capturing the essential information in sentences with modality, we also define modality features which include subjunctive sequence and category of modality. We also compare our method with some state-of-the-art alternative solutions including 1) traditional lexicon-based methods, 2) standard support vector machines, and 3) Naive Bayesian classification, to determine the improvement in classification performance due to the adaption of our proposal.

Since we only focus on sentences with modality, one may wonder whether our technique can be easily integrated into an existing sentiment analysis or opinion mining system. The answer is positive because those sentences can be easily detected using part-of-speech (POS) tag information (they contain at least one modal auxiliary verb, whose POS tag is *"MD"*), and our technique can be applied specifically to the detected sentences.

The main contributions of our work can be summarized as follows:

- We are the first to conduct a focused investigation into the problem of sentiment analysis of sentences with modality, and provide a detailed linguistic analysis of modalities, which forms the basis of studies on this topic;
- We identify some sophisticated linguistic features ,i.e., modality sequence and category of modality, which can be extracted and used for sentiment classification;
- We build an SVM classifier to perform sentiment classification of sentences with modality; the experimental results on real datasets demonstrate that the linguistic features are useful, and our classifier outperforms traditional lexicon-based, and unigram-based SVM and Naive Bayes classifiers.

The rest of our paper is organized as follows. We first introduce some related works in Section 2. Section 3 provides a problem statement of our work. In Section 4, we present the linguistic perspective of modality. We take a supervised learning approach to classify the sentiment orientation of sentences by integrating two groups of features in Section 5. In Section 6, we introduce three baseline strategies that can be used to compare with our proposal. Section 7 reports on the experimental results. We conclude the paper in Section 8.

2   Related Work

In this section, we survey briefly on some areas that are closely related to our topic.

2.1   Subjectivity Classification

Sentiment analysis has been widely studied in recent years, and there are many research directions along this line. One of the main directions is subjectivity classification, which focuses on determining whether a sentence is subjective or objective[24]. Most researchers apply supervised learning methods to process the classification. For example, Wiebe[24] use the Naive Bayes classifier with many binary features including the presence of adjectives, pronouns, cardinal numbers, etc. In contrast, Wiebe[23] develops an unsupervised method, which use the presence of subjective expressions for discrimination. Research in the same vein also includes [3,16,8,25].

2.2 Sentiment Classification

Another important direction is sentiment classification at different granularities, which is used to determine whether an opinion is positive, negative, or neutral. Some researchers have processed sentiment classification at document level (determining the overall sentiment orientation expressed in an opinion document)[20]. For instance, Pang[19] apply supervised methods, e.g., Naive Bayes classification, and SVM (support vector machine), to classify movie reviews as positive or negative. Turney[22] proposes an unsupervised learning method using some fixed syntactic patterns. Meanwhile, some research has been conducted at a finer level, i.e., for sentences or aspects. As another example, Yu[26] study sentence level classification to determine whether the sentiment expressed in a sentence is positive, negative or neutral. For aspect/feature level analysis, aspect/feature is considered as product attribute on which an opinion is expressed. Their goal is to extract features or topics in sentences and determine their corresponding sentiment orientations [4,1,13].

2.3 Sentiment Analysis on Special Sentences

However, all works listed above deal with general problems. Several researches argue that their effectiveness may be poor in some special cases, e.g., for conditional sentences[15]. They also believe that there is unlikely a one-technique-fits-all solution and a divide-and-conquer approach is needed to deal with some special sentences with unique characteristics. Sentiment analysis of some special sentences has been studied with the help of their linguistic characteristics. In [5] and [2], the researchers analyze sentiment analysis on comparative sentences, whose goal is to identify comparative sentences in customer reviews and determine whether they are positive, negative or neutral. Narayanan[15] investigate sentiment analysis of conditional sentences. Inspired by these works, we make an attempt to handle sentiment analysis on another special type of sentences, i.e., sentences with modality. Again, no focused study has been done on such sentences. As such sentences can be easily identified, our method can be readily fused with any general system and improve its efficiency.

3   Problem Statement

We focus on sentiment classification of a special type of sentences: sentences with modality. Here sentences with modality refer to those sentences containing at least one modal auxiliary word. Modality has been well studied linguistically in [18] and [17]. These sentences have their own characteristics, which render general sentiment analysis strategies unsuitable. For example, the sentence: *this phone would be perfect if a back-up battery were supplied.* expresses a negative opinion although the only opinion word it contains, "perfect", is usually used to express positive opinion. Sentences like that may reduce the performance of general sentiment analysis systems.

Our methodology is to first find out the principles of the usage of modalities, and then use those principles to guide the construction of features to be used in a classifier, which can be used to determine whether opinions expressed in sentences with modality are positive, negative, or neutral.

4   Linguistic Analysis of Modality

This section presents the linguistic perspective of modality. Modality is a linguistic device that indicates the degree to which an observation is possible, probable, likely, certain, permitted, or prohibited. In English, these notions are most commonly expressed by modal auxiliaries, sometimes combined with *not*. A variety of classification methods have been propounded by linguists; we follow the one by Palmer. There are two basic distinctions in the way languages deal with the category of modality: *mood* and *modal system* [18]. Each kind has its own characteristics and will be detailed below.

4.1   Mood

Mood has three sub-categories: indicative, imperative and subjunctive. Indicative is used to state fact, for example,*this is a perfect cellphone.* Imperative is used to state command, require or absolutely necessary, e.g.,*go and wash your hand.* Subjunctive is typically used to express various states of unreality such as wish, emotion, possibility, judgment, opinion, necessity, or action that has not yet occurred, e.g., *this phone would be perfect if a back-up battery were supplied,* and *inventor should have studied American vacuums a little closer.*

A general classification strategy, such as a lexicon-based one, usually base its classification decision on the opinion words contained in the sentence. However, this approach is problematic in processing sentences in subjunctive mood, as there may be no opinion words in such sentences. To make things worse, subjunctive sentences may sometimes express sentiment orientation opposite to that of the opinion words. For example, consider the sentence, *It would be nice if the screen were a little bigger.* Although it contains a typical positive word *"nice"*, it expresses a negative opinion. This usage usually appears in patterns like *"would/should be + adjective + that/if... "*, which can be captured via POS tag.

There are three types of constructions that have been deemed subjunctive in English [18]. One is the use of *were* in *I just wish that it were longer, so that I could get to the rest of our 14-foot vaulted ceiling.* The second construction is in a clause following the verbs like *insist, demand* and so on, such as *I demand the turbo be more powerful.* The third is the use of some modal verbs like *should* and *would*. For example, *I think they should offer an electrically powered brush for stair cleaning.* With the help of POS tags, subjunctives can be easily distinguished form the other moods.

4.2   Modal Systems

There are two main types of modality: propositional modality and event modality. In most cases, propositional modality is concerned with the speaker's attitude to the true-value factual status of the proposition, while event modality is used to refer to events that are not actualized, events that have not taken place but are merely potential[18]. They will be separately detailed below.

Propositional Modality. Epistemic modality and evidential modality are the two most important types of propositional modality. The main difference between them is that with epistemic modality, speakers express their judgments about the factual status of the proposition, while with evidential modality they indicate the evidence they have for its factual status[18]. For example,

- *This vacuum may be the best that costs less than $400.* (epistemic)
- *Manufacturer must have enhanced the turbo power.* (evidential)

Usually, the main modal auxiliary verbs used in epistemic modality are *may* (and its past tense *might*), *can* (and its past tense *could*), and *will*, which can be paraphrased as *"a possible conclusion is that..."*. The past tense of the modal verbs is usually used to express weakly possible. *Must* is the marker of evidential modality. It is not strictly possible to simply paraphrase *must* as *"it is certain that..."*, which merely indicates the strength of the speaker's belief [17]. The most suitable paraphrase is *"the only conclusion is that..."*.

Event Modality. Event modality is concerned with events that have potential to or may take place. Two main classes of event modality are deontic modality and dynamic modality. The basic difference between them is that with deontic modality, speakers try to get others to do something, while with dynamic they just express their ability or willingness to do something. For example,

- *You may choose iPhone 4s rather than iPhone 4.* (deontic)
- *I will recommend this vacuum to my friends.* (dynamic)

Many modal auxiliary verbs can be used in deontic modality, such as *may*, *can*, and *should*. It can be paraphrased as *"you were committed to..."* or *"you got my commitment that..."*. There are two kinds of dynamic modality, expressing ability and willingness, which are expressed in English by *can* and *will* [18], respectively.

What one should pay attention to is that a modal auxiliary verb can be used in more than one type of modalities. For example,

- *This vacuum can lose suction at any minute.* (epistemic)

– *This vacuum can pick up the dirt easily.* (dynamic)

Therefore, we cannot simply rely on the existence of a certain auxiliary verb to determine the sentiment orientation of a sentence. We must also take the contextual information into account, which motivates the proposal of the features used in classification.

5   Feature Construction

We take a supervised learning approach to classifying the sentiment orientation of sentences. To build a feature vector for each sentence, we construct two groups of features, i.e., general linguistic features, and features that are directly related to modality analysis.

5.1   General Linguistic Features

Some general features, such as opinion words, and negation words, have proved useful in sentiment analysis of all kinds of sentences. We also use these features as part of our feature set.

1. *Opinion words*
   Opinion words are words that are primarily used to express speakers' subjective opinions on an object, an event, etc. Typically, opinion words are adjectives, adverbs, or nouns. In general, the appearances of opinion words in a sentence have a big impact on its sentiment orientation. We identify those opinion words from sentences with POS tag information.
2. *Negation words*
   The presence of a negation word (such as *not, never, hardly, no*, etc.) usually changes the sentiment orientation of a sentence. For example, the sentence *you will not be disappointed with this vacuum* expresses a positive opinion on this vacuum; without the negation word *not*, it will express a negative one. In the experiment, we use a binary feature to indicate whether there is a negation word or not. We use a sliding-window of 4 to detect whether there is a negation word or not near an opinion word, i.e., any negation word appearing with a distance less than 4 from an opinion word is counted.
3. *Relative location with adversative conjunctions*
   Adversative conjunctions are used to connect two clauses with opposite meanings. Sentiment orientations of these two clauses are usually different from each other. In most cases, it is likely that the posterior clause is the one that is emphasized. For example, in the compound sentence *this vacuum does cost a lot, but I will recommend it to my friends*, the prior clause expresses a negative opinion on this product's price, but the whole sentence expresses a positive opinion and the speaker considers the product well worth its price. As we deal with sentence-level sentiment analysis, we only need to pay attention to the sentiment orientation of the posterior clause. In some cases,

if the sentiment of the posterior clause is difficult to analyze, but the sentiment orientation of the prior clause can be easily determined, the sentiment orientation of the whole sentence can be taken as opposite to that of the prior clause.

4. *Punctuations*

The presence of some punctuations, such as '!' and '?', also bears some useful information. In general, a sentence ended with "?" just expresses the speaker's question or query, whose sentiment may be neutral, e.g., *"is iphone4s worth having?"*. A sentence ended with "!" almost always expresses the speaker's strong emotion, e.g., *"what a amazing phone!"* This feature is also incorporated into our classification strategy.

5.2 Modality Features

Besides those general features listed above, our feature set also involves some types of features that are directly correlated with the linguistic analysis of modality presented in Section 4.

1. *Modality sequence*

Our analysis in the Section 4 shows that a modal verb can be used in many different modalities with different implications on the sentiment. It is therefore difficult to judge the sentiment orientation of a sentence merely based on the appearance of modal verbs. We must also take into consideration the context in which the modal verb resides.

For this purpose, we introduce the notion of *modality sequence* to refer to the sequence constituted by a modal auxiliary verb and the opinion bearing word that immediately follows it. For example, in the sentence *I would recommend this vacuum to anyone*, the modality sequence is *would recommend*. The opinion bearing word considered here could be a verb, an adjective, an adverb, or a noun. In fact, we find that the majority of opinion bearing words that follow model auxiliary verbs are verbs rather than the others. Only in some cases when the verb following a modal auxiliary verb is a linking verb that does not have sentiment orientation, such as "be", "get", and "become", does the modality sequence become one composed of a modal verb and the predicative word (may be an adjective, or a noun) right after the verb. For example, in the sentence, *you will be disappointed to buy this cellphone*, the modality sequence is *will disappointed*. We identify modal verbs, verbs, and predicative words with the help of POS tag information, i.e., the POS tag of a modal verb is *MD*, the POS tag of a noun may be *NN* or *NNS*, and the POS tag of a verb may be *VB*, *VBN* or *VBP*.

2. *Category of modality*

Category of modality also has an effect on the opinion expressed in a sentence with modality. For example, the sentiment orientation of a sentence with subjunctive mood may be opposite to the sentiment orientation of the opinion word in the sentence. In this paper, we handle three types of modality, i.e., subjunctive mood, deontic modality, and dynamic modality. We use three binary features, each one of which indicates whether a sentence with modality belongs to a particular category or not. We build some rules to recognize these categories using both words and their POS tag information. Rules we use are detailed below:

(a) *could(would, should) + past perfect*

This rule is used to recognize one type of subjunctive. The presence of the tag sequence *MD+VB+VBN* is regarded as the mark of this type of subjunctive. A sentence matching this rule will be treated as a subjunctive. For example, the sentence below will be regarded as a subjunctive due to the presence of tag sequence *MD+VB+VBN*.

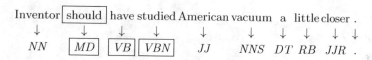

(b) *if + past tense → would + bare infinitive*

This rule is used to recognize one type of conditional subjunctive, whose condition clause is in the past tense (POS tag is "VBD") and consequent clause is in the present tense (POS tag is "VB"). For example, the sentence below will be regarded as a subjunctive for matching this rule.

<table>
<tr><td>If</td><td>it</td><td>had</td><td>a</td><td>larger</td><td>battery</td><td>,</td><td>I</td><td>would</td><td>buy</td><td>one</td><td>.</td></tr>
<tr><td>↓</td><td>↓</td><td>↓</td><td>↓</td><td>↓</td><td>↓</td><td>↓</td><td>↓</td><td>↓</td><td>↓</td><td>↓</td><td>↓</td></tr>
<tr><td>IN</td><td>PRP</td><td>VBD</td><td>DT</td><td>JJR</td><td>NN</td><td>,</td><td>PRP</td><td>MD</td><td>VB</td><td>CD</td><td>.</td></tr>
</table>

(c) *Singular noun (third person singular, first person singular) + were*

This rule is used to recognize another type of subjunctive using the knowledge that in the subjunctive, the past form of the verb *to be* is *were*, regardless of the subject. For example, the sentence below is recognized by this rule.

<table>
<tr><td>I</td><td>wish</td><td>it</td><td>were</td><td>a</td><td>little</td><td>lighter</td><td>.</td></tr>
<tr><td>↓</td><td>↓</td><td>↓</td><td>↓</td><td>↓</td><td>↓</td><td>↓</td><td>↓</td></tr>
<tr><td>PRP</td><td>VB</td><td>PRP</td><td>VBD</td><td>DT</td><td>RB</td><td>JJR</td><td>.</td></tr>
</table>

(d) *You should(may) + VB*

This rule is used to recognize deontic modality, with which the speaker tries to get others to do something. Deontic modality recognized by this rule is used to give advice to the audience "you". For example, the sentence below is a deontic modality matching this rule.

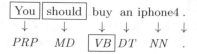

(e) *I will(would) + VB*

This rule is used to recognize dynamic modality, with which the speaker expresses his/her willingness to do something. The subject of this type of sentence is "I", and the modality auxiliary word is "will" or "would". For example, the sentence below is recognized as dynamic modality using this rule.

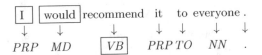

These rules may be too strict to identify all types of modality, but we argue that precision is much more important than recall in the task of features extraction.

6   Classification Strategy

In this section, we will introduce three baseline strategies that have been used in prior studies to determine the sentiment orientation of a sentence, as well as the construction of our own strategy that incorporates the features identified in the previous section. The first baseline method is a lexicon-based approach, which only uses opinion word as features, whereas the second is a standard SVM classifier, which is an improved method. We also use a Naive Bayes classifier as a baseline.

6.1   Lexicon-Based Strategy

The lexicon-based strategy is probably the most straightforward way to determine the sentiment orientation of a sentence. In a given sentence s, let p be the number of positive opinion words in this sentence, and n be the number of negative words. If p is greater (less) than n, the sentiment orientation of s will be designated as positive (negative), and neutral or mixed in the case of a tie. The opinion lexicon we use contains 4,783 negative opinion words and 2,006 positive opinion words[1]. This method has some drawbacks. For example, some sentences may not contain those opinion words, and the sentient orientation of some words depends on the context.

6.2   Standard SVM Classifier

SVM (Support Vector Machine) classifiers are widely used in some domains such as document classification and sentiment classification. In this paper, we use a Unigram-based (also called bag-of-words-based) SVM classifier. Typically, a unigram-based classifier builds a feature vector for a sentence based on flat term

---

[1] Available at http://www.cs.uic.edu/~liub/FBS/sentiment-analysis.html

frequencies or TF-IDF scores. TF-IDF score, which measures both how frequent a term appears in a document and how rare this term is in the other documents, is a classic quantity used in information retrieval and document classification. However, when dealing with sentiment classification, it has a serious drawback for not measuring how biased a term is to positive and negative corpora. Based on this observation, an improved TF-IDF method, called Delta TF-IDF, has been proposed by Martineau[11]. Delta TF-IDF calculates a term's feature value for a given document by computing the difference of that word's TF-IDF scores in the positive and negative training sets. We use this method in both the standard SVM classifier and our own method to be described in the sequel. The weight of a given term i in a sentence S is calculated as:

$$W_{i,S} = C_{i,S} * \log_2 \left(\frac{|P|}{P_i} \right) - C_{i,S} * \log_2 \left(\frac{|N|}{N_i} \right)$$
$$= C_{i,S} * \log_2 \left(\frac{|P|}{P_i} \frac{N_i}{|N|} \right)$$

where $W_{i,S}$ is the weight of i in a sentence S; $C_{i,S}$ is the number of times item i occurs in S; $|P|$ and $|N|$ are the numbers of sentences in the positive and negative corpora respectively; P_i and N_i are the numbers of positive and negative sentences containing term i respectively.

6.3   Naive Bayes Strategy

Naive Bayes is a widely used probabilistic classifier and performs efficiently in text classification [12]. Due to its solid statistical foundation, it has been successfully adopted to distinguish opinions from facts for sentences [26], and to discriminate sentiment polarities for documents [20]. Similar to SVM, Naive Bayes also share the benefit of being an unigram model. However, unlike SVM, it only considers word occurrence due to its multinomial distribution assumption. In this work, to answer the call of selecting the most appropriate classifier in modality classification, we also implement a Naive Bayes classifier with Weka[2]. We will compare its effectiveness with SVM-based strategy through experiments.

6.4   Our Strategy

Our new classification strategy expands on the feature vector used in the standard SVM classifier. In addition to that vector, we also incorporate all the features identified in Section 5, except for the opinion words as they are already built into the feature vector used for the unigram-based SVM strategy. Similar to the unigram-based SVM strategy, we also calculate the delta TF-IDF score for each modality sequence and use it as the feature value rather than the raw TF-IDF score. The category of modality features are also used to built vectors.

---

[2] Available at http://www.cs.waikato.ac.nz/ml/weka/

7   Experimental Evaluation

7.1   Data Sets

We carry out the experiments using customer reviews of four categories of products: vacuum, cellphone, haircare and mattress. There are 9,152 sentences in total. We tag these sentences using the Stanford Log-linear Part-Of-Speech Tagger, which gives a 97.24 percent accuracy on the Penn Treebank WSJ [21]. Using the POS tag information, we extract 1,635 sentences from the data set, each of which contains at least one modal verb. We also annotate their sentiment orientation (positive, negative or neutral) manually in the same way as described in (Hu and Liu, 2004). During our annotation, we found that most of these sentences carry users' opinions, i.e., around 72% sentences express opinions. Table 3 shows the class distribution of this data set. We also manually correct a few erroneous tags, for example, *"can"* is a noun but has been tagged as a modal auxiliary verb (*garbage/NN can/MD*) for 4 times; *"need"* is a noun but has been tagged as a modal auxiliary verb once(*your/PRP need/MD*).

Our method to annotate the data follows that in [15]. At first, three graduate students annotate the sentiment orientation of each sentence separately. When all finished, we gather the data to check if there is any conflict. If there is, the annotators discuss to reach an agreement. In the end, there is no sentence that they can not reach an agreement on, and we can assume that they are all annotated correctly. We measure the agreement of the sentiment annotation using the Kappa score. In our annotation, we achieve a Kappa score of 0.67, which indicates good agreements between annotators.

Table 3. Distribution of classes

Sentiment orientation	# of sentences
Positive	699
Negative	456
Neutral	480

7.2   Experiment Results and Discussion

We now present the experimental results for different classification strategies. Because we focus on sentence-level sentiment analysis, some sentences like *this vacuum works well, but its too noisy for me* are treated as negative in the system. In what follows, we discuss the results for two-class and three-class classification respectively. For each type of classification, our experiments focus on two aspects: (1) comparing the performance of our strategy with three baseline methods, and (2) studying the effect of each type of features on the performance of classification.

Two-Class Classification. We first present the results for two-class classification where only positive and negative sentences are used in the experiment;

the neutral sentences are not included (29.35% of the total). For the lexicon-based method, if the orientation of a sentence equals 0, we classify it as positive or negative randomly. In the standard SVM method and our method, we use SVMlight[3] with linear kernel [6] for classification. We use Weka to process Naive Bayes based classification. We calculate the accuracy, the arithmetic average precision and recall of each class, and the F-score. The results are shown in Figure 1(a). As illustrated in Figure 1(a), the performance of the lexicon-based method is quite poor. It proves that general classification strategies can not deal with some special types of sentences effectively, e.g., sentences with modalities, which may contain no apparent opinion words or may express an opinion opposite to that of the opinion word. Another observation is that our strategy outperforms the standard method and Naive Bayes by a significant margin; indeed, there is a statistically significant difference(p<0.001), paired t-test, actually the t score equals 6.69, which means the confidence is more than 99.9 percent), which means that the linguistic features used in our method are reliably helpful.

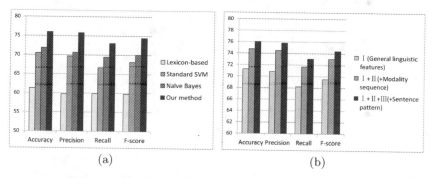

(a)   (b)

Fig. 1. Two-class classification - positive and negative

Figure 1(b) shows the performance of our strategy when some types of features are withheld. In { I }, we only use general linguistic features listed in Section 5.1, i.e., opinion words, negation words, relative location with adversative conjunctions, and punctuations, as features. This is regarded as the baseline. We can observe that it performs better than Unigram-based SVM method, which indicates that general linguistic features listed ahead besides opinion words are also helpful. In { I + II }, we use both general linguistic features and modality sequence features. The classifier's performance improves a lot. In { I + II + III }, we use all these three features, i.e. general linguistic features, modality sequence features, and category of modality features. The performance improves further.

Three-Class Classification. We now move to the more challenging and practical classification strategy that involves all three classes: positive, negative and neutral. For training three-class classifiers, we use SVM^{multiclass}[4] with the

---
[3] Available at http://svmlight.joachims.org/
[4] Available at http://svmlight.joachims.org/svm_multiclass.html

regularization parameter c set to 1.0 for the best result [7]. We also calculate the accuracy, the arithmetic average precision and recall, and F-score. The results are shown in Figure 2. Figure 2(a) shows the comparison of the results of the three baseline methods and our strategy, from which we can observe that ours significantly outperforms the others($p<0.001$, paired t-test). Another observation is that all these methods have a lower performance in three-class classification than in two-class classification. This is because the three-class classification is more difficult than a two-class one in general. For example, the expected accuracy of a randomly classifier for two-class classification is 50%, while for three-class classification it is only 33%.

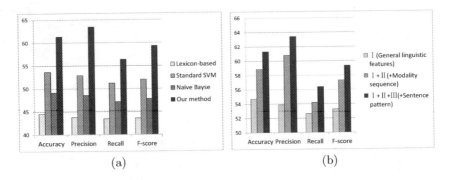

(a)   (b)

Fig. 2. Three-class classification - positive, neutral, and negative

Figure 2(b) shows the performance of our proposed strategy when some features are taken away. The conclusion is the same as two-class classification, which illustrates that both modality sequence and category of modality are useful in sentiment analysis of sentences with modality.

From the results of both two-class classification and three-classification, we can draw a conclusion that our method demonstrates significant improvement over the other three traditional methods. The linguistic features of modality sequence and sentences are important contributing factors.

8   Conclusion and Future Work

Due to the language diversity (each type of sentence has its own characteristic), it is too hard to find a one-size-fit-all approach to handle sentiment classification on all types of sentences accurately. Narayanan[15] argued that a divide-and-conquer approach is needed, which calls for different strategies to deal with different sentences. Motivated by his work, we developed a method to handle one particular type of sentences, i.e., sentences with modality. In this paper, we first performed linguistic analysis for sentences, and then develop two groups of features, i.e., general features and modality features, to construct a classifier. In order to select the most appropriate modality features, we carefully studied the use of different modal auxiliary verbs in different types of modalities, and discovered a series of patterns based on their linguistic characteristics. In addition, we built a classifier based on SVMs to determine whether an opinion expressed in a sentence with modality is positive, negative or neutral. Results of our experiment proved the effectiveness of our approach. Moreover, our method can be easily integrated into a general classification system as sentences with modality can be filtered easily using POS information.

For future work, we will continue to improve the performance of our method. One possible solution is to build a modality lexicon that can recognize the type of modality more accurately. It may incur a larger computational cost; however, it might well worth the effort when these sentences constitute a high proportion of all customer reviews corpus.

Acknowledgement. A preliminary version of this paper appeared as a workshop paper [10]. This work was supported in part by the National Natural Science Foundation of China Grant (No. 61272092), the Natural Science Foundation of Shandong Province of China Grant (No. ZR2012FZ004), the Independent Innovation Foundation of Shandong University (No. 2012ZD012), the Taishan Scholars Program, and NSERC Discovery Grants. Bing Liu was supported in part by a grant from National Science Foundation (NSF) under grant No. IIS-1111092.

References

1. Ding, X., Liu, B., Yu, P.S.: A holistic lexicon-based approach to opinion mining. In: WSDM, pp. 231–240 (2008)
2. Ganapathibhotla, M., Liu, B.: Mining opinions in comparative sentences. In: COLING, pp. 241–248. Association for Computational Linguistics (2008)
3. Hatzivassiloglou, V., Wiebe, J.M.: Effects of adjective orientation and gradability on sentence subjectivity. In: COLING, pp. 299–305. Association for Computational Linguistics (2000)
4. Hu, M., Liu, B.: Mining and summarizing customer reviews. In: KDD, pp. 168–177. ACM (2004)
5. Jindal, N., Liu, B.: Identifying comparative sentences in text documents. In: SIGIR, pp. 244–251. ACM (2006)
6. Joachims, T.: Making large scale svm learning practical (1999)
7. Joachims, T., Finley, T., Yu, C.: Cutting-plane training of structural svms. Machine Learning 77(1), 27–59 (2009)
8. Kim, S., Hovy, E.: Determining the sentiment of opinions. In: COLING, p. 1367. Association for Computational Linguistics (2004)
9. Liu, B.: Sentiment analysis and opinion mining. Synthesis Lectures on Human Language Technologies 5(1), 1–167 (2012)
10. Liu, Y., Yu, X., Chen, Z., Liu, B.: Sentiment analysis of sentences with modailities. In: Proceedings of the International Workshop on Mining Unstructured Big Data using Natural Language Processing (UnstructureNLP 2013), pp. 39–44 (2013)
11. Martineau, J., Finin, T.: Delta tfidf: An improved feature space for sentiment analysis. In: Proceedings of the 3rd AAAI International Conference on Weblogs and Social Media, pp. 258–261 (2009)

12. McCallum, A., Rosenfeld, R., Mitchell, T., Ng, A.: Improving text classification by shrinkage in a hierarchy of classes. In: ICML, pp. 359–367 (1998)
13. Mei, Q., Ling, X., Wondra, M., Su, H., Zhai, C.X.: Topic sentiment mixture: modeling facets and opinions in weblogs. In: WWW, pp. 171–180. ACM (2007)
14. Mihalcea, R., Banea, C., Wiebe, J.: Learning multilingual subjective language via cross-lingual projections. In: ACL, vol. 45, p. 976 (2007)
15. Narayanan, R., Liu, B., Choudhary, A.: Sentiment analysis of conditional sentences. In: EMNLP, pp. 180–189. Association for Computational Linguistics (2009)
16. Nasukawa, T., Yi, J.: Sentiment analysis: Capturing favorability using natural language processing. In: Proceedings of the 2nd International Conference on Knowledge Capture, pp. 70–77. ACM (2003)
17. Palmer, F.R.: The English Verb. Longman London (1974)
18. Palmer, F.: Mood and Modality. Cambridge University Press (2001)
19. Pang, B., Lee, L.: Opinion mining and sentiment analysis. Foundations and Trends in Information Retrieval 2(1-2), 1–135 (2008)
20. Pang, B., Lee, L., Vaithyanathan, S.: Thumbs up? sentiment classification using machine learning techniques. In: EMNLP, pp. 79–86 (2002)
21. Toutanova, K., Klein, D., Manning, C., Singer, Y.: Feature-rich part-of-speech tagging with a cyclic dependency network. In: NAACL, pp. 173–180. Association for Computational Linguistics (2003)
22. Turney, P.D.: Thumbs up or thumbs down?: semantic orientation applied to unsupervised classification of reviews. In: ACL, pp. 417–424. Association for Computational Linguistics (2002)
23. Wiebe, J.: Learning subjective adjectives from corpora. In: Proceedings of the National Conference on Artificial Intelligence, pp. 735–741 (2000)
24. Wiebe, J., Bruce, R., O'Hara, T.: Development and use of a gold-standard data set for subjectivity classifications. In: ACL, pp. 246–253 (1999)
25. Wiebe, J., Wilson, T.: Learning to disambiguate potentially subjective expressions. In: Proceedings of the 6th Conference on Natural language Learning, pp. 1–7. Association for Computational Linguistics (2002)
26. Yu, H., Hatzivassiloglou, V.: Towards answering opinion questions: separating facts from opinions and identifying the polarity of opinion sentences. In: EMNLP, pp. 129–136 (2003)

Word-Level Emotion Recognition
Using High-Level Features

Johanna D. Moore, Leimin Tian, and Catherine Lai

University of Edinburgh
School of Informatics
Informatics Forum, EH8 9AB Edinburgh, UK
J.Moore@ed.ac.uk, s1219694@sms.ed.ac.uk, clai@inf.ed.ac.uk

Abstract. In this paper, we investigate the use of high-level features for recognizing human emotions at the word-level in natural conversations with virtual agents. Experiments were carried out on the 2012 Audio/Visual Emotion Challenge (AVEC2012) database, where emotions are defined as vectors in the Arousal-Expectancy-Power-Valence emotional space. Our model using 6 novel disfluency features yields significant improvements compared to those using large number of low-level spectral and prosodic features, and the overall performance difference between it and the best model of the AVEC2012 Word-Level Sub-Challenge is not significant. Our visual model using the Active Shape Model visual features also yields significant improvements compared to models using the low-level Local Binary Patterns visual features. We built a bimodal model By combining our disfluency and visual feature sets and applying Correlation-based Feature-subset Selection. Considering overall performance on all emotion dimensions, our bimodal model outperforms the second best model of the challenge, and comes close to the best model. It also gives the best result when predicting Expectancy values.

1 Introduction

Affective Computing, the study of recognizing, understanding, and synthesising human emotions using computational technologies, has shown great potential both in academic studies of human behaviour as well as industrial applications. For example, by detecting affective states, such as boredom, an Intelligent Tutoring System can improve student learning and increase user satisfaction [1]. Multimodal emotion recognition has recently become a focus of affective computing. However, this task remains challenging, especially with respect to spontaneous spoken dialogue. Much of the early work on this topic was based on acted expressions of emotions [2], leading to models with good performance when the test and training data are similar, but which perform poorly when applied to a system working in a more natural environment. Moreover, differences in data collection and annotation style make it difficult to compare results across studies.

To address these issues, recent studies have focused on recognizing emotions in more realistic dialogues while shared tasks such as the annual Audio/Visual

A. Gelbukh (Ed.): CICLing 2014, Part II, LNCS 8404, pp. 17–31, 2014.
© Springer-Verlag Berlin Heidelberg 2014

Emotion Challenge (AVEC) have been held with the goal of comparing different approaches on common datasets of spontaneous speech. Despite these steps forward, the performance of existing multimodal emotion recognition models leaves much room for improvement. Predicted values from the top competitors in AVEC2012 [3], for example, exhibit relatively weak correlations for both the frame and word level subchallenges. As a regression task, the average correlation-coefficients over all test sessions for the best Fully-Continuous Sub-Challenge (FCSC) model [4] and the best Word-Level Sub-Challenge (WLSC) model [5] are 0.456 and 0.280 respectively, i.e., weak to moderate correlations. A possible reason for this poor performance is that the lexical, acoustic, and visual features often examined in these tasks are too low-level to predict emotions.

In this paper, we investigate the predictiveness of high-level features in the word-level emotion recognition task. These features include six disfluency features and locations of facial landmarks. We extracted our features from the AVEC2012 database and compared the predictiveness of our high-level features with that of the conventional lower-level audio and visual features used by the AVEC2012 WLSC baseline models. These include spectral and prosodic (SP) features and Local Binary Patterns (LBP) [6]. We compare our models to the corresponding AVEC2012 WLSC baseline models, as well as the three best performing models from the AVEC2012 WLSC. We find that our high-level features are more predictive than the low-level features, and the performance of our best bimodal model is competitive with the highest scoring models from the AVEC challenge, while using at most 22 features.

1.1 Background

Previous approaches to emotion prediction based on the AVEC data work with a high dimensional space of low-level features (1842 SP features and 5908 LBP features in the baseline model). However, the results from the top performing WLSC model [5] show that significant gains can be made by including lexical features. In this paper, we investigate whether other higher level features can be used to reduce feature space dimensionality and improve performance for this task.

Studies of both human cognition [7] and natural language processing [8] suggest that disfluencies are powerful clues for recognizing the emotional states of a speaker. Thus, disfluency features may have a stronger relationship with emotions than SP features or more general lexical features extracted from content words, and may contain less noise. Therefore, we conjecture that a unimodal emotion recognition model using disfluency features will outperform models using SP features or more general lexical features, and may contain less noise. Therefore, we conjecture that a unimodal emotion recognition model using disfluency features will outperform models using SP features or more general lexical features.

Both the best [4] and the second best [9] performing FCSC models of AVEC2012 chose high-level visual features that describe the facial expressions of the speaker using positions of facial landmark points, instead of the LBP features that describe the orientation of pixels. Their results suggest that high-level visual features may also improve performance when recognizing emotions at word level.

Studies in cognitive science [10] and affective computing [11] show that the audio and visual modalities have different strengths and weaknesses when predicting different emotion dimensions. Therefore, combining the modalities should lead to improved performance, at least compared to the lower performing modality. We also expect to see better performance by combining our high-level features as opposed to combining low-level features.

However, differences in performance may arise depending on how the modalities are combined. For example, the bimodal model of Savran et al. [5] uses Decision-Level (DL) fusion, in which unimodal models are built separately and their individual predictions are then combined. This bimodal model outperforms both unimodal models. However, the WLSC baseline model [3] uses Feature-Level (FL) fusion, in which the audio and visual features are concatenated and a single classifier is built from this combined feature set. This bimodal model only outperforms the worse performing unimodal model (the visual model). This suggests that applying feature engineering methods (e.g., Principal Components Analysis (PCA) and Correlation-based Feature-subset Selection (CFS)), to the concatenated feature set may improve the performance of the bimodal model by reducing the drawbacks of the less predictive features and increasing the benefits given by the more predictive features.

To sum up, in this work, we test the following three hypotheses:

1. *Using high-level features will improve the performance of emotion recognition compared to using low-level features.*
2. *Fusing modalities by concatenating feature sets will give better results compared to unimodal models.*
3. *Applying feature engineering to the concatenated feature set will improve performance of the bimodal model.*

The rest of this paper is organised as follows: In Section 2, we introduce the AVEC2012 database and our experimental setup. Section 3 presents the results of regression experiments using different feature sets. Section 4 provides general discussion and future directions for our work. Section 5 contains the conclusions.

2 Method

2.1 The AVEC2012 Challenge

The Database and Its Definition of Emotions

We use the AVEC2012 database [3] in the following experiments. It includes audio-visual recordings and manual transcriptions with word timings of 24 subjects conversing with 4 virtual agents, which were collected as part of the SE-MAINE corpus [12]. Each agent is designed with a different personality, namely even-tempered Prudence, happy Poppy, angry Spike, and depressive Obadiah. Conversations were conducted in a Wizard-of-OZ setup. Topics of conversation varied from daily life to political issues. The 24x4 recordings are divided into training set, development set, and test set, each of which contains 32 dialogue sessions. In the AVEC database, subjects in the test set are different people from those in the training and development sets. For the WLSC, each word spoken by a subject is a data instance. The number of instances contained in the training set, development set, and test set are 20169, 16300, and 13405, respectively.

The AVEC2012 database uses real-value vectors in the Arousal-Expectancy-Power-Valence (AEPV) space to represent emotions. Arousal represents the activeness of the subject; Expectancy represents the predictiveness the subject feels towards the conversation; Power represents the degree of dominance the subject feels over the conversation; Valence represents the positiveness the subject feels towards the conversation [13]. For example, using this representation, we may describe the emotional state of someone who has just been informed that she has won the best paper award as $a = (0.6, -0.3, -0.1, 0.9)$, which means she is excited (A = 0.6) about this great (V = 0.9) news, and cannot stop herself (P = -0.1) from jumping up and down at this surprise (E = -0.3). The AEPV emotional space is capable of describing most of our everyday emotions [13]. The original emotion annotations in the AVEC2012 database had different value ranges for the four dimensions. In our work, we rescale all the AEPV values into the range [-1, 1].

Baseline Audio and Visual Features Provided by the Challenge

The AVEC2012 baseline SP feature set provides 1842 audio features, including pitch, energy, voicing, spectral related low-level descriptors, and voiced/unvoiced duration features, which were extracted from the users' speech using OpenS-MILE [14] over words.

The AVEC2012 baseline LBP feature set provides 5908 features related to the size and position of the facial regions, as well as LBP descriptors. Faces in frames are detected using OpenCV's Viola and Jones face detector [15], then aligned by eye locations.

The Best Performing WLSC Model

The model that won the 2012 WLSC, proposed by Savran et al. [5], uses a subset of the baseline audio and visual features, together with lexical features they extracted from the transcripts provided.

Their lexical features were computed using Pointwise Mutual Information (PMI) values, which are measurements of the correlations between words and binarized emotion dimensions. They extracted two types of lexical features, i.e., sparse PMI features using a 1000-dimension bag-of-words approach, and non-sparse PMI features using word counts. These were the only lexical features used in the AVEC2012. Their experiments showed that the sparse PMI features gave significantly better results than the non-sparse PMI features, and were the most predictive features. Since our disfluency features are also extracted from the transcripts, performance of our disfluency feature model will be compared to their lexical model.

2.2 Disfluency Features

In our work, we extract 6 novel disfluency features from the manual transcripts and word-timings provided by the challenge. Each of our disfluency features describes one type of disfluency as described in the following list:

1. **Filled pauses:** non-lexical sounds people make when speaking. For example, "Hmm" in the utterance "Hmm... Maybe we should try another road." The three most common filled pause words in the AVEC database are "em", "eh", and "oh".
2. **Fillers:** phrases used by speakers when they pause to think but they still want to hold the turn. For example, "you know" in the utterance "I just want to, you know, get a drink and forget all about it.". The three most common fillers in the AVEC database are "well", "you know", and "I mean".
3. **Stutters:** words or parts of words the speaker involuntarily repeats when speaking. For example, "Sa" in the utterance "Sa Saturday will be fine.", or the first "I didn't" in the utterance "I didn't, I didn't mean it."
4. **Laughter:** sounds labelled as ⟨LAUGH⟩ in the transcripts provided with the AVEC challenge.
5. **Breath:** sounds labelled as ⟨BREATH⟩ in the transcripts.
6. **Sigh:** sounds labelled as ⟨SIGH⟩ in the transcripts.

We note that this is only a subset of the types of disfluencies that are studied in natural language. For example, content based repairs ("I went hiking on Saturday...no, Sunday.") were not annotated. We choose to use these 6 types because they are the most common disfluencies occurring in the corpus and they are relatively easy to detect from transcripts alone.

Filled pauses, fillers, and stutters were detected semi-automatically: we first ran a keyword search for known disfluency words on the transcripts, then manually checked the annotation results to reduce mistakes such as annotating the "well" in "It works well" as a filler. For laughter, breath, and sighs, we use the annotations provided by the challenge, which were manually labelled. In this paper, we use the manually corrected "gold standard" disfluency features to establish whether our disfluency features are good predictors for emotion recognition. If so, we plan to develop methods to detect disfluencies fully automatically.

To calculate the disfluency features, we used a moving window, which contains the target word and its 14 preceding words, as shown in Figure 1. Our window works on dialogue sessions, and it slides word by word, until it reaches the end of a session. The window for w_n contains w_{n-14} to w_n when $n > 15$. When $n \leq 15$, a window that contains w_1 to w_{15} is used. We chose a window length of 15 words because this is the average length of a speaker turn, and the emotional states of the words within a speaker turn are often highly correlated.

We compute our disfluency features using equation (1):

$$D_i = \frac{t_d}{T_i} \tag{1}$$

Here, D_i is the disfluency feature D of the i-th word; t_d is the total duration of disfluency type D within the window of this word; T_i is the total utterance length

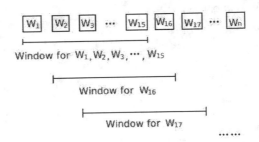

Fig. 1. The moving window

of all 15 words within this window. The reason we use durations of disfluencies instead of their counts is that duration of disfluency features also contains clues for emotions, e.g., saying "Hmm..." (longer duration) may indicate that the speaker feels more uncertain than saying "Hmm" (shorter duration).

One issue with our disfluency features is that some types of disfluency are very infrequent in the data. The percentages of non-zero values of our disfluency features are shown in Table 1 (before speaker normalization is applied). We can see that some types of disfluency are very sparse, with non-zero values occurring less than 10% of the time. However, these infrequent disfluency features may also contribute to emotion recognition by indicating where the also infrequent periods of strong emotions are located. This may help to predict the emotional values of time periods that are highly emotional, as well as those of time periods that are calm and neutral.

Table 1. Frequency of non-zero values of our disfluency features

Data set	Filled Pause	Filler	Stutter	Laughter	Breath	Sigh
train(%)	45.1	16.0	12.2	8.2	2.9	0.6
devel(%)	37.3	20.7	12.0	10.7	1.5	0.5
tests(%)	34.2	20.9	12.4	9.6	3.0	0.1

We examined two additional types of features: silent pause features, which are calculated by finding silent gaps between word timings, and window-based PMI features. However, our experiments showed that these features are not as predictive as our disfluency features. Therefore, these features and the experiments related to them will not be further discussed in this paper.

2.3   The ASM Visual Features

In our work, we use the horizontal and vertical positions of 77 facial landmarks as our visual features. The face detection and eye alignment procedure is similar to that employed in the AVEC2012 baseline visual feature extraction. To locate the facial landmarks on detected facial regions, the Active Shape Model (ASM) [16] is used. In ASM, a model for the shape of an object is first constructed from the training samples based on the geometric features calculated using PCA. This model is then applied to the test sample and iteratively fit to it. The reason we chose ASM instead of the Active Appearance Model (AAM) used by the best [4] and the second best [9] FCSC models, is that, in general, ASM works better than AAM when the test subjects are different from the training subjects [17]. In our case, we used an existing face model that is trained on the MUCT Face Database [18], so ASM is a more reasonable choice here.

The STASM (ASM with SIFT descriptors) tool [19] was used to automatically locate facial landmarks. Horizontal and vertical locations of these points are shown in Figure 2. This gives us a 154-dimension vector representing the facial expressions in each frame. We then compute a mean representation over all the frames within the duration of a word. The same moving window used in the disfluency feature extraction is applied again and the mean representation of each word within the window is concatenated. This leads to our ASM visual feature set of 2310 (154x15) features.

2.4   Speaker Normalization

We applied z-score speaker normalization to all our features to reduce the influence of speaker variance, as follows.

$$V_a' = \frac{V_a - \bar{V}_a}{Std_a} \qquad (2)$$

V_a' is the normalized value of an attribute a; V_a is the original value of attribute a; $\bar{V}_a$ is the mean value of attribute a over all the samples extracted from the speaker; Std_a is the standard deviation of values of attribute a. Speaker normalization is applied after grouping the data by speaker.

2.5   Modality Fusion and Feature Engineering

In our work, we applied a FL modality fusion method and concatenated our disfluency and ASM visual feature sets into one set. Simple concatenation without any further feature engineering is referred to as Basic-FL in the following. The feature set used by our Basic-FL model contains 2316 features.

We also study the influence of two feature engineering methods on the Basic-FL model. The first method is PCA, which maps the original features to a lower dimensional space, thus reducing the size and redundancy of the feature set. After reserving 99% of the total variance, the new feature set contains 59 transformed features.

The second method is CFS [21], which ranks features based on their predictiveness and correlation with other features. The predictiveness of features is evaluated by building single feature classifiers. Features with the best performance and low redundancy are iteratively added to our starting set containing

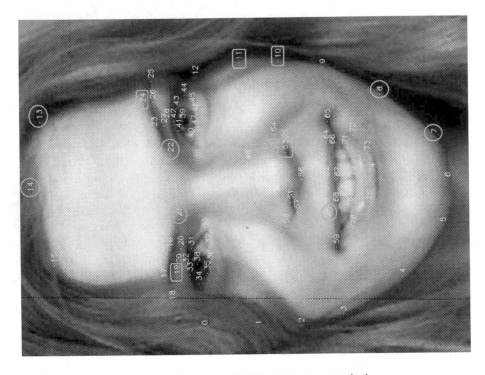

Fig. 2. Locations of 77 facial landmarks [20]

the 6 disfluency features, until the performance decreases. The CFS-FL model uses different feature sets when predicting different emotion dimensions, and there are at most 22 features (for Valence) in these subsets. The variance in the number and features contained in the subsets for predicting different emotion dimension also highlights the varying relationship between features and these dimensions.

Taking a closer look at the feature set selected and referring to the annotations of facial landmarks shown in Figure 2, we also find that the small number of visual features selected by the CFS method often represents facial expression changes within the moving window. For example, the 8 visual features selected when predicting Arousal values are w_1y_{21}, w_3y_{22}, w_5y_{21}, w_7y_{21}, w_9x_{24}, $w_{10}y_{22}$, $w_{12}y_{22}$, $w_{15}y_{21}$. Here, w_i represents the i-th word in the window. We use w_ix_j and w_iy_j to represent the averaged horizontal and vertical positions of the facial landmark numbered at j during word w_i. In Figure 2, we can see that the facial landmarks No.21 and No.22 are the inner corners of eyebrows. This subset of visual features mainly describes the vertical movements of these two key facial points within our window. Other important facial landmarks are also labelled in Figure 2. Those marked by circles indicate that the vertical movement of this point is used, and those marked by squares indicate that the horizontal movement is used.

2.6   Regression and Evaluation Metric

We use Support Vector Regression (SVR [22]) as our regression method for comparability with the AVEC2012 baseline and top performing models. Following the settings of the best WLSC model [5], we implemented epsilon-SVR with a linear kernel using the LibSVM [23] toolbox on the WEKA [24] interface. Before regression begins, all features are normalized to range [0, 1] in the regression models. The AEPV values are predicted independently. We use training, development, and test sets as set out by the AVEC guidelines.

In the AVEC2012 challenge, Cross-Correlation Score (CCS), which is the average of correlation-coefficients of all 32 test sessions, was defined as the evaluation metric. The value range of CCS is from 0 to 1, with higher scores representing better performance. In this paper, we evaluate significance of differences between CCS scores using a two-tailed z-test after Fisher's r-to-z transformation.

3   Experiments and Results

Results of our experiments are shown in Table 2. In Table 2, DF is our disfluency feature model; PMI is the sparse PMI feature model used by Savran et al. [5]; SP is the AVEC2012 WLSC baseline audio model using SP features; ASM is our ASM visual feature model; R-LBP is the dimensionality-reduced LBP model used by Savran et al. [5]; LBP is the AVEC2012 WLSC baseline visual model using LBP features; B-FL, P-FL, C-FL are our Basic-FL, PCA-FL, CFS-FL bimodal models; AV-B, AV-1, AV-2, AV-3 are the AVEC2012 WLSC baseline, the best [5], the second best [25], and the third best [26] audio-visual (AV) models.

3.1   The Disfluency Feature Model

As shown in Table 2, compared to all the other unimodal models, our disfluency feature model has the best performance when predicting all four emotion dimensions. Its overall performance is not significantly different from the best result of WLSC ($p = 0.424$).

As shown in Figure 3, our disfluency feature model outperforms the baseline model using SP features. Comparing our disfluency features with the PMI-based lexical features used by Savran et al. [5], our features look at data from a higher level, and give significantly better performance when predicting all emotion dimensions. This is consistent with our hypothesis that using high-level features will improve model performance.

Note that we only have 6 disfluency features, while there are 1842 SP features and 1000 PMI features. This huge difference in dimensionality will influence the efficiency of the emotion recognizer greatly, and lower dimensional features are often preferred, especially in real-time interactive systems. Therefore, our high-level disfluency features are more predictive and more efficient.

We also compared the predictiveness of different types of disfluency using the rank generated by the CFS method (see Section 2.5). Results are shown in

Table 2. Experimental results

Models	A	E	P	V	Mean
DF	0.250	0.313	0.288	0.235	0.271
PMI	0.131	0.285	0.254	0.188	0.214
SP	0.014	0.038	0.016	0.040	0.027
ASM	0.205	0.246	0.172	0.219	0.211
R-LBP	0.184	0.156	0.146	0.226	0.178
LBP	0.005	0.012	0.018	0.005	0.011
B-FL	0.205	0.274	0.223	0.207	0.227
P-FL	0.214	0.268	0.269	0.225	0.244
C-FL	0.274	0.258	0.266	0.215	0.253
AV-B	0.021	0.028	0.009	0.004	0.015
AV-1	0.302	0.194	0.293	0.331	0.280
AV-2	0.210	0.240	0.289	0.208	0.237
AV-3	0.267	0.241	0.223	0.138	0.192

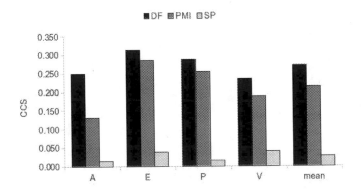

Fig. 3. The disfluency feature model

Table 3. The rank of a disfluency feature when predicting a particular emotion dimension is from 1.0 to 6.0, with 1.0 representing the highest predictiveness. The results indicate that filled pauses and laughter are the most predictive disfluency features in this task.

3.2 The ASM Visual Feature Model

As seen in Figure 4, for all four emotion dimensions our high-level ASM visual features are more predictive than the pixel-level LBP features extracted from the whole facial region. Our model also outperformed the feature-selected LBP model on most of the emotion dimensions. On the Valence dimension, our model has slightly lower CCS, but the difference is not significant ($p = 0.549$). These results verified our hypothesis that our high-level ASM features are more predictive then the low-level LBP features.

Table 3. Predictiveness rank of different disfluency features

Disfluency	A	E	P	V	mean
Filled pause	1.0	2.0	1.0	2.0	1.5
Filler	6.0	5.0	5.0	6.0	5.5
Stutter	5.0	6.0	6.0	3.0	5.0
Laughter	2.0	1.0	2.0	1.0	1.5
Breath	4.0	3.0	4.0	5.0	4.0
Sigh	3.0	4.0	3.0	4.0	3.5

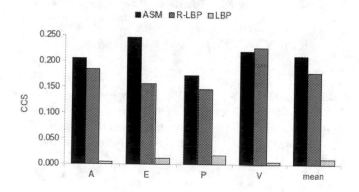

Fig. 4. The ASM visual feature model

3.3   The Bimodal Models

The performance of our unimodal models (DF and ASM) and our bimodal models (B-FL, P-FL, and C-FL) is shown in Figure 5. As we can see, our disfluency feature model outperforms our ASM visual model on all emotion dimensions. After simple concatenation of the feature sets, the increase on mean CCS of the lower-performing visual modality is not significant ($p = 0.168$). Recall that there are only 6 disfluency features, while there are 2310 ASM visual features. This suggests that the large visual feature set is dominating and introduces noise into the model. We can see that the PCA-FL model and the CFS-FL model perform better than the Basic-FL model in general, thus applying feature engineering to the concatenated feature set helps to reduce the influence of noisy visual features. These two feature-engineered bimodal models both give significant improvements on mean CCS compared to the lower-performing visual model. However, compared to the Basic-FL model, only CFS gives significant improvement on mean CCS ($p = 0.024$).

We also compared our best bimodal model, the CFS-FL model, with the baseline and the best three models of AVEC2012 WLSC. As shown in Figure 6, performance of our bimodal model is significantly better than the AVEC2012 WLSC baseline model when predicting all four dimensions of emotions, as expected.

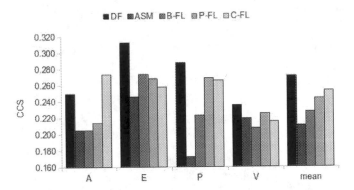

Fig. 5. Modality fusion and feature engineering

Comparing the overall performance (mean CCS), our model is significantly better than the third best WLSC model [26]. Our model also outperformed the second best WLSC model [25], but the difference is not significant ($p = 0.165$). When predicting Expectancy values, our model gives the highest CCS. The reason may be that Expectancy is the easiest dimension to predict for both our disfluency and ASM features, as shown in Figures 3 and 4.

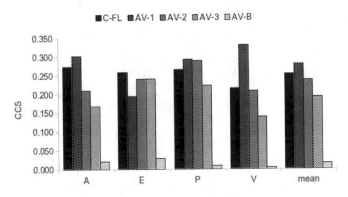

Fig. 6. Comparing our bimodal model with the AVEC2012 WLSC results

4   Discussion

Based on the experimental results, we verified our main hypothesis that using high-level features, namely the disfluency features and the ASM features, improves model performance compared to using low-level features.

The average performance of our disfluency model using only the 6 disfluency features is not significantly different from the best audio-visual model in AVEC2012 WLSC. This indicates that disfluency features are especially powerful in emotion prediction. Our disfluency model also outperformed the PMI model used by Savran et al. [5]. This suggests that learning other high level classes of lexical features may be useful for this task.

In the future, we plan to study whether the disfluency features are still highly predictive of emotion when using other corpora. The utility of disfluencies also depends on how well they can be detected. Further work will investigate performance using disfluencies detected from the output of an automatic speech recognizer, rather than manual transcription. Similarly, integrating work on developing a fully automatic disfluency detection method based on existing studies should be helpful for this task. For example, Liu et al [27] use a Hidden-Markov Model that combines textual and prosodic clues to detect disfluencies. The work of Niewiadomski et al. [28] also highlights the importance of automatic laughter detection.

The visual feature subset selected by the CFS method illustrates a way to compute visual features that also have longer duration. In the future, we will use the position changes of a subset of the 77 facial landmarks over the window as visual features, thus further reducing the dimensionality of the visual feature set.

Our experimental results also verify that fusing modalities can give improvements compared to the lower-performing unimodal model. However, it is difficult for FL fusion models to improve on the better-performing unimodal model. The fact that feature-engineered models do not provide huge gains may be due to a lack of control in feature weighting and, as such, DL fusion may be more appropriate for the task. Compared to FL fusion, DL fusion has the natural advantage of flexibility when weighting different modalities. In the future, we plan to apply DL fusion to our model and study whether or not it improves performance.

Finally, the low CCS of all models in the AVEC2012 may indicate that CCS is not the best evaluation metric for this task. CCS evaluates average performance of the classifier for predicting values of all data in the corpus. However, occurrences of strong emotions are relatively rare in conversations, which makes the values of a large portion of the data unsuitable for classifiers that are designed to predict such emotions. Therefore, a more appropriate evaluation metric is needed. One possible alternative would be to detect emotionally strong events first, using methods such as those previously used for the detection of hot spots in meetings [29], and only evaluate model performance on these segments.

5   Conclusions

In this paper, we introduced a new emotion recognition approach that used a small number of human-interpretable high-level features. Our unimodal and bimodal models built using these features have significantly better performance compared to the baseline models, which used a large number of low-level audio, visual, or lexical features. In fact, our models have the best performance for predicting the Expectancy dimension of emotion compared to all AVEC2012 competitors. Using only 6 disfluency features, we built a model with performance not significantly different from the best AVEC2012 bimodal model overall. Previous studies on automatic disfluency detection also give us reason to believe that these features can be computed automatically in real or near-real time with reasonable accuracy. This in turn would allow the development of a fast and accurate emotion classifier which holds promise for future applications in interactive systems.

Acknowledgments. We are grateful to Dr. Ani Nenkova for discussion about her research [5], and to Dr. Michel Valstar for answering questions about the AVEC2012 challenge and database.

References

1. D'Mello, S., Jackson, T., Craig, S., Morgan, B., Chipman, P., White, H., Person, N., Kort, B., el Kaliouby, R., Picard, R., et al.: AutoTutor detects and responds to learners affective and cognitive states. In: Workshop on Emotional and Cognitive Issues at the International Conference on Intelligent Tutoring Systems (2008)
2. Busso, C., Deng, Z., Yildirim, S., Bulut, M., Lee, C.M., Kazemzadeh, A., Lee, S., Neumann, U., Narayanan, S.: Analysis of emotion recognition using facial expressions, speech and multimodal information. In: Proceedings of the 6th International Conference on Multimodal Interfaces, pp. 205–211. ACM (2004)
3. Schuller, B., Valster, M., Eyben, F., Cowie, R., Pantic, M.: AVEC 2012: the continuous audio/visual emotion challenge. In: Proceedings of the 14th ACM International Conference on Multimodal Interaction, pp. 449–456. ACM (2012)
4. Nicolle, J., Rapp, V., Bailly, K., Prevost, L., Chetouani, M.: Robust continuous prediction of human emotions using multiscale dynamic cues. In: Proceedings of the 14th ACM International Conference on Multimodal Interaction, pp. 501–508. ACM (2012)
5. Savran, A., Cao, H., Shah, M., Nenkova, A., Verma, R.: Combining video, audio and lexical indicators of affect in spontaneous conversation via particle filtering. In: Proceedings of the 14th ACM International Conference on Multimodal Interaction, pp. 485–492. ACM (2012)
6. Ojala, T., Pietikainen, M., Maenpaa, T.: Multiresolution gray-scale and rotation invariant texture classification with local binary patterns. IEEE Transactions on Pattern Analysis and Machine Intelligence 24, 971–987 (2002)
7. Scherer, K.R.: Expression of emotion in voice and music. Journal of Voice 9, 235–248 (1995)
8. Devillers, L., Vidrascu, L., Lamel, L.: Challenges in real-life emotion annotation and machine learning based detection. Neural Networks 18, 407–422 (2005)
9. Soladié, C., Salam, H., Pelachaud, C., Stoiber, N., Séguier, R.: A multimodal fuzzy inference system using a continuous facial expression representation for emotion detection. In: Proceedings of the 14th ACM International Conference on Multimodal Interaction, pp. 493–500. ACM (2012)
10. Silva, L.C., Miyasato, T.: Degree of human perception of facial emotions based on audio and video information. IEICE Technical Report. Image Engineering 96, 9–15 (1996)
11. Chen, L.S., Huang, T.S., Miyasato, T., Nakatsu, R.: Multimodal human emotion/expression recognition. In: Proceedings of Third IEEE International Conference on Automatic Face and Gesture Recognition, pp. 366–371. IEEE (1998)

12. McKeown, G., Valstar, M.F., Cowie, R., Pantic, M.: The SEMAINE corpus of emotionally coloured character interactions. In: IEEE International Conference on Multimedia and Expo (ICME), pp. 1079–1084. IEEE (2010)
13. Fontaine, J.R., Scherer, K.R., Roesch, E.B., Ellsworth, P.C.: The world of emotions is not two-dimensional. Psychological Science 18, 1050–1057 (2007)
14. Eyben, F., Wöllmer, M., Schuller, B.: OpenSMILE: the munich versatile and fast open-source audio feature extractor. In: Proceedings of the International Conference on Multimedia, pp. 1459–1462. ACM (2010)
15. Viola, P., Jones, M.: Rapid object detection using a boosted cascade of simple features. In: Proceedings of the 2001 IEEE Computer Society Conference on Computer Vision and Pattern Recognition, CVPR 2001, vol. 1, pp. I–511. IEEE (2001)
16. Cootes, T.F., Taylor, C.J., Cooper, D.H., Graham, J.: Active shape models-their training and application. Computer vision and image understanding 61, 38–59 (1995)
17. Cootes, T.F., Edwards, G.J., Taylor, C.J.: Comparing active shape models with active appearance models. BMVC 99, 173–182 (1999)
18. Milborrow, S., Morkel, J., Nicolls, F.: The MUCT landmarked face database. Pattern Recognition Association of South Africa 201 (2010)
19. Milborrow, S., Nicolls, F.: Active shape models with sift descriptors and mars 1, 5 (2014)
20. Milborrow, S.: Stasm User Manual (2013), http://www.milbo.users.sonic.net/stasm
21. Hall, M.A.: Correlation-based Feature Subset Selection for Machine Learning. PhD thesis, University of Waikato, Hamilton, New Zealand (1998)
22. Drucker, H., Burges, C.J., Kaufman, L., Smola, A., Vapnik, V.: Support vector regression machines. Advances in Neural Information Processing Systems, 155–161 (1997)
23. Chang, C.C., Lin, C.J.: LIBSVM: a library for support vector machines. ACM Transactions on Intelligent Systems and Technology (TIST) 2, 27 (2011)
24. Hall, M., Frank, E., Holmes, G., Pfahringer, B., Reutemann, P., Witten, I.H.: The WEKA data mining software: an update. ACM SIGKDD Explorations Newsletter 11, 10–18 (2009)
25. Ozkan, D., Scherer, S., Morency, L.P.: Step-wise emotion recognition using concatenated-HMM. In: Proceedings of the 14th ACM International Conference on Multimodal Interaction, pp. 477–484. ACM (2012)
26. van der Maaten, L.: Audio-visual emotion challenge 2012: a simple approach. In: Proceedings of the 14th ACM International Conference on Multimodal Interaction, pp. 473–476. ACM (2012)
27. Liu, Y., Shriberg, E., Stolcke, A., Hillard, D., Ostendorf, M., Harper, M.: Enriching speech recognition with automatic detection of sentence boundaries and disfluencies. IEEE Transactions on Audio, Speech, and Language Processing 14, 1526–1540 (2006)
28. Niewiadomski, R., Hofmann, J., Urbain, J., Platt, T., Wagner, J., Piot, B., Cakmak, H., Pammi, S., Baur, T., Dupont, S., et al.: et al.: Laugh-aware virtual agent and its impact on user amusement. In: Proceedings of the 2013 International Conference on Autonomous Agents and Multi-agent Systems, pp. 619–626. International Foundation for Autonomous Agents and Multiagent Systems (2013)
29. Lai, C., Carletta, J., Renals, S.: Detecting summarization hot spots in meetings using group level involvement and turn-taking features. In: Proceedings of Interspeech 2013, Lyon, France (2013)

Constructing Context-Aware Sentiment Lexicons with an Asynchronous Game with a Purpose

Marina Boia, Claudiu Cristian Musat, and Boi Faltings

École Polytechnique Fédérale de Lausanne
Switzerland
firstname.lastname@epfl.ch

Abstract. One of the reasons sentiment lexicons do not reach human-level performance is that they lack the contexts that define the polarities of words. While obtaining this knowledge through machine learning would require huge amounts of data, context is commonsense knowledge for people, so human computation is a better choice. We identify context using a game with a purpose that increases the workers' engagement in this complex task. With the contextual knowledge we obtain from only a small set of answers, we already halve the sentiment lexicons' performance gap relative to human performance.

1 Introduction

Sentiment analysis identifies expressions of subjectivity in texts, such as sentiments or emotional states. We consider the sentiment classification task, which determines whether the sentiments expressed in a text are positive or negative. This task requires commonsense knowledge about the polarities of sentiment words.

The relative ease of construction led early researchers in the field toward corpus-based sentiment classification [1–3]. These methods aggregate statistical, syntactic, and semantic relations between words. A significant downside is that the classifiers that result are efficient only on narrow domains. This may be the reason why the competing, lexicon-based approach is currently the backbone of sentiment classification. Several sentiment lexicons [4–6] have been available for a significant period of time. However, multiple lexicons continue to appear [7, 8], showing that a satisfying solution has not yet been found.

The most successful methods perform syntactic preprocessing to extract relevant words, and then consider the resulting set of independent words as features of the text. Sentiment classification is performed on these features, by adding word polarity scores compiled in sentiment lexicons or learned with statistical methods. These models obtain from 60% to 80% accuracy [2, 1]. Better results can sometimes be achieved by training domain-specific classifiers, but only at the expense of narrow coverage. This performance is lower than that of people, who can extract sentiment with 80% to 90% agreement [9], depending on the domain of the texts.

A reason why these classifiers cannot reach human-level performance is that the words' polarities are influenced by context: a *small hotel room* is negative, while a *small digital camera* is positive. By representing texts as independent words, context

A. Gelbukh (Ed.): CICLing 2014, Part II, LNCS 8404, pp. 32–44, 2014.

is ignored. In narrow domains, words mostly occur in a single context, thus high accuracy can be achieved. For broad domains, it is necessary to enrich the feature set with contexts, by including word combinations. However, the complexity of the resulting models would explode, and it would no longer be feasible to acquire them from data. Nevertheless, the polarity of most words has only a few exceptions, so the size of these models could be manageable if these exceptions are identified. This is very difficult to do by statistical methods, but easy for people. We thus investigate how we can obtain contextual knowledge through human computation.

Most human computation tasks consist of simple tasks, such as object labeling [10]. We ask workers to find the contexts that influence the polarities of words. Because our task is more difficult, workers can become demotivated and give sloppy answers. Moreover, there is considerable freedom in indicating context, thus quality assurance by agreement with peers is unfeasible.

To increase engagement, we develop a novel task design that combines the entertaining aspect of games with the large worker pool available on a paid crowdsourcing platform. Workers play a game in rounds, where in each round they increase their score by submitting answers that contain sentiment words, contexts, and polarities. At the end, workers receive a payment proportional to their score.

To solve the quality assurance issue, we develop a scoring mechanism that elicits useful answers. The scoring rewards answers using a model that we derive from existing sentiment knowledge and the workers' input. We thus create the illusion of a game played with others, which ensures quality and also makes the task fun.

To further boost quality, we improve task understanding with tutorials. We develop a static tutorial with textual information, as well as an interactive one with quizzes. We show that the latter greatly increases worker performance.

The output of the game is a context-dependent sentiment lexicon. Our contextual knowledge refines the polarities of sentiment words with contexts, so it naturally complements the standard, context-independent lexicons. We thus assess how our lexicon improves several standard ones. Even with a small number of answers, we considerably improve the performance of two context-independent lexicons. For the lexicon that performs best on our corpus, we increase the accuracy from 68% to almost 75%, halving its gap relative to human performance.

This paper thus makes several contributions:

- To the best of our knowledge, we are the first that use human computation to acquire contextual knowledge for sentiment classification.
- We obtain a context-dependent lexicon that significantly improves sentiment classification accuracy.
- We develop a game that increases the workers' motivation in a complex task.
- We create the illusion of synchronous worker interaction using a model that rewards useful answers, thus ensuring quality.
- We analyze the effect of tutorials on worker performance.

Section 2 overviews the related work, Section 3 defines the context, and Section 4 presents our game. Subsequently, Section 5 discusses quality assurance, Section 6 presents our experiments and results, and Section 7 draws conclusions.

2 Related Work

There are two main directions in sentiment analysis. Lexicon-based methods use sentiment lexicons, which are out-of-the-box lists of sentiment words paired with polarities. Several lexicons [4–6, 11] have been available for a significant period of time. However, multiple ones continue to appear. [7] combined ANEW [12] and WordNet [13] to create a superior lexicon. [8] developed a domain-specific lexicon using a random walk algorithm, while [14] extended a lexicon with additional knowledge, such as parts of speech and word senses. The continual emergence of similar methods for a known problem shows a satisfactory solution has still not been found. A key problem of these lexicons is that they do not consider the influence of context on the polarities of words.

Corpus-based methods learn sentiment words and their polarities from text corpora. This is typically done with machine learning algorithms applied to annotated datasets [1], or by aggregating syntactic and semantic relations between words [2, 3]. These methods do not explicitly take context into account. However, when the texts target narrow enough domains, the resources learned become domain-specific, with sentiment words that occur in only a single context. This is why the corpus-based methods typically perform much better than their lexicon-based counterparts, but only in the domain of the corpus.

Context is important in sentiment analysis: [15] noted that words may have different polarities, depending on the features they describe. [16] learned a taxonomic organization of the entities, features, and sentiment words in a domain. They used this taxonomy to represent the contextual variations of sentiment words. Similarly, [17] generated context-dependent lexicons of sentiment words paired with features. However, expanding these approaches beyond a limited set of features is hard without a priori domain knowledge. We believe context can be more effectively identified through human computation.

Previous human computation applications to sentiment analysis have been limited to eliciting simple polarity annotations. Training data have been obtained through tasks that required polarity or emotion annotation of texts [18]. Several lexicons have been produced through polarity or emotion annotation of some predefined vocabularies [19]. The tasks in [20, 21] were more complex and required humans to select both sentiment words and polarities. Most of these annotation tasks had a simple design that motivated workers with payment, whereas others were designed as games [19–21]. We no longer focus on basic annotations and selections. We require workers to complete a more complex task, by characterizing the contexts in which sentiment words have certain polarities.

The outcome of a task is highly sensitive to the workers' motivation. One way to inspire motivation is through payment, as in online labor markets, such as Amazon Mechanical Turk [1]. Another way is through enjoyment - tasks can be packaged as games with a purpose, when players submit answers and are rewarded with points, reputation, badges etc. In human computation, the two have been mutually exclusive so far, and games have not yet taken advantage of the large worker pools available on

---

[1] www.mturk.com

crowdsourcing platforms. Instead, we combine enjoyment and payment, and obtain a game played for money, like poker or like the games on Swagbucks [2].

A further consideration is quality assurance, which can be done before, during, or after workers complete the task. During the task, output agreement setups place workers in teams and require them to agree on their answers [10]. Posterior measures are applied after the work is done, by filtering or aggregating data to remove answers that are irrelevant or malicious [22]. Preemptive measures are applied before collecting answers, by making workers aware of the desired level of performance. Tutorials aim to induce a basic understanding of the task and to explain what kind of answers are required [23]. We employ all three types of quality control.

3 Context Definition

We use human computation to obtain contextual knowledge for sentiment classification. We structure this knowledge using the following concepts:

- A phrase *phr* is a word construct that can carry sentiment.
- A context *con* is a word construct in the presence of which a phrase carries sentiment.
- A polarity *pol* is the positive *pos* or negative *neg* orientation of the sentiment conveyed by a word construct.

Depending on whether a context is needed to define its polarity, a phrase can be unambiguous or ambiguous. The unambiguous phrases have the same polarity in every context: the word *amazing* is always positive. The ambiguous phrases have different polarities depending on the context: the word *high* is positive in the context of *salary*, but negative in the context of *debt*.

The most common phrases are typically compiled in sentiment lexicons. Given a phrase vocabulary P, these lexicons pair phrases with their default polarities: $L = \{(phr, pol)|phr \in P, pol \in \{pos, neg\}\}$. This context-independent representation either includes ambiguous phrases with polarities that do not make sense, or excludes them altogether. Instead, we consider a more articulate representation that is sensitive to context. Given an additional context vocabulary C, context-dependent lexicons include the contexts that disambiguate the polarities of ambiguous phrases: $CL = \{(phr, con, pol)|phr \in P, con \in C, pol \in \{pos, neg\}\}$.

4 Human Computation Task

To build context-dependent lexicons, we ask workers to find the contexts that disambiguate the ambiguous phrases. Our task thus requires cognitive engagement, so workers might quickly lose interest in it. It is unclear if extrinsic motivators alone can keep workers interested. In previous experiments, we obtained poor results for a simple polarity annotation task where we incentivized colleagues with prizes. To motivate workers, we make our task fun by designing it as a game.

---

[2] www.swagbucks.com

4.1   Task Design

In our task, workers see text fragments from which they submit answers that contain a phrase, a context, and a polarity. Workers construct an answer by selecting a phrase and a context from a text fragment, and annotating the resulting word combination with a polarity. For instance, from the text *I don't like this camera. It has tiny buttons*, workers could submit the answer $(tiny, buttons, neg)$.

We motivate workers through enjoyment and payment. Enjoyment impacts intrinsic motivation. We entertain the workers with point rewards that reflect the usefulness of their answers. Payment targets extrinsic motivation. Once they finish playing, workers receive monetary rewards proportional to their scores. By combining enjoyment and payment, we obtain a game played for money.

We elicit useful answers with an intelligent scoring mechanism. An answer is useful if it has common sense and is novel. We consider an answer commonsensical if it agrees with the contextual knowledge acquired in the game, which means that it complies with the opinion of many workers. We consider an answer novel if it greatly impacts the contextual knowledge, which means that it is given early in the game or that it contains ambiguous phrases and disambiguating contexts. We reward answers with scores that are the sum of an agreement score and a novelty score (Section 4.3).

4.2   Gameplay

Guesstiment (Figure 1) is a round-based game. In each round, a worker sees a text fragment, from which she submits an answer. She then receives score and payment rewards, and starts a new round. The worker constructs an answer in three steps:

  – *Step 1: phrase selection*, when she selects a phrase from the text fragment.
  – *Step 2: context selection*, when she optionally selects a context.
  – *Step 3: polarity annotation*, when she annotates the resulting phrase and context combination with a polarity.

The worker has the option to skip to the next round without submitting an answer, if the text fragment does not contain one.

4.3   Scoring Mechanism

To compute score rewards, we use a model that contains our beliefs on the polarities of phrase and context combinations. To a phrase phr and a context con (possibly empty: $con = nil$), we associate a Beta distribution [24] from which we estimate the probabilities that, in the context con, the phrase phr has positive and negative polarities respectively: $\Pr(pos|phr, con)$ and $\Pr(neg|phr, con)$. We assign a Beta distribution to every word and word combination that appear in the sentences of our corpus $Train$ (Section 6.1). We initialize these distributions using word frequencies in positive and negative documents, as well as word polarities from context-independent lexicons. We incorporate incoming answers by modifying the distributions through a Bayesian update process. Therefore, the probabilities $\Pr(pos|phr, con)$ and $\Pr(neg|phr, con)$ assimilate the fractions of positive and negative answers respectively: (phr, con, pos) and (phr, con, neg).

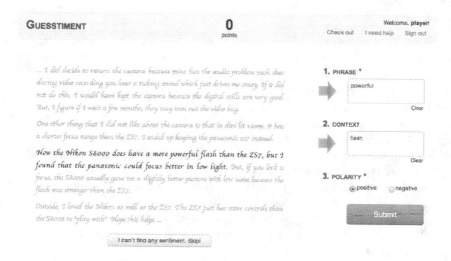

Fig. 1. The game interface

For an answer (phr, con, pol), we compute an agreement score $ag \in [0, ag_{max}]$ that reflects if it is commonsensical. We reward the answer with a high score if it agrees with the scoring model early in the game, because it substantially improves the model's confidence. We compute a low score if the answer disagrees with the model early in the game, because it substantially harms the model's confidence. We give a medium score if the answer comes late in the game, because it has less impact on the model. We use the entropy over $\Pr(pol|phr, con)$ to measure the model's uncertainty in the polarity of the phrase and context combination. The answer either decreases or increases entropy, depending on whether it agrees or disagrees with the model. Moreover, the change in entropy is bigger when the answer is submitted early in the game. We thus compute ag by linearly mapping the entropy update to $[0, ag_{max}]$.

For the answer (phr, con, pol), we also compute a context novelty score $cn \in [0, cn_{max}]$ that reflects if the answer contains an ambiguous phrase and a well-defined context. We use the model's uncertainty in the phrase's out of context polarity as an indicator for the phrases's ambiguity. If $con = nil$, we set $cn := 0$. Otherwise, we use the entropy over $\Pr(pol|phr, nil)$ to measure the ambiguity of phr. We thus compute cn by linearly mapping the entropy to $[0, cn_{max}]$.

We reward an answer with a score that is the sum of the agreement score and the context novelty score. Because we do not want to gather only unique answers, we give a bigger importance to agreement. We control this by making $ag_{max} > cn_{max}$, and we use $ag_{max} := 40$ and $cn_{max} := 10$.

5   Quality Assurance

We control quality before, during, and after the workers' activity. The scoring mechanism ensures answer quality during the game. Before the game, we introduce tutorials. After the game, we filter out the bad answers, and we aggregate the remaining ones.

5.1 Tutorials

With tutorials, we ensure workers understand the task. We create a static tutorial with text instructions only. We divide the instructions into sections that explain: the concepts of phrase, context, and polarity; how to construct an answer in a game round; and how the scoring mechanism works. We illustrate the scoring with examples of simple game rounds, along with potential answers and score updates.

We used the static tutorial when we first launched the game, but noticed that the quality of answers was not high. Therefore, we created another tutorial that teaches workers with text instructions and interactive quizzes. The first quiz presents several word constructs and asks which words are phrases and which words are contexts. This quiz also asks which phrases are ambiguous and which ones are unambiguous. The second quiz emulates the game interface (Figure 1) and asks workers to construct a simple answer. The final quizzes also emulate the game interface and ask workers to give specific types of answers: with an ambiguous or unambiguous phrase, with or without a context.

5.2 Answer Filtering

From the answers $A = \{(phr, con, pol) | phr \in P, con \in C, pol \in \{pos, neg\}\}$, we remove the ones that are not commonsensical. As filtering takes place only when all answers have been gathered, we can use a more precise evaluation than for the scoring mechanism (Section 4.3), based on actual agreement between workers:

Heuristic *wa*. An answer (phr, con, pol) is commonsensical if its phrase phr and context con have been included in answers by at least two workers. These answers form the *worker agreement* set: $wa(A)$.

We initially used the worker agreement heuristic, but noticed that a lot of answers were removed because workers could not agree on the same interpretation of context. Therefore, we introduced other heuristics in which we check whether answers contain proper phrases and contexts (Section 3), as workers are more likely to agree on well-constructed answers. We thus consider an answer (phr, con, pol) commonsensical if:

Heuristic *slr*. its phrase phr is a word from context-independent lexicons, and its context con is a noun or a verb. These answers form the *sentiment lexicon restrictive* set: $slr(A)$.

Heuristic *slp*. its phrase phr is at most two words with at least one from context-independent lexicons, and its context con is at most two words with at least one a noun or a verb. These answers form the *sentiment lexicon permissive* set: $slp(A)$.

Heuristic *psr*. its phrase phr is an adjective or an adverb, and its context con is a noun or a verb. These answers form the *part of speech restrictive* set: $psr(A)$.

Heuristic *psp*. its phrase phr is at most two words with at least one an adjective or an adverb, and its context con is at most two words with at least one a noun or a verb. These answers form the *part of speech permissive* set: $psp(A)$.

We combine these heuristics to obtain two higher-order ones, which consider an answer commonsensical if:

Heuristic *res*. it complies with the worker agreement heuristic or with the two restrictive ones. These answers form the *restrictive* set: $res(A)$.

Heuristic *per*. it complies with the worker agreement heuristic or with the two permissive ones. These answers form the *permissive* set: $per(A)$.

5.3   Answer Aggregation

After we filter out the bad answers, we aggregate the remaining ones to form the set: $AA = \{(phr, con, wkr^{pos}, wkr^{neg}) | phr \in P, con \in C, wkr^{pos}, wkr^{neg} \in \mathbb{N}\}$. An aggregate answer $(phr, con, wkr^{pos}, wkr^{neg})$ contains a phrase and a context that appear together in answers, along with the number of workers that included them in positive and negative answers respectively. We use the aggregate answers to obtain a context-dependent lexicon: $CL = \{(phr, con, pol)\}$. For each aggregate answer $(phr, con, wkr^{pos}, wkr^{neg}) \in AA$, we add (phr, con, pos) to CL if $wkr^{pos} > wkr^{neg}$. Conversely, we add (phr, con, neg) to CL if $wkr^{pos} < wkr^{neg}$.

6   Experiments and Results

6.1   Dataset

To construct the game rounds, we used Amazon [3] product reviews of four categories of vacuum cleaners and digital cameras respectively. We used the reviews' numerical ratings as a gold standard for their polarities, and we ensured each product category had equal class representation. We randomly split the positive and negative reviews from each product category into training and test data, in a ratio of 2 : 1. We used the training data $Train$ to obtain two sentiment classifiers: a sentiment lexicon and a support vector machine. We applied these classifiers on the test data $Test$, and identified the reviews $Test_{game}$ that both misclassified. We constructed the game rounds using 2000 sentences that we extracted from $Test_{game}$. To test our context-dependent lexicons, we removed $Test_{game}$ from $Test$ and obtained a new test set $Test_{\overline{game}}$.

6.2   Worker Participation

We deployed the game on Amazon Mechanical Turk. We ran a first HIT, with the static tutorial, for a week. We ran a second HIT, with the quiz tutorial, for another week. Both were visible to only United States residents and had a base payment of $0.1, with

---

[3] www.amazon.com

bonuses of $0.1 for every 100 and 500 points milestones reached. In the first HIT, the game was accessed by 71 workers, 75% of which completed the tutorial and played at least one round. In the second HIT, 279 workers accessed the game, 27% of which completed the tutorial and played at least one round.

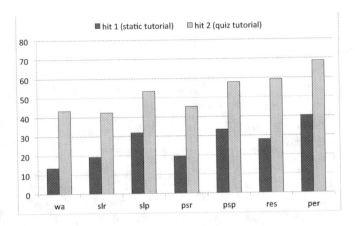

Fig. 2. The *number* of commonsensical answers per worker (according to the heuristics for common sense in Section 5.2)

On average, the workers in the first HIT played 95.51 rounds and gave 73.60 answers, while the ones in the second HIT played 105.52 rounds and gave 85.84 answers. Moreover, the workers in the first and second HITs earned 19.81 and 23.06 points per answer respectively. Two-tailed t-tests showed that these differences are significantly relevant at the 1% confidence level. The increase in performance shows that the workers that passed the quiz tutorial were more engaged, playing more and better.

6.3   Answer Usefulness Evaluation

We analyzed the frequency of useful answers. Figures 2 and 3 show that the workers in the second HIT submitted more commonsensical answers than the ones in the first HIT, according to the heuristics in Section 5.2. Two-tailed t-tests showed that these differences are significantly relevant at the 1% confidence level. Moreover, in the second HIT, workers submitted answers of which, on average, 4% paired phrases from context-independent lexicons with contexts that inverted their polarities; and 42% paired new phrases with well-defined contexts. Therefore, the workers who completed the quiz tutorial were more consistent about submitting answers with common sense, and they also identified new knowledge.

6.4   Context-Dependent Lexicon Evaluation

Context-Aware Classification. We classify documents by complementing a context-independent lexicon $L = \{(phr, pol^L)\}$ with a context-dependent one $CL = \{(phr\ ,$

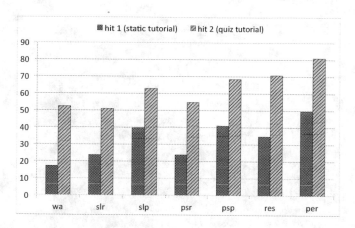

Fig. 3. The *percentage* of commonsensical answers per worker (according to the heuristics for common sense in Section 5.2)

$con, pol^{CL})\}$, thus assessing the impact of context. For each document d, we compute a sentiment score $\sigma(d)$, then we obtain a sentiment label by taking the sign of $\sigma(d)$. To compute the score $\sigma(d)$, we split d into sentences. For each sentence $s \in d$, we identify the phrases that are in CL or in L. For every phrase phr we find:

- We scan a window of t words centered around it to identify the contexts $con \neq nil$ for which we have $(phr, con, pol^{CL}) \in CL$. We use $t = 7$.
- For every $(phr, con, pol^{CL}) \in CL$ that we find, we update $\sigma(d)$ with one unit using the sign of pol^{CL}.
- If we cannot find any $(phr, con, pol^{CL}) \in CL$, we determine if $(phr, nil, pol^{CL}) \in CL$ and if $(phr, pol^{L}) \in L$:
  - If $(phr, pol^{L}) \in L$, we update $\sigma(d)$ with one unit using the sign of pol^{L}.
  - If $(phr, pol^{L}) \notin L$ but $(phr, nil, pol^{CL}) \in CL$, we update $\sigma(d)$ with one unit using the sign of pol^{CL}.

Context Impact. We analyzed the performance of our context-dependent lexicons. From the HITs with static and quiz tutorials, we obtained two lexicons[4]: CL^{static} and CL^{quiz}. We compared them with three context-independent lexicons: General Inquirer L^{gi} [4], OpinionFinder L^{of} [6], and the lexicon of Bing Liu L^{bl} [25]. We first used the context-independent lexicons individually, then we complemented them with our context-dependent ones. We tested on the review corpus $Test_{\overline{game}}$, averaging accuracies over all product categories.

Figure 4 shows the lexicons' performance. CL^{static} increased the accuracy of L^{gi} and L^{bl} by 6% and 3.5% respectively. Moreover, CL^{quiz} improved L^{gi} and L^{bl} by 10% and 6% respectively. Two-tailed t-tests showed that these improvements are significantly relevant at the 1% confidence level. Our contextual knowledge thus successfully complemented the context-independent lexicons.

---

[4] Available at http://liawww.epfl.ch/data/lexicons.zip

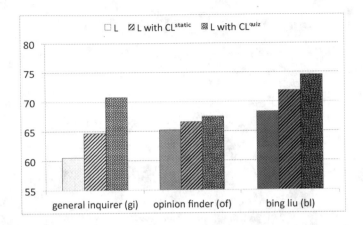

Fig. 4. Lexicon performance

Figure 5 illustrates how the lexicons' performance relates to the human performance of 83.50% [9]. For L^{gi}, CL^{quiz} reduced the gap relative to human performance by 44.52%. For L^{of}, CL^{quiz} reduced the gap by 12.51%. Finally, for L^{bl}, we decreased the gap by 41.23%. Our contextual knowledge thus halved the context-independent lexicons' deficit relative to humans.

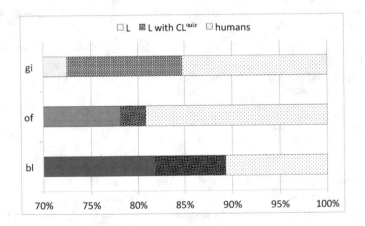

Fig. 5. Lexicon performance gap relative to human performance (83.50%)

7   Conclusions

Sentiment lexicons need commonsense knowledge about the contexts that impact the polarities of words. Because automatically extracting context would require a huge amount of data, human computation is a better alternative.

We are the first to acquire contextual knowledge with human computation. A first challenge was motivating workers to do a task that required cognitive engagement. We solved this by designing the task as a game that we launched on Amazon Mechanical Turk. A second challenge was controlling answer quality when workers were not simultaneously present and thus we could not use worker agreement. We overcame this with a scoring mechanism that assessed answer common sense and novelty using a model derived from the workers' activity.

We improved the workers' understanding of the task with two tutorials: one that used textual instructions and one that with interactive quizzes. We showed that the latter substantially improved worker performance.

We obtained contextual knowledge of good quality. Even with a small set of answers, our context-dependent lexicons substantially improved two established context-independent ones, halving their deficit relative to humans. We believe that a significantly larger number of answers could further improve performance.

References

1. Pang, B., Lee, L., Vaithyanathan, S.: Thumbs up?: Sentiment classification using machine learning techniques. In: Proceedings of the ACL Conference on Empirical Methods in Natural Language Processing, pp. 79–86 (2002)
2. Turney, P.D.: Thumbs up or thumbs down?: Semantic orientation applied to unsupervised classification of reviews. In: Proceedings of the 40th Annual Meeting on Association for Computational Linguistics, pp. 417–424 (2002)
3. Hatzivassiloglou, V., McKeown, K.R.: Predicting the semantic orientation of adjectives. In: Proceedings of the 35th Annual Meeting of the Association for Computational Linguistics, pp. 174–181 (1997)
4. Stone, P.J., Dunphy, D.C., Smith, M.S., Ogilvie, D.M.: The General Inquirer: A Computer Approach to Content Analysis. MIT Press (1966)
5. Whissell, C.M.: The dictionary of affect in language, vol. 4, pp. 113–131. Academic Press (1989)
6. Wilson, T., Hoffmann, P., Somasundaran, S., Kessler, J., Wiebe, J., Choi, Y., Cardie, C., Riloff, E., Patwardhan, S.: OpinionFinder: A system for subjectivity analysis. In: Proceedings of HLT/EMNLP on Interactive Demonstrations, pp. 34–35 (2005)
7. Loureiro, D., Marreiros, G., Neves, J.: Sentiment analysis of news titles the role of entities and a new affective lexicon. In: Antunes, L., Pinto, H.S. (eds.) EPIA 2011. LNCS (LNAI), vol. 7026, pp. 1–14. Springer, Heidelberg (2011)
8. Tan, C., Lee, L., Tang, J., Jiang, L., Zhou, M., Li, P.: User-level sentiment analysis incorporating social networks. In: Proceedings of the 17th ACM SIGKDD International Conference on Knowledge Discovery and Data Mining, KDD 2011, pp. 1397–1405. ACM (2011)
9. Musat, C.C., Faltings, B.: A novel human computation game for critique aggregation. In: AAAI (2013)
10. von Ahn, L., Dabbish, L.: Labeling images with a computer game. In: Proceedings of the SIGCHI Conference on Human Factors in Computing Systems, pp. 319–326 (2004)
11. Esuli, A., Sebastiani, F.: SentiWordNet: A publicly available lexical resource for opinion mining. In: Proceedings of the 5th Conference on Language Resources and Evaluation, pp. 417–422 (2006)
12. Bradley, M.M., Lang, P.J.: Affective norms for english words (ANEW): Instruction manual and affective ratings. Technical Report C1 The Center for Research in Psychophysiology (1999)
13. Miller, G.A.: WordNet: a lexical database for english. Commun. ACM 38, 39–41 (1995)

14. Maks, I., Vossen, P.: A lexicon model for deep sentiment analysis and opinion mining applications. Decision Support Systems 53, 680–688 (2012)
15. Ding, X., Liu, B., Yu, P.S.: A holistic lexicon-based approach to opinion mining. In: Proceedings of the International Conference on Web Search and Web Data Mining, pp. 231–240 (2008)
16. Lau, R.Y.K., Lai, C.C.L., Ma, J., Li, Y.: Automatic domain ontology extraction for context-sensitive opinion mining. In: Proceedings of the 30th International Conference on Information Systems, pp. 35–53 (2009)
17. Lu, Y., Castellanos, M., Dayal, U., Zhai, C.: Automatic construction of a context-aware sentiment lexicon: An optimization approach. In: Proceedings of the 20th International Conference on World Wide Web, pp. 347–356 (2011)
18. Brew, A., Greene, D., Cunningham, P.: Using crowdsourcing and active learning to track sentiment in online media. In: Proceedings of the 19th European Conference on Artificial Intelligence, pp. 145–150 (2010)
19. Hong, Y., Kwak, H., Baek, Y., Moon, S.: Tower of Babel: A crowdsourcing game building sentiment lexicons for resource-scarce languages. In: Proceedings of the 22nd International Conference on World Wide Web Companion, pp. 549–556 (2013)
20. Al-Subaihin, A., Al-Khalifa, H., Al-Salman, A.: A proposed sentiment analysis tool for modern arabic using human-based computing. In: Proceedings of the 13th International Conference on Information Integration and Web-Based Applications and Services, pp. 543–546 (2011)
21. Musat, C.C., Ghasemi, A., Faltings, B.: Sentiment analysis using a novel human computation game. In: Proceedings of the 3rd Workshop on the People's Web Meets Natural Language Processing, pp. 1–9 (2012)
22. Ipeirotis, P.G., Provost, F., Wang, J.: Quality management on Amazon Mechanical Turk. In: Proceedings of the ACM SIGKDD Workshop on Human Computation, pp. 64–67 (2010)
23. Sintsova, V., Musat, C.C., Pu, P.: Fine-grained emotion rrecognition in olympic tweets based on human computation. In: Proceedings of the 4th Workshop on Computational Approaches to Subjectivity, Sentiment and Social Media Analysis, pp. 12–20 (2013)
24. Gupta, A.K.: Beta distribution. In: Lovric, M. (ed.) International Encyclopedia of Statistical Science, pp. 144–145. Springer, Heidelberg (2011)
25. Hu, M., Liu, B.: Mining and summarizing customer reviews. In: Proceedings of the 10th ACM SIGKDD International Conference on Knowledge Discovery and Data Mining, pp. 168–177 (2004)

Acknowledging Discourse Function
for Sentiment Analysis

Phillip Smith and Mark Lee

University of Birmingham
School of Computer Science
Edgbaston, B15 2TT
{P.Smith.7,M.G.Lee}@cs.bham.ac.uk

Abstract. In this paper, we observe the effects that discourse function attribute to the task of training learned classifiers for sentiment analysis. Experimental results from our study show that training on a corpus of primarily persuasive documents can have a negative effect on the performance of supervised sentiment classification. In addition we demonstrate that through use of the Multinomial Naïve Bayes classifier we can minimise the detrimental effects of discourse function during sentiment analysis.

1   Introduction

In discourse, sentiment is conveyed not only when a speaker is expressing a viewpoint, but also when they are attempting to persuade. In this study we examine the influence of these two functions of discourse on sentiment analysis. We hypothesise that training a supervised classifier upon a document set of a single discourse function will produce errors in classification if testing on a document set of a different discourse function.

Ideally, we would have tested this theory on a currently used resource in sentiment classification in order to examine and compare the behaviour of the expressive and persuasive discourse functions in the overall classification process. However, no such resource is appropriately annotated with the expressive and persuasive labels that are needed to test our hypothesis. We have therefore developed a document set from the clinical domain, annotated on the document level with discourse function information. The document set used for our experiments contains 3,000 short documents of patient feedback, which we have made available online.[1]

We investigate our hypothesis by testing four supervised classifiers that are commonly used in both the machine learning and sentiment analysis literature [1]. The classifiers that we use are the simple Naïve Bayes (NB), multinomial Naïve Bayes (MNB), logistic regression (LR) and linear support vector classifier (LSVC). We use both binary presence and term frequency features for each classifier. We investigate our hypothesis by running four sets of experiments,

---

[1] http://www.cs.bham.ac.uk/~pxs697/datasets

A. Gelbukh (Ed.): CICLing 2014, Part II, LNCS 8404, pp. 45–52, 2014.
© Springer-Verlag Berlin Heidelberg 2014

varying the training and testing sets of each. We run two across the same discourse function, expressive to expressive and persuasive to persuasive, and we run two using the concept of transfer learning; expressive to persuasive and persuasive to expressive. Results of experiment exhibit decreases of up to 38.8% F_1 when training on the persuasive document set and testing on the expressive. We also show that the classifier with the least variability in macro-average F_1 is the MNB classifier, which suggests its robustness to the the effects of discourse function when performing supervised sentiment classification.

The remainder of this paper is structured as follows. Section 2 outlines the theory of discourse function, and describes the nature of expressive and persuasive utterances that we encountered. In Section 3 we describe the corpus used for experimentation. We describe how our experiments were set up in Section 4, and in Section 5 we discuss our results and their relative implications. Finally, we conclude and discuss avenues for future work in Section 6.

2   Discourse Function

Our study hinges on the premise that a difference in discourse function may be detrimental to the use of supervised machine learning classifiers trained for sentiment analysis. We base our definition of discourse function on that proposed by Kinneavy [2] who argues that the aim of discourse is to produce an effect in the average reader or listener for whom the communication is intended. This could be to share how one is feeling, or perhaps to persuade them. These two discourse functions fall into the *expressive* and *persuasive* categories, respectively. Kinneavy also includes two other discourse functions, informative and literary, in his theory of discourse [3].

To illustrate his theory, Kinneavy represents the components of the communication process as a triangle, with each vertex representing a different role in the theory. This is somewhat similar to the schematic diagram of a general communication system that is proposed by Shannon [4]. The three vertices in the triangle are labelled as the encoder, the decoder and the reality of communication. The signal, the linguistic product, is the medium of the communication triangle. The encoder is the writer or speaker of a communication, and the decoder is the reader or listener.

2.1   Expressive

In communication, when the language product is dominated by a clear design of the encoder to discharge his or her emotions, or to achieve his or her own individuality then it can be stated that the expressive discourse function is being utilised [3]. In this paper, we take expression to be communicated through text. Since the discourse function is in effect the personal state of the encoder, there is naturally an expressive component in any discourse. We however narrow this definition to only observe explicit examples of the expressive discourse function in text.

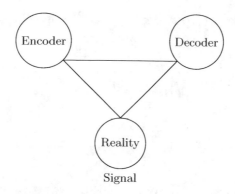

Fig. 1. Kinneavy's Communication Triangle [2]

We decompose the general notion of emotions that are conveyed to be valenced reactions, as either a positive or negative polarity based label. There is little consensus as to the set of emotions that humans exhibit, however methods have been put forward to extend these polarities into the realm of emotions [5, 6], so future work could extend this where needs be.

The components of expressive discourse when explicitly expressed are often trivial to identify. Utterances beginning with the personal pronoun *I* followed by an emotive verb often pertain to the expressive discourse function being utilised if they are succeeded by an additional emotion bearing component. Much research in sentiment analysis has observed the expressive discourse function [7–9].

2.2   Persuasive

Persuasion attempts to perform one or more of the following three actions: to change a decoder's belief or beliefs, to gain a change in a decoder's attitude, and to cause the decoder to perform a set of actions [10].

Sentiment can be viewed as a key component in persuasion, yet it is no trivial feat to define what a positive persuasive utterance is. We define what we shall call contextual and non-contextual persuasive utterances. First, let us observe the non-contextual persuasive utterances. An example of a positive persuasive utterance is: *You should give him a pay rise.* Taking this utterance alone, it is clear that the encoder of the signal is attempting to persuade the decoder to give someone more money for their work, which can be understood to be attempting to elicit a positive action from the decoder, for the benefit of the subject of the utterance.

To contrast this, we must demonstrate a non-contextual negative persuasive utterance. For example, take the utterance *Please fire him.* Here the encoder is attempting to stop the subject of the utterance from working, by persuading the decoder to ensure they cease working, which is typically seen as something negative (at least in Western societies).

Table 1. Persuasive & expressive corpus statistics

Corpus	D_N	W	$D_{avglength}$	$W_{uniq.}$
Expressive				
Positive	750	47875	62	4869
Negative	750	50676	67	5411
Persuasive				
Positive	750	44527	59	4587
Negative	750	97408	129	7391

Now, we must also consider the class of persuasive utterances that we describe as 'contextual'persuasive utterances. An example of such an utterance is: Please give me a call. At first glance, this utterance lacks a clear sentiment. However if we precede this with the sentence *Great work!*, the above persuasive utterance becomes positive. However, if we precede out initial persuasive utterance with the sentence *You've messed up.* our seemingly emotionless persuasive utterance becomes negative. This agrees with the view of Hunston [11], that indicating an attitude towards something is important in socially significant speech acts such as persuasion and argumentation.

3   Corpus

The corpus which we use in our experiments is the NHS Choices Discourse Function Corpus, introduced in [12]. This is a corpus of patient feedback from the clinical domain. Patients were able to pass comments on hospitals, GPs, dentists and opticians through an online submission form. Whilst there were many fields to fill in, the fields that were of relevance to sentiment analysis were those labelled as 'Likes', 'Dislikes'and 'Advice'. These blanket labels help to define individual documents, and made the automatic extraction for experimentation a straightforward process. There was also no need to hand-label the likes and dislikes for sentiment, as the labels presupposed this. Annotation was required for the advice, as to whether a positive or negative sentiment was conveyed. This was undertaken by two annotators, and inter-annotator agreement was measured.

Typically, sentiment analysis concentrates on the positive and negative aspects of a review; the likes and dislikes. However, the literature [3] has shown that these expressive aspects of discourse function are not alone in communicating sentiment. As shown in earlier sections, the persuasive discourse function also conveys sentiment when it is employed. Advice comes under the umbrella term that is persuasion. When offering advice, the intention is often to persuade the decoder of the advice to act in a certain manner, or to acquire a certain belief set. This can be rephrased by saying that when we use the persuasive discourse function, we often use advice to successfully perform this action. Therefore, in this corpus, the comments of the likes and dislikes section of the corpus form the expressive subsection, and the comments that the patients submitted under the advice header form the persuasive subsection of the corpus.

In this paper we concentrate on a 3,000 document subset of the corpus. This is divided into two 1,500 document sets for the documents that primarily used the expressive and persuasive discourse functions respectively. The corpus can be further divided into two 750 document subsets with documents communicating a positive sentiment, and a negative sentiment. Table 1 outlines the token counts, average document length, and the number of unique tokens present in each section of the corpus that we used for experimentation. We should note that there were at least 750 contributors to this corpus, and with the data being mined from an online source, there were no stipulations as to the qualifications of the poster, so the language model that would be learnt by the classifier would have great linguistic variation.

4 Method

In our experiments, we wanted to explore how supervised machine learning algorithms are able to generalise across discourse function. In particular we examine the transferability of learned models trained on corpora of different discourse function. Our hypothesis is that differences in discourse function will detract from the transferability of learned models when detecting sentiment, if they are tested on datasets of differing discourse function. We experiment across all pairwise combinations of the training and testing document set. We initiate the experiments in this way so that the directionality of discourse function could be tested, along with the transferability of the learned models across discourse function. We use scikit-learn [13] and NLTK [14] Python packages for our classifiers.

When training our models (NB, MNB, LR and LSVC), we used the same data for each algorithm. The training set consisted of 1,000 training documents, 500 positive and 500 negative, for both of the respective discourse functions. The test set consisted of 500 documents, 250 positive and 250 negative, from each discourse function. These were randomly selected from the NHS Choices Discourse Function corpus [12], however ensuring that there is no overlap between the sets. For the expressive to expressive and persuasive to persuasive experiments, 10-fold cross validation of the machine learning methods was used.

5 Results and Discussion

Figure 2 shows the macro-average F_1 values for each experimental setup. Classifiers that are trained upon the expressive document set perform better than those trained on the persuasive document set, irrespective of classifier choice or feature set used. The NB classifier shows greatest variability in classifier performance, with a peak F_1 of 0.826 and a minima of 0.438. The LR and LSVC models also exhibit a degree of variation in F_1. The MNB classifier minimises the variability in performance, and is the most robust classifier that we tested. Where other classifiers struggle when training on the persuasive document set, MNB achieves a macro-average F_1 of 0.802.

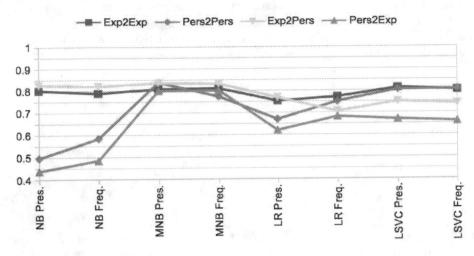

Fig. 2. Macro-averaged F_1 results for the cross-validation and transfer learning experiments

The results show the relative ease with which the expressive document set is able to create learned models of sentiment, and apply both to test sets of either an expressive or persuasive discourse function. When comparing cross-validation results to those of the transfer learning experiments, results exhibit minimal disturbance in macro-average F_1 score when models are trained on a corpus of expressive documents. This does not therefore support our hypothesis in the instance where we use the expressive document set to train our classifiers. However, this is only for the expressive function.

These results suggest that if there were a hierarchy of discourse functions, then persuasion is perhaps a subset of expression, and it inherits elements of the expressive vocabulary in order to carry out its role. We base this on the results of classification from the expressive to persuasive, and the poor adaptation of any classifiers trained on the persuasive document set. Consequently, we are inclined to believe that the persuasive discourse function cannot fully function without expressive elements. Examples of this are appeals to emotional elements, such as in congressional debates [15], where persuasion through fact alone are not the sole tactics used to sway the voters.

There is a clear drop in classifier performance when training on the persuasive corpus, and performing transfer learning. This supports our hypothesis for all classifiers, where each classifier trained in this way underperforms, sometimes to a considerable degree. We believe that this could be due to the implicit nature of sentiment that the persuasive discourse function conveys, and could be attributed to the structure of a text, in particular the interface between syntax and lexical semantics [16]. Further work is required to examine the differences in structure between documents of the respective discourse functions in order to confirm this assumption.

One interesting classifier is the MNB classifier. This performed consistently well during our study, and was even able to cope with the effects of cross-discourse classification to a high degree, performing well on the difficult persuasive to expressive classification experiments. We believe that this is due to the minimization in error rate that it has previously been shown to achieve, as it is able to deal with overlapping vocabularies and variable document length [17]. This performs considerably better than the simple NB classifier, and we believe that this is due to the difference in feature distribution that is observed in the models.

6   Conclusion

This paper has observed the effects of discourse function on supervised machine learning approaches to sentiment analysis. The effects of classification across the expressive and persuasive discourse function were recorded, and we found that despite both discourse functions conveying sentiment, the corpus with documents primarily utilising the expressive discourse function was preferable to train learned models upon, in comparison to a document set of primarily persuasive documents. In empirical results on a corpus of patient feedback containing documents of both discourse function testing across discourse, we found that there was an average improvement in accuracy of up to 38.8% when using the expressive subcorpus instead of the persuasive as a training set. We also find that the MNB classifier is preferable to others in order to minimise the effects of discourse function on sentiment classification.

In future work we will investigate further the effects of discourse function on other learned classifiers in order to determine if any others are able to minimize its effects on supervised machine learning models.

References

1. Liu, B.: Sentiment Analysis and Subjectivity. Handbook of Natural Language Processing 2, 568 (2010)
2. Kinneavy, J.E.: The Basic Aims of Discourse. College Composition and Communication 20, 297–304 (1969)
3. Kinneavy, J.L.: A Theory of Discourse: The Aims of Discourse. Norton (1971)
4. Shannon, C.E.: A Mathematical Theory of Communication. Bell Systems Technical Journal 27, 379–423 (1948)
5. Ortony, A., Clore, G.L., Collins, A.: The Cognitive Structure of Emotions. Cambridge University Press, Cambridge (1988)
6. Smith, P., Lee, M.: A CCG-based Approach to Fine-Grained Sentiment Analysis. In: Proceedings of the 2nd Workshop on Sentiment Analysis where AI meets Psychology, The COLING 2012 Organizing Committee, pp. 3–16 (2012)
7. Mullen, T., Collier, N.: Sentiment Analysis using Support Vector Machines with Diverse Information Sources. In: Lin, D., Wu, D. (eds.) Proceedings of EMNLP 2004, pp. 412–418. Association for Computational Linguistics (2004)

8. Bloom, K., Garg, N., Argamon, S.: Extracting Appraisal Expressions. In: Human Language Technologies 2007: The Conference of the North American Chapter of the Association for Computational Linguistics; Proceedings of the Main Conference, Association for Computational Linguistics, pp. 308–315 (2007)

9. Dermouche, M., Khouas, L., Velcin, J., Loudcher, S.: AMI&ERIC: How to Learn with Naive Bayes and Prior Knowledge: An Application to Sentiment Analysis. In: Second Joint Conference on Lexical and Computational Semantics (*SEM), Proceedings of the Seventh International Workshop on Semantic Evaluation (SemEval 2013), vol. 2, pp. 364–368. Association for Computational Linguistics (2013)

10. Miller, G.R.: 1. In: The Persuasion Handbook: Developments in Theory and Practice, pp.3–17. Sage (2002)

11. Hunston, S.: Corpus Approaches to Evaluation. Routledge (2011)

12. Smith, P., Lee, M.: Cross-discourse Development of Supervised Sentiment Analysis in the Clinical Domain. In: Proceedings of the 3rd Workshop in Computational Approaches to Subjectivity and Sentiment Analysis, pp. 79–83. Association for Computational Linguistics (2012)

13. Pedregosa, F., Varoquaux, G., Gramfort, A., Michel, V., Thirion, B., Grisel, O., Blondel, M., Prettenhofer, P., Weiss, R., Dubourg, V., Vanderplas, J., Passos, A., Cournapeau, D., Brucher, M., Perrot, M., Duchesnay, E.: Scikit-learn: Machine learning in Python. Journal of Machine Learning Research 12, 2825–2830 (2011)

14. Bird, S., Loper, E., Klein, E.: Natural Language Processing with Python. O'Reilly Media Inc. (2009)

15. Guerini, M., Strapparava, C., Stock, O.: Resources for Persuasion. In: Proceedings of LREC 2008, European Language Resources Association, pp. 235–242 (2008)

16. Greene, S., Resnik, P.: More than Words: Syntactic Packaging and Implicit Sentiment. In: Proceedings of Human Language Technologies: The 2009 Annual Conference of the North American Chapter of the Association for Computational Linguistics, pp. 503–511. Association for Computational Linguistics (2009)

17. McCallum, A., Nigam, K., et al.: A Comparison of Event Models for Naive Bayes Text Classification. In: AAAI 1998 Workshop on Learning for Text Categorization, vol. 752, pp. 41–48. AAAI (1998)

A Method of Polarity Computation of Chinese Sentiment Words Based on Gaussian Distribution

Ruijing Li, Shumin Shi, Heyan Huang, Chao Su, and Tianhang Wang

School of Computer Science and Technology BIT Beijing, China
{lrj,bjssm,hhy63,suchao,2120131070}@bit.edu.cn

Abstract. Internet has become an excellent source for gathering consumer reviews, while opinion of consumer reviews expressed in sentiment words. However, due to the fuzziness of Chinese word itself, the sentiment judgments of people are more subjective. Studies have shown that the polarities and strengths judgment of sentiment words obey Gaussian distribution. In this paper, we propose a novel method of polarity computation of Chinese sentiment words based on Gaussian distribution which can analyze an analysis of semantic fuzziness of Chinese sentiment words quantitatively. Furthermore, several equations are proposed to calculate the polarities and strengths of sentiment words. Experimental results show that our method is highly effective.

Keywords: Semantic fuzziness, sentiment words, polarity calculation, Gaussian distribution.

1 Introduction

As the number of product reviews grows rapidly, the identification of emotions (including sentimental orientation and polarity strength) in text becomes hot field. Generally speaking, opinions are divided into three categories: positive, neutral and negative. At present, there exists three levels of sentiment analysis of text, word-level, sentence-level and document-level.

Word-level sentiment analysis mainly includes construction of emotion dictionary, identification of candidate words and judging the sentiment polarities and strengths of candidate words [1]. Sentence-level sentiment analysis to determine the sentiment polarities and strengths of sentence [1][2]. Document-level sentiment analysis refers to recognize the sentiment polarities and strengths of the document on something [3]. Although most of current studies focus on sentence-level and document-level sentiment analysis, word-level sentiment analysis is the fundamental research, which is of great significance and has more challenge.

According to the characteristics of Chinese language itself, the minimum size of sentiment words word can best embody semantic fuzziness of the language. In analyzing semantic fuzziness of Chinese text, the polarity of sentiment words are identified, but the polarity strengths are continuous changeable. For ease calculation and statistics, the polarity strengths of sentiment words need to be quantified. Determining the

A. Gelbukh (Ed.): CICLing 2014, Part II, LNCS 8404, pp. 53–61, 2014.

polarity quantification of sentiment words become a typical problems of semantic fuzziness. Then in this paper, we only focus on word-level sentiment analysis. For the semantic fuzziness of Chinese word, we draw on ideas from Zadeh's fuzzy set theory [4]. And, Studies have shown that the polarities and strengths judgment of sentiment words obey Gaussian distribution [5],[6],[7]. So we take advantage of Gaussian distribution model on sentiment analysis. Furthermore, several equations are proposed to calculate the polarities and strengths of sentiment words.

To detect the polarities and strengths of sentiment words, a Chinese polarity lexicon is indispensable. At present, many scholars have done a lot of works in constructing the polarity lexicon. English sentiment analysis mainly uses the lexicon GI resource. It's a perfect polarity lexicon with manual annotation information of sentiment words, which building a foundation for text orientation analysis. The construction of the Chinese polarity lexicon is mainly based on the statistical method or semantic dictionary. But the effect of modifiers on polarity words has not been researched. This paper not only analyzes the sentiment words, but also its modifiers.

The rest of this paper is organized as follows. Section 2 describes the traditional research on polarity computation of Chinese sentiment words. We describe our algorithm in Section 3, experimental results and discussion in Section 4, and conclude in Section 5.

2   Related Works

Word-level sentiment analysis has been got a rapid development in recent years. There are mainly two ways:

One is sentiment polarity lexicon extended by some existing electronic dictionaries or word knowledge database. H and M in [8] got the sentiment orientation information of words from the synonyms and antonyms words set of WordNet and General Inquirer (GI). Kamps in [9] got the sentiment orientation of words by calculating the semantic distance between test words and benchmark words in WordNet [10] synonym structure chart. Ku in [11] first count the characters information of sentiment words in NTUSD polarity lexicon, and then get the polarities and strengths of sentiment words from its characters. Zhu in [12] determined the polarities and strengths of sentiment words by calculating the similarity between test word and benchmark word with semantic similarity provided by HowNet.

The other is methods based on machine learning. Turney and Littma in [13] proposed two algorithms for extraction of semantic orientations of words. Kim in [14] first got the synonym set of candidate words whose emotions are unknown. Yao and Lou in [15] combined associative-word processing which connected connection context based on PMI. Zhang in [16] calculated the relationship between the words and the context template (extracted from emotion tagging corpus), and then judged the sentiment orientation of the words.

Aiming at the deficiency of previous work, we propose a novel method of polarity computation of Chinese sentiment words based on Gaussian distribution and proposed several equations to calculate the polarities and strengths of sentiment words.

3   Polarity Computation Method based on Gaussian Distribution

3.1   Definitions

Fundamental sentiment words and compound sentiment words in this paper are defined as follows:

Fundamental Sentiment Word. The sentiment word which has no more than two characters and does not contain negative word and modifier.

Compound Sentiment Word. The sentiment word which has more than 2 characters or contain negative word or modifier.

3.2   Construction of Polarity Lexicon

We choose the sentiment words from HowNet and NTUSD as [10]. Table 1 shows the statistics of the sentiment words from these two resources.

Table 1. The statistics of the sentiment words

Lexicon	Positive	Negative
HowNet	728	933
NYUSD	2810	8276
Total	3538	9209

Obviously, modifiers in expressing emotion are also important for polarity analysis, e.g., the positive degree of "非常好 (very good)" is higher than "好 (good)" [17]. We take modifiers into account on the polarity calculation of compound sentiment words. We choose the degree adverb lexicon from HowNet. In addition, we construct a frequency adverb lexicon from [18].

3.3   Fundamental Sentiment Words Analysis

Since only determining the sentiment polarity of word is unable to meet the needs of text sentiment analysis. In this section we not only judge the polarity of word but also its polarity strength. Character is the smallest component units of the word. Some philologists and scholars believe that the same characters are often found in the words of the same polarity [11, 17]. Our method refers to [11] partly, while we pay more attention to semantic fuzziness of Chinese. We determine the polarity strengths of sentiment words from an interval period through Gaussian distribution model base on fuzzy set theory [4]; no longer represent it with a specific value. We later compare our method with [11].

Our methods are described as follows:

First we get the characters' polarity in test words through statistical and quantitative analysis of the distribution of characters in polarity lexicon. Considering that there are more negative sentiment words than positive sentiment words, we choose the Equations (1) and (2) to represent its sentiment tendency.

$$P_{ci} = \frac{fp_{ci}/\sum_{j=1}^{n} fp_{cj}}{fp_{ci}/\sum_{j=1}^{n} fp_{cj} + fn_{ci}/\sum_{j=1}^{m} fn_{cj}} \quad (1)$$

$$N_{ci} = \frac{fn_{ci}/\sum_{j=1}^{m} fn_{cj}}{fp_{ci}/\sum_{j=1}^{n} fp_{cj} + fn_{ci}/\sum_{j=1}^{m} fn_{cj}} \quad (2)$$

fp_{ci} and fn_{ci} denote the frequencies of a character ci in the positive and negative words respectively. We should calculate fp_{ci} and fn_{ci} of each character emerging in polarity lexicon first. P_{ci} and N_{ci} denote the weights of ci as positive and negative characters respectively; n and m denote the total number of unique characters in positive and negative words respectively. P_{ci}- N_{ci} in Equation (3) determines the sentiment tendency of character ci. If it is a positive value, then this character appears more often in positive Chinese words; and vice versa. A value close to 0 means that it is not a sentiment character or it is a neutral sentiment character.

$$S_{ci} = Random(\eta_s((P_{ci} - N_{ci}), \delta)) = Random(\frac{1}{\sqrt{2\pi}\delta} e^{-\frac{(x-(P_{ci}-N_{ci}))^2}{2\delta^2}}) \quad (3)$$

S_{ci} denotes the sentiment tendency of character ci. If it is a positive value, then this character appears more often in positive Chinese words; and vice versa. A value close to 0 means that it is not a sentiment character or it is a neutral sentiment character. $\eta_s((P_{ci} - N_{ci})$ is a Gaussian density function while $(P_{ci} - N_{ci})$ and δ is the corresponding mean and covariance. $Random(\eta_s((P_{ci} - N_{ci}))$ means to choose a value from this density function randomly as the sentiment tendency of character ci.

Equation (4) defines: a sentiment strength of a Chinese word w is the average of the sentiment tendency value of the composing characters $c1, c2, ..., cp$.

$$S_w = \sum_{j=1}^{p} S_{cj} / p \quad (4)$$

In this way, S_w representing the polarity and strength of word w has been calculated. If some characters without polarity degree appear, take the default value as zero.

3.4   Compound Sentiment Words Analysis

The analysis of compound sentiment words is so complex that we should make the classified discussion [19]. According to the characteristics of Chinese words and the thoughts of Sigmoid function, we designed some specific formulae to analyze the compound sentiment words from seven forms. They are as follows.

1. Degree Adverb Connects Fundamental Sentiment Word

Such as sentiment word w is "非常漂亮(very beautiful)", we first obtained the modifier strength of degree adverb "非常(very)"from degree adverb lexicon as S_{x2}; then calculate the polarity strength of fundamental sentiment word "漂亮(beautiful)"as S_{x1}; finally utilize Equation (5) to calculate the polarities and strengths of this kind of words.

$$S_w = \left(1 - \left(\frac{1-S_{x_1}}{1+S_{x_1}}\right)^{S_{x_2}+\Delta\mu}\right)\Big/\left(1 + (\frac{1-S_{x_1}}{1+S_{x_1}})^{S_{x_2}+\Delta\mu}\right) \qquad (5)$$

$\Delta\mu$ is displacement parameter valued 0.5, as the same follows.

2. Frequency Adverb Connects Fundamental Sentiment Word

Such as sentiment word w is "偶尔马虎(sloppy occasionally)", we first obtained the modifier strength of frequency adverb "偶尔(occasionally)"from frequency adverb lexicon as S_{x2}; then calculate the polarity strength of fundamental sentiment word "马虎(sloppy)"as S_{x1}; finally utilize Equation (5) to calculate the polarities and strengths of this kind of words.

3. Fundamental Sentiment Word Connects Fundamental Sentiment Word

Such as sentiment word w is "简单大方(simple and generous)", we first calculate the polarity strength of fundamental sentiment words "简单(simple)"and" 大方 (generous)"as S_{x1} and S_{x2} respectively; then utilize Equation (6) to calculate polarities and strengths of this kind of words.

$$S_w = \frac{(1+S_{x_1})(1+S_{x_2})-(1-S_{x_1})(1-S_{x_2})}{(1+S_{x_1})(1+S_{x_2})+(1-S_{x_1})(1-S_{x_2})} \qquad (6)$$

4. Reduplicated Word

Such as sentiment word w is "快快乐乐(happy-happy)", we first extract the fundamental sentiment word "快乐(happy)"and calculate its polarity strength S_{x1}; then utilize Equation (7) to calculate polarities and strengths of this kind of words.

$$S_w = \left(1 - \left(\frac{1-S_{x_1}}{1+S_{x_1}}\right)^2\right)\Big/\left(1 + \left(\frac{1-S_{x_1}}{1+S_{x_1}}\right)^2\right) \qquad (7)$$

5. Negative Word Connects Fundamental Sentiment Word

Such as sentiment word w is "不完善(imperfect)", we first calculate the polarity strength of fundamental sentiment word "完善(perfect)"and calculate its polarity strength S_{x1}; then utilize Equation (8) to calculate polarities and strengths of this kind of words.

$$S_w = -S_{x_1} \qquad (8)$$

6. Negative Word Connects Degree Adverb or Frequency Adverb and Reconnects Fundamental Sentiment Word

Such as sentiment word w is "不太认真(not too serious)", we first calculate the polarity strength of fundamental sentiment words "太(too)"and"认真(serious)"as S_{x2} and

S_{x1} with the above methods respectively; then utilize Equation (9) to calculate polarities and strengths of this kind of words.

$$S_w = \left(1 - \left(\frac{1-S_{x1}}{1+S_{x1}}\right)^{(1-S_{x2})+\Delta\mu}\right) \Big/ \left(1 + \left(\frac{1-S_{x1}}{1+S_{x1}}\right)^{(1-S_{x2})+\Delta\mu}\right) \tag{9}$$

7. Degree Adverb or Frequency Adverb Connects Negative Word and Reconnects Fundamental Sentiment Word

Such as sentiment word w is "太不认真(too serious)", we first calculate the polarity strength of fundamental sentiment words "太(too)"and"认真(serious)"as S_{x2} and S_{x1} with the above methods respectively; then utilize Equation (10) to calculate polarities and strengths of this kind of words.

$$S_w = -\left(1 - \left(\frac{1-S_{x1}}{1+S_{x1}}\right)^{S_{x2}+\Delta\mu}\right) \Big/ \left(1 + \left(\frac{1-S_{x1}}{1+S_{x1}}\right)^{S_{x2}+\Delta\mu}\right) \tag{10}$$

By now, the polarities and strengths of all kinds of sentiment words can be calculated.

4   Experiments

In this experiment, the degree of modification of degree/frequency adverb words is divided into five categories and valued 0.1, 0.3, 0.5, 0.7, 0.9 respectively. In addition, the accuracy of orientation value computation for new arrived sentiment words was mainly evaluated. The range of orientation value is from -1 to + 1. + 1 is the highest commendation, and -1 is the largest derogation.

We calculate the sentiment tendency of character use all the sentiment words that we collect. To build up testing sets, we construct a lexicon which includes 500 fundamental sentiment words and 300 compound sentiment words. The polarities and strengths of them are annotated by 3 annotators. The deviation threshold α of orientation value of artificial judgment for 0.3 on the condition of correct polarity judgment (because the sentiment judgments of people are more subjective, each person cannot be completely same, we allow the deviation in some extent.) is allowed.

4.1   Performance of Fundamental Sentiment Word Analysis

In order to validate our method, we refer to the method of Dr. Han [19]. His method is also based on Ku's works in [11] and only includes the polarity computation of fundamental sentiment words. The only difference between them is the Equation (4). It's as follows.

$$S_w = sign(S_{ci}) * Max(abs(S_{ci})) \tag{11}$$

Where $Max(abs(S_{ci}))$ is the maximum absolute sentiment tendency value of all characters; $sign(S_{ci})$ is the symbol value of character ci; If the sentiment tendency value greater than 0, valued +1; if less than 0, valued -1.

If we call our former method Ku with GD (Gaussian distribution), we will call it Han with GD which joined GD in [19].

The accuracy of orientation value for test words is shown in Table 2, including comparison among four methods. The widely used mean absolute error (MAE) and root mean square error (RMSE) in combination with the t-statistic are proposed as statistical indicators for the evaluation.

Table 2. Performance of fundamental sentiment word analysis

Methods	α=0.1	α=0.2	α=0.3	Average	MAE	RMSE
Ku's method	28.05%	48.74%	67.36%	48.05%	0.4439	0.5781
Ku's method with GD	35.63%	51.26%	69.89%	52.26%	0.4397	0.5740
Han's method	12.87%	28.74%	45.29%	28.97%	0.5666	0.6826
Han's method with GD	17.47%	31.03%	47.82%	32.11%	0.5545	0.6812

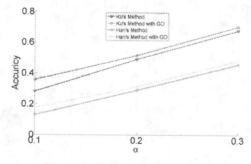

Fig. 1. Performance of four methods

We can visually see from Fig. 1 that methods which joining Gaussian distribution provide best performance in each kind of evaluation.

4.2 Performance of Compound Sentiment Word Analysis

We have conducted experiments on the compound sentiment word analysis. Table 3 and Fig. 2 below show the experiments results. As can be seen, our method is quite promising, and can be used in practical application.

Table 3. Performance of fundamental sentiment word analysis

Method	α=0.1	α=0.2	α=0.3	MAE	RMSE
Ku's method	24.06%	42.92%	59.43%	0.2922	0.3682
Our method	28.30%	44.81%	65.57%	0.2759	0.3453

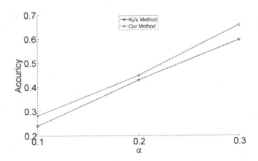

Fig. 2. Performance of two methods

4.3   Analysis

Through profound analysis and investigation, the results are not good enough for three reasons.

One is the scale of sentiment polarity lexicon. Although the scale of sentiment polarity lexicon is growing, it still cannot include all the emotional words. So there are deviations in the calculation of sentiment tendency of characters. Another is the fuzziness of Chinese words. There are no standard polarity and strength value of each Chinese word that we can only approximate to it. The last one is lack of sentence meaning analysis. For example, "骄傲(pride)"expresses positive opinion in the sentence of "古代四大发明是中国的骄傲。(The four great inventions of ancient times is Chinese pride.)", while negative opinion in the sentence of "骄傲自大是他的一个缺点。(Arrogance is one of his less attractive characteristics.)".

The current method cannot deal with these types of errors. We leave their solutions as future work.

5   Conclusions

This paper proposes a novel method to calculate the polarities and strengths of Chinese Sentiment Words, trying to analysis semantic fuzziness of Chinese. Compared with the conventional method, we use a probability value to represent the polarity strengths of sentiment word, rather than a determined value. The feasibility and effectiveness of our method with Gaussian distribution is proved. In view of the importance of this task, we should strive to explore better method to involve to our method.

In the future work, we will make further research on quantitative analysis of modifiers and analysis of the compound sentiment words with syntactical structure.

Acknowledgments. This work was supported by project of the National Natural Science Foundation of China (No. 61201352), the National Basic Research Program of China (973 Program) (2013CB329606, 2013CB329605), and Basic Scientific Research Fund of Beijing Institute of Technology partly.

References

1. Kim, S., Hovy, E.: Determining the Sentiment of Opinions. In: Proceedings of COLING, pp. 1367–1373 (2004)
2. Mao, Y., Lebanon, G.: Isotonic Conditional Random Fields and Local Sentiment Flow. In: Proceedings of NIPS, pp. 961–968 (2006)
3. Pang, B., Lee, L.: A Sentimental Education: Sentiment Analysis Using Subjectivity Summarization Based on Minimum Cuts. In: Proceedings of the 42nd Annual Meeting on Association for Computational Linguistics, pp. 271–278 (2004)
4. Zadeh, L.A.: Fuzzy sets. Information and Control 8(3), 338–353 (1965)
5. Zhang, G.B., Song, Q.H., Fei, S.M., et al.: Research on Speech Emotion Recognition. Computer Technology and Development 19(1), 92–96 (2009)
6. Chen, X.: Identification of Human Motion Emotions based on Gaussian Feature. Huazhong University of Science & Technology (2011)
7. Wang, J., Dou, R., Yan, Z., et al.: Exploration of Analyzing Emotion Strength in Speech Signal Pattern Recognition. In: Chinese Conference on CCPR 2009, pp. 1–4. IEEE (2009)
8. Hatzivassiloglou, V., McKeown, K.R.: Predicting the semantic orientation of adjectives. In: Proceedings of the 35th Annual Meeting of the Association for Computational Linguistics and Eighth Conference of the European Chapter of the Association for Computational Linguistics, pp. 174–181. Association for Computational Linguistics (1997)
9. Kamps, J., Marx, M.J., Mokken, R.J., et al.: Using wordnet to measure semantic orientations of adjectives (2004)
10. Fellbaum, C.: WordNet: An Electronic Lexical Database (1998)
11. Ku, L.W., Liang, Y.T., Chen, H.H.: Opinion Extraction, Summarization and Tracking in News and Blog Corpora. In: AAAI Spring Symposium: Computational Approaches to Analyzing Weblogs, pp. 100–107 (2006)
12. Zhu, Y.L., Min, J., Zhou, Y.Q., et al.: Research on semantic orientation calculation of words based on HowNet. Journal of Chinese Information Processing 20(1), 14–20 (2006)
13. Turney, P.D., Littman, M.L.: Measuring praise and criticism: Inference of semantic orientation from association. ACM Transactions on Information Systems (TOIS) 21(4), 315–346 (2003)
14. Kim, S.M., Hovy, E.: Determining the sentiment of opinions. In: Proceedings of the 20th International Conference on Computational Linguistics, p. 1367. Association for Computational Linguistics (2004)
15. Yao, T.F., Lou, D.C.: Research on the semantic orientation judgment of Chinese sentiment words. Research on computing technology and language problems. In: The Seventh International Conference on Chinese Information Processing (2007)
16. Zhang, Q., Qiu, X.P., Huang, X.J., et al.: Learning semantic lexicons using graph mutual reinforcement based bootstrapping. Acta Automatica Sinica 34(10), 1257–1261 (2008)
17. Wang, M., Shi, H.: Research on sentiment analysis technology and polarity computation of sentiment words. In: 2010 IEEE International Conference on Progress in Informatics and Computing (PIC), vol. 1, pp. 331–334. IEEE (2010)
18. He, S.B.: The study of modern Chinese frequency adverbs based on statistics. Nanjing Normal University (2006)
19. Shi, H.X.: Research on Fine-grained Sentiment Analysis. Soochow University (2013)

A Sentence Vector Based Over-Sampling Method for Imbalanced Emotion Classification

Tao Chen[1], Ruifeng Xu[1,*], Qin Lu[2], Bin Liu[1], Jun Xu[1], Lin Yao[3], and Zhenyu He[1]

[1] Key Laboratory of Network Oriented Intelligent Computation,
Shenzhen Graduate School, Harbin Institute of Technology, Shenzhen, China
{chentao1999,xuruifeng.hitsz,hit.xujun}@gmail.com,
bliu@insun.hit.edu.cn, zyhe@hitsz.edu.cn
[2] Department of Computing, The Hong Kong Polytechnic University, Hong Kong, China
csluqin@comp.polyu.edu.hk
[3] Peking University Shenzhen Graduate School, Shenzhen, Guangdong, China

Abstract. Imbalanced training data poses a serious problem for supervised learning based text classification. Such a problem becomes more serious in emotion classification task with multiple emotion categories as the training data can be quite skewed. This paper presents a novel over-sampling method to form additional sum sentence vectors for minority classes in order to improve emotion classification for imbalanced data. Firstly, a large corpus is used to train a continuous skip-gram model to form each word vector using word/POS pair as the unit of word vector. The sentence vectors of the training data are then constructed as the sum vector of their word/POS vectors. The new minority class training samples are then generated by randomly add two sentence vectors in the corresponding class until the training samples for each class are the same so that the classifiers can be trained on fully balanced training dataset. Evaluations on NLP&CC2013 Chinese micro blog emotion classification dataset shows that the obtained classifier achieves 48.4% average precision, an 11.9 percent improvement over the state-of-art performance on this dataset (at 36.5%). This result shows that the proposed over-sampling method can effectively address the problem of data imbalance and thus achieve much improved performance for emotion classification.

Keywords: Emotion classification, Imbalanced training data, Over-sampling, Sentence vector.

1 Introduction

With the rapid growth of social network, people have found a very effective media to share their emotions over the internet. This also motivated researchers to explore emotion analysis techniques to recognize and analyze emotions in the text. Generally speaking, text emotion classification techniques are camped into dictionary-based

---

* Corresponding author.

A. Gelbukh (Ed.): CICLing 2014, Part II, LNCS 8404, pp. 62–72, 2014.
© Springer-Verlag Berlin Heidelberg 2014

approaches [1,2] and machine learning based approaches [3,4]. Currently, machine learning based approaches, especially supervised learning based emotion classification, have achieved better performance. However, imbalanced training samples always poses a problem to supervised learning methods because skewed training data for different classes makes those training data for those so called majority classes dominate class predictions such that the minority classes can hardly be predicted. The imbalanced training data problem is more serious in emotion classification. In the two widely used Chinese emotion corpora, namely Ren-CECps [5] and NLP&CC2013 Chinese micro blog emotion classification dataset[1,2] , each having 8 and 7 emotion categories, the training samples corresponding to the largest emotion category are about 10 and 11 times the size of the smallest category, respectively.

There are some studies to deal with data imbalances in supervised learning. Major approaches include the data preprocessing approach [6], the algorithmic approach [7] and the feature selection approach [8]. The data preprocessing approach aims to generate relatively balanced data through adjustment of the imbalanced training data, which is has attracted much research interests. Typical methods include under-sampling which decreases the amount of training samples for the majority classes and over-sampling which increases the amount of training samples for the minority classes. The major issue for over-sampling is how to construct the appropriate new training samples.

In this paper, a sentence vector based over-sampling method is proposed to improve performance in emotion classification for imbalanced data. Considering that a word sometimes has more than one part-of-speech (POS) which leads to different lexical functions, in this paper, the word/POS pair is used as the basic lexical unit rather than the word. A large corpus is firstly used to train a continuous skip-gram model to construct the vector representation for word/POS unit. The sentence vectors of the training sentences are then constructed by the sum vector of the word/POS vectors in the sentences. Observations show that the sum vector of two sentence vectors in one class tends to be in the same class. Based on this observation, the new minority class training samples are generated by using the sum vector of two randomly selected sentence vectors in the same minority class. Such generation is done iteratively until its size is equal to that of the largest majority class. A multi-label k-nearest neighboring (ML-kNN) classifier is then trained on the generated fully balanced training samples for text emotion classification. Evaluations on NLP&CC2013 Chinese micro blog emotion classification dataset shows that the obtained classifier achieves 48.4% average precision, an 11.9% percent improvement over the state-of-art performance on this dataset(at 36.5%).

The rest of this paper is organized as follows. Section 2 briefly reviews related works. Section 3 presents the sentence vector based over-sampling method for classifier training. Evaluations and discussions are given in Section 4. Section 5 gives the conclusion and future directions.

---

[1] http://tcci.ccf.org.cn/conference/2013/dldoc/evsam02.zip
[2] http://tcci.ccf.org.cn/conference/2013/dldoc/evdata02.zip

2   Related Works

There are two major approaches in text emotion classification. The first is dictionary-based which uses emotion words and their associated attributes such as mutual information of keywords, semantic orientation of words [1], word emotional polarity in WordNet [2], and word tendency in HowNet as discriminate features for classification. The second is based on supervised learning which uses the classifiers such as Support Vector Machine [3], Hidden Markov Model [4], Naive Bayes, Maximum Entropy, and Conditional Random Fields for emotion classification.   Other approaches either use sentiment sentence framework [9], commonsense knowledgebase [10], association rules, or syntactic parse trees to analyze emotions in text.

Longadge et al. divided techniques for improving classifier training on the imbalance data into three major categories [11]: (1) Data-preprocessing approach which generates new training samples for minority classes or remove existing samples for majority classes in order to balance the training data for each class [6]; (2) Algorithmic approach which modifies classification algorithms to increase the cost of misclassification for minority classes or decrease the cost of misclassification for major classes [7] and further use ensemble learning to construct multiple classifiers for imbalance classification [12]; and (3) Feature selection approach which selects optimized set of features for imbalance classification [8].

The idea of vector-space based word representation comes from the field of information retrieval. Deerwester et al.[13] used Latent Semantic Indexing to represent word feature vectors based on the co-occurrence probability of these words in the same document. Bellegarda et al.[14] proposed the idea of continuous representation of word. Latent Dirichlet Allocation (LDA) model was also used to obtain continuous representation of words [15]. The idea of using neural networks to represent words in vector space was proposed by Riis and Krogh [16]. Mikolov et al.[17] pointed out that neural network based methods are better than Latent Semantic Indexing for preserving linear regularities because of   reduced computation complexity and the capability to handle large datasets. Vector representation of sentence is widely investigated in automatic summarization, topic discovery, information retrieval. Typical methods are bag-of-words model which uses the vector space model to represent the sentence vector [18]. Such methods always suffer from data sparseness problem as well as scalability to handle large-scale data.

3   Our Approach

3.1   Sentence Vector Representation Using Word/POS Vector

In this work, we first make use of a large micro blog corpus to learn the word/POS vector representation using word2vec[3] tools to train a continuous skip-gram model [17]. This model is a neural network model which is designed for learning word vectors from large-scale datasets. Compared to other word representation models, it has the advantage of high efficiency, without syntactic or semantic knowledge requirement and parallel training capability.

---

[3] https://code.google.com/p/word2vec/

Normally, the continuous skip-gram model is directly trained using words or hyphenated phrase vectors, i.e. one word generates one vector. Considering that a word with different part-of-speech (POS) always leads to different lexical functions and representations, in this study, the word/POS pair is used as the basic lexical unit. The dimension for each vector is set by various applications ranging from 80 to 640 in other works[17]. In this work, the vector dimensions is experimentally determined to be 200(denoted as 200-d to stand for 200 dimensions) is sufficient because experiment result shows that the classifier using 200-d vectors has very similar performance compared to 640-d vectors whereas the training cost for 640-dimensional vectors is about 3 times longer.

With the obtained word/POS vectors, two methods are applied to select words which we consider important to reflect the semantics of sentences in sentence vector constructions. The first method is called POS filtering and the second method is vector reversion for negation sentences as explained below:

- POS filtering: Filter out punctuations and some functional words. Only content words such as verbs, adjectives, nouns, conjunctions, and adverbs are used for generating sentence vectors.
- Vector Negation: The negation adverbs such as "不(no/not)"、"没有(without)"、"未(did not)"、"别(do not)"、"不必(need not)"、"不曾(never)" are identified to find negation sentences. The first verb/adjective/noun following the negation adverb is then obtained. The vector of this word is marked as negative to reflect the reversion in polarity.

Given the following example sentence:

Sentence 1: "等待不苦，苦的是，没有希望的等待……" (*Waiting is not painful, waiting without hope is painful……*)

After word segmentation and POS filtering, the following content words are obtained:

等待/v (*waiting*) 不/d (*not*) 苦/a (*painful*)苦/a (*painful*) 是/v (*is*) 没有/d (*without*) 希望/n (*hope*) 等待/v(*waiting*).

After the identification of the two negation adverbs,"不/d(*not*)" and "没有/d(*without*)", the following content words "苦/a (*painful*)" and "希望/n (*hope*)", are marked as negatives.

For a given sentence, its sentence vector, V_s, is defined as the sum of all the remaining word vectors, and the words that are marked negative are used as negatives in the summation   as given in Equation 1:

$$V_s = \sum_{i=1}^{n} v_i \qquad (1)$$

Where, V_s v_i represents the vector of the i-th content word in s, n is the total number of content words in the sentence. Details on how to obtain v_i is given in [17].

As an example, he sentence vector corresponding to Sentence 1 is defined as:

$$V_{\text{sentence1}} = v_{等待/v} - v_{苦/a} + v_{苦/a} + v_{是/v} - v_{希望/n} + v_{等待/v}$$

3.2   Sentence Vector Based Over-Sampling for Constructing Balanced Corpus

In this study, we investigate the use of over-sampling strategy. The key issue then becomes how to ensure the newly generated samples have good quality. In other words, they should also belong to the corresponding minority classes.

Let us take a look at two sentences from NLP&CC2013 Chinese microblog emotion classification datasets (in short NLP&CC2013 dataset) which all belong to the "happiness" class:

Sentence 2: "庆幸能看到这样一部上下集的大片... " (*Very lucky to have watched such a big blockbuster...*)

Sentence 3: "[大笑] [大笑] [大笑] " (*[Laught] [laught] [laught]*)
   Their corresponding 200-d sentence vectors are:
   [*0.0225006951251, -0.0774350013884, ..., 0.0143682950516, -0.185020536687*]
   [*-0.118491013725, -0.173214807502, ..., 0.0286399217377, -0.068343229329*]
   The sum vector of the two sentence vectors is:
   [*-0.095990318600, -0.250649808890, ..., 0.043008216789,-0.253363766016*]
   The following gives the top 3 similar sentence vectors based on the 200-d vector comparison:
   [*-0.143734412487, -0.113143146774, ..., 0.0673401045878, -0.102529027699*]
   [*0.0492763489994, -0.0518607813522,..., -0.0148164923998, -0.163521642871*]
   [*0. 0364673898094, -0. 028686698383,..., -0. 0708333786426, -0. 123988753054*]
   The corresponding sentences are:

Sentence 4: " 有木有? [大笑][猫咪] " (*Right? [Laugh][pussy cat]*)

Sentence 5:"...吃大餐收大礼[嘻嘻][偷笑][鼓掌][钱] " (*... had a big dinner and received a wonderful gift [gigle][titter][applause][money]*)

Sentence 6:"...在Macdrive听着周杰伦的一首歌，突然世界变得美好了... " (*Listening to a Jay Chou's song in Macdrive. Suddenly, the world is beautiful...*)

It is obvious that all these three sentences are also in the *"happiness"* class, the same as the original Sentence 2 and Sentence 3. In fact, it is generally true that the sum vector of two sentences in the same class always tends to be similar to the sentence vectors in the same class.

Based on this observation, we propose the following hypothesis:

Hypothesis 1: If the sentence vector V_1 and V_2 belong to a class c, their sum vector V_1+V_2 also belongs to class c, i.e.(LQ Note: please bracket V1 + V2 below: (V1 + V2))

$$V_1 \in c \wedge V_2 \in c \rightarrow V_1 + V_2 \in c \qquad (2)$$

Based on this hypothesis, an over-sampling method is designed which will eventually allow every class to have equal sample size. Let us assume the size of the maximum training samples of all classes is M. For any class c whose sample size is less than M, two sentence vectors in this class are randomly selected to construct a new sentence vector using the sum vector with the vector selection factor α for normalization. Their sum vector is labeled as the c class. This process repeats until the number of sample vectors in c reaches M. In this way, the training data of all different classes are fully balanced. The pseudo code of the algorithm is described as follows:

Algorithm 1. Over-sampling for Construction of Balanced Training Set

```
Input: sentence vectors of training set, class set.
  M ←max(sentence vector number in each class)
  For each class c
    loop
        v₁ ←randomly select a sentence vector in c
        v₂ ←randomly select a sentence vector in c
        v =  v₁ + (1- )v₂
        add v into class c
    until size(c) ≥ M
  End for
Output: balanced sentence vector set
```

where $size(c)$ represents the number of sentence in class c, α is a vector selection factor, $0 \leq \alpha \leq 1$.

3.3   Training Classifier Using Balanced Data

After the construction of the fully balanced training corpus, a multi-label k-nearest neighboring (ML-kNN) classifier is trained for emotion classification. In the nearest neighbor estimation, the similarity of any two sentences is the cosine of the angle between their corresponding sentence vectors, namely,

$$similarity(s_1, s_2) = cos(\theta) = \frac{V_{s_1} \cdot V_{s_2}}{\|V_{s_1}\| \times \|V_{s_2}\|} \tag{3}$$

Where, θ represents the angle between two sentence vectors V_{s_1} and V_{s_2}, $\| V_{s_1} \|$ represents the norm of the sentence vector of s_l.

4   Evaluation and Discussions

4.1   Experiment Setting

The NLP&CC2013 Chinese micro blog emotion classification dataset is used to evaluate the proposed over-sampling method. The emotional sentences are labeled up to 2 classes in 7 emotion classes including *happiness, like, surprise, fear, disgust, sadness, anger* while factual sentences are labeled as *neutral*. The training set consists of 13,250 sentences in 4,000 micro blogs including 4,949 emotional ones and 8,301 factual ones. The testing set consists of 32,185 sentences in 10,000 micro blogs. The numbers and percentages of each emotion class in the training set and testing set are listed in Table 1, respectively.

Table 1. Emotional Class Distribution in NLP&CC2013 dataset

Class	Training Set				Testing Set			
	Primary Emotion		Secondary Emotion		Primary Emotion		Secondary Emotion	
Like	1226	24.8%	138	21.6%	2888	27.6%	204	26.1%
Disgust	1008	20.4%	187	29.2%	2073	19.8%	212	27.1%
Happiness	729	14.7%	95	14.8%	2145	20.5%	138	17.6%
Anger	716	14.5%	129	20.2%	1147	10.9%	82	10.5%
Sadness	847	17.1%	45	7.0%	1565	14.9%	84	10.7%
Surprise	309	6.2%	32	5.0%	473	4.5%	43	5.5%
Fear	114	2.3%	14	2.2%	186	1.8%	20	2.6%

Note that in the training set, the size of the largest class, *like*, is about 4 times and 11 times of *surprise* and *fear*, respectively. Obviously, the training data is quite imbalanced.

A 3-billion-word Chinese micro blog corpus from weibo.com is used to train the continuous skipgram model for generating the word/POS vectors. Here, the context window size is set to 10 words while the vector dimension is set to 200, empirically. Finally, 454,071 word/POS vectors are obtained.

Average precision, the metric adopted in NLP&CC2013 emotion classification evaluation, is used in the experiment as the performance measure as is shown below:

$$Precision_{average} = \frac{1}{2n}\sum_{i=1}^{n}\sum_{k=1}^{2}\frac{|(emotion_k \in Y_i|rank(x_i,emotion_k)\leq rank(x_i,y))|}{rank(x_i,emotion_k)} \tag{4}$$

Where n is the total number of sentences in the test corpus, x_i is the *i-th* sentence, $emotion_k$ is the k-th emotion label, $emotion_k \in Y_i$ indicates the label is correctly identified, $rank(x_i,emotion_k)$ is the emotion label ranking of the *i*-th sentence.

There are loose measures and strict measures in the evaluation according to different scoring criteria for the secondary emotion, respectively. In loose measures, both the primary emotion and the secondary emotion can add 1 point when either is correctly recognized. In strict measures, the primary emotion can also add 1 point while the secondary emotion can add 0.5 point when it is correctly recognized.

4.2   Evaluation on Balanced Training Data Construction

As listed in Table 1, the top emotion class "*like*" has 1,226 training sentences, so the variable M in Algorithm 1 is set to 1,226. Through the over-sampling process, following Algorithm 1, the sizes of the other 6 categories are expanded to 1,226. So the fully balanced corpus has 9,808 sentence vectors.

To evaluate the effect of over-sampling, three measures are used to estimate the scatter or clustering degree of the sentence vectors before and after over-sampling. Let v^c be the center vector for class c, defined by $v^c = \frac{1}{n_c}\sum_{i=1}^{n_c} v_i^c$. Then the within-class scatter degree of a class c is defined as:

$$D_b^c = \frac{1}{n_c}\sum_{i=1}^{n_c} distance(v_i^c, v^c) \tag{5}$$

Where distance(v_i^c, v^c) is the cosine of the angle between the two vectors, n_c is the total of samples in class c.

Furthermore, the within-class scatter degree of all classes is defined as:

$$D_b^c = \frac{1}{N} \sum_{c \in C} \sum_{i=1}^{n_c} \text{distance}(v_i^c, v^c) \tag{6}$$

Where N is the total number of samples of the training set, C is the set of all classes.

The between-class scatter degree is defined as:

$$D_w = \sum_{c_1 \in C} \sum_{c_2 \in C} \text{distance}(v^{c_1}, v^{c_2}) \tag{7}$$

Where v^{c_1} and v^{c_2} are the centre vector of class c_1 and c_2. A larger D_b indicates a closer clustering of all classes and a smaller D_w indicates a better scatter between different classes. Table 2 show scatter measures corresponding to the original training corpus and the generated balanced corpus , respectively.

Table 2. Scatter Degree of the Training Data Before/After Over-sampling

	Raw Corpus	Balanced Corpus
D_b^c (c = Like)	0.402	0.402
D_b^c (c = Disgust)	0.443	0.445
D_b^c (c = Happiness)	0.404	0.405
D_b^c (c = Anger)	0.439	0.441
D_b^c (c = Sadness)	0.476	0.474
D_b^c (c = Surprise)	0.426	0.426
D_b^c (c = Fear)	0.470	0.474
D_b	0.431	0.438
D_w	17.529	17.534

Table 2 shows that the within-class scatter degree and between-class scatter degree are almost unchanged before/after over-sampling which indicates very good consistency of the generated balanced corpus.

4.3   Evaluations on Emotion Classification

A ML-kNN classifier is trained using the balanced training data. Table 3 shows the achieved performance. For comparison, the baseline which uses Bag-Of-Words as features using the original training data is also listed in Table 3. The performances of emotion classification using other vector representations and training sets are also given in Table 3. It should be noted that the duplicating instance is another over-sampling method by randomly duplicate existing training sentences for the minority classes. Here, k=41 is used for the ML-kNN classifier and the vector selection factor α=0.5.

As shown in Table 3, the classifier trained on word/POS vector and the balanced training set performs the best. It achieves 0.484 and 0.468 for the loose and strict average precisions, respectively. Compared to the 19 submitted systems in NLP&&CC2013

evaluation, which has the top performance of 0.365(loose) and 0.348(strict)[4], our system outperformed the top performer by 0.119 (loose) and 0.120 (strict), translating into 32.6% and 34.5% improvement, respectively. It also shows that all classifiers trained on balanced training set performs better than the original training set, which indicates that the proposed over-sampling method are effective in improving classifier training on imbalanced data. It should be pointed out that the classifier based on word/POS vectors gives 1.9% and 29.3% improvement over the one based on word vector, respectively, indicating the word/POS pair is a much better feature to use, especially for the strict measure.

Table 3. Emotion Classification Results with different vector representations and training set

Settings	Average Precision (Loose)	Average Precision (Strict)
Bag-Of-Words + training set	0.271	0.261
word vector + training set	0.320	0.315
word/POS vector + training set	0.334	0.325
word vector + balanced training set	0.475	0.362
word/POS vector + balanced training set	**0.484**	**0.468**
word/POS vector + duplicating instances	0.383	0.369
NLP&&CC2013 best performance	0.365	0.348

Details of the classification performance of our system by each emotion class compared to the same classifier trained by the original training set are given in Table 4.

Table 4. Performance of Classification on Each Emotion Class Before/After Over-Sampling

Class	Average Precision	
	Raw Training Set	Balanced Training Set
Like	0.424	0.634
Disgust	0.256	0.514
Happiness	0.280	0.332
Anger	0.208	0.315
Sadness	0.413	0.603
Surprise	0.009	0.139
Fear	0.024	0.278

It is observed that classification performance on all emotion classes are improved using the balanced training data. The precision for the minority classes for *surprise* and *fear,* in particular, improves about 16 and 11 times, respectively. It should also be pointed out that even for the majority class *like*, the improvement is 49.5% which is quite impressive and yet quite explainable. Table 4 is a clear indication that the proposed over-sampling method is very effective in improving the classification for imbalanced data.

---

[4] http://tcci.ccf.org.cn/conference/2013/dldoc/evres02.pdf

4.4   Error Analysis

Further examination and analysis of the results show several problems which need to be addressed in the future. Firstly, in the proposed method, continuous skip-gram model is used to train word/POS pair as the vector unit. We observed that the quality of the word/POS vectors on nouns is much better than that of verbs and adjectives. The top words that are similar to an adjective may be its synonyms, words within similar context, and sometimes even its antonyms. Obviously, the sentence vector constructed by such word/POS vectors could not accurately reflect the sentiment orientation of those sentences. Secondly, in this study, word vectors used in sentence vector construction have equal weights. However, different words play different roles in a sentence and their importance to the sentence, especially for emotion classification, are in general different. Further analysis shows that performance improvement for minority classes for *surprise* and *fear,* in particular, improves about 16 and 11 times, respectively. It should also be pointed out that even for the majority class *like*, the improvement is 49.5% which is quite impressive and yet quite explainable. Table 4 is a clear indication that the proposed over-sampling method is very effective in improving the classification for imbalanced data. Further directions on development of a good weighting scheme should be investigated in the future to improve the performance of emotion classification.

5   Conclusion

This paper presents a novel over-sampling method to form additional sum sentence vectors for minority classes in order to improve emotion classification for imbalanced data. Through iteratively generating new minority class training samples by the sum vector of two existing sentence vectors, the training dataset are fully balanced. Evaluations on NLP&CC2013 Chinese micro blog emotion classification dataset shows that the obtained classifier achieves 48.4% average precision, a 32.6% percent improvement over the state-of-art performance on this dataset (at 36.5%). This result shows that the proposed over-sampling method can effectively address the problem of data imbalance and thus achieve much improved performance for emotion classification. Detailed analysis also shows that performance improvement for minority classes can achieve up to 16 times for skewed data. Even for the majority class *like*, the improvement is 49.5%. These are all clear indications that the proposed over-sampling method is very effective in improving the classification for imbalanced data.

Future directions include the development of weighting schemes that can reflect the importance of different words to emotion classification. The proposed sentence vector based over-sampling method may be applied to other text classification tasks with imbalanced training data.

Acknowledgments. This research is supported by National Natural Science Foundation of China 61203378, 61370165, Natural Science Foundation of Guang Dong Province S2013010014475, MOE Specialized Research Fund for the Doctoral Program of Higher Education 20122302120070, Shenzhen Foundational Research Funding JCYJ20120613152557576 and JC201104210033, Shenzhen International

Cooperation Research Funding GJHZ2012 0613110641217, Technology Research and Development Project of Shenzhen CXZZ20120618155717337, Open Projects Program of National Laboratory of Pattern Recognition and The Hong Kong Polytechnic University (Z0EP).

References

1. Turney, P.-D.: Thumbs Up or Thumbs Down? Semantic Orientation Applied to Unsupervised Classification of Reviews. In: Proceedings of ACL 2002, pp. 417–424 (2002)
2. Kamps, J., Marx, M., Mokken, R.-J., de Rijke, M.: Using WordNet to Measure Semantic Orientation of Adjectives. In: Proceedings of LREC 2004, pp. 1115–1118 (2004)
3. Pang, B., Lee, L., Vaithyanathan, S.: Thumbs up? Sentiment Classification using Machine Learning Techniques. In: Proceedings of EMNLP 2002, pp. 79–86 (2002)
4. Gu, X.-J., Wang, Z.-L., Liu, J.-W., Liu, S.: Research on Modeling Artificial Psychology Based on HMM. Application Research of Computers 12, 30–32 (2006)
5. Quan, C., Ren, F.: Construction of a Blog Emotion Corpus for Chinese Emotional Expression Analysis. In: Proceedings of EMNLP 2009, pp. 1446–1454 (2009)
6. Chawla, N.V., Japkowicz, N., Kolcz, A.: Editorial: Special Issue on Learning from Imbalanced Data Sets. SIGKDD Explorations 6(1), 1–6 (2004)
7. Zhou, Z.-H., Liu, X.-Y.: Training Cost-sensitive Neural Networks with Methods Addressing the Class Imbalance Problem. Knowledge and Data Engineering 18(1), 63–77 (2006)
8. Ertekin, S., Huang, J., Bottou, L., Giles, C.-L.: Learning on the Border: Active Learning in Imbalanced Data Classification. In: Proceedings of CIKM 2007 (2007)
9. Chen, T., Xu, R., Wu, M., Liu, B.: A Sentiment Classification Approach based on Sentiment Sentence Framework. Journal of Chinese Information Processing 27(5), 67–74 (2013)
10. Ren, J.-W., Yang, Y., Wang, H., Lin, H.: Construction of the Binary Affective Commonsense Knowledgebase and its Application in Text Affective Analysis. China Science Paper Online (2013), http://www.paper.edu.cn/releasepaper/content/201301-158
11. Longadge, R., Dongre, S.-S., Malik, L.: Class Imbalance Problem in Data Mining Review. International Journal of Computer Science and Network 2(1), 1305–1707 (2013)
12. Wang, Z.-Q., Li, S.-S., Zhu, Q.-M., Li, P.-F., Zhou, G.-D.: Chinese Sentiment Classification on Imbalanced Data Distribution. Journal of Chinese Information Processing 26(3), 33–37 (2012)
13. Deerwester, S., Dumais, S.-T., Furnas, G.-W., Landauer, T.-K., Harshman, R.: Indexing by Latent Semantic Analysis. Journal of the American Society for Information Science 41(6), 391–407 (1990)
14. Bellegarda, J.-R.: A Latent Semantic Analysis Framework for Large–span Language Modeling. In: Proceedings of Eurospeech 1997, pp. 1451–1454 (1997)
15. Blei, D.-M., Ng, A.-Y., Jordan, M.-I.: Latent Dirichlet Allocation. Journal of Machine Learning Research 3, 993–1022 (2003)
16. Riis, S., Krogh, A.: Improving Protein Secondary Structure Prediction using Structured Neural Networks and Multiple Sequence Profiles. Journal of Computational Biology, 163–183 (1996)
17. Mikolov, T., Chen, K., Corrado, G., Dean, J.: Efficient Estimation of Word Representations in Vector Space. In: Proceedings of ICLR Workshop (2013)
18. Han, J., Kamber, M.: Data mining: Concepts and Technique. Morgan Kaufman, San Francisco (2006)

Reader Emotion Prediction Using Concept and Concept Sequence Features in News Headlines

Yuanlin Yao[1], Ruifeng Xu[1,*], Qin Lu[2], Bin Liu[1], Jun Xu[1], Chengtian Zou[1], Li Yuan[1], Shuwei Wang[1], Lin Yao[3], and Zhenyu He[1]

[1] Key Laboratory of Network Oriented Intelligent Computation,
Shenzhen Graduate School, Harbin Institute of Technology, Shenzhen, China
{yuanlinhappy,xuruifeng.hitsz,hit.xujun,chsky.zou,
yuanlisail,wswangshuwei}@gmail.com,
bliu@insun.hit.edu.cn, zyhe@hitsz.edu.cn
[2] Department of Computing, Hong Kong Polytechnic University, Hong Kong, China
csluqin@comp.polyu.edu.hk
[3] Peking University Shenzhen Graduate School, Shenzhen, Guangdong, China
yaolin@insun.hit.edu.cn

Abstract. This paper presents a method to predicate news reader emotions. News headlines supply core information of articles, thus they can serve as key information for reader emotion predication. However, headlines are always short which leads to obvious data sparseness if only lexical forms are used. To address this problem, words in their lexical forms in a headline are transferred to their concepts and concept sequence features of words in headlines based on a semantic knowledge base, namely HowNet for Chinese. These features are expected to represent the major elements which can evoke reader's emotional reactions. These transferred concepts are used with lexical features in headlines for predicating the reader's emotion. Evaluations on dataset of Sina Social News with user emotion votes show that the proposed approach which do not use any news content, achieves a comparable performance to Bag-Of-Word model using both the headlines and the news contents, making our method more efficient in reader emotion prediction.

Keywords: Emotion Prediction, Concept Feature, Concept Sequence Feature, HowNet.

1 Introduction

Reader's emotion prediction investigates methods to predict the most likely reader emotional reactions, such as *happy*, *disappointed*, after reading an article or a piece of text. In previous studies, this task was always treated as a text classification problem which classifies given documents to categories evoking different emotions[1-4]. The majority of works may be camped into either the Bag of Words (BOW) based approach[4] or the topic model based approach[1-3].The BOW approach explores the

---

* Corresponding author.

A. Gelbukh (Ed.): CICLing 2014, Part II, LNCS 8404, pp. 73–84, 2014.

role of words in headline and news content, especially emotional words, which may evoke reader's emotions. This approach can achieve good performance in a high dimensional feature space with sufficient training data. However, such training data are almost never available. The topic model based approach seeks to find the connection between reader's emotional reactions and the topic of news article. Since the topic model obviously can reduce the feature space, the topic model based approach achieves better predication performance. However, this approach only finds isolated words to represent the topic of a piece of news without considering word order. The lack of word sequence sometimes confuses the emotion predicator. For example, "人砸了手机(*A personbroke a mobile phone*)" and "手机砸了人(*A mobile phone brokea person*)" are two different events, which can trigger quite different emotional reactions. However, they are regarded as the same in a typical topic model, thus the emotion prediction model cannot come up with different emotional reactions. Such a problem also occurs in the BOWapproach.

Generally speaking, headline of a piece of news contains the key information of that news article. Thus, it provides the essential features for predicating reader emotions. Unfortunately, headlines are always short in length. So using lexical units alone would face data sparseness problem for reader emotion predication.

In this paper, we propose a novel approach to predict reader emotions using only headline of news article.To address the mentioned data sparseness problem, instead of using the article content for more features, a semantic dictionary, HowNet[1], is used as conceptual knowledgebase to augment the concepts and semantics to reveal the emotional reactions from reader's cognition perspective. By transfer headline words in lexical forms to concept representations, the feature space is much reduced which is helpful in reducing the influence of data sparseness. Ordered concept sequences are used as new features which are better abstractions for the event(s) described by the headlines. Based on the transferred concepts, concept sequences, as well as word features, a multi-label classifiers is developed to predicate the reader's emotion using concept, concept sequence and word features. The evaluations on Sina Social News Reader Emotion Corpus show that the proposed approach achieved a comparable performance to the BOW based approach using both headlines and contents. It indicates using headlines alone, can supply as sufficient information as using content, making our proposed approach much more efficient in emotion predication. Furthermore, the proposed approachis much more applicable to reader emotion predication for short text such as twitters.

The rest of this paper is organized as follows. Section 2 briefly reviews related works. Section 3 presents our proposed headline based emotion prediction approach using a combination of concept feature, concept sequence feature and word feature. Evaluations and discussions are given in Section 4. Section 5 is the conclusion.

2   Related Works

Reader emotion prediction is always treated as a text classification problem in which a piece of text are classified into the most likely triggered emotion category. Most of

---

[1] http://www.keenage.com

the emotion prediction techniques can be grouped into the BOW based approach and the topic model based approach.

To explore the connection between emotions and lexical semantics, the 4th International Workshop on Semantic Evaluations (SemEval-2007) held a shared task on Affective Text [5-7]. This task is mainly based on the headlines of news articles. The task proposes to classify reader emotions to two polarities, namely *positive* and *negative* or six emotion classes including *anger, fear, happiness, sadness, surprise* and *disgust*. In this evaluation, the submitted systems were mainly based on lexicon-based method and supervised BOW model. Using corpus from Yahoo! Kimo News with emotion votes, Lin et al. [8-10] used supervised machine learning method for reader emotions predication. The features they used include Chinese words, word bigrams, news Meta data such as the source and author of the news, the affix similarity and so on. Different from direct emotional words which describe feelings directly, implicit emotional words refer to thosewords that can trigger other people's emotions such as "*ghost*" and "*wicked*"[11].

In recent years, topic model based techniques are used in reader emotion prediction through identifying the link between topic and reader emotion. On the basis of Latent Dirichlet Allocation (LDA) model, Bao et al. [1] added another emotion layer into the standard LDA model which considersthe emotion factor in the process of generating text.The study simplifies emotion prediction by assuming that only topic words cantriggerreader emotions. Through reducing the feature space with less information loss, the topic model based approach generally achieves good performance. Xu et al. [2] proposed a Weighted LDA model on the basis of traditional LDA with weights for each word according to their importance to reader emotion categories. Through generate topic words using partitioned text corresponding to different reader emotions; Xu et al. [3] also proposed a Partitioned LDA model which transforms the unsupervised LDA model to a supervised model which achieves a better performance on reader emotion prediction.

Even though topic model based methods show advantages in reader emotion prediction, they still face some common problems. The first problem is that it is a supervised method which means getting a sufficiently large training data is very costly. Secondly, when a new training sample is added to the inference process, the topic based model must be re-inferred for all the parameters, so the algorithms are not efficient. When the text content is relatively complex and need more number of topics, the efficiency is particularly low. Thirdly, from the characteristics of the topic model itself, only isolated words can be used as features for training without methods to consider word/concept sequence.

It is important to pointout that both the BOW approach and the topic model approach always require sufficiently long text to generate enough discriminative features. Thus they are not suitable to be used to predict reader emotions for short text, such as twitters.

3   Our Approach

In this study, a novel approach is proposedto predicate reader emotionsby using short text such as news headlines. In order to solve the word sparseness problem, we transform words in headlines to semantically based concepts through the semantic knowledge base, HowNet. Two kinds of concept based features are designed o represent headlines: concept term feature and concept sequence feature. These two additional features are then used along with the regular word feature in a multi-label classifier for reader emotion prediction.

3.1 Concept Feature

In the semantic knowledge base HowNet, each word is described by set of concept linked sememes (labeled as DEF). The sememes are connected by eight types of relations, including hypernym, hyponym, synonymy, antonymy and so on. The basic sememesfor nouns and verbs then form a tree-like hierarchy. The sememes of some example words are listed in Table 1.

Table 1. Example Words and Their sememes (DEF) in HowNet

Word	DEF
亮点(highlight)	{Advantagel 利:host={Situationl 状况}{entityl 实体}{eventl 事件}}
服毒(take poison)	{eatl 吃:patient={physicall 物质:modifier={poisonousl 有毒}},purpose={suicidel 自杀}
清 明 (Tomb-sweeping Day)	{timel 时间:TimeFeature={festivall 节},TimeSect={dayl 日},{condolel 致哀: location={facilitiesl 设施:{buryl 埋入:location={~},patient={partl 部件:PartPosition={bodyl 身},whole={humanl 人:{diel 死:experiencer={~} }}}}}, time={~}}
杰出(outstanding)	{fantasticl非常好}
优异(excellent)	{fantasticl非常好}

There are two notable characteristic of DEFs in HowNet:

(1) Words with the same DEFs havethe same semantic meaning. If two words have similar semantic or related concepts, their DEFs in HowNet will share the same sememes. This characteristic is helpful to discover the conceptual connection between two headlines.

(2) The sememes in DEF of a word provide certain clues related to emotion triggers for human beings from the cognitive perspective. For example, the DEF of the word "服毒(*take poison*)" contains sememe of "自杀|suicide" and the DEF of "清明 (*Tomb-sweepingDay*)" contains "节日|festival, 致哀|mourn and 死|death". Because both DEF contains similar sememes such as "死|death", the emotional reactions evoked by reading these two words should also be similar.

Different from emotion classification which emphasizeson the recognition and classification of emotion expression related keywords; Reader emotion prediction emphasizes the relationship between the events in news articles which can trigger emotions. Using concept (DEF) feature is helpful to construct the relations between different specific events using different words but sharing the same or similar concepts. The events with the same or similar concepts tend to evoke the same or similar emotional reactions of their readers. The use of concept features is also expected to describe events using much lower dimensions compared to word features which are important to the emotion predication task.

In this work, the Chinese words in headlines are first converted to their corresponding DEFs in HowNet. For words having more than one DEF records in HowNet, we choose DEF that has the same POS as the original word when looking up HowNet.

For example, a headline "父亲6天未合眼寻找走失儿子(*Father did not sleep for six dayslooking for his lost son*)" is transferred as follows:

{*human*|人:*belong*={*family*|家庭},*modifier*={*lineal*|直系}{*male*|男}{*senior*|长辈}} {*cardinal*|基数} {*original*|原} {*FuncWord*|功能词:*adjunct*={*neg*|否},*modifier*={*past*|过去}} {*shut*|关闭:*patient*={*part*|部件:*PartPosition*={*eye*|眼},*whole*={*AnimalHuman*|动物}}} {*LookFor*|寻} {*disappear*|消失} {*human*|人:*belong*={*family*|家庭},*modifier*={*junior*|小辈}{*lineal*|直系}{*male*|男}}

3.2   Concept Sequence Feature

Although the above concept level representation is expected to reduce data sparseness, the augmented concepts are still used as isolated feature items. As discussed before, different event/word order may trigger different reader emotions. To clearly describe an event in a headline, concept sequence feature is also considered.

When transferring headline words to DEFs, the frequently associated sememe sequences formed by ordered sememes are extracted as concept sequence features. The process of extracting concept sequences can be treated as class sequentialrules (CSR) mining. A class sequential rule is combined toform a sequence pattern plusits class tag. The target of CSR mining is to find all sequence patterns that have highly class correlations. In this study, the Prefixspan [12] algorithm is used. This algorithm is a depth-first algorithm. It first segments all the sequence items with prefix that have a frequency higher than the threshold. A corresponding projection list is then constructed for different prefix according to the sequences whose length is longer than 1. The steps above are repeated for every projection until no new sequencepattern are detected. Here, CHI-square test is used to measurethe importance of the generated concept sequence. The greater CHI score indicates that this sequence has stronger association with one specific reader emotion. The top ranked extracted concept sequences and their CHI-square test value are listed in Table 2.

For example, the concept sequence "human|人，suicide|自杀"can be extracted from the follow headlines:

-初中女生偷尝禁果怀孕男友喝农药自杀

(*A junior school girl pregnant, her boyfriend drink pesticide to suicide*)

-武大34岁博士疑因论文难通过在家自杀身亡

(*A 34 years old doctor of WuHan University suicide at home.*)

-男子枪杀妻子后饮弹自杀

(*A man killed himself after he shot his wife.*)

Table 2. The Top Extracted ConceptSequences

Concept sequence	CHI score
human人，die死	772.71
human人，suffer遭受	642.58
RelatingToCountry与特定国家相关，图image	549.69
human人，suicide自杀	438.29
TakeCare照料，human人，FrequencyValue频度值	405.27
human人，force强迫，human人	403.95
Beast走兽，image图	393.08

Obviously, reader emotions triggered by this news are quite similar. It indicates that the concept sequence feature should be a good discriminative feature for predicting reader emotion with good generalization capability.

3.3   Reader Emotion Predication Using Multi-label Classifier

Since emotions are always combination of several base emotions, reader emotion predication is naturally a multi-label classification problem. In this study, we incorporate the words, concepts and concept sequences as features in a multi-label classifier to predicate the reader's emotion through classify given news to categories triggering specific emotions. As the classifier itself is not the emphasis of this work, we chose to use two classifiers, RAkEL [13] and MLkNN [14], as the base multi-label classifiers, in the experiments.

4   Evaluation and Discussion

4.1   Experiment Setting

The dataset used in this evaluation comes from Sina Social News[2]. The news articles and their corresponding reader votes of emotional feedback are collected, i.e., *Touched, Empathy, Boredom, Anger, Amusement, Sadness, Surprise* and *Warmness*. Only news articles with more than 20 votes are selected to construct a reader emotion predication corpus. The detail procedure using readers' emotional votes to generate emotion labels is presented in [4]. Table 3 shows the statistics on this dataset. It has 11,007 news documents in which 2,738 documents have two emotion labels and others have one emotion label. The digitsin diagonal elementsare the corresponding numbers of single emotion and otherdigits are the number of documents with two emotions.

In the following experiments, CHI square Test is used for feature selection. Binary weighting is usedfor document representation. RAkEL[13] and MLkNN[14] are employed as the base multi-label classifiers, respectively.

---

[2] http://news.sina.com.cn/society/moodrank/

Table 3. The Statistics of Reader EmotionCorpus

Emotion	Touched	Empathy	Boredom	Anger	Amusement	Sadness	Surprise	Warmness
Touched	1,387	60	2	27	0	9	5	118
Empathy		308	11	214	19	84	25	70
Boredom			426	51	241	11	496	115
Anger				2,878	468	417	32	40
Amusement					1,532	113	79	29
Sadness						603	10	41
Surprise							799	77
Warmness								84
Total	1,608	791	1,353	4,127	2,481	1,288	1,523	574

The commonly used metrics for multi-label classification are different from single-label classification due to the inherent differences of the classification problem. In this study, five evaluation metrics are adopted in the multi-label classification experiment include Average Precision (AVP), Coverage (COV),One-Error (OE),Ranking Loss (RL) and Hamming Loss (HL)[15].

The dataset is randomly split into 4 equal parts. 4 fold cross validation is performed. All the performances given below are the average of 4 folds.

4.2 Experimental Results and Discussions

Two multi-label classifier, RAkEL and MLkNN using BOW features in the headlines are treated as baselines, respectively. After filtering stop words, Chinese word unigrams ofthe headlines are used as discriminative features.Table 4 shows the performance achieved by the baselines.

Table 4. The Baseline Performance of Emotion Prediction (Unigram word features)

#Features	Classifier	AVP	COV	OE	RL	HL
800 words	MLkNN	0.609	2.070	0.577	0.239	0.153
	RAkEL	0.658	2.340	0.428	0.264	0.126
1,200 words	MLkNN	0.613	2.075	0.568	0.240	0.151
	RAkEL	0.678	2.214	0.410	0.248	**0.126**
1,600 words	MLkNN	0.611	2.070	0.505	0.306	0.155
	RAkEL	0.681	2.163	**0.391**	0.236	0.128
2,000 words	MLkNN	0.618	**2.022**	0.563	**0.232**	0.151
	RAkEL	**0.682**	2.151	**0.391**	0.237	0.131

Generally speaking, the performances of baselinesimprove with more word features. The RAkEL classifier with 2,000 word features achieves the best performance, with average precision reaching 0.682.

The second experiment evaluates the performance by using concept features individually. Table 5 shows the performance using different number of concept features.

Table 5. Emotion Prediction Performance Using Concept Features

#Features	Classifier	AVP	COV	OE	RL	HL
Concept 400	MLkNN	0.654	1.837	0.515	0.206	0.147
	RAkEL	0.660	2.364	0.427	0.269	**0.133**
Concept600	MLkNN	0.661	1.796	0.503	0.201	0.146
	RAkEL	0.673	2.250	0.418	0.252	0.135
Concept800	MLkNN	0.662	1.788	0.501	0.199	0.145
	RAkEL	0.679	2.196	**0.414**	0.250	0.137
Concept 1,000	MLkNN	0.663	**1.777**	0.501	**0.198**	0.145
	RAkEL	**0.682**	2.152	0.415	0.239	0.138

This experiment is conducted with much lower dimension compared to word unigram since the number of sememes used in HowNet is quite limited. Note thatthe more concept features used, the better performanceis achieved. Compare to the baselines, using concept features individually can achieve similar performance with only the One-Error slightly higher, while the feature dimension is much lower.

The third experiment evaluates the performance by combining word and concept features. Here, the word and concept feature are sorted by their Chi-test value. The performance achieved by the top word and concepts are given in Table 6.

Table 6. Emotion Prediction Performance Using Word and Concept Features

#Features	Classifier	AVP	COV	OE	RL	HL
Word+ Concept 800	MLkNN	0.655	1.835	0.508	0.205	0.145
	RAkEL	0.689	2.137	0.398	0.236	**0.129**
Word+ Concept1,200	MLkNN	0.660	1.807	0.502	0.201	0.145
	RAkEL	0.692	2.194	0.399	0.241	0.126
Word+ Concept1,600	MLkNN	0.663	1.786	0.499	0.199	0.145
	RAkEL	0.696	2.152	0.390	0.225	0.129
Word+ Concept2,000	MLkNN	0.661	**1.798**	0.501	**0.200**	0.144
	RAkEL	**0.715**	1.917	**0.384**	0.207	0.133

Table 6 shows that the performanceof using both features is better than using them alone. The average precision increases from 0.682 to 0.715with a 4.84% of improvement when the top 2,000 word and concept features are used. This result indicates that the combined use of word feature and concept feature are helpful to improve emotion predication.

The fourth experiment evaluates the performance of concept sequence features. For comparison, the individual use of word sequence features and concept sequence features are evaluated, respectively. The achieved performances are given in Table 7 and Table 8, respectively.

Table 7. Emotion Predication Performance Using WordSequence Features (WSF)

#Features	Classifier	AVP	COV	OE	RL	HL
WSF 200	MLkNN	0.581	2.307	0.612	0.268	0.155
	RAkEL	0.346	3.873	0.870	0.500	0.154
WSF 400	MLkNN	**0.584**	2.287	0.609	0.267	0.154
	RAkEL	0.367	3.783	0.839	0.486	**0.153**
WSF 600	MLkNN	0.583	2.286	0.611	0.266	0.155
	RAkEL	0.380	3.728	0.821	0.478	0.154
WSF 800	MLkNN	0.584	**2.274**	**0.610**	**0.265**	0.154
	RAkEL	0.391	3.663	0.807	0.469	0.154
WSF 1000	MLkNN	0.583	2.276	0.613	0.266	0.154
	RAkEL	0.398	3.601	0.790	0.463	0.154

Table 8. Emotion Predication Performance Using Concept Sequence Features (CSF)

#Features	Classifier	AVP	COV	OE	RL	HL
CSF 200	MLkNN	**0.609**	**2.090**	0.573	0.242	**0.153**
	RAkEL	0.470	3.329	0.684	0.416	0.148
CSF 400	MLkNN	0.607	2.090	0.575	0.242	0.154
	RAkEL	0.501	3.181	0.639	0.392	0.147
CSF 600	MLkNN	0.601	2.136	0.586	0.248	0.155
	RAkEL	0.523	3.075	0.609	0.376	0.147
CSF 800	MLkNN	0.603	2.126	0.579	0.246	0.154
	RAkEL	0.541	2.936	0.583	0.359	0.146
CSF 1000	MLkNN	0.605	2.113	0.572	**0.236**	0.154
	RAkEL	0.561	2.852	**0.566**	0.343	0.153

Table 7 shows that the individual use of word sequence features achieves low performance attribute to the feature sparseness. In fact, the sparseness for word feature is already shown in the first experiment in Table 4. The continuous word sequence features leads to much more sparseness, thus degrading performance even further. Table 8 shows that classifiers using concept sequence features alone do not achieve very good performance even though they are still better than using word sequence features. This is because feature dimension reduction by using concepts is helpful.

The fifth experiment evaluates the performance by using combined word feature, concept feature and concept sequence feature. Results are shown in Table 9.

Table 9 shows that the combined use of three types of features is the most effective. The RAkEL classifier with three types of features achieves the best average precision performance, 0.750, an increase of 6.80%, which translate to 9.97% improvement overthe baseline using headline BOW features.

Table 9. Emotion Predicative Using Word, Concept and Concept Sequence Feature

#Features	Classifier	AVP	COV	OE	RL	HL
Word+Concept+CSF900	MLkNN	0.693	1.640	0.445	0.179	0.135
	RAkEL	0.697	1.944	0.383	0.212	0.135
Word+Concept+CSF1,100	MLkNN	0.696	1.636	0.446	0.178	0.136
	RAkEL	0.716	1.894	0.375	0.203	0.131
Word+Concept+CSF1,400	MLkNN	0.687	1.629	0.448	0.179	0.139
	RAkEL	0.718	1.893	0.374	0.200	0.130
Word+Concept+CSF1,600	MLkNN	0.695	1.626	0.445	0.174	0.131
	RAkEL	0.729	1.840	0.369	0.198	0.129
Word+Concept+CSF1,800	MLkNN	0.680	1.660	0.449	0.182	0.138
	RAkEL	0.743	1.799	0.360	0.198	0.128
Word+Concept+CSF2000	MLkNN	0.694	**1.620**	0.442	**0.178**	0.131
	RAkEL	**0.750**	1.780	**0.349**	0.188	**0.124**

The sixth experiment is conducted to evaluate emotion prediction using headline and content using BOW features. The achieved AVP is 0.656 when the top 2,000 word features are used. The performance is even lower than our baseline at 0.682 given in Experiment 1. This result shows that use headline words alone are more productive for emotion prediction. It should be pointed out that when 9,000 word features are used, content using BOW achieved AVP of 0.749 comparable to our proposed approach using combined features in the headlines. However, our approach has much lower feature dimensions, thus more efficient and making is more applicable for emotion predication of shorter text such as Twitter text.

5   Conclusion

In this paper, we present a novel approach to predict reader emotionsusing only the headlines of news articles. The sememes in HowNet are used to augment the words in headlines on the concept level. Based on the augmented concepts, concept sequence features are also extracted for the headlines.These features and word features are then incorporated in a multi-label classifier for predicting reader emotion as a text classification task. Experiments on Sina Social News dataset show that the combined use of word feature, concept feature and concept sequence feature using the RAkEL classifier achieves about 10%improvement compared to the use of word features which shows the contribution of concept and concept sequence as features. Furthermore, the performance of the proposed approach is similar to that of the BOW approach using both word features from headline and content using much higher dimensions. This is a clear indication that the headlines of news article are very productive in emotion predication and our approach is much more efficient than using content for emotion prediction. More importantly, the proposed approach is much more suitable for emotion detection and prediction for short text, such as Twitter, whereas the BOW based approach and the topic model approach is not suited for shorter text.

Acknowledgement. This research is supported by National Natural Science Foundation of China 61203378, 61370165, Natural Science Foundation of Guang Dong Province S2013010014475, MOE Specialized Research Fund for the Doctoral Program of Higher Education 20122302120070, Shenzhen Foundational Research Funding JCYJ20120613152557576 and JC201104210033, Shenzhen International Cooperation Research Funding GJHZ2012 0613110641217, Technology Research and Development Project of Shenzhen CXZZ20120618155717337, Open Projects Program of National Laboratory of Pattern Recognition and The Hong Kong Polytechnic University (Z0EP).

References

1. Bao, S., Xu, S., Zhang, L., et al.: Mining Social Emotions from Affective Text. IEEE Transactions on Knowledge and Data Engineering 24(9), 1658–1670 (2012)
2. Xu, R., Ye, L., Xu, J.: Reader's Emotion Prediction Based on Weighted Latent Dirichlet Allocation and Multi-label k-nearest Neighbor Model. Journal of Computational Information Systems 9(6), 2209–2216 (2013)
3. Xu, R., Zou, C., Xu, J., Lu, Q.: Reader's Emotion Prediction Based on Partitioned Latent Dirichlet Allocation Model. In: Proceedings of International Conference on Internet Computing and Big Data, USA (July 2013)
4. Ye, L., Xu, R., Xu, J.: Emotion Prediction of News Articles from Reader's Perspective Based on Multi-label Classification. In: Proceedings of International Workshop on Web Information Processing, pp. 2019–2024 (2012)
5. Strapparava, C., Mihalcea, R.: Semeval-2007 Task 14: Affective Text. In: Proceedings of the 4th International Workshop on Semantic Evaluations in ACL 2007, pp. 70–74 (2007)
6. Kozareva, Z., Navarro, B., Vázquez, S., et al.: Ua-Zbsa: A Headline Emotion Classification through Web Information. In: Proceedings of the 4th International Workshop on Semantic Evaluations in ACL 2007, pp. 334–337 (2007)
7. Katz, P., Singleton, M., Wicentowski, R.: Swat-Mp: The Semeval-2007 Systems for Task 5 and Task 14. In: Proceedings of the 4th International Workshop on Semantic Evaluations in ACL 2007, pp. 308–313 (2007)
8. Lin, K.H.-Y., Yang, C., Chen, H.-H.: Emotion Classification of Online News Articles from the Reader's Perspective. In: Proceedings of the 2008 IEEE/WIC/ACM International Conference on Web Intelligence and Intelligent Agent Technology, pp. 220–226 (2008)
9. Lin, K.H.-Y., Chen, H.-H.: Ranking Reader Emotions Using Pairwise Loss Minimization and Emotional Distribution Regression. In: Proceedings of the Conference on Empirical Methods in Natural Language Processing, pp. 136–144 (2008)
10. Yang, C., Lin, K.H.-Y., Chen, H.-H.: Writer Meets Reader: Emotion Analysis of Social Media from Both the Writer's and Reader's Perspectives. In: Proceedings of the 2009 IEEE/WIC/ACM International Joint Conference on Web Intelligence and Intelligent Agent Technology, pp. 287–290 (2009)
11. Strapparava, C., Valitutti, A., Stock, O.: The Affective Weight of Lexicon. In: Proceedings of the 5th International Conference on Language Resources and Evaluation, pp. 423–426 (2006)
12. Han, J., Pei, J., Mortazavi-Asl, B., et al.: Prefixspan: Mining Sequential Patterns Efficiently by Prefix-projected Pattern Growth. In: Proceedings of the 17th International Conference on Data Engineering, pp. 215–224 (2001)

13. Zhang, M.-L., Zhou, Z.-H.: ML-kNN: A Lazy Learning Approach to Multi-label Learning. Pattern Recognition 40(7), 2038–2048 (2007)
14. Tsoumakas, G., Vlahavas, I.P.: Random k-Labelsets: An Ensemble Method for Multilabel Classification. In: Kok, J.N., Koronacki, J., Lopez de Mantaras, R., Matwin, S., Mladenič, D., Skowron, A. (eds.) ECML 2007. LNCS (LNAI), vol. 4701, pp. 406–417. Springer, Heidelberg (2007)
15. Schapire, R.E., Singer, Y.: Boostexter: A Boosting-Based System for Text Categorization. Machine Learning 39(2-3), 135–168 (2000)

Identifying the Targets of the Emotions Expressed in Health Forums

Sandra Bringay[1,2], Eric Kergosien[1,3], Pierre Pompidor[1], and Pascal Poncelet[1]

[1] 1-LIRMM UMR 5506, CNRS, University of Montpellier 2
{bringay,kergosien,pompidor,poncelet}@lirmm.fr
[2] AMIS, University of Montpellier 3
[3] Maison de la télédétection, IRSTEA

Abstract. In the framework of the French project Patients' Mind, we focus on the semi-automatic analysis of online health forums. Online health forums are areas of exchange where patients, on condition of anonymity, can talk about their personal experiences freely. These resources are a gold mine for health professionals, giving them access to patient to patient exchanges, patient to health professional exchanges and even health professional to health professional exchanges. In this paper, we focus on the emotions expressed by the authors of the messages and more precisely on the targets of these emotions. We suggest an innovative method to identify these targets, based on the notion of semantic roles and using the FrameNet resource. Our method has been successfully validated on real data set.

Keywords: Opinion mining, emotion analysis, health system applications.

1 Introduction

In the framework of the French project Patients' Mind[1], we focus on the semi-automatic analysis of online health forums. Online health forums are areas of exchange where patients, on condition of anonymity, can talk about their personal experiences freely. An example is the very active forum *healthforum*[2], which allows internet users (often non-health professionals) to exchange their opinions on their health situation. In [1], Hancock and al. demonstrated that the ability to communicate anonymously via computers facilitates the expression of affective states such as emotions, opinions, doubts, risk, fears, *etc.* These affective states are generally repressed in more traditional communication contexts, such as face to face interviews or when people answer surveys. These resources are a gold mine for health professionals, giving them access to patient to patient exchanges, patient to health professional exchanges and even health professional to health professional exchanges.

---

[1] Funded by the MSH-M (Maison des Sciences de l'homme de Montpellier, trans. Humanities Home of Montpellier) and the inter-MSH network.
[2] http://www.healthforum.com/

In literature, while many approaches have been proposed for text polarity analysis (*positive, neutral* and *negative*) [2], few ones focus on sentiment analysis (*joy, anger, sadness, etc.*) [3] or on the intensity of these two emotional states [4]. These approaches have been applied in various fields and can, for instance, be used to classify texts effectively. However, these methods rarely address the detection of targets or sources even if they provide relevant information. For example, lets consider the two following sentences: *I'm afraid of my doctors reaction* and *My doctor is afraid of my reaction to the drug.* Conventional methods detect that these two sentences are negative and that they contain the emotion *fear*. In the first case, they do not detect that this emotion involves the reaction of the doctor and that the *fear* is felt by the speaker. In the second case, they do not detect that the emotion deals with the patient's response to the drug and that the fear is felt by the doctor. In this paper, we focus on this level of precision in order to aggregate this type of information, in a second time, to multiple messages (*e.g.*, find the number of messages in which patients express their fear about a drug). In literature, some methods analyze the sources of affective states (*e.g.*, who feels the emotional state) [5], and others the targets (*e.g.*, what is the object of the emotional state?) [6]. In this article, we focus on the target, in the specific context of the emotions expressed in health forums.

Our method aims to identify traces of emotions in texts associated with a target or a context, if possible medical ones, facilitating the interpretation of the emotion. This method can be generalized to other affective states and other fields of application. In this purpose, we suggest as [7,8] do to incorporate the *Shallow semantic parsing* and use the FrameNet[3] lexical resource. Based on the concept of semantic role defined by Baker and al. [9], this resource describes situations schematically. Annotation based on this resource allows us to identify in sentences expressions of emotions and to explain their components. We propose a typology of these annotations dedicated to our specific context. To our best knowledge, there is no method based on such type of annotations, customised thanks to a typology for health forum application field. We compare this approach to the more classic ones, which are based on a distance computation in the dependency tree of the sentences, to identify associations between emotions and predefined targets. Our method has been successfully evaluated on real posts.

The rest of this paper is organized as follows: in Section 2, we motivate our work in the framework of the semi-automatic analysis of health forums and we define the referred task of sentiment analysis. In Section 3, we show the latest methods dedicated to this task. In Section 4, we describe the conducted experiments, then the main results in Section 5. Finally, in Section 6, we conclude and give the main perspectives associated to this work.

2   Motivations and Task Definition

As Siegrist pointed out [10], one of the greatest challenges for health professionals is to capture patients satisfaction. With this aim, he studied patients' feedback

---

[3] https://framenet.icsi.berkeley.edu/fndrupal/home

after their stay in the main American hospitals and turned them into raw data for decision making. Using the forums as an object of study, we are getting closer to the patients private sphere. Indeed, patients express things through posts they do not express in comments (even anonymous). However, precisely identifying the emotional state of patients through these messages is a difficult objective task and not always verifiable. However, we could consider using these large amounts of emotionally-charged texts to construct indicators that are relevant for health professionals. An example of such an application is *We feel fine* [11]. This tool scans the web with the aim of assessing users' moods. Every 10 minutes, the application considers sentences with emotional words and performs statistical calculations based on the type of feelings, age, gender, *etc.*

Sentiment analysis has been widely studied since the early 2000s. Many communities are interested in this area and their definitions and interpretations are highly varied (*e.g.*, psychology, social sciences, computational linguistics, natural language processing, data mining, *etc.*). Sentiment analysis involves the extraction of emotional states expressed or implied in texts. There are several models of the opinion [12,13], which vary according to the purpose of the study and the completed tasks. To generalize our work to any emotional state and any task, we consider the model described in Figure 1. Emotional state is experienced by a *source* (or experiencer). It refers to a *polarity*, that is to say, a judgment which can be positive if it is linked to a positive effect for the experiencer and in the opposite negative or in some cases neutral. The emotional state can also refer to an *emotion* such as *anger, joy, sadness, etc.* Generally, emotions are associated with a polarity. *Joy* is a positive example, *anger* is negative and *surprise* is neutral. Different levels of *intensity* can be associated to the emotional state (*e.g.*, very successful, a little sad, *etc.*). Finally, the emotional state has a *target* which is the receptacle of the opinion or emotion.

In this work, we focus on target identification. They are generally present in texts as named-entities, events, abstract concepts, features associated with these abstract concepts or general contexts [14,15]. Lets consider the examples in Table 1. In S1, S2 and S3 sentences, the emotion *fear* is related to an entity represented respectively by the general concept *drug*, by the event *beginning of the chemotherapy* and by the named-entity *IVEMED* (which is a drug name). In S4, the target of the opinion involves one *aspect*, a characteristic, *the tolerance rate to the drug*. In S5, only the aspect is present. In S6, there is no explicit target and the target refers to the general context the sentence is formulated in. In S7, the target is detailed in the rest of the sentence and it is not limited to medical entity *pain*. Sometimes, it is difficult to distinguish the target from the *circumstances* which cause the emotional state. This is the case for example in S8. These examples illustrate the complexity of the task because targets can be expressed in very different ways. Unlike most of the approaches of the literature, we take the position in this proposal not to limit the target to a few words, but rather to provide as much information as possible, like it is done in S7 and S8.

The (semi-)automatic analysis of forums is difficult from a technological standpoint. Most (semi)-automatic methods used in the health domain are applied to

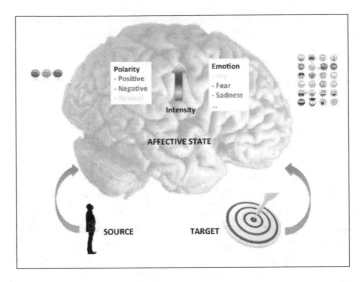

Fig. 1. Affective state model

Table 1. Examples of emotion expressions

S1: I am afraid of this drug.
S2: I am afraid to begin the chemotherapy.
S3: I am afraid to take the IVEMEND.
S4: The rate of tolerance for this drug is excellent!
S5: The rate of tolerance is excellent.
S6: I am afraid.
S7: I'm afraid to live in pain for another ten years.
S8: I was given a 10 chance of living for 10 years, announced real prospects of recurrence and the fear stayed with me all this time.

publications and hospital reports. Adapting these methods to messages from social media like forums is not at all simple. Such messages are written by patients in a quite loose style. They vary in size (between a hundred and a thousand characters). They contain non-standard grammatical structures, many misspellings, abbreviations, emotion-rich expressions as well as emotion-rich words (*I love, I hate*), unconventional lay-out, *e.g.*, repeated use of capital letters (*TIRED*), unconventional spelling (*enooooooough*), and punctuation repetitions (*!!!!!!!!*), slang words that are (or not) specific to the forum or the topic (*LOL vs. IVF*) and emoticons (*:-)*). Message volumes are generally very high (in the French forum dedicated to breast cancer on Doctissimo website, there are more than 3,300 threads, some of which containing more than 2,000 replies). Finally, the processing of health forum data based on semi-automatic information extraction methods is a significant technological challenge. The method proposed in this paper is efficient without special pretraitement such as misspelling and grammar correction, which are difficult to implement generically in the case of health forum because of the non standardised vocabulary which varies a lot from a forum to another.

3   State of the Art

Before associating an emotional state to a target, the first step consists in identifying in the texts some candidates for these two categories. **To identify marks of emotions,** many resources (list of words, phrases, idioms) are available. Most of these resources have been compiled for English texts and polarity analysis, *e.g.,* General Inquirer [16], Linguistic Inquiry and Word Count [17], MicroWNOp [18]. More specific resources, such as the DAL dictionary [19] or the lexicon elaborated by Mohammad and Turney [20] were created for emotional words. There are also approaches for extending these vocabularies for specific application domains by building manual rules [21] or identifying co-occurring terms with words already identified as denoting emotions through large corpora or the web [22,23]. There are also methods to identify emotions in text that are not limited to the use of lexicons such as [24,25] which implement machine learning methods. **To identify potential targets of opinions,** approaches are generally specific to the application domain. Hu and Liu [26] used an association rules algorithm to identify common characteristics mentioned in products reviews. Zhuang and al. [27] used for movie reviews, annotated data and grammatical patterns. In the case of health forums, target identification is *a priori* more difficult because the authors discuss many entities which are hardly comparable and listable in advance as we have shown in the examples of Table 1.

Once candidates for opinions and candidates for targets are generated, two families of approaches can be used to connect them. The first one considers essentially linguistic aspects, represented as rules [26,28] such as valence shifters (*e.g.,* not, only, very, *etc.*) or conjunctions (*e.g.,* the drug x is good but Y is better). These rules are very complex and some have been theorized in the context of the study of *compositional semantics*. Dowty and al. [29] consider that the meaning of an expression depends on the meaning of its components and composition rules. For example, *my pain was reduced significantly* is a positive expression composed of a negative term *pain* and a relationship *significantly reduced*. The effectiveness of this first family of methods is strongly related to language style that impacts on the linguistic rules to consider. In the context of health forums, they are difficult to implement because the styles vary from a forum to another. Although there are exceptions, in forums dealing with motherhood, addressing young women, the language is often familiar, close to SMS language while the language in forums dealing with back pain and addressing primarily elderly people is much more sustained. For our application, it is quite difficult to develop a generic method for any type of forums, relying solely on approaches based on lexicons and rules. The second family of methods is based on different distance computations between the emotional words and the potential targets. The most commonly used is the proximity: the chosen opinion candidate is the closest one to the target in number of words [26]. It is also possible to use the dependencies tree of a sentence. For example, Zhuang and al. [27] consider this tree as a graph and calculate the distance between a target candidate and an opinion candidate by a wide first search where the shortest path is calculated as the number of edges. Wu and al. [6] consider the distance in depth between the target candidate and the opinion candidate based on the lowest common parent in the dependencies tree. To improve the performance of these methods, some authors propose hybrid approaches and introduce linguistic rules for weighting the edges in the dependencies tree [30]. In our context, because of the langage pecularities (misspelling and gramatical errors...), we will show in the next section the importance of the robustness of the parser.

Table 2. Simplified example of the frame Experiencer Focus

FEs Definition: The words in this frame describe the **Experiencer**'s EMOTIONS with respect to some target. Although the target may refer to an actual, current state of affairs, quite often it refers to a general situation which causes the emotion. With certain verbs, the *circumstances* can also be expressed.
Annotations Examples: **My** ENJOYMENT of the movie was considerably spoiled by the seven-foot-tall guy sitting in front of me.   **Smithers** takes great PLEASURE in collecting matchboxes.   **I** HATE it *when you do that.*
LUs Examples: abhor.v, abhorrence.n, abominate.v, adoration.n, adore.v, afraid.a, agape.a, antipathy.n, apprehensive.a, calm.a, comfort.n, compassion.n, , *etc.*

In this article, we propose to incorporate the *Shallow semantic parsing* not only to identify traces of emotions in health forums but also to improve targets detection. The principle is to associate semantic roles to different components of a statement such as *Agent, Patient, Subject, etc.* We are particularly interested in the resource developed in the framework of the FrameNet project [9]. These authors define *Frames*, corresponding to schematic representations of situations. *Semantic roles*, called *Frame Elements* (FE) are exclusively associated with frames. They are expressed by *Lexical Units* (LU). Table 2 shows a simplified example of the frame *Experiencer Focus*. This frame has two main FEs: the *experiencer* and the *target*. Another FE can also be used: the *circumstances*. As we can see in the annotations examples, the same FEs can be raised by constituents with different types of syntax and grammar. This theory has been successfully applied to automatic translation [31] and question/answer systems [32]. To our knowledge, only Kim and Hovy [8] have used the FrameNet resource for the identification of opinions. In this article, we suggest to handle specifically the case of emotions and we refine the notion of target used by these authors to distinguish different types of frames in order to better capture the expressed emotional states in relation with health issues.

4   Initial Evaluations

Figure 2 describes the overall approach for identifying affective states and targets[4]. It is divided into three steps:

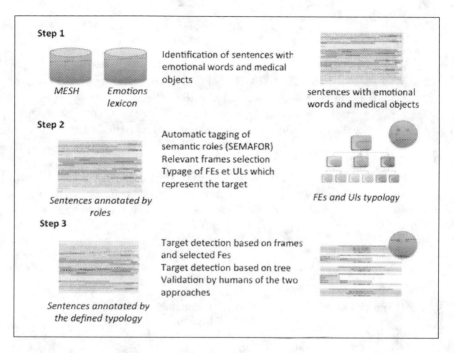

Fig. 2. Global approach

Step 1: Corpus definition. We built a corpus from $17,000$ messages collected in the English-language *Spine-health forum*[5]. We automatically annotated the corpus with the vocabulary of emotions described in [20]. This lexicon consists in more than $14,000$ entries characterized by their polarity and associated with 8 emotions. In this work, we consider only 6 emotions [33]: *anger, disgust, fear, joy, sadness and surprise*. This automatic annotation enabled us to filter 22% of the messages (not containing emotional words). In order to focus only on emotions associated with medical items, we used the classical medical thesaurus MeSH[6] to identify medical units, which allowed us to filter messages without any medical references (6% of messages). In a message, more than 6 emotions were usually expressed because the messages were relatively long. We therefore chose to segment the messages in sentences. We finally kept $1,000$ sentences.

---

[4] Data and annotated results are available at:
http://tatoo.lirmm.fr/~pompidor/cgi/cicling2014.html
[5] http://www.spine-health.com/forum
[6] http://www.ncbi.nlm.nih.gov/mesh

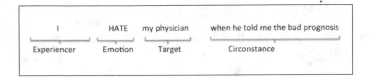

Fig. 3. Expression of emotions

Step 2: Identification of relevant frames for the emotions target detection. We have used the SEMAFOR tool [34] to annotate texts according to the elements provided by the FrameNet ressources. Figure 3 shows an example of the result. We studied the frames identified by this tool and chose manually those that are relevant to the target identification in the medical context. Among the existing 1164 frames, we selected 16 ones relative to the expression of emotional states, 7 explaining generally the expression of these emotional states and 5 which are specific to medical objects. For each frame, we chose the FEs that play the role of emotion or target. For example, for the frame FEAR, we used the FE *expressor* as *emotion* and *topic* as *target*. Table 3 lists these frames.

Table 3. Typology of frames of interest manually identified

Emotions	Targets explaining emotions
Experiencer focus	**General context**
Experiencer obj	Mental stimulus exp focus
Emotions	Partiality
Emotion active	Activity start
Emotion directed	Causation
Emotion heat	Awareness
Emotions by stimulus	
Emotions success or failure	**Medical context**
Complaining	Medical conditions
Contrition	Observable body parts
Desirability	Perception body
Desiring : Event	Cause harm
Fear	Intoxicants
Feeling	Cure
Sensation	Experience bodily harm
Tolerating	

Step 3: Evaluation. In the initial corpus, we kept only the sentences corresponding to the frames selected at the second step (i.e. 345 sentences). We transformed the output of the SEMAFOR tool according to the typology described in Table 3. We used one color to reveal emotions and another one for the context and any element that can help for interpreting the emotion. Finally, the output presented to the users corresponds to a list of typed information. We call this approach the *Roles approach*. We also suggested for each sentence a relationship

between emotions and medical objects based on the distance computation in the dependencies tree. To do this, we used the morphosyntactic tagger developed at Sandford[7]. To determine the distance between targets and emotion candidates, we chose the computation based on the shortest path in number of edges that had be proven to be the most effective (see Section 3). The output presented to the user corresponds to couples *Emotion - Medical Unit*. We call this approach the *Path approach*. The results of these two approaches were validated via a web interface (see Figure 4) used by 10 experts to verify if the information was: *correct, partially correct* (if a part of the target had been forgotten or added) or *incorrect*. At least three experts answered this question for each sentence.

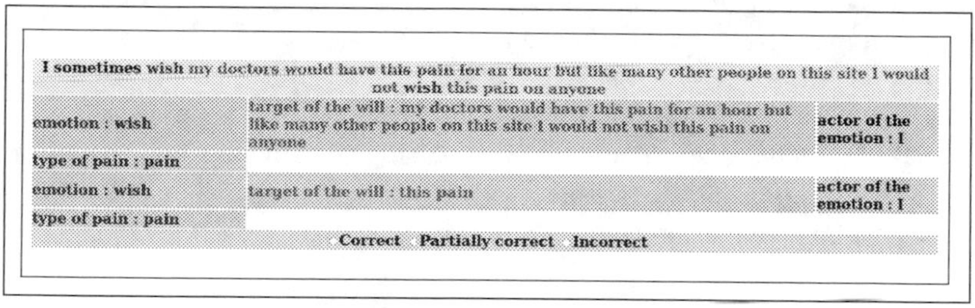

Fig. 4. Validation interface

5   Results

Table 4 presents the results obtained from the methodology described in Section 4. We studied the agreement between annotators using the *Fleiss Kappa* measure[8] and a measure based on Regular Equivalence Classes (REC) and obtained in both cases a moderate agreement.

As might be expected, the *Path Approach* is the least effective approach. The main reason is due to the fact that in our case study and contrary to other application domains, it is not possible to define in advance a comprehensive list of targets. For example, when studying opinions about products, targets are related to the products themselves or to their characteristics that can be easily identified because they appear frequently in the comments. In the forums, the targets are really diversified and they are not limited to medical objects identified using the MeSH. For example, in the sentence *I fear the long term tendency*, the *Path Approach* identifies *fear* as the emotion but it will not be able to associate it with tendency which is not an entry in the MeSH. A second limitation is related to the parser performances used to extract the dependency tree on the forums sentences because they are often poorly constructed (punctuation, spelling, *etc.*). The SEMAFOR parser used in the Role approach is more robust for langage pecularities. Finally, in step 2, the *Role Approach* efficiently excludes sentences

---

[7] http://nlp.stanford.edu/software/lex-parser.shtml
[8] http://en.wikipedia.org/wiki/Fleisskappa

(not excluded if we use only the *Path approach*) that do not contain emotion expressions in the context of health forums. They have been selected in the initial corpus because of the presence of a word of emotion issued from a general resource, which is not always representative in the specific medical context. For example, the sentence *If you want your curve progression halted, it can only be done by surgery* was chosen because of the presence of the word *progression* in the general emotions resource, which is not relevant in our context.

Roles Approach is the most effective one, especially when emotion is carried by a verb (*e.g., I fear the surgeon will be reluctant to continue helping control my pain*). However, it shows some limitations. The study of sentences for which human annotations were different, shows that the targets identified as relating to health are easier to interpret than the general targets. For example, *I hope that your injection starts to bring you raised soon* is simple to analyze. Moreover, the FrameNet resource does not fully integrate all emotional states we aim to identify in texts. For example, only two frames identify the uncertainty (*Certainty, Degree*) which are particularly interesting to consider in health forums. Finally, the generalization of this method to other languages is difficult. There is no resource, for French, as complete as FrameNet.

Table 4. Results

	Correct	Partially correct	Incorrect	Fleiss Kappa measure	REC measure
Role Approch	8%	43%	49%	0.48 (Moderate)	0.52
Path Approch	53%	34%	13%	0.58 (Moderate)	0.63

6   Conclusions and Perspectives

Forums represent a large and diverse base of knowledge about patients' perceptions of their illness and the care that are eventually provided to them. In this article, we described an approach to help a reader to identify traces of emotions in the health forum and to explain their components. We have shown that the use of a semantic roles tagger is quite effective to interpret these targets without any pretraitement of the messages.

Prospects associated with this work are numerous. First, in this work we focus only on the targets expressed in sentences and we now have to focus on inter-sentence relationships at paragraph or message level. In addition, identifying the opinion holder could represent some additional information that would be relevant for the forums analysis. It would be quite simple to capture this information as it is already present in the elements returned by the semantic roles tagger. Contrary to the analysis of products reviews that usually contain only the feelings of the commentators, patients in the forums relate emotions that are not their own (*e.g., my doctor is concerned that my blood glycerol increases*). Moreover, once the relationship between emotion and target is identified, it can be generalized to a set of messages in order to summarize the emotional states of different patients about a specific target. For example, in the case of the analysis is associated with a particular emotions treatment, characteristics associated with this medical purpose are well known (price, tolerance, side effects, *etc.*). It is then possible to make associations between target and opinion as it is done in the case of movies or products reviews [26,27].

Finally, the main limitation of this contribution is to restrict emotion identification only to the case in which an emotional word is present. As pointed out by [35], in most cases people express emotions implicitly without using these emotional words. An emotion can be limited to something a person feels about a fact and not the sentiment that a person expresses about this fact. Thus, it could be common to explicitly express sentiments about things, but it is more common to feel emotions without expressing them explicitly. In future work, we should take into account this fact and try to identify emotions beyond the explicit cases, because it is possible that in such implicit cases the identification of targets is different.

References

1. Hancock, J.T., Toma, C.L., Ellison, N.B.: The truth about lying in online dating profiles. In: Rosson, M.B., Gilmore, D.J. (eds.) CHI, pp. 449–452. ACM (2007)
2. Pang, B., Lee, L., Vaithyanathan, S.: Thumbs up?: Sentiment classification using machine learning techniques. In: Proceedings of the ACL 2002 Conference on Empirical Methods in Natural Language Processing, EMNLP 2002, vol. 10, pp. 79–86. Association for Computational Linguistics, Stroudsburg (2002)
3. Strapparava, C., Mihalcea, R.: Learning to identify emotions in text. In: Symposium on Applied Computing, pp. 1556–1560. ACM, New York (2008)
4. Wiebe, J.: Learning subjective adjectives from corpora. In: Proceedings of the Seventeenth National Conference on Artificial Intelligence and Twelfth Conference on Innovative Applications of Artificial Intelligence, pp. 735–740. AAAI Press (2000)
5. Choi, Y., Cardie, C., Riloff, E., Patwardhan, S.: Identifying sources of opinions with conditional random fields and extraction patterns. In: Proceedings of the Human Language Technology Conference and the Conference on Empirical Methods in Natural Language Processing, HLT/EMNLP (2005)
6. Wu, Y., Zhang, Q., Huang, X., Wu, L.: Phrase dependency parsing for opinion mining. In: Proceedings of the 2009 Conference on Empirical Methods in Natural Language Processing, EMNLP 2009, pp. 1533–1541. Association for Computational Linguistics, Stroudsburg (2009)
7. Ruppenhofer, J., Somasundaran, S., Wiebe, J.: Finding the sources and targets of subjective expressions. In: Calzolari, N., Choukri, K., Maegaard, B., Mariani, J., Odjik, J., Piperidis, S., Tapias, D. (eds.) Proceedings of the Sixth International Language Resources and Evaluation (LREC 2008). European Language Resources Association (ELRA), Marrakech (2008),
   http://www.lrec-conf.org/proceedings/lrec2008/
8. Kim, S.M., Hovy, E.: Extracting opinions, opinion holders, and topics expressed in online news media text. In: Proceedings of the Workshop on Sentiment and Subjectivity in Text, SST 2006, pp. 1–8. Association for Computational Linguistics, Stroudsburg (2006)

9. Baker, C.F., Fillmore, C.J., Lowe, J.B.: The Berkeley FrameNet project. In: Proceedings of COLING/ACL, pp. 86–90 (1998)
10. Siegrist, J.: Emotions and Health in Occupational Life: New Scientific Findings and Policy Implications: Inauguration Speech Belle Van Zuylen Professorship. Universiteit Utrecht (1994)
11. Kamvar, S.D., Harris, J.: We feel fine and searching the emotional web. In: ACM International Conference on Web Search and Data Mining, pp. 117–126. ACM, New York (2011)
12. Kim, S.M., Hovy, E.: Determining the sentiment of opinions. In: Proceedings of the International Conference on Computational Linguistics (COLING), pp. 1367–1373 (2004)
13. Kobayashi, N., Inui, K., Matsumoto, Y.: Extracting aspect-evaluation and aspect-of relations in opinion mining. In: Proceedings of the 2007 Joint Conference on Empirical Methods in Natural Language Processing and Computational Natural Language Learning (EMNLP-CoNLL), pp. 1065–1074 (2007)
14. Popescu, A.M., Etzioni, O.: Extracting product features and opinions from reviews. In: Proceedings of the Conference on Human Language Technology and Empirical Methods in Natural Language Processing, HLT 2005, pp. 339–346. Association for Computational Linguistics, Stroudsburg (2005)
15. Wilson, T., Wiebe, J., Hoffmann, P.: Recognizing contextual polarity in phrase-level sentiment analysis. In: Proceedings of the Human Language Technology Conference and the Conference on Empirical Methods in Natural Language Processing (HLT/EMNLP), pp. 347–354 (2005)
16. Stone, P.J., Hunt, E.B.: A Computer Approach to Content Analysis: Studies Using the General Inquirer System. In: AFIPS 1963 (Spring). ACM, New York (1963)
17. Tausczik, Y.R., Pennebaker, J.W.: The psychological meaning of words: Liwc and computerized text analysis methods 29, 24–54 (2010)
18. Cerini, S., Compagnoni, V., Demontis, A., Formentelli, M., Gandini, G.: Language Resources and Linguistic Theory: Typology, Second Language Acquisition, English Linguistics, Milano, IT, Franco Angeli Editore, Franco Angeli Editore (2007)
19. Whissell, C.: The dictionary of affect in language. Academic Press (1989)
20. Mohammad, S.M., Turney, P.D.: Emotions Evoked by Common Words and Phrases: Using Mechanical Turk to Create an Emotion Lexicon. In: Workshop on Computational Approaches to Analysis and Generation of Emotion in Text, pp. 26–34. ACL, Stroudsburg (2010)
21. Neviarouskaya, A., Prendinger, H., Ishizuka, M.: Affect analysis model: Novel rule-based approach to affect sensing from text, vol. 17, pp. 95–135. Cambridge University Press, New York (2011)
22. Harb, A., Plantié, M., Dray, G., Roche, M., Trousset, F., Poncelet, P.: Web Opinion Mining: How to extract opinions from blogs? Categories and Subject Descriptors. In: International Conference on Soft Computing as Transdisciplinary Science and Technology, pp. 211–217 (2008)
23. Kozareva, Z., Navarro, B., Vazquez, S., Montoyo, A.: UA-ZBSA: a headline emotion classification through web information. In: 4th International Workshop on Semantic Evaluations, pp. 334–337. ACL, Stroudsburg (2007)
24. Strapparava, C., Mihalcea, R.: Learning to identify emotions in text. In: Proceedings of the, ACM Symposium on Applied Computing, SAC 2008, pp. 1556–1560. ACM, New York (2008)

25. Pardo, F.M.R., Rosso, P.: On the identification of emotions and authors' gender in facebook comments on the basis of their writing style. In: Proceedings of the First International Workshop on Emotion and Sentiment in Social and Expressive Media: Approaches and perspectives from AI (ESSEM 2013) A workshop of the XIII International Conference of the Italian Association for Artificial Intelligence (AI*IA 2013), Turin, Italy, December 3, pp. 34–46 (2013)

26. Hu, M., Liu, B.: Mining opinion features in customer reviews. In: Proceedings of the 19th National Conference on Artifical Intelligence, AAAI 2004, pp. 755–760. AAAI Press (2004)

27. Zhuang, L., Jing, F., Zhu, X.Y.: Movie review mining and summarization. In: Proceedings of the 15th ACM International Conference on Information and Knowledge Management, CIKM 2006, pp. 43–50. ACM, New York (2006)

28. Mudinas, A., Zhang, D., Levene, M.: Combining lexicon and learning based approaches for concept-level sentiment analysis. In: Proceedings of the First International Workshop on Issues of Sentiment Discovery and Opinion Mining, WISDOM 2012, pp. 5:1–5:8. ACM, New York (2012)

29. Dowty, D.R., Wall, R.E., Peters, S.: Introduction to Montague Semantics, vol. 11. D. Reidel, Dordrecht (1989)

30. Ding, X., Liu, B.: The utility of linguistic rules in opinion mining. In: Proceedings of the 30th Annual International ACM SIGIR Conference on Research and Development in Information Retrieval, SIGIR 2007, pp. 811–812. ACM, New York (2007)

31. Boas, H.C.: Bilingual FrameNet dictionaries for machine translation. In: Rodríguez, M.G., Araujo, C.P.S. (eds.) Proceedings of the Third International Conference on Language Resources and Evaluation, Las Palmas, vol. IV, pp. 1364–1371 (2002)

32. Narayanan, S., Harabagiu, S.: Question answering based on semantic structures. In: Proceedings of the 20th International Conference on Computational Linguistics, COLING 2004. Association for Computational Linguistics, Stroudsburg (2004)

33. Ekman, P.: An argument for basic emotions 6, 169–200 (1992)

34. Das, D., Schneider, N., Chen, D., Smith, N.A.: Probabilistic frame-semantic parsing. In: Human Language Technologies: The 2010 Annual Conference of the North American Chapter of the Association for Computational Linguistics, HLT 2010, pp. 948–956. Association for Computational Linguistics, Stroudsburg (2010)

35. Osherenko, A., André, E.: Lexical affect sensing: Are affect dictionaries necessary to analyze affect? In: Paiva, A.C.R., Prada, R., Picard, R.W. (eds.) ACII 2007. LNCS, vol. 4738, pp. 230–241. Springer, Heidelberg (2007)

Investigating the Role of Emotion-Based Features in Author Gender Classification of Text

Calkin Suero Montero, Myriam Munezero, and Tuomo Kakkonen

School of Computing, University of Eastern Finland
{calkins,myriam.munezero,tuomo.kakkonen}@uef.fi

Abstract. Research has shown that writing styles are influenced by an extensive array of factors that includes text genre and author's gender. Going beyond the analysis of linguistic features, such as n-grams, stylometric variables and word categories, this paper presents an exploratory study of the role that emotions expressed in writing play to aid discriminating author gender in different text genres. In this work, the gender classification task is seen as a binary classification problem where discriminating features are taken from a vectorial space that includes emotion-based features. Results show that by exploiting the emotional information present in personal journal (diary) texts, up to 80% cross-validation accuracy with support vector machine (SVM) algorithm can be reached. Over 75% cross-validation accuracy is reached when classifying the author gender of blog texts. Our findings show positive implications of emotion-based features on assisting author's gender classification.

Keywords: Text analysis, sentiment analysis, gender classification, emotion-based features.

1 Introduction

In society, emotions are a two-way channel through which individuals experience the social world, and in turn, the social world is what constrains or regulates emotions [23]. Gender-biased, stereotypical expectations for the expression of emotions create, as well, a strong constrain on the way individuals express their emotions [41]. In fact, the manner in which emotions are expressed has been identified as an illustrator of the individual's gender from early age [7]: from facial expression and general body language [16], to written text ([37], [40]) women and men express emotions differently.

Relevant to the interests of the natural language processing (NLP) community is the analysis of emotions expressed in written text. In accordance with sterotypic emotion expression, research has shown that female writing style is in general more contextual, personal and emotional than male writing, in which the style is typically more impersonal, formal and judgmental [17]. These differences have been observed throughout a variety of text genres ranging from online personal diaries to fiction and non-fiction novels ([34], [36]).

The fact that the expression of emotion is so readily gender-differentiated and can be observed in varied situations and text genres has strong implications for the

A. Gelbukh (Ed.): CICLing 2014, Part II, LNCS 8404, pp. 98–114, 2014.

identification of author's gender. That is, it is reasonable to expect that the likelihood of correctly identifying the author's gender increases if the type of emotions present in, for instance, a customer review, is considered in addition to other features, such as stylometry [33].

The gender information, alongside the type of emotions that mark the gender distinction, could prompt a company or a political candidate to redefine the targets of its marketing campaigns, to shape and maintain its public image, to evaluate future trends and so on. Furthermore, not only market analysis could benefit from a domain-robust and informative approach to author gender classification, but also the problem of authorship attribution [22] could make use of such gender information.

To this end, we explore the impact that classification features based on emotions have on the identification of the author's gender for two types of genre: personal journals and blogs. Within the NLP community, literature abounds on the sentiment classification of text, i.e., classifying opinions, users reviews and personal blogs, either by using positive/negative valence or by assigning the text to emotional classes (angry/happy/sad) [20], [48] but the impact that emotions per se have as features for author gender classification has been overlooked thus far. The main contributions of our work can be summarized as follows:

—Our work shows that by using emotion-based features to aid author gender classification, the need for feature space dimensionality reduction is avoided, along with requirements of computationally expensive feature selection algorithms.

—Our work demonstrates that emotion-based features have an effective positive impact on the classification of the author's gender of personal journals. This highlights the significance of using a small and informative feature set for efficient author gender classification.

—Our work also provides insights for the analysis of the writer's preferences since the emotions used as gender discriminants by the classifiers are easily extracted. These emotions can give information, for instance, into the writer's satisfaction with a product or their political orientation. Such information is highly valuable.

2   Related Work

2.1   Gender Classification

Author gender identification has been explored as a text classification problem and as an authorship detection problem ([21]; see the works of Juola [18], and Koppel et al., [22] for detailed surveys in authorship attribution; Rangel et al., [43] provide an overview of the PAN 2013 age and gender author profiling task for authorship detection). We focus on the text classification aspect of the problem.

As a text classification problem, social media and other online resources provide the gender classification task with a rich source of data against which systems and algorithms can be tested. Analyzing web blogs, Argamon et al. [3] identified stylometric (i.e. writing style) differences between genders in that male bloggers used more personal pronouns, auxiliary verbs and conjunctions than their female counterparts. Interestingly, the same authors had previously identified similar type of features as author gender discriminators in fiction and non-fiction writing ([21], [2]), suggesting that some writing characteristics are constant for females and males across text genres.

In a more comprehensive approach, Murkherjee and Liu [33] reported 88% accuracy in gender classification of web blogs authors using a very large and compounded set of features. These features included: contextuality measure, as given by the F-measure of Heylighen and Dewaele [17]; stylometric information, using what they referred to as "blog words" (for example, "lol, hmm and smiley"); gender preferential features taken from Corney et al.'s work [10]; factor analysis features, i.e., groups of related words that appear in similar documents [9]; POS sequence patterns; and word class lists containing positive, negative and emotional connotations. Zhang et al. [53] also used a large set of features combining content-free (lexical, syntactic, and structural) and content-dependent (uni-, bi-grams) features to classify the gender of web forum messages' authors.

Author gender classification task has also been explored for emails ([51], [10]) with up to 91% accuracy in gender prediction; and university students' essays [34] with discriminant analysis prediction of 72.5% assertion. These approaches rely on large number of features (in some cases in the order of thousands) in order to yield high levels of accuracy. Hence, in order to use these approaches, a large number of genre-specific training data is needed: the larger the feature set is, the more training examples are needed for training a reliable classification model.

2.2   Emotion Analysis

Emotions have been long investigated in various fields of research ranging from social psychology to computational linguistics [24]. Lists of primary or "basic" emotions have been put forward in the psychological field prominently by Frijda [13], Ekman [11] and Plutchik [42] among others. The basic emotion categories used in these lists include anger, sadness, joy, love, surprise, happiness, fear and disgust (see Ortony et al. [39], and Shaver et al. [44], for a detailed compilation of primary emotion lists). Within the NLP research community, more often than not researchers use Ekman's six basic emotion categories [11]: anger, disgust, fear, happiness, surprise and sadness ([1], [46]).

Broadly speaking, two main methods exist for the analysis of emotions within the NLP community: word lists-based and machine learning-based. Word list based methods use lexical resources such as lists of emotion-bearing words, lexicons or affective dictionaries ([38], [12]); and databases of commonsense knowledge [26]. The General Inquirer (GI) [45], the Affective Norms for English Words (ANEW) [5], the WordNet-Affect ([47], [52]), and more recently the NRC word-emotion association lexicon ([31], [30]) are well-known lexical resources. Emotion analysis systems such as UPAR7 [8] and SentiProfiler [19], among others, have been developed using WordNet-Affect as lexical resource.

Machine learning based models, on their part, treat the problem as a multi-classification task and require a significant amount of annotated data to represent each one of the emotions to classify. Applications include automatic affective feedback for email clients [25], text-to-speech emotion-aware storyteller system [1], among others.

Regardless of the method used, emotion analysis has proven to be a valuable asset when processing text. From marketing analysis of products and services consumers' satisfaction [4] to political party or candidate popularity [28], the emotional information in written text is a gold mine to exploit.

2.3   Emotion Expression as Features vs. Author's Gender

Emotion expressivity has been identified in a variety of genres as one of the author gender discriminators [36]; and genre indeed influences the preferred writing style of men and women [15]. There is also evidence that suggests that certain emotions such as fear, disgust, love and happiness are expressed more openly by females whereas emotions such as anger and pride are more readily expressed by males [41]. Although in certain context, for instance some social network sites, negative comments representing a minority of the total are expressed equally by both genders; females are more likely to express positive comments than males are [50]. Research has also shown that in work-place email correspondences women tend to use words expressing joy and sadness, whereas men use words related to fear and trust [32]. Based on these gender-wise stereotypic emotion expression behaviors, it is reasonable to expect distinctive patterns in the emotions expressed in personal journals and other genres in which the writing style is not formal.

Some previous author gender classification studies have used emotion words as a part of the feature set. In a relevant work, Murkherjee and Liu [33] included "word classes implying positive, negative and emotional connotations" as part of a larger feature set for the classification of author gender in blogs. An original feature selection algorithm was put forward for dimensionality reduction and feature discrimination. Our work differs from their approach in that we investigate the impact that a feature set mainly influenced by emotion features has (described in Section 3). Also, in our work there is no need for feature selection algorithms since we are handling a smaller set of features.

Another relevant work is that of Newman et al. [35], in which they used the Linguistic Inquiry and Word Count (LIWC) on a set of over 14,000 documents, and found differences in emotion expression between genders. The LIWC analyzed seven dimensions within the emotion category: positive emotions, optimism, positive feelings, negative emotions, anxiety, sadness and anger. The study by Newman et al. showed that women expressed positive feelings (e.g., happy, joy) more frequently than men, along with negative emotions, anxiety and sadness; whereas men used more swear words. Our method differs from LIWC in that we apply a richer set of emotion classes, beyond the seven emotion dimensions analyzed by LIWC. Moreover, beyond linguistic characteristics (e.g., word count, negations, etc) or "social words" (e.g., share, mom, group, etc) as the LIWC utilizes, in our approach emotions are the basis of the features used.

The work of Mohammad [30] also relates to our work, in that emotion features were used for sentence classification. In Mohammad's work, newspaper headlines were given scores from 0 to 100 to indicate how strongly the headline expressed each one of the 6 emotions categories given by Ekman [11], i.e., anger, disgust, fear, joy sadness and surprise. Mohamad showed that, generally, affective features improved the performance of ngram features; and that affective features were more robust than ngram features for classification in previously unseen domains. Our work differs from Mohammad's work in that we use a deeper representation of emotion expression beyond Ekman's emotion categories. Furthermore, in our work there is not a predetermined classification of the analyzed text into emotion categories, but the emotions expressions that are found in the text are the ones used as features for gender classification.

3   Method

3.1   Emotions for Gender Classification

In order to identify emotion words in text, we use an ontology of emotion categories. Each category contains a set of emotion classes and emotion words. Figure 1 illustrates the positive emotion category with samples from the liking class.

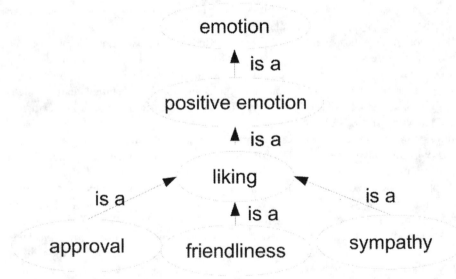

Fig. 1. Sample from the positive-emotion category branch of the ontology

The emotion ontology along with contextual disambiguation rules are used for tagging the input texts. A contextual disambiguation rule would distinguish between the usage of "like" in "PRP (personal pronoun) + likes", as an emotion bearing word (verbal meaning); and "like + WDT (determiner)" as a preposition, for example. We used contextual rules to disambiguate different cases of expressions using: like, good, blue, down and get to.

The emotion ontology was developed based on WordNet-Affect. As a lexical resource, WordNet-Affect contains the representation of affective knowledge based on WordNet [29]. While WordNet-Affect was a good starting point for creating our emotion ontology, preliminary experiments pointed out some issues related to tagging documents for categorization. First, we found that the number of words in WordNet-Affect was rather limited (1,316 words divided into 250 classes). Second, we found the WordNet-Affect hierarchy too detailed for many practical purposes. For instance, the class positive-emotion/joy/merriment has three sub-classes: hilarity, jocundity and jollity; and similarly the class negative-emotion/general-dislike/anger/annoyance has five sub-classes associated to it. Such subtle differences in emotions are very challenging to automatically detect, hence we opted to modify the ontology.

Our emotion ontology consists of 85 classes. The number of classes was reduced by combining classes representing similar emotions in WordNet-Affect, and removing leaf classes in cases such as the ones described above. Moreover, new words were added into our ontology to improve its coverage. Hence, our emotion ontology contains 1,499 words. After adding new emotion words and removing redundant emotion classes, this emotion ontology has 17.6 words per emotion class on average. The wider coverage of emotion classes and emotions words makes the ontology more suitable to use as the basis for tagging documents for author gender classification, which is our aim in this work.

It is important to notice that the process of pruning the ontology structure was performed before running any experiments on our development and test data, i.e., the ontology was not fitted to any particular dataset. Table 1 shows examples from the emotion ontology.

We used two types of emotion-based features for classification: ontology-dependent and ontology-independent emotion features.

Table 1. Emotion classes and attributes

Emotion Category	Emotion Class	Emotion Words
Negative-emotion	Anxiety	distress, edginess, impatience
	Despair	cynicism, pessimism, discouragement
	Sadness	depression, sorrow, misery
Positive-emotion	Love	adore, amorous, intimacy
	Liking	admiration, approval, friendliness
	Enthusiasm	avid, ardor, eager

Note: An emotion class consists of a number of emotion words, as represented in our ontology.

3.2 Ontology-Dependent Emotion Features

These features are obtained through an emotion tagging process using the emotion ontology.

Aggregated Value over Ontology Classes

This aggregated value is calculated using the following formula:

$$x.AggValue = \sum_i (FEC_i)_x; \quad FEC_i = \frac{(eWords)_i}{aWords} \tag{1}$$

where x.AggValue represents the sum of all the relative frequencies of the emotion classes (FEC_i) that belong to the emotion categories represented in the ontology. The frequency of an emotion class i, FEC_i, is calculated as the number of emotion words belonging to class i, $(eWords)_i$, that are found in the document, divided by the total number of words in the document $(aWords)$. This feature holds the emotional representation of the document. Taking into account that females tend to use a more varied vocabulary than males when expressing emotions [37] we expect to find more emotion classes represented in female writing.

Positive to All Emotion Ratio

This feature represents the ratio between the positive emotions and all the emotions in the text. It is calculated by:

$$PosToAllRatio = \frac{pos.AggValue}{pos.AggValue + neg.AggValue} \tag{2}$$

where pos.AggValue and neg.AggValue are given by formula (1). This feature offers information on how positively oriented a document is. With this feature, we capture differences in the rate of positive/negative emotion expressions. Based on gender stereotypic expression of emotions, with this feature we explore whether female writings are distinguishably more positive than male writings.

Document Emotion Ratio

This feature shows the portion of the document that contains emotional value. It is based on the ratio of emotion words (eWords) to all words in the document:

$$eRatio = \frac{emotionWordsInDocument}{totalWordsInDocument} \tag{3}$$

With this feature, we capture differences in the rate of emotion expressions between genders. Based on the assumption that females express their emotions more freely than men, it is interesting to explore whether higher emotion ratio is found in female-authored texts.

3.3   Ontology-Independent Emotion Features

In order to cover a wider aspect of emotion expression analysis, we adopted two additional features outside the emotion ontology.

Emotion Strength

This feature is obtained from the SentiStrength system [49], a dictionary-based sentiment analysis system that provides sentiment strength measures for text. The SentiStrength measures for a piece of text are given from -5 to +5. SentiStrength is able to detect a range of non-standard spellings of emotion-bearing words, which makes it particularly useful for analyzing the type of data of our study. According to the literature [41], women are perceived to express some emotions more intensively than men. We, hence, try to capture with this feature differences in the expressed emotion's strength and expect to find female authors having higher sentiment strength measures for some emotions.

F-score

The F-score [17][1] has shown to be useful for describing the contextuality/formality of language. This score is based on the frequencies of specific POS categories in the analyzed document. It is given by the following formula:

$$Fscore = 0.5 * [(nounFreq+adjFreq+prepFreq+artFreq) \\ - (pronFreq+verbFreq+advFreq+intFreq) + 100] \tag{4}$$

In their analysis of a speech corpus with known speakers gender, Heylighen and Dewaele [17] found that female used a "more contextual" style (i.e. relatively low F-score) than their male counterparts, concluding that "(A) formal style of expression is characterized by detachment, precision, and objectivity, but also rigidity and cognitive load; a contextual style is much lighter in form, more flexible and involved, but correspondingly more subjective, less accurate and less informative".

This F-score has been used successfully to highlight gender differences in various writing contexts including emails and blogs [36]; and as a feature for gender classification [33]. We consider contextuality or involvement to be linked to the frequency of emotional expressions, and hence we also used the F-score as a classification feature. We expect to find a clear distinction between genders in the level of involvement expressed in their writing.

4   Experimental Design

4.1   Datasets

We used two sets of annotated corpora for our study: personal journals and weblogs.

Personal Journals Corpora

Newman et al. [35] analyzed gender differences in language usage from 14,000 text samples collected from 70 studies that were carried out around the world. We obtained the corpora of 58 of those studies[2] with a total of 21,814 documents. We distri-

---

[1] This F-score is different from the F-score used to assess the accuracy of text classification, which is based on precision and recall.

[2] Personal communication with James W. Pennebaker.

buted these documents according to their content into personal journal (10,029 documents) and general content (11,785 documents).

Since we are using emotions as features, we worked with documents from the personal journal content. We based this choice on the premise that people are less inhibited when it comes to expressing their emotions in personal diaries [27].

Table 2 shows a detailed description of the personal journals used in our study. These sub-corpora were chosen for their reasonable size and their relatively high number of documents authored by females and males respectively. Age information was available only in PJ1 (female participants' average age, 39.4; male participants, 41.5) and PJ4 (females, 33.9; males 33.2).

Table 2. Personal journal (PJ) documents description by gender with average word count (Avg), and standard deviation (STD)

Sub-corpus	Female			Male			NA[a]
	Files	Avg	STD	Files	Avg	STD	
PJ1	88	89	203	54	112	161	84
PJ2	54	459	119	51	384	105	10
PJ3	321	73	59	317	63	51	0
PJ4	3653	71	142	1747	57	135	208
PJ5	338	185	77	266	128	76	0
PJ6	563	201	112	215	126	97	0

[a] no gender information available

Topics in these personal journal corpora varied:

— PJ1: work life and relationships with important people in their lives
— PJ2: feelings and experiences of going to college
— PJ3: relationship with parents
— PJ4: feelings regarding the USA 9/11/2001 attacks
— PJ5: stories about lives and daily routines
— PJ6: feelings while enrolled in addiction treatment.

In these six sub-corpora, the participants contributed to the study several times within the same context. We merged all documents written by the same author into a single file (similar to Newman et al. [35]). An exception to this was PJ6, which had 11 female participants and only four males. In this case, for each gender we combined five files together into one document. When the difference in the number of documents for each gender was large, we balanced the sub-corpus such that similar number of files for both genders was randomly selected for analysis[3]. See Table 3 for a summary of the merged documents and their emotional content.

Weblogs Corpora

---

[3] Hence, out of the 3,812 documents written by female in PJ4, we used only 1,884 randomly selected documents; and similarly for PJ1 and PJ6.

To test the applicability of our emotion-based classification method to genres other than personal journals, we looked into the author gender classification of weblogs. Weblogs as a genre have "blog types" [15]: they are not only about personal and emotional matters ("diaries"), but also about "events external to the author" ("filters"). This type of data provides us the opportunity to test the feasibility of using emotion-based features for the author gender classification of general informal text.

For this part of the study, we used the weblog corpora by Mukherjee and Liu [33][4]. This corpus consists of 3,217 blogs (1,674 by male authors and 1,543 by female authors respectively). This blog corpus was collected from various sources (blogger.com, technorati.com, among others), and it contains a variety of topics (daily life situations, advice on specific matters, etc). For analysis, 50 blogs were combined into 1 document, resulting in a total of 31 female and 34 male files.

4.2   Feature Value Assignments

The value of each feature was assigned depending on the type of the feature. Fscore, eRatio and PosToAllRatio were assigned binary values depending on the following thresholds:

$$FscoreFeature = \begin{cases} 1 & if\ Fscore \geq 50 \\ 0 & if\ Fscore < 50 \end{cases}$$

$$eRatioFeature = \begin{cases} 1 & if\ eRatio \geq 3 \\ 0 & if\ eRatio < 3 \end{cases} \tag{5}$$

$$PosToAllRatioFeature = \begin{cases} 1 & if\ PosToAllRatio \geq 0.5 \\ 0 & if\ PosToAllRatio < 0.5 \end{cases}$$

These thresholds were defined empirically by using a random sample of 20 weblogs (equal distribution female/male authors) during the system development phase. The SentiStregth and the aggregated value over ontology classes features used their nominal values as given by their relative calculations. This is similar to the term frequency (TF) feature value assignment.

4.3   Classifiers

We used the WEKA API [14] to implement two types of classifiers: Support Vector Machine (SVM) and C4.5 decision tree (J48). These classifiers are well-known in the field and are a good representation of both statistical (J48) and function-based (SVM) machine learning approaches. The vectorial representation of each document to analyze was given by:

$$DocVector = [SSPos, SSNeg, PosToAllR, Fscore, eRatio, OntoClassAggValue] \tag{6}$$

---

[4] Retrieved on Dec 22nd, 2011. Online at
http://www.cs.uic.edu/~liub/FBS/sentiment-analysis.html.

where SS(Pos/Neg) are the positive and negative SentiStrength scores respectively; PosToAllR is the positive to negative emotions ratio of the document; Fscore is the F-score for the whole document; eRatio is the overall emotion ratio of the document; and OntoClassAggValue represents the aggregated frequency value over each one of the 85 emotion classes in the ontology. Any given document vector is, hence, represented using emotion-based features by a minimum of 85 features (only the ontology emotion classes), and a maximum of 90 features (all the emotion-based features described).

Table 3. Number of documents (Docs) analyzed for each gender. The number of words, emotion words (eWords), emotion categories (eCat) and positive to all emotion ratio (pToall) shown are the averages per document.

	Gender	Docs	Words	eWords	eCat	pToall
PJ1	Female	20	1054.4	25.4	8.0	0.56
	Male	18	1028.5	22.9	7.6	0.51
PJ2	Female	18	2391.0	110.7	15.3	0.44
	Male	17	2147.2	73.4	10.8	0.47
PJ3	Female	39	583.5	15.2	5.9	0.57
	Male	41	491.3	12.7	4.6	0.55
PJ4	Female	47	2828.8	146.7	26.0	0.20
	Male	47	2284.2	109.3	23.0	0.19
PJ5	Female	106	555.2	12.9	4.6	0.41
	Male	78	372.3	10.9	3.3	0.35
PJ6	Female	43	917.7	44.0	14.1	0.37
	Male	43	648.9	23.9	8.1	0.52
Blogs	Female	31	20771.8	443.9	36.1	0.66
	Male	34	20998.1	376.3	35.7	0.62

5   Results

Table 3 shows a summary of the aggregated documents analyzed in each personal journal dataset. We found significant difference of emotion expressivity (i.e., the usage of emotion words) between genders. Unpaired t-test for PJ2 ($t(33)=2.32$, $p<0.013$), PJ4 ($t(92)=5.12$, $p<0.0001$) and PJ6 ($t(84)=7.42$, $p<0.0001$) showed that female authors used more emotion words in their writings than male authors. No significant difference was found for PJ1, PJ3, PJ5 or blogs.

Similarly, female authors used more emotion categories than their male counterparts in their writings in PJ2 ($t(84)=7.89$, $p<0.0001$), PJ3 ($t(78)=1.98$, $p<0.05$), PJ4 ($t(92)=3.24$, $p<0.001$), PJ5 ($t(182)=3.29$, $p<0.001$) and PJ6 ($t(84)=7.89$ $p<0.0001$). PJ1 and blogs showed no significant difference. Significant difference was also found in the positive to all emotion ratio but only for PJ6 ($t(84)=3.6$, $p<0.0005$), where male authors had more positively oriented text; and blogs ($t(63)=2.42$, $p<0.01$), where female authors showed more positively oriented writing.

The results of the classification experiments are summarized in Table 4. The Accuracy, Precision and Recall shown are for 10-fold cross validation for the SVM algorithm with RBF-kernel, optimal grid-search γ, and default C settings. We compared the performance of the emotion-based features to that of bag-of-words (BoW) features, and to that of a combined emotions+BoW features set.

Using SVM classification method, BoW performance was improved when adding emotion features (Emo+BoW) in PJ1 - PJ5, achieving the best classification accuracy of all feature sets in PJ3 - PJ5. Emotion-based features alone had the best accuracy in PJ1 and PJ2, whereas BoW alone had the best accuracy classifying PJ6 and the blogs corpus. Using emotion-based features, the blogs author gender classification accuracy (75.4%) was similar to that of BoW, in spite of the number of features being significantly less: only 90 emotions features were used to represent the blogs, whereas BoW features ranged over a thousand. This result is comparable to that reported by Mukherjee and Liu [33], where 68% accuracy was achieved for the same corpus with a feature set of about 130 features.

Observing the classification results of the BoW feature sets of PJ4 - PJ6, the classifier's performance appears to be better than that achieved with an emotion-based feature set alone. However, a closer look at the descriptors used by BoW in PJ6, for instance, revealed that the SVM classification algorithm exploited differences in usage frequency of some linguistic features, i.e., great, but. In contrast, a closer look at the emotion features used by SVM for discrimination showed that expressions of "negative-fear" and "resentment", among other negative-emotions were found in the top descriptors of the text (Table 4). Hence, albeit its lower accuracy, the information given by an emotion-based feature set is more valuable and useful in a variety of

Table 4. SVM 10-fold cross-validation accuracy results. *Emotion* represents the optimal emotion-based feature space using only emotion features, *BoW* represents a bag-of-words feature space and *Emo+BoW* represents a combined feature space for each sub-corpus in our study. The Top Descriptors shown are from the 10 best attributes for each feature set.

	Features	Accuracy	Precision	Recall	Top Descriptors
PJ1	**Emotion**	**84.2**	**84.5**	**84.2**	**approval, humility, horror**
	BoW	78.9	81.9	78.9	radio, five, drain, afraid, state
	Emo+BoW	81.6	83.7	81.6	sit, five, afraid, radio, respect
PJ2	**Emotion**	**54.3**	**54.7**	**54.3**	**love, jealousy, horror, fScore**
	BoW	45.7	37.0	45.7	boyfriend, into, everyone
	Emo+BoW	45.7	43.0	45.5	upset, boyfriend, found, kept
PJ3	Emotion	57.5	62.1	57.5	negative-fear, disgust, sadness
	BoW	58.7	60.2	58.8	run, anymore, reason, their
	Emo+BoW	**61.2**	**64.0**	**61.3**	**their, away, negative-fear**
PJ4	Emotion	67.0	68.0	67.0	love, gratitude, anxiety, fury
	BoW	82.9	83.0	83.0	ones, don, Islamic, lost, punish
	Emo+BoW	**84.0**	**84.2**	**84.0**	**know, act, distress, gratitude**
PJ5	Emotion	67.9	69.0	67.9	humility, horror, melancholy, love
	BoW	75.5	79.3	75.5	talk, embarrassed, score, she, year
	Emo+BoW	**76.6**	**80.1**	**76.6**	**embarrassed, she, death, love**
PJ6	Emotion	79.2	80.1	79.3	negative-fear, SSneg, resentment
	BoW	**91.8**	**92.4**	**91.9**	**-, great, fear, group, but**
	Emo+BoW	91.8	92.4	91.9	talk, tired, however, fear
Blogs	Emotion	75.4	76.1	75.4	love, gratitude, joy, forlornness
	BoW	**75.4**	**80.0**	**75.4**	**sauce, performance, wedding**

Features	Accuracy	Precision	Recall	Top Descriptors
Emo+BoW	72.3	79.0	72.3	love, gratitude, version, baby
Mukherjee and Liu 2010*	86.2	-	-	-

applications than the purely linguistic information given by the BoW approach. That is, it is more interesting to know, from the marketing point of view for instance, that the female users "hate" a product rather than the number of times they use "what" in their feedback. A combined Emo+BoW features set takes advantage of the linguistic features given by the BoW plus the insightful information given by the emotion features.

Using the J48 classification algorithm, BoW feature sets and Emo+Bow feature sets had similar performances but their accuracy was substantially lower than that achieved with SVM. Using the emotion feature sets alone, however, the J48 algorithm was sensitive to the choice of emotion features. Figure 2 shows the impact of the different emotion features on PJ author gender classification. Apart from the ontology aggregated value, the addition of other emotion features had little impact on the SVM classification performance, however, except for PJ2, the J48 algorithm yielded best results when the feature set contained all emotion features.

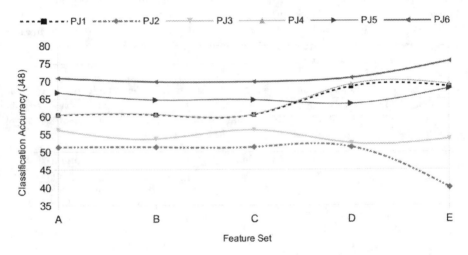

Fig. 2. Impact of emotion-based features sets on PJ classification accuracy with J48. A=*OntoClassAggValue* (only); B= A + *eRatio*; C= B + *PosToAllRatio*; D= C + *SS(Pos/Neg)*; E= D + *Fscore* (all features)

6   Discussion and Implications

Several observations can be drawn from the results of our experiments:

— As we expected and in accordance to earlier reports (for a description of recent advances see Brody and Hall [6]), female authors used a wider range of emotion categories (as represented by emotion words) compared to male authors in our datasets. The emotion-based features positively affected the performance of BoW features in most of cases.

— The positive to all emotion ratio did not yield a marked distinction between the genders. Exceptions were PJ6 in which female writing was found significantly more negative than male writing; and blogs, where female writing was found significantly more positive (as also reported by Thewall et al. [50]).

— Similarly, we found the stereotypical tendency of women expressing more emotions than men to be more context-related than a norm. That is, when asked to describe their feelings (PJ2, PJ4, and PJ6), female participants showed significantly more emotion expressivity. However, contrary to intuition, the emotion ratio of the document did not have a strong impact as a classification feature. This implies that a similar emotion ratio was generally found in male and female writing, including both positive and negative emotions.

— The SentiStrength feature did not have a marked positive impact on classification. Perhaps the sentiment strength was similar for both genders because within the analyzed genres all participants described mostly very personal matters.

— In contrast to previous studies in gender classification, the F-score lacked distinctive impact as a classification feature in our work. However, this feature, along with the SentiStrength, seemed to influence the performance of the J48 algorithm. This could be due to the information gain that they represent, as features external to the ontology.

— Adding emotions-based features positively influenced the performance of a BoW features set. An exception to this was seen in the Blogs corpus, were an Emo+BoW features set had a slightly less accuracy than that of a BoW features set alone. Since emotions features give interesting insights on the analyzed text, a slightly less accuracy in performance could be a trade-off worth considering.

— It is important to note that the classification accuracy using the different feature sets varies widely between corpora (e.g., PJ2 has 54.3% accuracy, whereas PJ6 has 91.8%). Hence, within the personal diaries genre, the topic of the writing seems to influence how well the classification features perform. In our study, PJ2 and PJ3 had low classification accuracy which suggests that when describing their feeling and experience of going to college, or their relationship with their parents, female and male authors used similar emotion words and expressions, however more in-depth analysis is needed to validate this. Also, the age of the authors may influence their writing style and emotion expression, but this variable was not controlled for in the datasets. Such cases represent a limitation to an emotion-based classifier.

— However, the experiments support our intuition that emotions features aid the performance of BoW features for author gender classification as compared to BoW features alone for the same corpus. Our contribution, moreover, showed that the behavior of emotions as features for author gender classification in different informal text genres yielded an acceptable performance, particularly when using SVM. Investigating the performance of different parameter settings and other types of classifiers will be an interesting vein to pursue.

7   Conclusion

We reported on a study in which the emotions detected from texts were used as features to aid author gender classification. We compared the performance of classification algorithms when using emotions as sole features to that of the BoW features, as well as to that of a combined BoW and emotion-based features. Our experimental results indicated that the accuracy of the BoW features is positively influenced by the addition of emotion-based features.

The following benefits of using emotion-based features can be highlighted from our work: 1) a relatively small and informative set of emotion-based features can be used to aid author's gender classification with positive impact in performance. The set size avoids the need for feature reduction/selection algorithms. 2) Discriminatory emotions, based on which classification is made, are easily identified. And 3) the emotion features are easily extracted for further analysis on the writers preferences.

Future directions will explore the applicability and reliability of emotions features to mark literary genres differences, and to detect cyberbullying in online discussion forums.

Acknowledgement. This work was supported by "Detecting and Visualizing Emotions and their Changes in Text" grant No. 14166, Academy of Finland.

References

1. Alm, C.O., Roth, D., Sproat, R.: Emotions from text: machine learning for text-based emotion prediction. In: Proceedings of HLT/EMNLP, Vancouver, Canada, pp. 579–586 (2005)
2. Argamon, S., Koppel, M., Fine, J., Shimoni, A.R.: Gender, genre, and writing style in formal written texts. Text - Interdisciplinary Journal for the Study of Discourse 23(3), 321–346 (2003), http://www.cs.biu.ac.il/~koppel/papers/male-female-text-final.pdf (accessed online on May 05, 2012)
3. Argamon, S., Kopel, M., Pennbaker, J.W., Schler, J.: Mining the Blogosphere: Age, Gender and the varieties of self-expression. First Monday 12(9), 3 (2007), http://firstmonday.org (accessed online on May 01, 2012)
4. Bougie, R., Pieters, R., Zeelenberg, M.: Angry Customers Don't Come Back, They Get Back: The Experience and Behavioral Implications of Anger and Dissatisfaction in Services. Journal of the Academy of Marketing Science 31(4), 377–393 (2003)
5. Bradley, M.M., Lang, P.J.: Affective norms for English words (ANEW). Instruction Manual and Affective Ratings. Technical report, The Center for Research in Psychophysiology, University of Florida (1999)
6. Brody, L.R., Hall, J.A.: Gender and Emotion in Context. In: Lewis, M., Haviland-Jones, J.M., Barrett, L.F. (eds.) Handbook of Emotions, ch. 24, 3rd edn., pp. 395–408. The Guilford Press (2008)
7. Chaplin, T.M., Cole, P.M., Zahn-Waxler, C.: Parental Socialization of Emotion Expression: Gender Differences and Relations to Child Adjustment. Emotion 5(1), 80–88 (2005)
8. Chaumartin, F.-R.: UPAR7: a knowledge-based system for headline sentiment tagging. In: Proceedings of the 4th International Workshop on Semantic Evaluations, pp. 422–425. ACL (2007)
9. Chung, C.K., Pennebaker, J.W.: Revealing dimensions of thinking in open-ended self-descriptors: An automated meaning extraction method for natural language. Journal of Research in Personality 42, 96–132 (2008)
10. Corney, M., de Vel, O., Anderson, A., Mohay, G.: Gender-Preferential Text Mining of E-mail Discourse. In: Proceedings of the ACSAC 2002, pp. 282–289 (2002)
11. Ekman, P.: Facial Expression and Emotion. American Psychologist 8(4), 376–379 (1993)
12. Elliott, C.: The affective reasoner: A process model of emotions in a multi-agent system. Ph.D. thesis, Institute for the Learning Sciences, Northwestern University (1992)

13. Nico, H.: Emotional Behavior. In: The Emotions. Studies in Emotion and Social Interaction, ch. 2. Cambridge University Press (1986)

14. Hall, M., Frank, E., Holmes, G., Pfahringer, B., Reutemann, P., Witten, I.H.: The WEKA Data Mining Software: An Update. SIGKDD Explorations 11(1) (2009)

15. Herring, S.C., Paolillo, J.C.: Gender and genre variation in weblogs. Journal of Sociolinguistics 10(4), 439–459 (2006)

16. Hess, U., Adams Jr., R.B., Kleck, R.E.: Facial Appearance, Gender, and Emotion Expression. Emotion 4(4), 378–388 (2004)

17. Heylighen, F., Dewaele, J.-M.: Variation in the Contextuality of Language: An Empirical Measure. Context in Context, Special issue of Foundations of Science 7(3), 293–340 (2002)

18. Juola, P.: Authorship Attribution. Foundations and Trends in Information Retrieval 1(3), 233–334 (2006)

19. Kakkonen, T., Kakkonen, G.G.: SentiProfiler: Creating Comparable Visual Profiles of Sentimental Content in Texts. In: Proceedings of the LaTeCH Workshop, Associated with the RANLP 2011 Conference, Hissar, Bulgaria (2011)

20. Kim, S.-M., Hovy, E.: Determining the Sentiment of Opinions. In: Proceedings of the 20th International Conference on Computational Linguistics, Article 1367 (2004)

21. Koppel, M., Argamon, S., Shimoni, A.R.: Automatically Categorizing Written Text by Author Gender. Literary and Linguistic Computing 17(4), 401–412 (2002)

22. Koppel, M., Schler, J., Argamon, S.: Computational methods in authorship attribution. Journal of the American Society for Information Science and Technology 60(1), 9–26 (2008)

23. Leach, C.W., Tiedens, L.Z.: A World of Emotions. In: Tiedens, L.Z., Leach, C.W. (eds.) The Social Life of Emotions, pp. 1–16. Cambridge University Press (2004)

24. Liu, B.: Sentiment Analysis and Subjectivity. In: Indurkhya, N., Damerau, F.J. (eds.) Handbook of Natural Language Processing, 2nd edn. (2010)

25. Liu, H., Lieberman, H., Selker, T.: Automatic Affective Feedback in an Email Browser. MIT Media Laboratory Software Agents Group Technical Report (2002), http://larifari.org/writing/

26. Liu, H., Lieberman, H., Selker, T.: A Model of Textual Affect Sensing using Real-World Knowledge. In: Proc. of the 2003 IUI, pp. 125–132 (2003)

27. McNeil, L.: Teaching an Old Genre a New Trick: The Diary on the Internet. Biography: An Interdisciplinary Quarterly 26(1), 24–48 (2003)

28. Melville, P., Wojciech, G., Lawrence, R.D.: Sentiment analysis of blogs by combining lexical knowledge with text classification. In: Proceedings of the 15th ACM SIGKDD. ACM (2009)

29. Miller, G.A.: WordNet: A Lexical Database for English. Communications of the ACM 38(11), 39–41 (1995)

30. Mohammad, S.M.: Portable Features for Classifying Emotional Text. In: Proceedings of the 2012 NAACL HLT, pp. 587–591 (2012)

31. Mohammad, S.M., Turney, P.D.: Emotions Evoked by Common Words and Phrases: Using Mechanical Turk to Create an Emotion Lexicon. In: Proceedings of the NAACL HLT 2010 Workshop on Computational Approaches to Analysis and Generation of Emotion in Text, pp. 26–34 (2010)

32. Mohamad, S.M., Yang, T.(W.): Tracking Sentiment in mail: how genders differ on emotional axes. In: Proceedings of the 2nd Workshop on Computational Approaches to Subjectivity and Sentiment Analysis, pp. 70–79. ACL, USA (2011)

33. Mukherjee, A., Liu, B.: Improving gender classification of blog authors. In: Proceedings of the 2010 EMNLP, pp. 207–217. ACL, USA (2010)
34. Mulac, A., Lundell, T.L.: Effect of Gender-Linked Language Differences in Adult's Written Discourse: Multivariate Tests of Language Effects. Language and Communication 14(3), 299–309 (1994)
35. Newman, M.L., Groom, C.J., Handelman, L.D., Pennebaker, J.W.: Gender Differences in Language Use: An Analysis of 14,000 Text Samples. Discourse Processes 45, 211–236 (2008)
36. Nowson, S., Oberlander, J., Gill, A.J.: Weblogs, Genres and Individual Differences. In: Proceedings of The 27th Annual Conference of the Cognitive Science Society, pp. 1666–1671 (2005)
37. O'Kearney, R., Dadds, M.: Developmental and gender differences in the language for emotions across the adolescent years. Cognition and Emotion 18(7), 913–938 (2004)
38. Ortony, A., Clore, G.L., Foss, M.A.: The Referential Structure of the Affective Lexicon. Cognitive Science 11, 341–364 (1987)
39. Ortony, A., Clore, G.L., Collins, A.: The Structure of the Theory. In: The Cognitive Structure of Emotions, ch. 2, pp. 15–33. Cambridge University P. (1994)
40. Picard, D., Boulhais, M.: Sex differences in expressive drawing. Journal of Personality and Individual Differences 51, 850–855 (2011)
41. Ashby Plant, E., Hyde, J.S., Keltner, D., Devine, P.G.: The Gender Stereotyping of Emotions. Psychology of Women Quarterly 24, 81–92 (2000), doi:10.1111/j.1471-6402.2000.tb01024.x
42. Plutchik, R.: The Nature of Emotions. American Scientist 89(4), 344–350 (2001)
43. Rangel, F., Rosso, P., Koppel, M., Stamatatos, E., Inches, G.: Overview of the Author Profiling Task at PAN 2013. In: Forner, P., Navigli, R., Tufis, D. (eds.) Notebook Papers of CLEF 2013 LABs and Workshops, CLEF (2013)
44. Shaver, P., Schwartz, J., Kirson, D., O'Connor, C.: Emotion Knowledge: Further Exploration of a Prototype Approach. In: Parrot, G.W. (ed.) Emotions in Social Psychology: Key Readings, pp. 26–56. Taylor & Francis, USA (2001)
45. Stone, P.J., Dunphy, D.C., Smith, M.S., Ogilvie, D.M.: The General Inquirer: A Computer Approach to Content Analysis. The MIT Press, Cambridge (1966)
46. Strapparava, C., Mihalcea, R.: Semeval-2007 task 14: Affective text. In: Proceedings of SemEval 2007, Prague, pp. 70–74 (2007)
47. Strapparava, C., Valitutti, A.: WordNet-Affect: an Affective Extension of WordNet. In: Proc. of the 4th LRE, pp. 1083–1086 (2004)
48. Strapparava, C., Mihalcea, R.: Learning to Identify Emotions in Text. In: Proc. of the ACM SAC 2008, pp. 1556–1560 (2008)
49. Thelwall, M., Bucley, K., Paltoglou, G., Cai, D.: Sentiment Strength Detection in Short Informal Text. Journal of The American Society for Information Science And Technology 61(12), 2544–2558 (2010)
50. Thelwall, M., Wilkinson, D., Uppal, S.: Data mining emotion in social network communication: Gender differences in MySpace. Journal of the American Society for Information Science and Technology 61(1), 190–199 (2010)
51. Thompson, R., Murachver, T.: Predicting gender from electronic discourse. British Journal of Social Psychology 40(2), 193–208 (2001)
52. Valitutti, A., Strapparava, C., Stock, O.: Developing Affective Lexical Resources. PsychNology 2(1), 61–83 (2004)
53. Zhang, Y., Dang, Y., Chen, H.: Gender Classification for Web Forums. IEEE Transactions on Systems, Man and Cybernetics, Part A: Systems and Humans 41(4), 668–677 (2001)

A Review Corpus for Argumentation Analysis

Henning Wachsmuth[1], Martin Trenkmann[2], Benno Stein[2],
Gregor Engels[1], and Tsvetomira Palakarska[2]

[1] Universität Paderborn, s-lab – Software Quality Lab, Paderborn, Germany
{hwachsmuth,engels}@s-lab.upb.de
[2] Bauhaus-Universität Weimar, Weimar, Germany
{benno.stein,martin.trenkmann,tsvetomira.palakarska}@uni-weimar.de

Abstract. The analysis of user reviews has become critical in research and industry, as user reviews increasingly impact the reputation of products and services. Many review texts comprise an involved argumentation with facts and opinions on different product features or aspects. Therefore, classifying sentiment polarity does not suffice to capture a review's impact. We claim that an argumentation analysis is needed, including opinion summarization, sentiment score prediction, and others. Since existing language resources to drive such research are missing, we have designed the *ArguAna TripAdvisor corpus*, which compiles 2,100 manually annotated hotel reviews balanced with respect to the reviews' sentiment scores. Each review text is segmented into facts, positive, and negative opinions, while all hotel aspects and amenities are marked. In this paper, we present the design and a first study of the corpus. We reveal patterns of local sentiment that correlate with sentiment scores, thereby defining a promising starting point for an effective argumentation analysis.

1 Introduction

Argumentation is a key aspect of human communication and cognition, consisting in a regulated sequence of speech or text with the goal of providing persuasive arguments for an intended conclusion or decision. It involves the identification of relevant facts about the topic or situation being discussed as well as the structured presentation of pros and cons [3]. In terms of text, one of the most obvious forms of argumentation can be found in reviews. Reviews provide facts and opinions about a product, service, or the like in order to justify a particular overall rating or sentiment, as in the following example: *"This was truly a lovely hotel to stay in. The staff were all friendly and very helpful. The location was excellent. The atmosphere is great and the decor is beautiful."*

In the last decade, the vast amount of user reviews in the web has become a primary influence factor of the reputation of products and services. As a consequence, research and industry put much effort into approaches and resources for the automatic analysis of reviews. Most approaches classify sentiment polarity at the text-level [12]. However, the facts, pros, and cons in review texts have proven beneficial for more complex tasks, such as summarizing opinions on different product features [7], interpreting local sentiment flows [9], or predicting sentiment scores [19]. Still, there has been no publicly available linguistic resource

A. Gelbukh (Ed.): CICLing 2014, Part II, LNCS 8404, pp. 115–127, 2014.

until now that makes it possible to jointly analyze the different types of information involved in the argumentation of reviews (cf. Section 2 for details).

In this paper, we present our design of the annotated *ArguAna TripAdvisor corpus* for analyzing the argumentations of web user reviews. The corpus consists of 2,100 English hotel reviews from an existing *TripAdvisor* dataset [17,19], evenly distributed across seven hotel locations. Such a review comprises a text, a set of ratings, and some metadata. In each text, we let experts manually annotate all hotel aspects and amenities as product features. In addition, we segmented the texts into subsentence-level statements. Then, we used crowdsourcing to classify every statement as a fact, a positive, or a negative opinion. In total, the corpus comprises 24.5k product features and 31k statements, while it is balanced with respect to the reviews' overall ratings, i.e., sentiment scores from 1 to 5.

The corpus is freely available at http://www.arguana.com for scientific use. It serves as a linguistic resource for the development and evaluation of approaches to sentiment-related tasks. Some example tasks have been named above [7,9,19], but the corpus also enables research on novel tasks. For large-scale evaluations and semi-supervised learning [13], nearly 200k further reviews from [19] are given without manual annotations. In general, we think that an *argumentation analysis* of texts will provide new insights into the use of language and can improve effectiveness in several natural language processing tasks.

To show the benefit of our corpus, here we investigate how the argumentation of a review text relates to the review's global sentiment. We offer evidence for the importance of the distribution of local sentiment in a review text, both in general and regarding specific product features. Moreover, we reveal common patterns of changes in the flow of local sentiment and their correlations with global sentiment scores. Altogether, our main contributions are the following:

1. We present the design of a freely available text corpus for analyzing the argumentation of web user reviews in terms of sentiment (Section 3).
2. We analyze the corpus to obtain new findings on correlations between a hotel review's sentiment score and the local sentiment in the review's text, giving insights into the ways web users argue in reviews (Section 4).

2   Related Work

In his pioneer study of arguments, Toulmin [16] models the basic argumentation structure with facts and warrants justified by a backing, leading to a qualified claim unless a rebuttal counters the facts. An approach to infer similar structures from scientific articles is given by *argumentative zoning* [15]. Recently, research has started to generally address *argumentation mining*, which analyzes natural language texts to detect the different types of arguments that justify a claim as well as their interactions [10]. In the reviews we consider, however, the actual claim is often not explicit, but it is quantified in terms of a sentiment score.

The argumentation of reviews is related to the concept of discourse, but it differs from *conversational* discourse, where the participants present arguments to persuade each other [4]. A review comprises a *monological* and *positional* argumentation, where a single presenter collates and structures a choice of facts and opinions in order to inform the intended recipient about his or her beliefs [3]. Accordingly, the aim of our corpus is not to check whether a claim is well argued, but to analyze what information is chosen and how arguments are structured to justify the claim, assuming the claim holds.

Following [10], an *argumentation analysis* enables a better understanding of discourse, intentions, and beliefs. This helps analyzing the sentiment of reviews, which in turn benefits the reputation management of products and services [12]. Different recent approaches exploit discourse structure on the subsentence-level to improve sentiment polarity classification, e.g. [11,21]. Others extract and summarize opinions [7] or they infer scores for several aspects from reviews [19]. All these approaches capture review argumentation to some extent. However, while sentiment corpora exist for several tasks and domains (cf. [12] for a selection), to our knowledge our corpus is the first that enables a combination of the approaches. The *MPQA corpus* [20] contains phrase-level annotations of opinions and other private states, but it is not meant for analyzing argumentations.

Below, we analyze review texts with respect to the flow of local sentiment. Our work resembles [9] where a sequential model first classifies the sentiment of each sentence in a text. The resulting flow is then used to predict the global sentiment of the text. In contrast, we focus on the identification of abstract argumentation patterns and we provide a corpus for related research.

3   Design of a Corpus for Argumentation Analysis

We now present our main design decisions in the compilation, annotation, and formatting of the *ArguAna TripAdvisor corpus* for the argumentation analysis of web user reviews. The corpus serves the scientific development and evaluation of approaches to tasks like sentiment score prediction [19] and opinion summarization [7]. It can be freely accessed at http://www.arguana.com.

3.1   Balanced Sampling of Web User Reviews

The ArguAna TripAdvisor corpus is based on a carefully chosen subset of a dataset originally used for aspect-level rating prediction [19]. The original dataset contains nearly 250k crawled English hotel reviews from *TripAdvisor* [17] that cover 1,850 hotels from over 60 locations. Each review comprises a text and a set of numerical ratings. The text quality is not perfect in all cases, certainly due to crawling errors: Some line breaks have been lost, which hides a number of sentence boundaries and, sporadically, word boundaries. In our experience, however, such problems are typical for web contexts. We rely on this dataset because its size, the quite diverse hotel domain, and the restriction to English serve as a suitable starting point for analyzing argumentations. We computed the distributions of locations and sentiment scores in the dataset, as shown in Figure 1. The latter should be representative for TripAdvisor in general.

Our sampled subset consists of 2,100 texts balanced with respect to both location and sentiment score. In particular, we selected 300 texts of seven of the 15 most-represented locations in the original dataset, 60 for each sentiment score

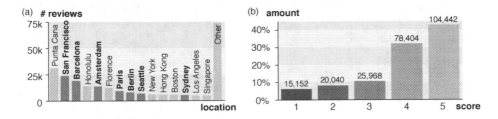

Fig. 1. (a) Distribution of the locations of the reviewed hotels in the original dataset from [19]. The ArguAna TripAdvisor corpus contains 300 annotated texts of each of the seven marked locations. (b) Distribution of sentiment scores in the original dataset.

Table 1. The number of reviewed hotels of each location in the ArguAna TripAdvisor corpus as well as the number of texts for each sentiment score from 1 to 5 and in total

Set	Location	Hotels	1	2	3	4	5	Σ
training	Amsterdam	10	60	60	60	60	60	300
	Seattle	10	60	60	60	60	60	300
	Sydney	10	60	60	60	60	60	300
validation	Berlin	44	60	60	60	60	60	300
	San Francisco	10	60	60	60	60	60	300
test	Barcelona	10	60	60	60	60	60	300
	Paris	26	60	60	60	60	60	300
complete	**all seven**	**120**	**420**	**420**	**420**	**420**	**420**	**2100**

between 1 (worst) and 5 (best). This supports an optimal training for learning approaches to sentiment score prediction. For opinion summarization, we ensured that the reviews of each location cover at least 10 but as few as possible hotels. To counter location-specific bias, we propose a corpus split with a training set containing the reviews of three locations, and both a validation set and a test set with two of the other locations. Table 1 lists details about the compilation.

3.2 Tailored Annotation Scheme for Argumentations

The reviews in the original dataset from [19] include optional ratings for seven aspects of hotels, namely, *value, room, location, cleanliness, front desk, service,* and *business service,* as well as a mandatory overall rating. We interpret the latter as the review's *sentiment score*. Besides, there is metadata about each review text (the username of the *author* and the creation *date*) and the reviewed hotel (*ID* and *location*). We maintain this data as text-level annotations in our corpus. In addition, we have enriched the corpus with annotations of local sentiment and product features to allow for an analysis of review argumentation.

Researchers have observed that reviews often contain *local sentiment* on the subsentence-level [21]. A common approach to handle this level is to divide a text into discourse units according to the *rhetorical structure theory* [8]. However, parsing discourse tends to be error-prone on noisy text [11] while being computationally expensive, which can be critical in web contexts. Also, not all discourse

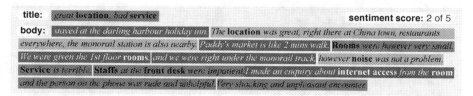

Fig. 2. Illustration of a text from the ArguAna TripAdvisor corpus. Each text is segmented into *positive opinions* (light green background), *negative opinions* (medium red), and objective *facts* (dark gray). All annotated aspects and amenities are marked in bold.

units are meaningful on their own, as in the following example, where the first unit depends on the context of the second one:

Statement 1. [*Although we had the suite,*]$_{unit1}$ [*our room was small,*]$_{unit2}$
Statement 2. [*but everything in the room was great.*]$_{unit3}$

Therefore, we have segmented each text into single *statements* instead, where we define a statement to be at least a clause and at most a sentence that is meaningful on its own. We assume each statement to have only one sentiment, even though this might be wrong in some cases. For reproducibility, the segmentation was done automatically using a rule-based algorithm provided with the corpus. The algorithm relies on lexical and syntactic clues derived from tokens, sentences, and part-of-speech tags. To classify the sentiment of all statements, we used crowdsourcing (see below). Our classification scheme follows approaches like [5], which see sentiment as a combination of subjectivity and polarity: We distinguish objective *facts* from subjective *opinions*. The latter are either *positive* or *negative*.

With the term *product features*, on the one hand we refer to *aspects*, such as those given above or others like "atmosphere". On the other hand, a product feature can be anything that is called an *amenity* in the hotel domain. Examples are facilities, e.g. "coffee maker" or "wifi", and services like "laundry". All mentions of such product features have been manually annotated in the corpus.

Figure 2 shows a sample text from the corpus, exemplifying the typical writing style often found in web user reviews: A few grammatical inaccuracies (e.g. inconsistent capitalization) and colloquial phrases (e.g. "like 2 mins walk"), but easily readable. More importantly, Figure 2 illustrates the corpus annotations. Each text has a specified title and body. In this case, the body spans nine mentions of product features, such as "location" or "internet access". It is segmented into 12 facts and opinions, which reflect the review's rather negative sentiment score 2 while e.g. showing that the internet access was not seen as negative.

The general numbers of corpus annotations are listed in Table 2 together with some statistics. The corpus includes 31,006 classified statements and 24,596 product features. On average, a text comprises 14.76 statements and 11.71 product features. Figure 3(a) shows a histogram of the text length in the number of statements, grouped into intervals. As can be seen, over one third of all texts span less than 10 statements (intervals 0-4 and 5-9), whereas less than one fourth spans 20 or more. Figure 3(b) visualizes the distribution of sentiment scores for all intervals that cover at least 1% of the corpus. Most significantly, the fraction

Table 2. Statistics of the tokens, sentences, manually classified statements, and manually annotated product features in the ArguAna TripAdvisor corpus

Type	Total	Average	Std. dev.	Median	Min	Max
tokens	442,615	210.77	171.66	172	3	1823
sentences	24,162	11.51	7.89	10	1	75
statements	**31,006**	**14.76**	**10.44**	**12**	**1**	**96**
facts	6,303	3.00	3.65	2	0	41
positive opinions	11,786	5.61	5.20	5	0	36
negative opinions	12,917	6.15	6.69	4	0	52
product features	**24,596**	**11.71**	**10.03**	**10**	**0**	**180**

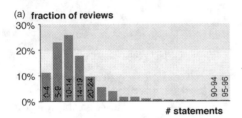
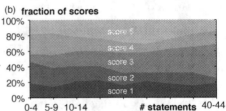

Fig. 3. (a) Histogram of the number of statements in the texts of the ArguAna TripAdvisor corpus. The numbers are grouped into intervals. (b) Interpolated curves of the fraction of sentiment scores in the corpus depending on the numbers of statements.

of reviews with sentiment score 3 increases under higher numbers of statements. This matches the intuition that long reviews may indicate so-so experiences.

3.3 Annotation by Web Users and Review Experts

Most hotel reviews are written by regular travelers and hence reflect the argumentation of average web users rather than review experts. Consequently, the classification of a statement as being a fact, a positive, or a negative opinion is in general a straightforward task. For this reason, we let web users annotate the sentiment of all 31,006 statements in our corpus using crowdsourcing. In particular, we relied on *Amazon Mechanical Turk* [1] where so called *workers* can be requested to perform *Human Intelligence Tasks* (HITs) and are paid a small amount of money in case their results are approved by the requester.

The HIT that we assigned to the workers involved the classification of 12 statements. To make the task as simple as possible, we experimented with different task descriptions. The main question of the final description was the following:

"When visiting a hotel,
are the following statements positive, negative, or neither?"

Below, we added notes: (1) to pick "neither" only for facts, not for unclear cases, (2) to pay attention to subtle statements where sentiment is expressed implicitly or ironically, and (3) to pick the most appropriate answer in controversial cases. A carefully chosen set of examples was given to illustrate the different cases.

The workers were allowed to work on a HIT at most 10 minutes and were paid $0.05 for an approved HIT. To assure quality, we assigned the HITs only to workers with over 1,000 approved HITs and an average approval rate of at least 80%. Moreover, we always put two hidden check statements with known and unambiguous classification among the 12 statements in order to recognize faked or otherwise flawed answers. The workers were informed that HITs with incorrectly classified check statements are rejected. For a consistent annotation, we assigned each statement to three workers and then applied majority voting to obtain the final classifications. Rejected HITs were reassigned to other workers.

Altogether, we received 14,187 HITs from 328 workers with an approval rate of 72.8%. On average, a worker spent 75.8 seconds per HIT. We measured the inter-annotator agreement for all statements, resulting in the value 0.67 of *Fleiss' Kappa* [6], which is interpreted as "substantial agreement". 73.6% of the statements got the same classification from all workers and 24.7% had a 2:1 vote (4.8% with opposing opinion polarity). The remaining 1.7% mostly referred to controversial statements, e.g. *"nice hotel, overpriced"* or *"It might not be the Ritz"*. So, we classified them ourselves in the context of the associated review.

Compared to the statement classifications, the annotation of product features is more complex since it requires to mark zero or more appropriate spans within a given text fragment. Moreover, the concept of a product feature is not clear by itself in the hotel domain. This renders crowdsourcing problematic, as it opens the door to ambiguities. In fact, a preliminary study produced very unsatisfying answers with a rejection rate of 43.3%. Thus, we decided to let two experts with linguistic background annotate the corpus, one from a university and one from our partner *Resolto Informatik GmbH*. We gave them the following guideline:

> *"Read through each review text. Mark all product features of the reviewed hotel in the sense of hotel aspects, amenities, services, and facilities."*

For clarity, we specified (1) to omit attributes of product features, e.g. to mark "location" instead of "central location" and "coffee maker" instead of "in-room coffee maker", (2) to omit guest belongings, and (3) not to mark the word "hotel" or brands like "Bellagio" or "Starbucks". Again, we gave a set of examples.

Based on 30 initial texts, we discussed and revised the annotations produced so far with each expert. Afterwards, the experts annotated all other texts from the corpus, taking about 5 minutes per text on average. To measure agreement, 633 statements were annotated twice. In 546 cases, the experts marked the same set of product features, which results in the value $\kappa = 0.73$ for *Cohen's Kappa* [6], assuming a chance agreement probability of 0.5.

3.4   Standard Corpus Format and Tool Support

The ArguAna TripAdvisor corpus comes as an 8 MB packed zip archive (28 MB uncompressed), which contains XMI files preformatted for the *Apache UIMA* framework, the industry standard for natural language processing applications [2]. Such an XMI file stores a text followed by its annotations, while the possible types of annotations are specified in a global *type system descriptor file*.

In addition, we converted all 196,865 remaining reviews of the original dataset with a correct text and a correct sentiment score between 1 and 5 into the same

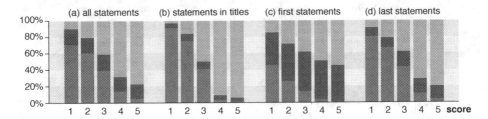

Fig. 4. (a) The fractions of positive opinions (light green), negative opinions (medium red), and objective facts (dark gray) in the texts of the ArguAna TripAdvisor corpus, separated by sentiment score. (b–d) The fractions for specific positions of statements.

format without manual annotations but with all TripAdvisor ratings and meta-data. This unannotated dataset (265 MB; 861 MB uncompressed) can be used both for semi-supervised learning techniques similar to [13] and for a large-scale evaluation of sentiment score prediction and the like. Also, we attached some software tools and UIMA-compliant text analysis algorithms with associated *UIMA analysis engine descriptor files* to the corpus. They can be executed to conduct the following analyses, thereby demonstrating how to process the corpus.

4   Analysis of Review Argumentation on the Corpus

In this section, we report on statistical analyses of the ArguAna TripAdvisor corpus. In particular, we focus on the questions how and to which extent the local sentiment in a review text determines the review's global sentiment.

4.1   The Impact of the Local Sentiment Distribution

First, we investigate how the distribution of local sentiment in a review text affects the review's global sentiment score. Intuitively, the larger the fraction of positive opinions, the better the sentiment score, and vice versa. More precisely:

Hypothesis 1. *The global sentiment score of a hotel review correlates with the ratio of positive and negative opinions in the review's text.*

As can be seen in Figure 4(a), Hypothesis 1 turns out to be true statistically for our corpus. On average, a review with sentiment score 1 contains 71% negative and 9.4% positive opinions. This ratio decreases strictly monotonously under increasing sentiment scores down to 5.1% negative and 77.5% positive opinions for sentiment score 5. Interestingly, the fraction of facts remains quite stable close to 20% in all cases. To further analyze the connection of local and global sentiment, we computed the distributions of opinions and facts in the review titles as well as in the first and last statements of the review's bodies. Based on the results shown in Figure 4(b–d), we checked for evidence for or against Hypothesis 2:

Hypothesis 2. *The global sentiment score of a hotel review correlates with the polarity of opinions at certain positions of the review's text.*

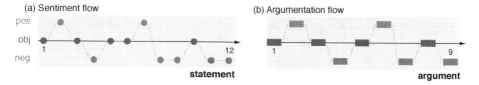

Fig. 5. Illustrations of the local sentiment in the sample text from Figure 2: (a) The *sentiment flow*, i.e., the sequence of all statement sentiments. (b) The *argumentation flow*, where consecutive statements with the same sentiment belong to the same argument.

Compared to Figure 4(a), the distributions for titles in Figure 4(b) entail much stronger gaps in the above-mentioned ratio with a rare appearance of facts, suggesting that the sentiment polarity of the title often reflects the polarity of the whole review. Conversely, over 40% of all first statements denote facts, irrespective of the sentiment score (cf. Figure 4(c)). This number may originate in the introductory nature of first statements. It implies a limited average impact of the first statement on a review's sentiment score. So, both the titles and first statements support Hypothesis 2. In contrast, the distributions in Figure 4(d) do not differ clearly from those in Figure 4(a). A possible explanation is that last statements often serve as summaries, but they may also simply reflect the average.

4.2   The Impact of the Local Sentiment Flow

Knowing that both the distribution and the positions of local sentiment have an impact, we next look at the importance of the structure of review texts. For generality, we do not consider the title of a review text as part of its structure, since unlike in our corpus many review texts do not have a title.

To quantify the impact of the structure, we analyze the flow of local sentiment in review texts. In accordance with [9], we define the *sentiment flow* of a text as the sequence of all statement sentiments in the body of the text, where by *sentiment* we either mean the *positive* or *negative* polarity of an opinion or the *objective* nature of a fact. As an example, we visualize the sentiment flow of the text from Figure 2 in Figure 5(a). Our hypothesis is the following:

Hypothesis 3. *The global sentiment score of a hotel review depends on the flow of local sentiment in the review's text.*

Our method to test Hypothesis 3 is to first determine common *flow patterns* in the corpus, i.e., flows of local sentiment that occur in a significant fraction of all texts in the corpus. Then, we check how much these patterns correlate with certain sentiment scores. From an analysis perspective, the two quantifications underlying these steps can be viewed as measuring recall and precision: We define the *recall* **R** of a flow pattern in a given corpus as the fraction of all texts in the corpus where the flow pattern occurs. The *precision* **P(s)** of a flow pattern with respect to a sentiment score **s** is the fraction of texts with sentiment score **s** under all texts in the given corpus where the flow pattern occurs.

However, the only five sentiment flow patterns with a recall of at least 1% in our corpus (i.e., more than 20 texts) are trivial without any change in local sentiment.

Table 3. The 13 argumentation flow patterns with the highest recall **R** in the ArguAna TripAdvisor corpus and their precision **P(s)** with respect to each sentiment score s

#	Argumentation flow	R	P(1)	P(2)	P(3)	P(4)	P(5)
1	(pos)	7.7%	1.9%	3.1%	7.5%	31.1%	**56.5%**
2	(obj)	5.3%	3.6%	13.6%	20.0%	**33.6%**	29.1%
3	(neg)	3.5%	**58.9%**	30.1%	9.6%	1.4%	–
4	(pos, obj, pos)	3.0%	–	–	6.5%	35.5%	**58.1%**
5	(obj, pos)	2.7%	–	1.8%	7.0%	31.6%	**59.6%**
6	(pos, neg, pos)	2.1%	–	15.9%	11.4%	**56.8%**	15.9%
7	(obj, pos, obj, pos)	1.9%	–	–	5.1%	35.8%	**59.0%**
8	(pos, neg)	1.7%	11.1%	**36.1%**	33.3%	19.4%	–
9	(neg, obj, neg)	1.7%	**88.9%**	8.3%	2.8%	–	–
10	(obj, pos, neg, pos)	1.5%	–	3.2%	**32.3%**	**32.3%**	**32.3%**
11	(neg, pos, neg)	1.5%	35.5%	**51.6%**	12.9%	–	–
12	(obj, neg, obj, neg)	1.1%	**77.3%**	18.2%	4.5%	–	–
13	(obj, neg)	1.1%	**83.3%**	16.7%	–	–	–

In [9], improvements are obtained by ignoring the objective facts. Our according experiments did not yield new insights except for a higher recall of the trivial patterns. We thus omit to present their results here, but the results can be easily reproduced using our software tools. The problem lies in the high variance of the reviews' lengths (cf. Figure 3(a)). While a solution is to length-normalize sentiment flows, a reasonable normalization is not straightforward. Instead, here we propose to move from statements to *arguments*, where we take the very simplyfing view that a single argument is a sequence of consecutive statements with the same sentiment. The following example shows the rationale behind:

Argument 1. [*I love that hotel!*]stmt1 [*Huge rooms, great location...*]stmt2
Argument 2. [*but it's so expensive!!!*]stmt3

Though the first two statements discuss different topics, the second can be seen as an *elaboration* of the first one in the discourse sense [8]. The third statement contrasts the others, thus denoting a different argument. Based on the notion of arguments, we define the *argumentation flow* of a text as the sequence of all argument sentiments in the body of the text, as illustrated in Figure 5(b).

In total, 826 different argumentation flows exist in our corpus. Table 3 lists the flow patterns with a recall of at least 1%. They cover 34.8% of the corpus texts. The highest-recall pattern *(pos)* represents all 161 fully positive texts (7.7%). Patterns with a high precision **P(5)** are made up only of objective and positive arguments (table line 4, 5, and 7). Quite intuitively, typical patterns of reviews with sentiment score 2 and 4 are *(neg, pos, neg)* and *(pos, neg, pos)*, respectively, whereas none of the listed patterns clearly indicates sentiment score 3. The highest correlation is observed for *(neg, obj, neg)*, which results in sentiment score 1 in 88.9% of the cases. While such correlations offer strong evidence for Hypothesis 3, all 13 patterns cooccur with more than one sentiment score. Consequently, the structure of a review text does not decide the global sentiment alone.

Table 4. A selection of the 25 product features with highest recall **R** in the ArguAna TripAdvisor corpus, the fractions of their positive (**pos**) and negative (**neg**) mentions, and the precision with respect to sentiment score 1 and 5 depending on these polarities

#	Feature	R	pos	$P_{pos}(1)$	$P_{pos}(5)$	neg	$P_{neg}(1)$	$P_{neg}(5)$
1	room	**80.3%**	36.9%	7.4%	31.1%	47.8%	38.4%	3.5%
2	staff	43.4%	62.9%	4.3%	38.0%	34.1%	**50.3%**	1.5%
3	location	42.2%	**84.7%**	5.7%	35.9%	11.8%	32.5%	1.6%
8	service	18.4%	38.9%	7.4%	44.1%	55.0%	45.1%	–
17	food	7.6%	52.3%	**9.9%**	34.7%	37.3%	45.8%	1.4%
20	towels	5.3%	**27.1%**	7.9%	**21.1%**	**67.1%**	35.1%	3.2%
24	parking	**5.1%**	30.6%	–	**46.3%**	56.0%	**25.3%**	**12.0%**

4.3 The Impact of the Local Sentiment regarding Product Features

Finally, we quantify the impact of the content of a hotel review, which is represented by the product features discussed within the review's text:

> **Hypothesis 4.** *The global sentiment score of a hotel review correlates with the polarity of opinions on certain product features in the review's text.*

To investigate the hypothesis, we consider the 25 product features with the highest recall **R** in the corpus. Similar to above, here *recall* means the fraction of all texts where the product feature occurs. First, we compute the fractions of positive and negative mentions of each product feature. For simplicity, we assume that an opinion always refers to the product features it contains. Then, we quantify the correlation between the polarity of a mention and the sentiment score of the respective review by reusing the concept of *precision* from Section 4.2 accordingly. In Table 4, we present a selection of the 25 product features.

The general importance of the *room* is reflected by a recall of 80.3%. The *location* appears most often in positive opinions (84.7%) and *towels* in negative ones (67.1%). However, other aspects and amenities seem to have a larger impact on a review's global sentiment: When e.g. the *staff* is seen as negative, this results in sentiment score 1 in 50.3% of the cases. Even more obvious, a negative mention of *service* never cooccurs with sentiment score 5 (interestingly, *staff* is used more in positive and *service* more in negative contexts). Conversely, we see that a positive *food* experience alone does not make a good hotel ($P_{pos}(1) = 9.9\%$), and 12% of all negative opinions on *parking* occur in reviews with the highest sentiment score. A good *parking* situation seems to be appreciated, though.

To summarize, our corpus reveals large differences in the impact of product features on a review's global sentiment, which supports Hypothesis 4. We hence conclude that an argumentation analysis of reviews should cover both structure and content. To this end, our results define a promising starting point.

5 Conclusion

The facts and opinions within the argumentation of a review text impact the reputation of products and services. To analyze argumentations, we have designed the freely available ArguAna TripAdvisor corpus based on a balanced collection of hotel reviews. Each review text is annotated with respect to local sentiment and the mentioned hotel aspects and amenities. We have explored the corpus to reveal argumentation patterns that correlate with the reviews' sentiment scores. While the corpus is restricted to hotel reviews, in future work we will investigate to what extent the patterns generalize to other domains. Generally, we believe that an argumentation analysis of texts allows for more effective approaches to sentiment-related tasks. At the same time, it implies new ways to explain obtained results, as it mimics the way humans interpret texts. Currently, we work on an approach that learns argumentation patterns in order to predict and explain sentiment scores. Apart from sentiment, our findings on argumentation may be transferrable to other natural language processing tasks, such as authorship attribution [14] or language function analysis [18]. For this purpose, we will need further resources that cover more domains and types of annotations.

Acknowledgments. This work was funded by the German Federal Ministry of Education and Research (BMBF) under contract number 01IS11016A.

References

1. Amazon Mechanical Turk, http://www.mturk.com
2. Apache UIMA, http://uima.apache.org
3. Besnard, P., Hunter, A.: Elements of Argumentation. The MIT Press (2008)
4. Cabrio, E., Villata, S.: Combining Textual Entailment and Argumentation Theory for Supporting Online Debates Interactions. In: Proc. of the 50th ACL: Short Papers, pp. 208–212 (2012)
5. Esuli, A., Sebastiani, F.: SENTIWORDNET: A Publicly Available Lexical Resource for Opinion Mining. In: Proceedings of the 5th LREC, pp. 417–422 (2006)
6. Fleiss, J.L.: Statistical Methods for Rates and Proportions, 2nd edn. John Wiley & Sons (1981)
7. Hu, M., Liu, B.: Mining and Summarizing Customer Reviews. In: Proc. of the Tenth SIGKDD, pp. 168–177 (2004)
8. Mann, W.C., Thompson, S.A.: Rhetorical Structure Theory: Toward a Functional Theory of Text Organization. Text 8(3), 243–281 (1988)
9. Mao, Y., Lebanon, G.: Isotonic Conditional Random Fields and Local Sentiment Flow. Advances in Neural Information Processing Systems 19, 961–968 (2007)
10. Mochales, R., Moens, M.F.: Argumentation Mining. AI and Law 19(1), 1–22 (2011)
11. Mukherjee, S., Bhattacharyya, P.: Sentiment Analysis in Twitter with Lightweight Discourse Analysis. In: Proc. of the 24th COLING, pp. 1847–1864 (2012)
12. Pang, B., Lee, L.: Opinion Mining and Sentiment Analysis. Foundations and Trends in Information Retrieval 2(1-2), 1–135 (2008)
13. Prettenhofer, P., Stein, B.: Cross-Language Text Classification using Structural Correspondence Learning. In: Proc. of the 48th ACL, pp. 1118–1127 (2010)
14. Sapkota, U., Solorio, T., Montes-y-Gómez, M., Rosso, P.: The Use of Orthogonal Similarity Relations in the Prediction of Authorship. In: Gelbukh, A. (ed.) CICLing 2013, Part II. LNCS, vol. 7817, pp. 463–475. Springer, Heidelberg (2013)

15. Teufel, S.: Argumentative Zoning: Information Extraction from Scientific Text. Ph.D. thesis, University of Edinburgh (1999)
16. Toulmin, S.E.: The Uses of Argument. Cambridge University Press (1958)
17. TripAdvisor, http://www.tripadvisor.com
18. Wachsmuth, H., Bujna, K.: Back to the Roots of Genres: Text Classification by Language Function. In: Proc. of the 5th IJCNLP, pp. 632–640 (2011)
19. Wang, H., Lu, Y., Zhai, C.: Latent Aspect Rating Analysis on Review Text Data: A Rating Regression Approach. In: Proc. of the 16th SIGKDD, pp. 783–792 (2010)
20. Wiebe, J., Wilson, T., Cardie, C.: Annotating Expressions of Opinions and Emotions in Language. Language Resources and Evaluation 1(2) (2005)
21. Zirn, C., Niepert, M., Stuckenschmidt, H., Strube, M.: Fine-Grained Sentiment Analysis with Structural Features. In: Proc. of the 5th IJCNLP, pp. 336–344 (2011)

Looking for Opinion in Land-Use Planning Corpora

Eric Kergosien[1,2], Cédric Lopez[4],
Mathieu Roche[3], and Maguelonne Teisseire[1,2]

[1] LIRMM - CNRS, Univ. Montpellier 2,
161 rue Ada, 34095 Montpellier Cedex 5, France
eric.kergosien@lirmm.fr
[2] Irstea, UMR TETIS, 500 rue J.F. Breton, 34093 Montpellier Cedex 5, France
maguelonne.teisseire@teledetection.fr
[3] Cirad, UMR TETIS, 500 rue J.F. Breton, 34093 Montpellier Cedex 5, France
mathieu.roche@cirad.fr
[4] Viseo - Objet Direct, 4 av. du Doyen Louis Weil, 38000 Grenoble, France
clopez@objetdirect.com

Abstract. A great deal of research on opinion mining and sentiment
analysis has been done in specific contexts such as movie reviews, commercial evaluations, campaign speeches, etc. In this paper, we raise the
issue of how appropriate these methods are for documents related to
land-use planning. After highlighting limitations of existing proposals
and discussing issues related to textual data, we present the method
called OPILAND (OPinion mIning from LAND-use planning documents)
designed to semi-automatically mine opinions in specialized contexts.
Experiments are conducted on a land-use planning dataset, and on three
datasets related to others areas highlighting the relevance of our proposal.

Keywords: Land-use planning, Text-Mining, Opinion-mining, Corpus,
Lexicon.

1 Introduction

The notion of territory, and more specifically of land-use-planning, is complex
and refers to many concepts such as stakeholders, spatial and temporal features,
opinions, politics, history, etc. Hence, territories reflect both economic, ideologi-
cal and political appropriation of space by groups who provide a particular view
of themselves, of their history, and their uniqueness. The characterization and
understanding of perceptions of a territory by different users is complex but
needed for land-use planning and territorial public policy. In this paper, we focus on on political and administrative territories (eg. local or regional territory)
and we propose an original approach to build specific vocabularies of opinions
related to our domain.

A. Gelbukh (Ed.): CICLing 2014, Part II, LNCS 8404, pp. 128–140, 2014.
© Springer-Verlag Berlin Heidelberg 2014

Opinion mining has been intensively studied in various fields such as movie reviews, political articles, tweets, etc. Methods are based on statistics or Natural Language Processing (NLP). A lexicon or a dictionary of opinions (with or without polarity) is often used. In the context of land-use planning, even if information published on the web (blogs, forums, etc.) and in media expresses a feeling, the traditional opinion mining approaches fail to extract opinion due to the context specificity (small or medium-size and specialized corpus). We propose to tackle this issue by defining a new approach, called OPILAND (OPInion mining for LAND-use planning documents), in order to semi-automatically mine opinion in specific contexts. This approach uses specialized vocabularies to compute a polarity score for documents.

This paper is structured as follows. In Section 2, an overview of opinion mining methods is presented. In Section 3, the OPILAND method is detailed. Section 4 reports experiments, firstly on the land-use planning corpus and secondly on three other corpus. The paper ends with our conclusions and future work.

2   State-of-the-Art

In **opinion mining** and NLP, the analysis of subjectivity and opinions expressed by people in texts (newspapers, technical documents, blogs, reviews, letters, etc.) is called opinion analysis [12]. Recognition polarity attempts to classify texts according to positivity or negativity with respects to the expressed opinions therein. Two main approaches can be identified: one based on the frequency of positive and negative words in each text [19], and the other one based on machine learning techniques from annotated texts [6]. Hybrid approaches would appear to offer the best results [9, 10]. In all these approaches, several features are used, including words, n-grams of words [13], the shifted words [8], and so on. These features can be exploited using machine learning methods based on an annotated corpus. Such corpora are made available in text analysis challenges such as TREC (Text Retrieval Conference), or DEFT (DÉfi Fouille de Textes) for assesment by the French community. However, only a few are annotated according to opinion and polarity. In addition, several classification methods can be grouped into voting systems proposed by [17] or applying reinforcement and bag of words methods [7]. Other approaches rely on incremental methods for opinion analysis [21]. Along with the classification of opinion texts, the team's research works focused on the automatic construction of opinion vocabularies [5]. The incremental approaches proposed are usually based on web-mining methods in order to learn an opinion vocabulary specifically linked to a topic or a sub-topic.

Concerning **analysis of user feelings** in the **land-use planning** domain, the overview of existing research reveals the involvement of several communities. For thirty years, the concept of territory, based on different definitions, has been widely used and discussed by ethologists and ecologists, geographers, sociologists, economists, philosophers, etc. In the French community, the geographer community has been particularly prolific, and their work has tended to adopt either a social or a political angle in their territorial analysis [20]. On the one hand, social geography analyzed the identity dimension of the territory, the membership reports and tighten reports [2]. On the other hand, political geographers have tried to represent dimensional aspects of the territory, through the analysis of public action initiatives [4]. To cope with its multiple definitions of territory, it often clarifies the words meaning by adding a qualifier: Biophysical territories (watershed, great landscape, etc.), politico-administrative territories (e.g. city, country, continent, etc.), large territories, suitable territories, mobile territories. According to the state-of-the-art, different types of approaches could be used. In our context, supervised approaches are not adapted because we process with a reduced size of labeled data (i.e data analysed by experts). Unsupervised approaches based on incremental methods using seed of polarized words to enrich are often used in sentiment analysis studies [16]. But they are not adapted in our context. Generally the enrichment is based on global information (e.g. hits of Web pages) returned by general search engines (for example, Google, Yahoo, Exalead, and so forth). But this is too general in our context.

However, to the best of our knowledge, there is no automatic or semi-automatic method to mine opinion in the land-use planning context. In this paper, the proposed approach defines a specific vocabulary of opinion related to the domain using general lexicons of opinion.

3   Towards a Specialized Vocabulary of Opinion Related to Land-Use Planning

Our aim is to automatically identify opinion in texts related to a domain application. The project SENTERRITOIRE refers to the territory of the Thau lagoon described in our corpus. Classical methods of opinion mining fail when applied on this corpus as the traditional lexicons are not appropriate to the associated domain (See Section 4.2, Table 1). We also tested a "bag of words" approach by applying classical classification algorithms for supervised learning. Unfortunately, due to the low proportion of data training and the diversity of topics covered, the results were also unsatisfactory (between 50-55% well-classified). We have thus defined the new and generic approach OPILAND which is detailed in the following subsections.

3.1   Overview

At a first step, there is no need for an annotated corpus contrary to conventional methods of opinion mining [18]. However, for assessment purpose, part of the corpus has been annotated by an expert (with positive and negative polarity). To extract opinion from texts, we propose to define a specialized vocabulary of opinions in three steps (See Figure 1): (1) *General vocabulary of pivot opinion (GVPO)*: list of generalized and polarized pivot words (i.e. words extracted from the corpus and existing in at least one of the traditional lexicons of opinion used); (2) *Contextualized vocabulary of opinion (CVO)*: list of words polarized with respect to their context in documents; (3) *Specialized vocabulary of opinion (SVO)*:list of polarized words related to the domain.

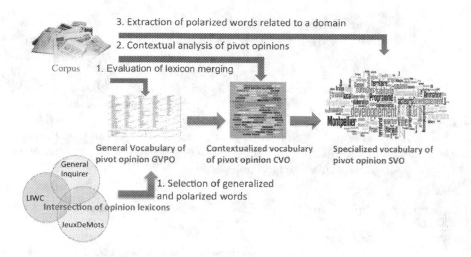

Fig. 1. OPILAND Approach

In the following subsections, in order to illustrate all steps involved in the OPILAND approach, we will use 100 extracts of documents (40 are negative and 60 are positive) related to land-use planning in the vicinity of Sète (See Figure 2). The extracts 1, 2 and 4 are positive and the extracts 3 and 100 are negative.

1 : positive	The Thau basin is made up of the *magnificent* and *natural* Thau lagoon, its watershed and marine front.
2 : positive	*Increasing* urbanisation means greater competition with vine-growing and *natural* space linked with the *beautiful* lagoon.
3 : negative	However, the SMBT *fears* that fishermen wich live in areas of Sète will move elsewhere to find work. The *heavily* criticized General Fisheries Council for the Mediterranean is working in order to *find* a *solution*.
4 : positive	The Thau basin is a *natural* marine environment, unrivalled in France, so *leisure* activities are strictly controlled to *protect* this eco-sensitive zone.
...	
100 : negative	From 1880 to 1884, seven concessions were granted in the vicinity of Sète. The consequences of these installations in *heavily* polluted waters did not take long to manifest, with *gastrointestinal* intoxications and even cases of *typhoid fever*.

Color legend:
- Negative opinion from GVPO ; *Positive opinion from GVPO*
- Negative opinion from CVO   *Positive opinion from CVO*
- Negative opinion from SVO ; *Positive opinion from SVO*

Fig. 2. Document extracts related to the vicinity of XXX

The first stage consists in automatically extracting pivot opinions such as: "*magnificent*", "*beautiful*", "*increasing*", "*solution*", "*criticized*", "*fears*", "*leisure*" and "*protect*". To this end, we propose a NLP extraction process built on the basis of opinion lexicons produced by the scientific community. A "Pivot Opinion" PO is a word extracted from the corpus existing in at least one of the three lexicons of opinion used. The second stage involves extracting words which contain pivot opinions in their context (CVO) such as, for example, the words "find" which is located in the same context of the positive pivot word "*solution*" and "*heavily*", located in the same context of the negative pivot opinion "*criticized*". The third stage resides in semi-automatically define a specialized vocabulary of opinion (SVO) related to the domain such as the word "*natural*". Therefore, we propose to adapt a text mining method to highlight a specific vocabulary of opinion related to our domain using the contextualized vocabulary of opinion (CVO).

3.2   Construction of the Specialized Vocabulary of Opinion

General vocabulary of pivot opinion
Opinion lexicon are often created manually, semi-automatically, or in a contributive way. In our approach, we use and evaluate three French lexicons:

- Lexicon 1: The General Inquirer lexicon in French [1], is a translated version of the General Inquirer4 which contains syntactic, semantic and pragmatic information about a list of polarized words. For each word, the polarity indicates that a word is positive or negative. This list is available in French after translation, stemming and validation done by two judges [1]. Finally, the lexicon contains 1246 positive words and 1527 negative words.
- Lexicon 2: The LIWC lexicon in French [14] is the translation from the English lexicon Linguistic Inquiry and Word Count5 (LIWC). The words are not in stemming form. In addition, verb conjugations, noun and adjective in sections increase the size of the lexicon (13626 polarized words) . There are also words describing positive and negative emotions.
- Lexicon 3: The lexicon JeuxDeMots [11] is a French lexicon extended to all part-of-speech (noun, verb, adjective, and adverb), and also to a large number of named entities (people, places, brands, events). The lexicon is composed of more than 250000 words obtained with the serious game JeuxDe-Mots[1]. Moreover an associated system, called LikeIt6, catches polarity information given by users. Currently, 27529 words have been polarized. We consider JeuxDeMots as the other lexicons : we define a value of 1 for positive words and -1 for negative ones.

862 words are ambiguous, i.e. present in one or more selected lexicons in both the positive and negative portion. So we decide to delete them. Our challenge is to find the best combination to obtain a relevant list of linguistic features for opinion called Pivot Opinion *PO*. We firstly propose to merge these different lexicons in a General Vocabulary of Pivot Opinion (GVPO):

- $GeneralInquirer \cap LIWC \cap JeuxDeMots = S_1$;
- $GeneralInquirer \cap LIWC = S_2$;

---

[1] http://www2.lirmm.fr/~lafourcade/JDM-LEXICALNET-FR/?C=M;O=D

- $GeneralInquirer \cap JeuxDeMots = S_3$;
- $LIWC \cap JeuxDeMots = S_4$;
- $GeneralInquirer = S_5$: words appearing only in the GeneralInquirer lexicon;
- $LIWC = S_6$: words appearing only in the LIWC lexicon;
- $JeuxDeMots = S_7$: words appearing only in the JeuxDeMots lexicon.

Three types of reliability scores are defined: a high score (S_1) to the words contained in the three lexicons, an average score (S_1, S_2, S_3) to the words included in two lexicons, and a low score (S_5, S_6, S_7) to the words contained in only one lexicon[2]. These scores are compared in Section 4.2. From the document extracts presented in Figure 2, the following words are PO, i.e., present in at least one of the three lexicons of opinion used: "magnificent, beautiful, leisure, increasing; solution, intoxication, protect, polluted, fever ".

Contextualized vocabulary of pivot opinion
The words contained in the GVPO vocabulary are therefore Pivot Opinions PO. Next, other words located in the same context of these PO are polarized. In practice, if a word is close to a PO (i.e. in a window size, for a given sentence), it is polarized depending on where the PO is. The assigned polarity score is defined in formula 1 in which d is the position (in number of words) of the current word W relative to a PO.

$$NeighborWordScore(W) = \frac{\sum \frac{WordScore(PO)}{d}}{\sum PO} \qquad (1)$$

For each non-pivot word CW, all neighboring Pivot Opinions PO are identified (i.e. in a window of neighboring words). Next, for each PO selected, a *NeighborPS* polarity score is calculated by dividing its polarity score by the distance to CW. The *NeighborWordScore* score of CW is then the average of its *NeighborPS* score. The CVO vocabulary obtained at this step consists of PO from GVPO and other words which are in the same context as PO. We use the following concrete example in Figure 3 to illustrate this step. Concerning the word "*heavily* ", if we choose a window of four neighboring words, only the PO named "*criticized* " is selected. The *NeighborWordScore* score of "*heavily* " is equal to -1, indicating that this word is negative in this context.

Fig. 3. Syntagm from a document extract related to the vicinity of XXX

From the document extracts presented in Figure 2, the following words are some of the words added in the CVO: "natural, find, heavily, gastrointestinal ".

---

[2] These lexicons are available here:
http://ekergosien.free.fr/file/SenterritoireProject_S1toS7Lexicons.zip

Specialized vocabulary of opinion

Text mining methods allow us to highlight a specific vocabulary of opinion related to our domain. We have implemented a module for the extraction of relevant linguistic features from a corpus based on the contextualized vocabulary of opinion CVO defined in Section 3.2. For each opinion feature O present in the CVO, the number of positive ($nbPos$) and negative ($nbNeg$) documents in which it occurs is counted. A first selection criterion is used to remove features (low presence with respect to $nbTDocs$, i.e, the corpus size) (See Formula 2). We have empirically tested the commonly measures (support, natural logarithm, logarithm to the base 10, tf-idf, etc.) to filter features on the basis of our corpora, and we have chosen to use the measure logarithm to the base 10. Therefore, our results indicate that this measure is a less restrictive than the others when used on small or medium sized corpora.

$$nbPos(O) + nbNeg(O) \leq log(nbTDocs) \tag{2}$$

From the document extracts presented in Figure 2, the words *"natural"* and *"heavily"* are, for example, retained as candidate for inclusion in the SVO using this selection criterion as shown below:

natural	*heavily*	*leisure*
$nbPos = 3$	$nbPos = 0$	$nbPos = 1$
$nbNeg = 0$	$nbNeg = 2$	$nbNeg = 0$
$log(100) = 2$	$log(100) = 2$	$log(100) = 2$
$3+0 \geq 2$	$2+0 \geq 2$	$1+0 \leq 2$

In contrast, the word *"leisure"* is not retained.

We assign to the remaining features O a weighting score WScore (See Formula 3) based on their discriminating factor and the proportion of positive and negative documents in the corpus, $nbTDocNeg$ and $nbTDocPos$ being the total number of negative and positive documents, respectively. The function max with the second parameter 1 is used to avoid that the denominator or the numerator is equal to 0, i.e. the feature O is not present positive or negative documents.

$$WScore(O) = \frac{max(nbPos(O),1)}{max(nbNeg(O),1)} \times \frac{nbTDocNeg}{nbTDocPos} \times nbTDocs \tag{3}$$

From the document extracts presented in Figure 2, the WScore for the example *"natural"* would be:

$$WScore(natural) = \frac{max(3,1)}{max(0,1)} \times \frac{40}{60} \times 100 = \frac{3}{1} \times \frac{200}{3} = 200$$

In the same way, the WScore for the word *"heavily"* would be 100/3:

$$WScore(heavily) = \frac{max(0,1)}{max(2,1)} \times \frac{200}{3} = \frac{1}{2} \times \frac{200}{3} \approx 33$$

This method is used to process an unbalanced corpus in terms of the number of polarized documents. A feature is deemed to be representative if its occurrence in a document class (positive or negative) is more significant than in the other. Therefore, ambiguous features, which are not representative of a class, are removed. For the remaining features O, we define a representativeness score RS related to the represented class (See Formula 4). The feature is assigned to a positive polarity if its weighting value is greater than, or equal to, T_r, and negative if its weighting value is lower than, or equal to, $1\text{-}T_r$.

$$
\begin{aligned}
if\ WScore(O) &\geq T_r \times nbTDocs\ Then \\
RS_{pos}(O) &= WScore(O) \\
if\ WScore(O) &\leq (1 - T_r) \times nbTDocs\ Then \\
RS_{neg}(O) &= 1 - WScore(O)
\end{aligned}
\tag{4}
$$

The threshold used in our experiments is $T_r = 65\%$, indicating that all features that do not have at least 65% of the distribution in one of the two classes are removed. Manual validation of these features generates a specialized vocabulary (SVO). For example, the word *"natural"* has been associated with the specialized vocabulary of positive elements related to land-use planning. Indeed, as shown below, the RS score is positive.

$$
WScore(natural) \geq 0,65 \times 100\ Then\ RS_{pos}(O)\ =\ 200
$$

In contrast, the RS score of the opinion *"heavily"* is assigned as negative:

$$
WScore(heavily) \leq 0,35 \times 100\ Then\ RS_{neg}(O)\ =\ 35
$$

3.3 Assigning Opinion Score to Documents

Once the different vocabularies are defined, the next step assigns a polarity score to each document. Two types of preprocessing are previously performed: (1) Statistical preprocessing: Removing not discriminant words based on IDF score (Inverse Documents Frequency), (2) Linguistic preprocessing: A final preprocessing effort consists in weighting pivot words according to their part-of-speech category using Tree-Tagger (i.e. *Cat*).

Based on parameters and preprocessing, an overall polarity score is assigned to each textual object (e.g. word, sentence, and document). Firstly, the Opinions (O) score is calculated according to their presence in the positive lexicon or the negative one (See formula 5, with *polarity (O)* $\in \{$-1, 1$\}$). Thereafter, this polarity is weighted using a reliability score S_i (See Section 3.2).

$$
WordScore(O) = S_i \times polarity(O)
\tag{5}
$$

From the document extract number 2 presented in Figure 2, the WordScore of the opinions *"Increasing"* and *"beautiful"* are both equals to 1 because they are extracted from positive lexicons. Then, the word *"natural"*, contained in the CVO has a WordScore equal to 1/3 (See Section 3.2 for the NeighborWordScore description). In this simple example, $S_i=1$. The score of a sentence S is obtained regarding to the scores assigned to each opinion O (See formula 6). A weight is assigned to the words according to their part-of-speech (i.e. *Cat* parameter). Here, the Cat parameter gives different weights to words according to their grammatical categories (e.g. adjectives, nouns).

$$SentenceScore(S) = \frac{\sum Cat \times WordScore(O)}{\sum Cat \times O} \quad (6)$$

From the document extract number 2 presented in Figure 2, the SentenceScore is about 2,33/3, indicating that it is a positive sentence. In this simple example, $Cat=1$. Finally, the overall score of the analyzed document is defined as the average of the scores of its associated sentences. In order to affect a score to document by using contextual information (See Section3.2), each word of the context is taken into account (See formula 6).

4   Experiments

4.1   Description of the Corpus

We selected newspapers related to land-use planning of a specific lagoon in Thau Agglomration. The corpus consists of 100 documents, called SENT_100. It is divided into two classes of opinion: positive and negative. The opinion is related to the formation of a new metropolitan area gathering several cities. The corpus was validated by geographer experts. It does not contain (i) ambiguous texts presenting different opinions, and (ii) texts without polarity information. The next section presents experimental results with the OPILAND approach on SENT_100.

4.2   Results

Which lexicon? The classification score of polarized document is obtained using the three selected lexicons (i.e. *General Inquirer, LIWC,* and *JeuxDeMots,* see Table 1). We note that the *General Inquire* lexicon, more complete than *LIWC,* gives better results (i.e. 57,5%). *JeuxDeMots* lexicon sounds less effective. These results rely to the construction of this resource that does not focus on opinion domain.

Table 1. Overall score from general lexicons of opinion

Lexicons	Scores of correct classification
General Inquirer	57,5%
LIWC	54,5%
JeuxDeMots	51,5%

Which combination? Furthermore, we merge the three lexicons in order to identify a general vocabulary of pivot opinion. To perform the optimal merging method, we vary all reliability scores S_1, S_2, S_3, S_4, S_5, S_6, and S_7 from 0 to 3^3 (See Section 3.2). The results are presented in Table 2. We observe that the scores are better using *GeneralInquirer* and/or *LIWC* in addition with the intersection of the three lexicons. In the experiments, tests are performed by taking into account Contextual Windows *CW* (See Table 2). The results show that the identification of polarity are improved when we extend the context window with 4 words preceding and following a target (See Test 7, Table 2).

Table 2. Scores with different combinations of lexicons

Test	CW	S_1	S_2	S_3	S_4	S_5	S_6	S_7	Score
1	0	1	1	1	1	1	1	1	52,5%
2	0	1	1	1	1	0	0	0	59,6%
3	0	3	0	2	0	1	0	1	54,5%
4	0	3	0	0	2	0	1	1	54,5%
5	0	3	2	0	0	1	1	0	63,6%
6	0	3	2	0	0	1	0	0	64,6%
7	4	3	2	0	0	1	0	0	65,6%

Which context? The fourth step defines a specialized vocabulary of opinion SVO from the vocabulary CVO and representative features identified by cross-validation. Firstly, every noun, verb, and adjective words are candidates to be features. Then, we apply both filters described in Section 3.2 in order to select relevant features. We consider a feature as representative if its occurrence in a polarized document is at least 65% in favor of one polarity (positive or negative). This threshold enables to select the most discriminating features. By selecting the whole 527 features obtained by cross-validation (see formulas 2 and 3), we obtain a vocabulary of opinion that significantly improves the identification of document polarity with a value at 81,8% (See Table 3, test a).

Table 3. Overall score with representative features

Test	Representativeness Score	Scores of correct classification
a	1	81,8%
b	RS(O)	91,9%

By weighting the features with their representativeness score, we improve significantly the identification of polarity scores with a value at 91,9% (see Table 3,

---

[3] A larger amplitude by weighting the scores from 0 to 10 was not experimentally relevant.

test b). Note that adding linguistic information (by part-of-speech weighting and by taking into account negation) does not improve classification results.

With these proposed parameters of OPILAND approach, classification of land-use planning documents is relevant. Now the question is: are these parameters suitable for other corpora?

4.3   Is the Opiland Approach Generic?

To show the genericity of our approach, we tested the OPILAND method on three French corpora associated to the DEFT'07 challenge (1) *CorpusP:* 300 anonymous interventions of politicians, (2) *CorpusV:* 994 reviews of video games, (3) *CorpusM:* 3000 reviews related to movies, books, shows, and comics. Similar to results based on land-use planning corpus, General Inquirer lexicon brings better results than the two others. Moreover, the combination of lexicons based on SVO gives best results (see Table 4). Then, these results underline the genericity of OPILAND approaches.

Table 4. Experiments of OPILAND on three corpora

	GVPO	CVO	SVO
CorpusP	54,0%	55,0%	69,8%
CorpusV	67,3%	68,4%	73,7%
CorpusM	77,6%	78,8%	82,6%

4.4   Opiland Approach vs Supervised Approaches

In order to gauge our work, we compared our approach with supervised methods. Table 5 presents results obtained using a bag-of-words approach (without stop words) with Naive Bayes and 10-cross-validation. Note that SVM method provides similar results. The results indicate that supervised methods are inefficient for corpus SENT_100 due to its specificity, the complexity of the used vocabulary, and its small size. Actually, small data sets are still challenging for classical learning approaches. In that way, dedicated approaches such OPILAND are more adapted for this kind of corpus.

Table 5. Naive Bayes classifier results

SENT_100	CorpusP	CorpusV	CorpusM
51,5%	78,6%	90,5%	88,9%

5   Conclusions and Future Work

OPILAND approach based on a general vocabulary of opinion combines three traditional lexicons of opinion. Our proposal improves identification of document polarity. Indeed OPILAND enables the identification of specialized vocabulary, in particular for land-use planning opinion. Experiments show our method has a good behavior for processing of small data sets.

In future work, we plan to experiment OPILAND approach on different kinds of documents such as blogs and websites that contain feelings about territorial planning. Indeed the information regarding the opinion is often insufficient. We plan to detail different types of sentiment [3] on multilingual corpora. For instance, the sentiment model of Hourglass [15] is based on four independent dimensions representing the emotional state of the mind (i.e., Sensitivity, Aptitude, Attention, Pleasantness). Each of the four affective dimensions is characterized by six levels which determine the intensity of the expressed/perceived emotion. Note that this model enables the different affective sentiments to co-exist as compound emotions (e.g., love and aggressiveness).

Acknowledgments. The authors thank Pierre Maurel (IRSTEA, UMR TETIS) for his expertise on the corpus. This work was partially funded by the labex NU-MEV and the Maison des Sciences de l'Homme de Montpellier (MSH-M).

References

1. Bestgen, Y.: Building affective lexicons from specific corpora for automatic sentiment analysis. In: Proceedings of LREC, Trento, Italy, pp. 496–500 (2008)
2. Buléon, P., Méo, G.D.: L'espace social. Armand Colin, Annales de la recherche urbaine (2005)
3. Cambria, E., Havasi, C., Hussain, A.: Senticnet 2: A semantic and affective resource for opinion mining and sentiment analysis. In: FLAIRS Conference. AAAI Press (2012)
4. Debarbieux, B., Vanier, M.: Ces territorialités qui se dessinent. Editions de l'Aube, 267 pages, Datar (2002)
5. Duthil, B., Trousset, F., Roche, M., Dray, G., Plantié, M., Montmain, J., Poncelet, P.: Towards an automatic characterization of criteria. In: Hameurlain, A., Liddle, S.W., Schewe, K.-D., Zhou, X. (eds.) DEXA 2011, Part I. LNCS, vol. 6860, pp. 457–465. Springer, Heidelberg (2011)
6. Esuli, A., Sebastiani, F.: Sentiwordnet: A publicly available lexical resource for opinion mining. In: 5th Conference on Language Resources and Evaluation, pp. 417–422 (2006)
7. Fan, W., Sun, S., Song, G.: Sentiment classification for chinese netnews comments based on multiple classifiers integration. In: Proc. of the Int. Joint Conf. on Comp. Sciences and Optimization, pp. 829–834 (2011)
8. Joshi, A., Balamurali, P., Bhattacharyya, P., Mohanty, R.: C-feel-it: a sentiment analyzer for microblogs. In: Proc. of HLT, pp. 127–132 (2011)
9. Kennedy, A., Inkpen, D.: Sentiment classification of movie reviews using contextual valence shifters. Computational Intelligence 22(2), 110–125 (2006)

10. Klebanov, B., Beigman, E., Diermeier, D.: Vocabulary choice as an indicator of perspective. In: Proceedings of the ACL 2010, Conference Short Papers, ACLShort 2010, pp. 253–257. Association for Computational Linguistics, Stroudsburg (2010)
11. Lafourcade, M.: Making people play for lexical acquisition. In: Proc. 7th Symposium on Natural Language Processing (SNLP 2007), pp. 13–15 (2007)
12. Liu, B.: Sentiment analysis and opinion mining, p. 167. Morgan and Claypool Publishers (2012)
13. Pak, A., Paroubek, P.: Microblogging for micro sentiment analysis and opinion mining. TAL 51(3), 75–100 (2010)
14. Piolat, A., Booth, R., Chung, C., Davids, M., Pennebaker, J.: La version française du dictionnaire pour le liwc: modalités de construction et exemples d'utilisation. Psychologie Française 56(3), 145–159 (2011)
15. Plutchik, R.: The nature of emotions. American Scientist 89(4), 344–350 (2001)
16. Rice, D.R., Zorn, C.: Corpus-based dictionaries for sentiment analysis of specialized vocabularies. In: Proceedings of NDATAD 2013: New Directions in Analyzing Text as Data Workshop 2013, London, England (2013)
17. Plantié, M., Roche, M., Dray, G., Poncelet, P.: Is a voting approach accurate for opinion mining? In: Song, I.-Y., Eder, J., Nguyen, T.M. (eds.) DaWaK 2008. LNCS, vol. 5182, pp. 413–422. Springer, Heidelberg (2008)
18. Torres-Moreno, J., El-Beze, M., Bechet, F., Camelin, N.: Thumbs up or thumbs down? semantic orientation applied to unsupervised classification of reviews. In: Proc. of ACL, pp. 417–424 (2009)
19. Turney, P.: Thumbs up or thumbs down? semantic orientation applied to unsupervised classification of reviews. In: Proc. of ACL, pp. 417–424 (2002)
20. Vanier, M.: Territoires, territorialité, territorialisation - controverses et perspectives. In: PUR, pp. 417–424 (2002)
21. Wiebe, J., Riloff, E.: Finding mutual benefit between subjectivity analysis and information extraction. IEEE Transactions on Affective Computing 2(4), 175–191 (2011)

Cross-Lingual Product Recommendation Using Collaborative Filtering with Translation Pairs

Kanako Komiya, Shohei Shibata, and Yoshiyuki Kotani

Tokyo University of Agriculture and Technology,
2-24-16 Naka-cho, Koganei, Tokyo, 184-8588 Japan
{kkomiya,kotani}@cc.tuat.ac.jp,
50009268024@st.tuat.ac.jp
http://www.tuat.ac.jp/~kotani/

Abstract. We developed a cross-lingual recommender system using collaborative filtering with English-Japanese translation pairs of product names to help non-Japanese buyers visiting Japanese shopping Web sites who speak English. The customer buying histories at an English shopping site and those at another Japanese shopping site were used for the experiments. Two kinds of experiments were conducted to evaluate the system. They were (1) two-fold cross validation where the half of the translation pairs was masked and (2) experiments where the whole of the translation pairs were used. The precisions, recalls, and mean reciprocal rank (MRR) of the system were evaluated to assess the general performance of the recommender system in the former experiment. On the other hand, what kinds of items were recommended in more realistic scenario was shown in the later experiments. The experiments revealed that masked items were found more efficiently than bestseller recommender system and showed that items only at the Japanese site that seemed to be related to buyers' interests could be found by the system in more realistic scenario.

1 Introduction

Japanese pop culture such as that exemplified by manga, anime, and gaming has gained popularity with young generations in recent years. Also, e-commerce has become widely used throughout the world and had enabled people to purchase products from abroad. However, there are some cases where non-Japanese buyers are unable to find products they want through Japanese shopping Web sites because they require Japanese queries. It is particularly difficult to translate product names such as titles of anime or movies using machine translation. We developed a cross-lingual recommender system using collaborative filtering with English-Japanese translation pairs of product names to alleviate this problem and compared the performances between the data of the two languages. We expected that even if non-Japanese buyers could not make Japanese queries to Japanese shopping Web sites, the recommender system would capture their interests from their buying histories and recommend various products that they wanted.

A. Gelbukh (Ed.): CICLing 2014, Part II, LNCS 8404, pp. 141–152, 2014.
© Springer-Verlag Berlin Heidelberg 2014

Since we had no customer buying histories of shopping sites in both English and Japanese, the customer buying histories at an English shopping site and those at another Japanese shopping site were used together for the experiments. The precision, recall, and mean reciprocal rank (MRR) of four types of systems were evaluated to investigate the performance of the cross-lingual recommendation. Two kinds of experiments were conducted to evaluate the system: (1) two-fold cross validation where the half of the translation pairs was masked and (2) experiments where the whole of the translation pairs were used. The experiments revealed that the system with collaborative filtering outperformed bestseller recommender system and products only at the Japanese site that seemed to be related to buyers' interests could be found by the system.

This paper is organized as follows. Section 2 reviews related work on recommendations and other trials to solve the problem where non-Japanese buyers cannot buy some products via Japanese shopping sites. Section 3 explains the outline of the recommendation system that we developed. Section 4 describes the data we used and Section 5 explains the experimental settings. We present the results in Section 6 and discuss them in Section 7. Finally, we conclude the paper in Section 8.

2   Related Work

Recommender systems, which involve recommending items such as products, pages, and articles, have been studied intensively in recent years. Methods of recommendation can be categorized into two types, i.e., content based and collaborative filtering based recommendations. We focused on the latter recommendations in this paper.

There has been much work on similarities between users for collaborative filtering. For example, [Symeonidis et al.2010] defined similarities that took into consideration the strength of links between users on Social Network Services (SNSs) and they used a graph based on similarities between links on SNSs and customer buying histories.

In addition, [Kawamae et al.2009] proposed collaborative filtering based on relationships between users who tended to purchase products ahead of other users by considering changes in users' interests. [Chang and Quiroga 2010] proposed a method of recommendation using bookmark data and content from Wikipedia to consider the serendipity of recommendations.

Our purpose is to help non-Japanese users of Japanese shopping sites who speak English purchase products that are sold only at Japanese shopping sites. [Tsuji et al. 2012] proposed transliteration from Japanese product names to alphabetical queries to alleviate the same problem. We developed a cross-lingual recommender system and exploited another way of recommending products that were only sold at a Japanese shopping site to non-Japanese users. We think that these two methods could be used together.

As far as we know, the closest work to ours is the cross-lingual paper recommender system proposed by [Uchiyama et al. 2011]. However, their system recommended papers rather than products such as books and movies, and they did not use translation pairs but used keywords in English abstract in Japanese papers. We carried out cross-lingual recommendation experiments with simple collaborative filtering by using customer buying histories from Japanese and English shipping sites and translation pairs to assess the general performance of the system. We also discuss the results.

3   Recommendation System to Help Non-japanese Buyers

The recommender system using collaborative filtering with translation pairs consists of six steps. After this, we call buyers from Japanese and English shopping sites, Japanese and English users.

1. The translation pairs are generated in advance.
2. Product names are formatted to identify the items in two sites and the translation pairs more efficiently.
3. Vectors of English and Japanese users are generated. The features of the user vectors are the indices of products and their values are binary: bought or not[1].
4. Similarities between English and Japanese users are calculated. Cosine similarities are used for measures.
5. Japanese users who are similar to English users are selected.
6. Items that Japanese users bought are recommended to the English users. (All the items bought by a Japanese user who is the most similar to an English user are recommended before any items bought by the Japanese user who is the second most similar to him/her.)

3.1   Formatting

The same items at the two sites are identified in the second step using English-Japanese translation pairs because they have different product names according to the language. However, the formats of product names are varied; "The Godfather" in the translation pair list is "Godfather, The" in the customer buying histories of the English shopping site. This kind of difference often causes problems where the products are not correctly identified. Therefore, we formatted the product names to improve product identification.

The procedures are as follows.

1. Expressions within brackets are removed to remove expressions such as "【送料無料】" (freight free), "（予約）" (pre-order), and "(Book 2)".
2. All the Japanese one-byte characters are converted into two-byte characters.
3. All the two-byte Latain characters used in Japanese shopping site are converted into one-byte characters.

---

[1] They were better than assessed-values.

4. All the Latin characters are converted into lower case.
5. Articles are removed.
6. Punctuations, marks such as apostrophes, and marks used in Japanese shopping site such as " $*$ " and " $♪$ " are removed.
7. Expressions like "special edition", "Blu-ray Disc Video", and "DVD コレクターズ BOX1" (DVD collectors BOX1) are removed using stop words list, which included 65 words.

4 Data

Two kinds of data, i.e., customer buying histories at English and Japanese shopping sites and English-Japanese translation pairs, were necessary to recommend items at a Japanese shopping site to non-Japanese buyers. Since we had no customer buying histories of a shopping site in both English and Japanese, customer buying histories from an English shopping site and those from another Japanese shopping site were used together in the experiments.

Rakuten Data Release[2] and GroupLens Research[3] were used for the customer buying histories of Japanese and English shopping sites, respectively. GroupLens Research had data on movies. However, Rakuten Data Release had data on products in various genres. Therefore, the reviews of "Rakuten Books"[4], which included data on movies and books, were extracted for the experiments. We used the reviews as the customer buying histories because they could only be written by users who bought items. We not only used movie data but also book data because the original stories of some movies were written in books and movies were often novelized. The customer buying histories we used were transactions that consisted of three data: user IDs, product names, and their assessed values. The values were graded by each user on a scale from one to five. After this, we call the items that were bought at both the English and Japanese sites, common items. They are important because our system uses collaborative filtering.

The numbers of transactions, users, and transactions per user of the original data and the data of common items according to the languages, are listed in Table 1.

The English-Japanese translation pairs were collected from Japanese Wikipedia[5] using regular expressions. Since they were Japanese movie titles extracted with their original titles, they included not only English translations but also translations of other languages such as French, Spanish, and Chinese. We used a total of 14,327 translation pairs. Three thousand seven hundred and fifteen products in the translation pairs were appeared in the original data of the two languages. However, 496 of them were common items. Finally, there were

---

[2] http://rit.rakuten.co.jp/rdr/index_en.html

[3] http://www.grouplens.org/

[4] http://books.rakuten.co.jp/

[5] http://ja.wikipedia.org/wiki/

Table 1. Number of transactions, users, and averaged items per user of original data and data of common items

Language		Japanese	English
Original	Transactions	346,104	442,845
Original	Users	81,160	62,839
Original	Transactions/user	4.26	7.05
Common Items	Transactions	2,553	32,623
Common Items	Users	1,935	7,372
Common Items	Transactions/user	1.32	4.43

196,327 types of items in the original customer buying histories of English and Japanese shopping sites.

Figures 1, 2 and 3 have examples of the Japanese customer buying histories, the English customer buying histories, and the translation pairs with and without formatting. The ones under lines are formatted.

1　【予約】【楽天限定（国内）】 Crystal Ball × NYLON JAPAN ドリームコラボ・ラブリー トートバッグ付 （ホワイト×ブラック　）　4
1　TSUMORI CHISATO 2010-11 AUTUMN & WINTER COLLECTION　5
1　【楽天管理商品】　4
2　ななちゃんとどうぶつさんワンワンワン　2
2　【送料無料】チャレンジミッケ！(5)　4
2　【送料無料】【ポイント4倍対象商品】Mr.Children / SENSE　1

1　crystalballnylonjapan ドリームコラボラブリートートバッグ付　4
1　tsumorichisato201011autumnwintercollection　5
2　ななちゃんとどうぶつさんワンワンワン　2
2　チャレンジミッケ　4
2　mrchildrensense　1

Fig. 1. Examples of Japanese customer buying histories

5   Experiment

Since the purpose of our research was to recommend items at a Japanese shopping site to non-Japanese buyers, we wanted to recommend items that were only sold at a Japanese shopping site to non-Japanese buyers. However, we could not automatically evaluate the performance of the recommender system using customer buying histories of products that were sold only at a Japanese shopping site because there were no customer buying histories where English users bought at the Japanese site.

```
1   Chasing Amy   5
1   Mr. Smith Goes to Washington   3
1   Return of the Jedi   5
1   Nadja   2
1   Weekend at Bernie's   3
1   Brothers McMullen, The   3

1   chasingamy   5
1   mrsmithgoestowashington   3
1   returnofjedi   5
1   nadja   2
1   weekendatbernies   3
1   brothersmcmullen   3
```

Fig. 2. Examples of English customer buying histories

```
ナイトライダー        Knight Rider
最後の猿の惑星        BATTLE FOR THE PLANET OF THE APES
バラッド        The Elephant's Child
猿の惑星・征服        CONQUEST OF THE PLANET OF THE APES
東京オリンピック        Tokyo Olympiad
宇宙の旅        2001: A Space Odyssey

ナイトライダー        knightrider
最後の猿の惑星        battleforplanetofapes
バラッド        elephantschild
猿の惑星征服        conquestofplanetofapes
東京オリンピック        tokyoolympiad
宇宙の旅        2001spaceodyssey
```

Fig. 3. Examples of translation pairs

Therefore, two kinds of experiment were carried out to evaluate the system. They were (1) two-fold cross validation where the half of the translation pairs was masked and (2) experiments where the whole of the translation pairs were used. The former experiment assessed the general performance of the recommender system. On the other hand, the later experiments showed what kinds of items were recommended in more realistic scenario. The system in both kinds of experiment recommended items from the Japanese shopping site to English users. The bestseller recommender system, where the bestsellers at a Japanese shopping site were recommended in order of the number of times they were sold, was developed and compared. Cross validation of items in English vectors was conducted; half of the common items (defined in Section 4) were masked and checked if they were recommended by the system. In other words, the half of the items was used for the calculation of similarities and the remaining half was used for the check. The Japanese customer buying histories we used in these two kinds of experiments were summarized in Table 1. Here, all the items that were translated from a Japanese product name were regarded as bought if it had more than one translation. The items that were in input vector were not recommended because we thought most users did not purchase the items they already had.

5.1   Cross Validation of Translation Pairs

In this experiment, two-fold cross validation where the half of the translation pairs was masked was conducted. (In other words, four validations were conducted for the combinations of cross validation of English items and that of translation pairs.) The items whose translation pair was masked were assumed to be the items sold only at a Japanese site. Only common items are used both in Japanese and English customer buying histories because the items in the two sites are too different from each other to assess the system using cross validation. The experiment consists of five steps.

1. Only common items are collected for the experiment.
2. Half of the translation pairs is masked for the recommendation phase.
3. The half of the common items, whose translation was masked, is masked for the test in English vectors.
4. The system recommends items at a Japanese shopping site to English users.
5. The system checks if the recommended items have been bought by the English user using uncovered translation pairs.

Here, the system recommends not only items whose translation pair is masked but also items whose translation pair is unmasked. The item-based precisions and recalls at 1^6 and at 10^7 , the user-based precisions at 1 and at 10, and mean reciprocal rank (MRR)[8] of the system were evaluated to assess the general performance of the recommender system.

5.2   More Realistic Scenario

This experiment shows what kinds of items were recommended in more realistic scenario using all the translation pair. Therefore, not only common items but also the items bought only by Japanese were recommended. We will show the examples and discuss them in Section 7 because the items only sold at a Japanese shopping site could not be automatically checked.

---

[6] The system recommended one item for one masked item.
[7] The system recommended ten items for one masked item.
[8] The system recommended any number of items for one masked item.

In addition, common items in English vector were masked for the test and item-based precisions at 1 and recalls are evaluated as reference. Two kinds of methods were used for the evaluations: strict and lenient evaluation methods. The system output is only correct in the strict evaluation when the masked item was recommended to users who bought it. However, the system output is correct in the lenient evaluation when the recommended item was bought by any users in the test set.

6   Results

Tables 2, 3, and 4 summarize item-based precision, recall, and F-measure, user-based precision, and MRR of experiments with cross validation of translation pairs. In addition, Table 5 shows item-based precision, recall, and F-measure of experiment in realistic scenario. "CF" in the tables means collaborative filtering. Bestseller (English) in Table 5 means the case that only English items recommended. (This system could not recommend items sold only at a Japanese site.)

Table 2. Item-based precision, recall, and F-measure of experiment with cross validation of translation pairs

Method	N of Answers	Precision	Recall	F-measure
CF	1	3.03% (326 / 10772)	2.79% (326 / 11690)	2.90
Bestseller	1	1.84% (215 / 11690)	1.84% (215 / 11690)	1.84
CF	10	3.11% (935 / 30017)	8.00% (935 / 11690)	4.48
Bestseller	10	1.71% (2024 / 118462)	17.31% (2024 / 11690)	3.11

Table 3. User-based precision of experiment with cross validation of translation pairs

Method	N of Answers	Precision
CF	1	3.73%
Bestseller	1	0.82%
CF	10	9.47%
Bestseller	10	7.57%

Table 4. MRR and rank of experiment with cross validation of translation pairs

Method	MRR	Rank
CF	0.0836	12th
Bestseller	0.0344	29th

Table 5. Item-based precision, recall, and F-measure of experiment in realistic scenario

Method	Strictness	Precision	Recall	F-measure
CF	Strict	1.58% (74 / 4678)	1.48% (74 / 4998)	1.53
Bestseller	Strict	0.00% (0 / 4998)	0.00% (0 / 4998)	0.00
Bestseller (English)	Strict	0.90% (45 / 4998)	0.90% (45 / 4998)	0.90
CF	Lenient	14.92% (698 / 4678)	13.97% (698 / 4998)	14.43
Bestseller	Lenient	0.00% (0 / 4998)	0.00% (0 / 4998)	0.00

We defined the item-precision, user-based precusion, and recall as follows:

$$R = \frac{N_{item}}{Np_{item}} \tag{1}$$

$$R = \frac{N_{user}}{Np_{user}} \tag{2}$$

$$R = \frac{N_{item}}{Nr} \tag{3}$$

where N_{item}, Np_{item}, N_{user}, Np_{user}, and Nr correspond to the number of the correct items that the system output, the number of items that the system output, the number of the users the correct items are recommended, the number of all the users whom items are recommended by the system, and the number of all the masked items. Here, $\frac{1}{k}$ is added to N_{user} when one item is correct (k is the number of the items that are masked for the user in question). Thus, user-based precision can be 100 % only when all the items of all the users correctly recommended.

In addition, MRR was calculated as:

$$MRR = \frac{1}{N} \sum_{i=1}^{N} \frac{1}{rank(i)}, \tag{4}$$

where, N denotes the number of test data (users) and $rank(i)$ denotes the highest ranking of correct products that are recommended to user i. The $rank(i)$ is zero when none of the items recommended to user i are correct.

Rank of Table 4 shows the rank the correct items are generally recommended.

7   Discussion

7.1   Cross Validation of Translation Pairs

Tables 2, 3, and 4 show the collaborative filtering system outperformed the bestseller recommender system in all the criteria except recall at 10 in the experiments with cross validation of translation pairs. In addition, although the recall at 10 of bestseller recommender system (17.1 %), much outperformed that of the collaborative filtering system (8.00 %), it is only because the bestseller recommender system had recommended many items; it had recommended three times as many as the collaborative filtering system recommended. Consequently, the precision (1.71 %) got much smaller than that of collaborative filtering (3.11 %) and the same held for F-measures. All the difference is significant according to a chi square test. The level of significance in the test was 0.05.

Table 4 shows that the 12th items the collaborative system recommends was generally correct although 29th items the bestseller recommender system recommends generally correct. This means that the bestseller recommender system has to wait more than twice as long as the collaborative filtering system has to.

7.2   More Realistic Scenario

Table 5 shows that the performance of the collaborative filtering system is low. In particular, the low values for lenient evaluations of recommendations from the Japanese shopping site to English users indicate how different they are. The reason for the poor performances of recommendations from the Japanese shopping site is the difference in the items the two shopping sites sell.

However, none of the bestseller recommender system recommended was bought by any English users. In addition, not only the strict precision and recall but also those of lenient evaluation were 0 %. Moreover, the bestseller recommender system could recommend correctly only 45 items even if the system recommended only items that have translation pairs. These results revealed that the naive collaborative filtering could recommend the items that the English buyers bought comparing with the bestseller recommender system.

Since the purpose of our research was to recommend items at a Japanese shopping site to non-Japanese buyers, the system would be successful if the system could recommend products only at the Japanese site that seemed to be related with buyers' interests, even though the precision or recalls were low.

Therefore, we will finally present examples of recommender system outputs. For users who bought "Fantasia", "Wizard of Oz, The", "Pinocchio", "Alice in Wonderland", "Snow White and the Seven Dwarfs", "Beauty and the Beast", and "Cinderella", the system recommended "The Aristocats", "The Little Mermaid", "Dumbo", "ねむれるもりのびじょ (Sleeping Beauty)", "Alice's Adventures in Wonderland", "101 ぴきわんちゃん (101 Dalmatians)", "しらゆきひめ (Snow White)", "風の谷のナウシカ 7 巻セット(Nausicaä of the Valley of the Wind Set: Volumes 1 to 7)", "Peter Pan", "Toy story 3", and "プーさんをさがせ (Look and Find Pooh)". Here, the English names were the products that had translation pairs and the Japanese names were those that did not have them. In addition, "バック・トゥ・ザ・フューチャー Part 3 (Back to the Future Part 3)" or "バックトゥザフューチャー 25th アニバーサリー (Back to the Future 25th Anniversary)" were recommended to users who bought "Back to the Future". These results indicated that even if the recommended items were determined to be incorrect by the system, they included many correct answers.

In addition, "Nausicaä of the Valley of the Wind Set: Volumes 1 to 7" and "Back to the Future 25th Anniversary" were products that were sold only at the Japanese shopping site. They also seemed to be related to buyers' interests. Therefore, we think that simple collaborative filtering between two shopping sites in different languages could provide many effective recommendations.

These items could not be recommended by the bestseller recommender system at all[9].

However, "Back to the Future Part 3" was also sold at the English shopping site, as was "Back to the Future Part III". The system could not identify they were the same products. We think that the more accurate translation pairs, the fewer there will be of these kinds of errors. Formatting using the edit distance was tried but it increased erroneous translations; "天使と悪魔(Angels & Demons)" was identified as "天使と小悪魔 (Literally meaning angels and demons, but they are totally different work)". More precise formatting of product names should be investigated in the future.

In addition, we plan to perform domain adaptation in a scenario of using a small amount of English customer buying histories in addition to Japanese customer buying histories in the future.

8   Conclusion

We developed a system that recommend products at a Japanese shopping site to non-Japanese users using naive collaborative filtering. The same items in Japanese and English sites were identified using translation pairs because the items had the different product names according to the language. Our experiments revealed that the collaborating filtering system outperformed bestseller recommender system. They also revealed that products only at the Japanese site that seemed to be related to buyers' interests could be found by the system. In addition, even if the recommended items were determined to be incorrect by the system they included many answers that seemed to be correct. We think that simple collaborative filtering between two shopping sites in different languages could provide many effective recommendations.

References

[Chang and Quiroga 2010] Chang, P.-C., Quiroga, L.M.: 2010. Using wikipedia's content for cross-website page recommendations that consider serendipity. In: Proceedings of International Conference on Technologies and Applications of Artificial Intelligence, pp. 293–298 (2010)

---

[9] For example, the bestsellers were 1. One Piece (Japanese manga), 2. This Is It (CD) 3. Cath Kidston "HELLO !" FROM LONDON, 4. 鋼の錬金術師 (Japanese manga, English title is "Fullmetal Alchemist"), 5. 体脂肪計タニタの社員食堂 (Japanese cookbook), and so on.

[Kawamae et al.2009] Kawamae, N., Sakano, H., Yamada, T., Ueda, N.: Collaborative
    filtering focusing on the dynamics and precedence of user preference. The Trans-
    actions of the Institute of Electronics, Information and Communication Engineers
    D J92-D(6), 767–776 (2009) (in Japanese)
[Symeonidis et al.2010] Symeonidis, P., Tiakas, E., Manolopoulos, Y.: Transitive node
    similarity for link prediction in social networks with positive and negative links.
    In: Proceedings of the ACM Conference Series on Recommender Systems 2010,
    pp. 183–190 (2010)
[Tsuji et al. 2012] Tsuji, R., Nemoto, Y., Luangpiensamut, W., Abe, Y., Kimura, T.,
    Komiya, K., Fujimoto, K., Kotani, Y.: The transliteration from alphabet queries
    to japanese product names. In: Proceedings of PACLIC 2012, pp. 490–496 (2012)
[Uchiyama et al. 2011] Uchiyama, K., Nanba, H., Aizawa, A., Sagara, T.:
    Osusume: Cross-lingual recommender system for research papers. In: Proceedings
    of the 2011 Workshop on Context-awareness in Retrieval and Recommendation,
    pp. 39–42 (2011)

Identifying a Demand Towards a Company in Consumer-Generated Media

Yuta Kikuchi[1], Hiroya Takamura[2], Manabu Okumura[2], and Satoshi Nakazawa[3]

[1] Interdisciplinary Gradient School of Science and Engineering,
Tokyo Institute of Technology,
4295 Nagatsuta Midori-ku Yokohama, Japan, 226-8503
kikuchi@lr.pi.titech.ac.jp
[2] Precision and Intelligence Laboratory,
Tokyo Institute of Technology,
4295 Nagatsuta Midori-ku Yokohama, Japan, 226-8503
{takamura,oku}@lr.pi.titech.ac.jp
[3] Knowledge Discovery Research Laboratories,
NEC Corporation,
47 8916 Takayama Ikoma Nara, Japan, 630-0101
s-nakazawa@da.jp.nec.com

Abstract. Demands in consumer-generated media (CGM) with regard to a product are useful for companies because these demands show how people want the product to be changed. However, there are many types of demand, and the demandee is not always the company that produces the product. Our objective in this study is to identify the demandees of demands in CGM. We focus on the verbs representing the requested actions and collect them using a graph-based semi-supervised approach for use as the features of a demandee classifier. Experimental results showed that using these features improves the classification performance.

1 Introduction

The values and behavioral patterns of people have become increasingly diversified in modern society. For example, there is increased attention on CGM such as blogs, Internet forums, and social networking services (SNS). Through CGM, we can directly hear the voices of the people. There is a growing need for companies to grasp the interests and opinions of ordinary people from their subjective expressions so that they can measure consumer satisfaction and utilize it for their advertising strategies or product development. The techniques for analyzing such information are becoming ever more important.

There are several types of opinion, such as evaluation, sentiment, demand, criticism, and so on. In this paper, we focus on demand. Demands do not simply indicate a positive or negative sentiment on behalf of a person: they also show how people want a product to be changed.

One of the most important components of a demand is its *demandee*, i.e., the person or company who is asked to perform the action. Demands are valuable for

A. Gelbukh (Ed.): CICLing 2014, Part II, LNCS 8404, pp. 153–163, 2014.

demandees because they (the demandees) can improve their products or services by taking heed of them. However, the demandee is seldom explicitly specified in the actual demand. Let us consider two example sentences pertaining to a mobile phone[1].

- 充電 機能 を   改善 して 欲しい.
  recharging function (accusative) improve   want

  (I want the recharging function to be <u>improved</u>.)

- おすすめ なので , 購入を   検討 して 欲しい .
  recommend so   buying (accusative) think   want

  (I recommend it, so <u>think</u> about buying it.)

The demandee of the first sentence is the company that produces the phone and that of the second sentence is whoever happens to be reading the sentence. In a situation in which a company is collecting demands about their particular product, they want to extract only sentences of the first type. Sentences of the second type also represent a demand, but they are not so important for the company because they do not carry any immediately useful information, e.g., details on what part of the product should be changed. While there are several studies about extracting demands or identifying the topic of demands, the task of identifying demandees has thus far never been addressed.

In this paper, for a given demand sentence that is written about a particular product (the *product in question*), we address the task of determining whether the demandee of the sentence is the company that produces the product in question. Demand sentences are collected from blog articles mentioning the product in question.

Previous studies have shown that modal verbs can be helpful in collecting demand sentences [7,10]. However, when we want to collect demand sentences that request some action of their demandees, we have to use more specific patterns. In Japanese, we can use the phrase "*verb* te-hoshii", which is used when the writer wants the demandees to do something. The phrase "te-hoshii" roughly corresponds to the English phrase "want something to be *verb*-ed (*past particle*)".

In the next section, we give a brief overview of related work. In Section 3, we describe the various types of demand sentences in order to clarify on which part of the demand sentence processing we are going to work. In Section 4, we describe our approach. We construct a demandee classifier that determines whether or not the demandee of a demand sentence is the company that produces the product in question. In constructing the classifier, we focus on verbs representing the requested actions. We collect these verbs using a graph-based semi-supervised approach and use them as features for the demandee classifier. In Section 5, we describe experiments we performed to determine the effectiveness of our approach.

---

[1] In this paper, we work with Japanese text data. In order for readers to understand the Japanese examples, we will add the word level translation and the sentence level translation to the examples throughout this paper.

2   Related Work

There have been many studies on opinion mining and sentiment analysis, and there are a couple of good books ([5] and [4]) that provide a helpful overview of these fields.

By using the technology of sentiment analysis, we can obtain a consumer's subjective sentiment for a particular product. In sentiment analysis, the key points are to identify the sentiment orientation of the words (positiveness or negativeness) [9,6] and to extract the topic (target) of the sentiment [8,6]. Demands take it one step further: they not only indicate a (negative) sentiment for a product but also concretely describe how people want the product to be changed.

Yamamoto et al. proposed an approach for identifying a demand sentence and the reason for the demand using an open-ended questionnaire [11]. Kanayama and Nasukawa tackled the problem of detecting demands as a product aspect that a user wants [3] and proposed a pattern induction method to increase the coverage of the automatic detection of that aspect. Goldberg et al. built the WISH corpus, derived from New Year's wishes [1], and then the *wish detector*, which leverages wish templates learned from the corpus. Ramanand et al. showed that modal verbs are the key to extracting demands [7], and Wu and He utilized modal verbs and pattern mining to improve the precision [10].

While there have been several studies on extracting demands or identifying the topic of demands, identifying demandees has thus far never been addressed. Furthermore, the task of identifying demandees is different from the task of topic (target) identification in that demandees are often not explicitly stated in the demand sentences, while topics in a topic identification task usually are. As we will describe in the next section, the task of identifying demandees hinges on determining whether or not the demandee of a demand sentence for a given product is a company that produces or is related to the product.

3   Types of Demand Sentences

In this section, we describe the various types of demand sentences. Let us consider blog articles written about product **A** of company **X**. There are several types of demands:

(i)   充電 機能 を   改善 して 欲しい.
     recharging function (accusative) improve   want

(I want the recharging function to be <u>improved</u>.)

(ii)   製品 A も   良い   が ,   製品 B の   新作 も
     product A (nominative) good although   product B (genitive) new version (nominative)
発売 して 欲しい.
release   want

(Although product **A** is good, I also want the new version of product **B** to be <u>released</u>.)

(iii) 企業 X に は　　もっと　　　情報 を　　　　提供　して　欲しい
company X (dative)　more　information (accusative)　provide　　　want

(I want company **X** to <u>provide</u> more information.)

(iv) 企業 Y に も　,　　製品 A の　　　　ような もの を　　　発売　して
company Y (dative)　product A (genitive)　something like (accusative)　release
欲しい.
want

(I hope company **Y** will also <u>release</u> something like product **A**.)

(v) おすすめ な の で ,　　購入 を　　検討 して 欲しい .
recommend　so　　buying (accusatve)　think　　　want

(I recommend it, so <u>think</u> about buying it.)

(vi) 早く　返して !! 欲しい .
quickly　return　want

(<u>Return</u> it quickly!!)

The example sentence of type (i) mentions product **A**; the demandee is company **X**. The demandee of the example sentence of type (ii) is also company **X**, but it mentions another product, **B**. The example sentence of type (iii) does not have any target product because the demand is towards **X** itself regarding its general behavior. The demandee of the example sentence of type (iv) is another company, **Y**. The demandee of the example sentences of types (v) and (vi) is readers of the article or friends of the author.

We consider two different situations here. In the first situation, where company **X** wants to collect only demands that directly mention their product **A**, (i) is the only type that **X** needs. In the second situation, where company **X** also wants to collect demands other than those for product **A**, they need the sentences of types (i)-(iv). The example sentences of types (ii)-(iv) are roughly related to product **A**, and the sentences may also be informative for **X**. We run experiments with settings corresponding to these two situations in Section 5.

We found that the verbs in demand sentences are useful for identifying the demandees. For example, the verb *improve* is often used for companies, and the verb *bring* is often used for others. Therefore, we use a verb in a demand sentence as a feature for classifying a demandee. However, there are many possible verbs appearing in a sentence. We have to consider arbitrary verbs to build a good classifier with a wide coverage. Therefore, we used a semi-supervised method to construct a large verb lexicon with demand orientation as the solution (described in Section 4.1).

4   Our Model

In this section, we describe our model. It determines whether or not the demandee of a demand sentence is the company that produces the product in question. We use a support vector machine (SVM), which is a well-known supervised machine learning method.

4.1 Building a Demand Orientation Lexicon by Spin Model

In Section 3, we described how the verb in a demand sentence is useful for identifying the demandee. However, there are many possible verbs appearing in a sentence, and we cannot cover all of them using only training data. In this study, we therefore rely on a graph-based semi-supervised method, which is often used for constructing linguistic resources including polarity lexicons for sentiment analysis. In the construction of a polarity lexicon, positive and negative seed words are given and the positive/negative labels are propagated on the graph in which nodes are associated with words. The graph (lexical network) is often constructed on the basis of a thesaurus, a corpus, and/or glosses in a dictionary.

Among many other alternatives, we chose a graph-based model proposed by Takamura et al. called the spin model [9,2]. The spin model is similar to the label propagation proposed by [12] but differs in that the spin model allows negative weights on the edges. We used the same lexical network used in the previous work [9] and the same algorithm but with different seed words: the verbs likely to be used in demands towards a company, and the other verbs. Henceforth, this graph-based algorithm discriminates between these two types of verbs.

Update Rule of Spin Model. Here, we describe the update rule of the orientations in the spin model. Variable x_i indicates the class. $x_i = 1$ means that i is a verb likely to be used in demands towards a company, and $x_i = -1$ means otherwise. Matrix $W = w_{ij}$ represents the weights between two words, i and j. The update rule for $x_i (i \in L)$ is

$$\bar{x}_i^{new} = \frac{\sum_{x_i} x_i \exp\left(\beta x_i s_i^{old} - \alpha(x_i - a_i)^2\right)}{\sum_{x_i} \exp\left(\beta x_i s_i^{old} - \alpha(x_i - a_i)^2\right)}, \tag{1}$$

where $s_i^{old} = \sum_j w_{ij} \bar{s}_i^{old} \cdot \bar{x}_i^{old}$ and $\bar{x}_i^{new}$ are the respective averages of x_i before and after the update. L is the set of seed words, a_i is the given class label of seed word i, and α is a positive constant. This expression means that if $x_i (i \in L)$ is different from a_i, the state is penalized. For the non-seed words $(i \in L)$, $\alpha(x_i - a_i)^2$ in the exponentials of both the numerator and the denominator is ignored.

Initially, the averages of the seed words are set according to their given class label. The other averages are set to 0.

The words with high final average values are classified as being likely to be used in the demands towards a company. The words with low final average values are classified as other verbs.

Seed Words. In this study, we leveraged product review texts to collect the seed verbs likely to be used in demands towards a company. Note that we are attempting to extract demands from general CGM text, i.e., we are not limited to just review text. Our preliminary investigation showed that, in review texts, demands that ask for some action are often directed towards a company. We can thus collect seed words efficiently by manually choosing verbs out of the ones found in product reviews. Note that we used product reviews only for collecting seed words. In the actual experiments, we use blog articles that contain various types of demand sentences (described in Section 3). For seed words of the opposite class, we manually chose verbs out of the ones that do not mention any product in blogs. We show the English translations of the original Japanese seed words that we obtained.

- Demands towards companies [2]
improve, upgrade, sell, develop, conceive, consider, advertise, refine, substitute, refer to, supplement, strive, discontinue, increase, produce, release, initiate, be inspired by, bear (in mind)

- Other demands [3]
forgive, invite, presume, agree, understand, finish, live together, explain, be relieved, study, judge, sing, say, go, leave, talk to oneself, write, see, return, praise, look, forget

We used these words as the seeds of the graph-based model. After applying the graph-based model, we can obtain a lexicon consisting of verbs with *demand orientations*. The orientation in the obtained lexicon is used as the feature in our model.

Note that the orientations in this lexicon will distinguish sentence types (i)-(iv) from (v)-(vi) in Section 3. Therefore, we have to leverage contextual information to distinguish sentence type (i) from the others.

4.2   Features

The features used in SVM are as follows. We use the letters a, b, d, p, s, t, and u next to feature names to represent feature combinations in the experiments in the next section.

- Demand orientation (s)
This feature stands for the demand polarity of the verb in a demand sentence. This is a three-dimensional binary feature. Each dimension corresponds to a demand towards a company[4], a demand towards another individual [5], and verbs that do not exist in the obtained lexicon.

- Bag of words (t,d,a,b)
To leverage contextual information, we used the bag-of-words features of sentences. There are four types of information for bag-of-words features: the title of a blog article (t), context before a demand sentence (b), the demand sentence itself (d), and context after a demand sentence (a). We compare combinations of these four features in the experiments.

---

[2] Corresponding Japanese verbs are 改善, 改良, 販売, 開発, 考案, 検討, 宣伝, 工夫, 代替, 真似, 付ける, 頑張る, やめる, 増やす, 作る, 出す, 始める, 見習う, and 持つ, respectively.

[3] Corresponding Japanese verbs are 勘弁, 招待, 推理, 同意, 理解, 終了, 同居, 解説, 安心, 勉強, 判定, 歌う, 言う, 行く, 出る, 呟く, 書く, 見る, 返す, 褒める, 調べる, and 忘れる, respectively.

[4] Sentence types (i)-(iv) in Section 3.

[5] Sentence types (v)-(vi) in Section 3.

- **Terms representing the product in question** (p) We use the occurrence of the terms representing the product in question for each of the regions t, d, a, b. For example, a feature associated with t is 1 if the term representing the product in question appears in region t, otherwise 0. The product in question is given beforehand while creating the dataset, as we will describe in Section 5.
- **Terms representing the other products** (u)
  We use the occurrence of the terms representing the other products (i.e., other than the product in question) for each of the regions t, d, a, b. We regard unknown alphabetical or katakana[6] strings as other products.

5   Experiments

We evaluated our model by comparing feature combinations. The dataset was built using sentences extracted from blog articles. We collected demand sentences with their contextual information from blog articles. First, we selected a list of products in question and collected blogs by *Google blog search*, mentioning one of these products. Next, each demand sentence including the term "te-hoshii" was extracted with its contextual information, as discussed in Section 1. The contextual information consists of the title of the article, up to four sentences before the demand sentence, and up to two sentences after the demand sentence.

We used precision, recall, and the F1 score as evaluation metrics:

$$precision = \frac{tp}{tp+fp}, \tag{2}$$

$$recall = \frac{tp}{tp+fn}, \tag{3}$$

$$F1 = 2 \cdot \frac{precision \cdot recall}{precision + recall}, \tag{4}$$

where tp denotes the number of sentences that the system correctly judged as a demand towards a company, fp denotes the number of sentences that the system incorrectly judged as a demand towards a company, and fn denotes the number of sentences that the system incorrectly judged as a demand towards others.

Table 1. Dataset statistics for Experiment 1

Product in question	No. of true sentences	No. of false sentences	Total
Finepix	67	113	180
Walkman	34	130	164
iPhone	32	150	182
iPod	73	144	217
MacBook Air	15	141	156
Total	221	678	899

---

[6] In Japanese, katakana strings are used mainly for representing loanwords from foreign languages. Therefore, the products with loanwords for their names are written in katakana.

5.1   Experiment 1: Identifying Sentence Type (i)

We annotated demand sentences with a label indicating whether the demandee of the demand sentence is the company that produces the product in question. Table 1 shows the statistics of the dataset. We show the results in Table 2.

Table 2. Results of Experiment 1

Feature combination	Precision	Recall	F1
td	0.604	0.421	0.496
dab	0.549	0.281	0.371
tdab	0.580	0.394	0.469
td+spu	0.616	0.443	0.516
dab+spu	0.595	0.326	0.421
tdab+spu	0.589	0.403	0.478
td+s	0.624	0.421	0.503
td+p	0.603	0.425	0.499
td+u	0.595	0.398	0.477

The letters *a, b, d, p, s, t, and u* refer the features in Section 4.2. The top three rows show the performance when we used only the bag-of-words features. According to these results, the title (*t*) and the demand sentence (*d*) are most important. And it also supposes that after and before context information are similar between true and false sentences. We collect the sentences indicating a demand. Hence, whether the label of the demand sentences is true or false, the topic of them are almost same , i.e., impression of products.

The middle three rows show the performance when we added the other three features. These results show that the additional features, including the demand orientation of verbs, improve the performance. However, this improvement is not statistically significant. The bottom three rows show the performance when we individually added three features to the best performed bag-of-words feature (*td*).

5.2   Experiment 2: Identifying Sentence Types (i)-(iv)

With Experiment 1, we assumed that the companies want to collect only demands that directly mention the product in question. In other words, Experiment 1 aims to identify sentence type (i) from Section 3. However, companies sometimes want to collect not only demands directly mentioning the product in question but also other types of demands that may be useful. In this section, our model tries to identify sentence types (i)-(iv).

Sentences will be annotated with the "true" label if they are sentence types (i)-(iv) and "false" otherwise. Table 3 shows the statistics of the annotated dataset. We show the results in Table 4.

Table 3. Dataset statistics for Experiment 2

Product in question	No. of true sentences	No. of false sentences	Total
Finepix	126	54	180
Walkman	121	43	164
iPhone	92	90	182
iPod	148	69	217
MacBook Air	80	76	156
Total	567	332	899

Table 4. Results of Experiment 2

Feature combination	Precision	Recall	F1
td	0.797	0.887	0.840
dab	0.784	0.859	0.823
$tdab$	0.793	0.871	0.830
$td+spu$	0.811	0.884	0.846
$dab+spu$	0.798	0.862	0.829
$tdab+spu$	0.789	0.889	0.836
$td+s$	0.805	0.875	0.839
$td+p$	0.797	0.885	0.839
$td+u$	0.805	0.891	0.846

The overall results are roughly similar to those of Experiment 1. The top three rows show that the bag-of-words features of the sentences before and after the demand sentence are more effective than in Experiment 1. This is because, in Experiment 2, the classifier can focus only on whether the context is related to a company or not; therefore, the scores including ab are improved. The improvement from adding three features (s,p,u) is not statistically significant.

The main difference from Experiment 1 is in the bottom three rows. These results show that the most effective feature is not the demand orientation. We think the reason might be the same as above and simply boil down to effective contextual information for this experiment.

Since the classifier can focus only on whether the demand is related to a company or not, the appearance of unknown alphabetical or katakana strings (other products, and sometimes other companies) becomes a more useful feature than the demand orientation.

Since we only classify demand sentences about products, the corresponding product names seem to be useful information. However, both results ($td+p$ in Tables 2 and 4) show that feature p does not improve the performance. Table 5 shows the occurrence ratio of the product in question in each region (t,d,a,b), where the occurrence ratio is a proportion of the number of sentences that include the product name to the total number of sentences in the region. There is no significant difference between in true and false sentences. It suggests that occurrence of a product name does not help identify demandees, even when we only tackle the classification of demand sentences.

Table 5. Occurrence ratio of a product name in each region

	Experiment 1				Experiment 2			
	t	d	a	b	t	d	a	b
true sentence	0.84	0.22	0.20	0.34	0.63	0.15	0.19	0.30
false sentence	0.52	0.12	0.19	0.25	0.53	0.14	0.19	0.23

We observed that the feature u degrades the performance in experiment 1, but improves the performance in experiment 2. In experiment 1, the feature u occured almost same times in both true and false sentences. Whereas in experiment 2, the occurence of u was less in false sentences than in true sentences. Since this feature indicates occurrences of other products, the results show that the sentences where the demandee is a company[7] contain more product names than the sentences where the demandee is not a company[8].

6   Conclusion

In this study, our objective was to identify the demandee of a demand. We built a classifier that identifies the demandee of a demand sentence. We leveraged the demand orientation of verbs in a demand sentence. To obtain the orientations for verbs, we used a graph-based semi-supervised model. We used contextual information by simple bag-of-words features.

We only collected sentences including the phrase "te-hoshii" because this explicitly indicates a demand for some action. However, only focusing on this phrase has some limitations in terms of coverage. Tackling the problem of collecting other demand sentences from CGM will be the focus of our future work.

References

1. Goldberg, A.B., Fillmore, N., Andrzejewski, D., Xu, Z., Gibson, B., Zhu, X.: May all your wishes come true: a study of wishes and how to recognize them. In: Proceedings of Human Language Technologies: The 2009 Annual Conference of the North American Chapter of the Association for Computational Linguistics, pp. 263–271 (2009)
2. Inoue, J., Carlucci, D.M.: Image restoration using the Q-Ising spin glass. Physical Review E 64(3), 036121 (2001)
3. Kanayama, H., Nasukawa, T.: Textual demand analysis: detection of users' wants and needs from opinions. In: Proceedings of the 22nd International Conference on Computational Linguistics, vol. 1, pp. 409–416 (2008)
4. Liu, B.: Sentiment Analysis and Opinion Mining. In: Synthesis Lectures on Human Language Technologies, Morgan & Claypool Publishers (2012)
5. Pang, B., Lee, L.: Opinion mining and sentiment analysis. Foundations and Trends in Information Retrieval 2(1-2), 1–135 (2008)

---

[7] The sentence types (i)-(iv) in Section 3.
[8] The sentence types (v)-(vi) in Section 3.

6. Qiu, G., Liu, B., Bu, J., Chen, C.: Opinion word expansion and target extraction through double propagation. Computational Linguistics 37(1), 9–27 (2011)
7. Ramanand, J., Bhavsar, K., Pedanekar, N.: Wishful thinking - finding suggestions and 'buy' wishes from product reviews. In: Proceedings of the NAACL HLT 2010 Workshop on Computational Approaches to Analysis and Generation of Emotion in Text, pp. 54–61. Association for Computational Linguistics, Los Angeles (2010)
8. Stoyanov, V., Cardie, C.: Topic identification for fine-grained opinion analysis. In: Proceedings of the 22nd International Conference on Computational Linguistics (Coling 2008), Manchester, UK, pp. 817–824. Coling 2008 Organizing Committee (August 2008)
9. Takamura, H., Inui, T., Okumura, M.: Extracting semantic orientations of words using spin model. In: Proceedings of the 43rd Annual Meeting on Association for Computational Linguistics, ACL 2005, pp. 133–140. Association for Computational Linguistics, Stroudsburg (2005)
10. Wu, X., He, Z.: Identifying wish sentence in product reviews. Journal of Computational Information Systems 7(5), 1607–1613 (2011)
11. Yamamoto, M., Inui, T., Takamura, H., Marumoto, S., Otsuka, H., Okumura, M.: Extracting demands and their reasons in answers to open-ended questionnaires. In: Proceedings of the 13th Annual Meeting of The Association for Natural Language Processing (2007) (in Japanese)
12. Zhu, X., Ghahramani, Z.: Learning from labeled and unlabeled data with label propagation. Technical report, Carnegie Mellon University (2002)

Standardizing Tweets with Character-Level Machine Translation

Nikola Ljubešić[1], Tomaž Erjavec[2], and Darja Fišer[3]

[1] University of Zagreb, Faculty of Humanities and Social Sciences, Zagreb, Croatia
`nikola.ljubesic@ffzg.hr`
[2] Jožef Stefan Institute, Department of Knowledge Technologies, Ljubljana, Slovenia
`tomaz.erjavec@ijs.si`
[3] University of Ljubljana, Faculty of Arts, Ljubljana, Slovenia
`darja.fiser@ff.uni-lj.si`

Abstract. This paper presents the results of the standardization procedure of Slovene tweets that are full of colloquial, dialectal and foreign-language elements. With the aim of minimizing the human input required we produced a manually normalized lexicon of the most salient out-of-vocabulary (OOV) tokens and used it to train a character-level statistical machine translation system (CSMT). Best results were obtained by combining the manually constructed lexicon and CSMT as fallback with an overall improvement of 9.9% increase on all tokens and 31.3% on OOV tokens. Manual preparation of data in a lexicon manner has proven to be more efficient than normalizing running text for the task at hand. Finally we performed an extrinsic evaluation where we automatically lemmatized the test corpus taking as input either original or automatically standardized wordforms, and achieved 75.1% per-token accuracy with the former and 83.6% with the latter, thus demonstrating that standardization has significant benefits for upstream processing.

Keywords: twitterese, standardization, character-level machine translation.

1 Introduction

This paper deals with the problem of processing non-standard language for smaller languages that cannot afford to develop new text processing tools for each language variety. Instead, language varieties need to be standardized so that the existing tools can be utilized with as little negative impact of the noisy data as possible. Slovene, the processing of which is difficult already due to its highly inflecting nature, is even harder to process when orthographic, grammatical and punctuation norms are not followed. This is often the case in non-standard and less formal language use, such as in the language of tweets which is becoming a predominant medium for the dissemination of information, opinions and trends and as such an increasingly important knowledge source for data mining and text processing tasks. Another important characteristics of twitterese is that it is rich

A. Gelbukh (Ed.): CICLing 2014, Part II, LNCS 8404, pp. 164–175, 2014.
© Springer-Verlag Berlin Heidelberg 2014

in colloquial, dialectal and foreign-language elements, causing the standard text processing tools to underperform.

This is why we propose an approach to standardizing Slovene tweets with the aim of increasing the performance of the existing text processing tools by training a character-level statistical machine translation (CSMT) system. CSMT has recently become a popular method for translating between closely related languages, modernizing historical lexicons, producing cognate candidates etc. The specificity of CSMT is that the translation and language model are not built from sequences of words, but characters. In all experiments we use the well-known Moses system[1] with default settings if not specified differently. In order to minimize the human input required, we explore the following strategy: we produce a manually validated lexicon of the 1000 most salient out-of-vocabulary (OOV) tokens in respect to a reference corpus, where the lexicon contains pairs (original wordform, standardized wordform). We also annotate a small corpus of tweets with the standardized wordform and use the lexicon resource for training the CSMT system and the corpus for evaluating different settings. We compare the efficiency of normalizing a lexicon of most-salient OOV tokens to the standard approach of normalizing running text. Finally, we also manually lemmatize our test corpus in order to evaluate how much the standardization helps with the task of lemmatization. The datasets used in this work are made available together with the paper[2].

The rest of this paper is structured as follows: Section 2 discusses related work, Section 3 introduces the dataset we used for the experiments, Section 4 gives the experiments and results, while Section 5 concludes and gives some directions for future work.

2   Related Work

Text standardization is rapidly gaining in popularity because of the explosion of user-generated text content in which language norms are not followed. SMS messages used to be the main object of text standardization [1,2] while recently Twitter has started taking over as the most prominent source of information encoded with non-standard language [3,4].

There are two main approaches to text standardization. The unsupervised approach mostly relies on phonetic transcription of non-standard words to produce standard candidates and language modeling on in-vocabulary (IV) data for selecting the most probable candidate [4]. The supervised approach assumes manually standardized data from which standardization models are built.

Apart from using standard machine learning approaches to supervised standardization, such as HMMs over words [2] or CRFs for identifying deletions [5], many state-of-the-art supervised approaches rely on statistical machine translation which defines the standardization task as a translation problem. There has been a series of papers using phrase-based SMT for text standardization [1,3]

---

[1] http://www.statmt.org/moses/
[2] http://www.cicling.org/2014/data/156/

and, to the best of our knowledge, just two attempts at using character-level SMT (CSMT) for the task [6,7]. Our work also uses CSMT but with a few important distinctions, the main one being data annotation procedure. While [6,7] annotate running tweets, we investigate the possibility of extracting a lexicon of out-of-vocabulary (OOV) but highly salient words with respect to a reference corpus. Furthermore, we apply IV filters on the n-best CSMT hypotheses which proved to be very efficient in the CSMT approach to modernizing historical texts [8]. Finally, we combine the deterministic lexicon approach with the CSMT approach as fallback for tokens not covered by the lexicon.

3   Dataset

The basis for our dataset was the database of tweets from the now no longer active aggregator sitweet.com containing (mostly) Slovene tweets posted between 2007-01-12 and 2011-02-20. The database contains many tweets in other languages as well, so we first used a simple filter that keeps only those that contain one of the Slovene letters č, š or ž. This does not mean that there is no foreign language text remaining, as some closely related languages, in particular Croatian, also use these letters. Also it is fairly common to mix Slovene and another language, mostly English, in a single tweet. However, standard methods for language identification do not work well with the type of language found in tweets, and are also bad at distinguishing closely related languages, especially if a single text uses more than one language. In this step we also shuffled the tweets in the collection so that taking any slice will give a random selection of tweets, making it easier to construct training and testing datasets.

In the second step we anonymized the tweets by substituting hashtags, mentions and URLs with special symbols (XXX-HST, XXX-MNT, XXX-URL) and substituted emoticons with XXX-EMO. This filter is meant to serve two purposes. On the one hand, we make the experimental dataset freely available and by using rather old and anonymized tweets we hope to evade problems with the Twitter terms of use. On the other, tweets are difficult to tokenize correctly and by substituting symbols for the most problematic tokens, i.e. emoticons, we made the collection easier to process.

We then tokenized the collection and stored it in the so called vertical format, where each line is either an XML tag (in particular, <text> for an individual tweet) or one token. With this we obtained a corpus of about half a million tweets and eight million word tokens which is the basis for our datasets.

3.1   Support Lexicons

As will be discussed in the following sections, we also used several support lexicons to arrive at the final datasets for our experiments. In the first instance, this is Sloleks[3] [9], a CC-BY-NC available large lexicon of Slovene containing

---

[3] http://eng.slovenscina.eu/sloleks/opis

the complete inflectional paradigms of 100,000 Slovene lemmas together with their morphosyntactic descriptions and frequency of occurrence in the Gigafida reference corpus of Slovene. We used only wordforms and their frequency from this lexicon, not making use of the other data it contains. In other words, to apply the method presented here to another language only a corpus of standard language is needed, from which a frequency lexicon, equivalent to the one used here, can then be extracted.

As mentioned, Slovene tweets often mix other languages with Slovene and, furthermore, the language identification procedure we used is not exact. As processing non-Slovene words was not the focus of this experiment, it was therefore useful to be able to identify foreign words. To this end, we made a lexicon of words in the most common languages appearing in our collection, in particular English and Croatian. For English we used the SIL English wordlist[4], and for Croatian the lexicon available with the Apertium MT system[5].

A single lexicon containing all three languages was produced, where each wordform is marked with one or more languages. It is then simple to match tweet wordforms against this lexicon and assign each such a word a flag giving the language(s) it belongs or marking it as OOV.

3.2   Lexicon of Twitterese

The most straightforward way to obtain standardizations of Twitter-specific wordforms is via a lexicon giving the wordform and its manually specified standardized form. If we choose the most Twitter-specific wordforms, this will cover many tokens in tweets and also take care of some of the more unpredictable forms.

To construct such a lexicon, we first extracted the frequency lexicon from the tweet corpus vertical file. We then used Sloleks to determine the 1,000 most tweet-specific words using the method of frequency profiling [10] which, for each word, compares its frequency in the specialized corpus to that in the reference corpus using log-likelihood. These words were then manually standardized, a process that took about three hours, i.e. on the average about 10s per entry, making it an efficient way of constructing a useful resource for standardization. This lexicon makes no attempt to model ambiguity, as a tweet wordform can sometimes have more than one standardization. We simply took the most obvious standardization candidate, typically without inspecting the corpus, which would have taken much more time. Sometimes one word is standardized to several standard words, i.e., a word is mapped to a phrase, so the relation between tokens in tweets and standardized ones is not necessarily one-to-one. Along with manual standardization, words were also flagged as being proper nouns (names), foreign words or errors in tokenization. The first are important as they can be OOV words as regards Sloleks, even though they are in fact standard words, the

---

[4] http://www-01.sil.org/linguistics/wordlists/english/
[5] http://wiki.apertium.org/wiki/
   Bosnian-Croatian-Montenegrin-Serbian_and_Slovenian

second as they are not really the subject of standardization, and the third as an error had been made in up-stream processing, so there is not much point in trying to standardize them. In this way we obtained a lexicon of 1,000 (195 of these flagged) of the most salient tweet-specific wordforms together with their standardized wordform, which constitutes part of the distributed dataset; we henceforth refer to this lexicon as the Training Lexicon, TL.

3.3 Manually Annotated Tweets

For development and testing various approaches we needed a collection of manually annotated tweets with typical Twitterese. We first filtered the vertical corpus file to select only interesting tweets, i.e., discarding those that are written in standard Slovene or have few Slovene words. The filter chooses tweets that have some Slovene words, less than half English words, more Slovene than Croatian words (note that each word can belong to more than one language), and at least a fifth of OOV words. We then took a sample of 10,000 lines from this collection and manually standardized and lemmatized it (the lemmatization was done in order to be able to use perform extrinsic evaluation of our standardization approach on this task as will be explained in Section 4.5). In the process of annotation, certain uninteresting Tweets, in particular the remaining ones in standard or foreign language, were discarded.

This gave us a manually corrected corpus of about 500 tweets and 7.500 tokens. The corpus was then split, one half to serve as the development set (TWEET-DEV), and the other as the test set (TWEET-TEST), both of which are part of the distributed dataset. The non-annotated remainder of tweets from our corpus was used to construct a resource for language modeling and CSMT hypothesis filtering containing in-vocabulary (IV) tokens with frequency higher than 10 only. We refer to this resource as TWEET-IV.

4   Experiments and Results

Our overall approach to tweet standardization is based on standardizing only OOV tokens by applying transformations on them with the goal of producing wordforms identical to the ones produced during manual corpus standardization. Therefore we evaluate our approaches with two types of accuracy on the corpus:

1. ACC-ALL – accuracy on all word tokens in the corpus
2. ACC-OOV – accuracy on OOV word tokens in the corpus

The first measure reports how well we do on the level of complete texts, and the second one how well we do on the tokens we perform our transformations on.We perform all together five sets of experiments.

4.1   CSMT Datasets

The first set of experiments attempts to identify the best subset of our TL lexicon for building the character-level translation model and the best target-language dataset for the character-level language model, along with the order of that language model. We perform evaluation on the TWEET-DEV dataset.

We experiment with all TL entries (ALL) and with TL entries where the original and standardized forms are different (DIFF). The results in Table 1 show that using all entries proves to be more informative than using just the entries where the original and standardized forms differ.

Table 1. Evaluation of the two TL subsets for building the translation model

	ACC-ALL	ACC-OOV
ALL	**0.766**	**0.481**
DIFF	0.754	0.443

Additional experiments with filtering the TL showed slight improvements when removing foreign words and errors in tokenization from the lexicon.

Regarding the order of the language models, we experiment with levels from 2 to 6. The best results are obtained with models of order 6, order 5 consistently producing slightly worse results, while lower-order LMs produce significantly worse results. We use Witten-Bell smoothing while constructing the language models.

We experiment with the following datasets for learning the character-level language model:

1. SLOLEKS – the inflectional lexicon of Slovene language
2. TWEET-IV-TOKEN – tokens from the non-annotated set of tweets with frequency above 10, confirmed in SLOLEKS
3. TWEET-IV-TYPE – types from the TWEET-IV-TOKEN dataset

The results in Table 2 show that significantly better results are obtained when using the TWEET-IV dataset than the SLOLEKS dataset which shows the benefits of using in-domain data. Using tokens rather than types, and thereby giving more probability mass to character sequences found in more frequent words improves the overall accuracy for 2.3 percent.

Table 2. Evaluation of different datasets for the character-level language model

	ACC-ALL	ACC-OOV
SLOLEKS	0.720	0.335
TWEET-IV-TOKEN	**0.766**	**0.481**
TWEET-IV-TYPE	0.743	0.410

4.2 Lower and Upper Bounds

The second set of experiments sets the lower (baseline) and upper bound (best possible performance, given the starting assumptions) of the remaining experiments calculated on the TWEET-DEV dataset.

We define two lower bounds, LB as the accuracy obtained without any intervention in the data while the second one, LB-TOP1, is the result of using the first hypothesis of the CSMT system obtained with the best performing settings from the first set of experiments.

We measure various upper bounds by inspecting n-best hypotheses from the CSMT system. We calculate UB-TOP5, UB-TOP10, UB-TOP20 and UB-TOP50. We calculate an overall upper bound UB as the accuracy if all OOV tokens were correctly standardized. Note that our method only standardizes OOV tokens and so cannot give perfect results, as some IV tokens are sort-of-false-friends between standard and non-standard language.

The results of calculating the lower and upper bounds are presented in Table 3. The LB lower bound shows that 26.6% of all tokens and 62% of OOV tokens require standardization. The LB-TOP1 lower bound, which applies the first CSMT hypothesis on OOV tokens, improves the overall accuracy by 3.2 points and OOV accuracy by 10.1 points.

The upper bounds calculated by taking into account n-best CSMT hypotheses show that most of the remaining correct hypotheses are positioned very high. We get an improvement of 5.4 points when taking into account the next four hypotheses and 3.8 points when inspecting the remaining 45 hypotheses.

The overall upper bound UB shows that the maximum overall accuracy, if all OOV tokens are standardized correctly, is 93.1%. There is a 7.3% gap between the UB-TOP50 and UB upper bound showing that the CSMT approach performs quite well (the difference between LB and UB-TOP50 is 12.4%), but that there is still room for improvement, probably by constructing a larger lexicon, i.e., producing more parallel data. There are 6.9% of tokens $(1-UB)$ that are IV but require standardization, showing that future effort will have to be made in identifying and standardizing those tokens as well.

Table 3. Different lower and upper bounds on TWEET-DEV

	ACC-ALL	ACC-OOV
LB	0.734	0.380
LB-TOP1	0.766	0.481
UB-TOP5	0.820	0.651
UB-TOP10	0.838	0.707
UB-TOP20	0.848	0.739
UB-TOP50	0.858	0.770
UB	0.931	1.0

4.3 CSMT Extensions

In the third set of experiments we compare the results of applying the TL only with different extensions of the basic CSMT approach on the TWEET-DEV dataset:

1. LEXICON – applying the TL only
2. CSMT-TOP1 – using the first hypothesis from the CSMT system (identical to the LB-TOP1)
3. CSMT-FILTER – using the first CSMT hypothesis if confirmed in the TWEET-IV dataset
4. CSMT-TOP5-FILTER –using the first of top 5 hypotheses confirmed in the TWEET-IV dataset
5. LEXICON-CSMT-FILTER – applying the TL and using the CSMT system with the TWEET-IV hypothesis filter as fallback for wordforms not covered in TL

The results of this set of experiments are presented in Table 4. Applying the TL only (LEXICON) performs significantly better than applying the first hypothesis of CSMT (CSMT-TOP1). By taking the first CSMT hypothesis only if confirmed in the TWEET-IV dataset (CSMT-FILTER), the CSMT approach does outperform the LEXICON approach with a small increase in accuracy of less than one point on all tokens and by 1.5 points on OOV tokens. Although LB-TOP1 and UB-TOP5 show that among hypotheses on positions 2-5 for 5.4% of tokens correct standardized wordforms can be found, choosing among the top 5 hypotheses the first one confirmed in the IV filter (CSMT-TOP5-FILTER) does outperform CSMT-TOP1, but underperforms regarding the simpler CSMT-FILTER. A possible explanation could be the fact that we work with a highly inflected language and that producing CSMT hypotheses covered by IV filters, but with wrong endings is pretty easy.

When combining the lexicon and the CSMT approach by using CSMT with the hypothesis filter as fallback on tokens not covered in the lexicon, we obtain the best results that outperform the LEXICON approach by 1.7 points on all tokens and 5.6 points on OOV tokens. With this joint setting we obtain an overall accuracy improvement on the TWEET-DEV dataset of 9.9 points on the whole corpus and 31.3 points on OOV tokens.

Table 4. Evaluation of various approaches to standardizing OOV tokens on TWEET-DEV

	ACC-ALL	ACC-OOV
LEXICON	0.816	0.637
CSMT-TOP1	0.766	0.481
CSMT-FILTER	0.820	0.652
CSMT-TOP5-FILTER	0.789	0.554
LEXICON-CSMT-FILTER	**0.833**	**0.693**

We performed additional experiments with using token-level LMs for reweighting CSMT hypotheses as performed in [6], but without any accuracy improvement. The probable reason is that we are already quite near the CSMT upper bounds calculated in the previous set of experiments. Namely, with the LEXICON-CSMT-FILTER setting we already obtained 80% of the maximum possible improvement in accuracy regarding the 50 best hypotheses produced by CSMT.

4.4   Lexicon vs. Corpus Standardization

The fourth set of experiments compares our lexicon approach of data preparation (LEX) to the standard approach of standardizing running text (COR) as performed in [6]. While all previous sets of experiments were evaluated on the TWEET-DEV dataset, here we use the TWEET-DEV dataset for building the lexicon from running text (the COR approach) and test both the LEX and COR approach on the TWEET-TEST dataset. We also consider this evaluation as final intrinsic evaluation of the LEXICON-CSMT-FILTER procedure constructed on the development set in the first three sets of experiments.

We construct the COR lexicon by taking a comparable amount of pairs of original and standardized forms from the TWEET-DEV dataset to the amount of forms in the TL by counting each entry where the original and standardized forms are identical as 0.5 and each entry where the forms differ as 1. We consider this to be a good estimate of the amount of effort necessary to inspect and possibly standardize each token in both approaches.

We present the results of the CSMT-FILTER and the LEXICON-CSMT-FILTER settings on both approaches in Table 5. We report the lower bound LB again, now calculated on the TWEET-TEST dataset.

The results show that in both approaches the lexicon approach (LEX) outperforms the corpus approach (COR). While the difference when using CSMT-FILTER is below one point, it does get more substantial when combining the lexicon and CSMT.

Comparing the CSMT-only and the joint approach of the two data annotation approaches we observe that, as one would expect, bigger improvement with the joint approach is achieved through the lexicon approach (1.5 points) than through the corpus approach (0.4 points). Nevertheless, using the deterministic approach for exact matches from the training corpus yields improvements on the corpus approach as well and should therefore be practiced.

It is important to note that, when constructing the lexicon from the corpus for the LEXICON-CSMT-FILTER approach, for each original form the most frequent (original form, standardized form) pair is used. The CSMT system is trained on all entries for an original form.

Last but not least, we compute the learning curves for both approaches as depicted in Figure 1. The left figure shows the results on using the CSMT setting with the TWEET-IV filter while the right figure shows the setting which uses filtered CSMT as fallback for the lexicon. The curves show that the LEX approach outperforms the COR approach on all sizes of the training data with

Table 5. Comparison of lexicon standardization to corpus standardization on TWEET-TEST

	ACC-ALL	ACC-OOV
LB	0.750	0.430
LEX-CSMT-FILTER	**0.841**	**0.716**
COR-CSMT-FILTER	0.836	0.701
LEX-LEXICON-CSMT-FILTER	**0.856**	**0.763**
COR-LEXICON-CSMT-FILTER	0.840	0.707

the COR approach slowly catching up as the training set size increases, more significantly in the CSMT-FILTER setting. All learning curves show room for additional improvement by annotating more data.

Annotating just one tenth of the data (100 tokens, below 20 minutes of annotation work) with the LEX approach and applying filtered CSMT already produces a significant improvement of 6.9 points to the LB lower bound which comprises 66% of the overall improvement obtained by the best performing setting of 10.4 points. The difference in accuracy between the LEX and the COR approach at that point is 4.1 points.

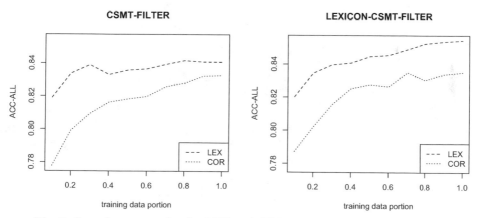

Fig. 1. Learning curves for the LEX and COR approach to data annotation

4.5 Lemmatization Experiment

In order to extrinsically test the effect of our best scoring standardization, we performed a small experiment on a basic but very important task, at least for languages with rich inflectional morphology, namely lemmatization. Lemmatization abstracts away from inflectional variation and is useful for full-text search and dictionary lookup but is at the same time quite complex since Slovene words exhibit a complicated system of endings and stem alternations, dependent on morphological and syntactic features of the word, which makes learning Slovene inflections one of the more daunting tasks for foreign speakers.

Lemmatization was performed with ToTrTaLe [11], a program that tokenizes, transcribes, PoS tags and lemmatizes a Slovene text. The transcription module was developed to standardize historical Slovene texts and uses hand-constructed transcription patterns for this task. For the current experiment we either removed this step (in order to determine how well the system works with the original wordforms) or substituted it with our module for standardization. It should be noted that the PoS tagging step already receives the standardized wordforms, just as lemmatization, and that lemmatization makes use of the PoS tags because it is impossible to determine the lemma of a (at least OOV) Slovene word without it.

Table 6 shows the results of the experiment. The accuracy of lemmatization directly on "raw" words is just over 75%, while lemmatization accuracy on manually (i.e., perfectly) standardized words is almost 92%, so twitterese does indeed have a significant impact on processing of such texts. With automatically standardized words, the accuracy is almost 83.6%, which, as the last two columns show, is 8.5% better than on original data and just about equally worse than with perfect standardization, which should be taken as the upper accuracy bound we could achieve with the standardization. In other words, automatically standardizing the words cuts the absolute error rate by half.

Table 6. Comparison of lemmatization accuracy on original, manually standardized and automatically standardized wordforms from TWEET-TEST

RAW	AUTO	MANUAL	AUTO - RAW	AUTO - MAN
0.750	**0.836**	0.919	0.085	-0.083

5   Conclusions

In this paper we have presented a method for standardizing non-standard text, more specifically Slovene tweets, by using character-level SMT as fallback to lexicon lookup.

We compared the approach of manually standardizing most salient OOV tokens with respect to a reference corpus to the approach of standardizing running text. We have shown that the former produces significantly better results, especially on small training sets. This is an interesting finding given that we work with a highly inflected language with many possible forms that heavily depend on the context. For character-level language models we have shown that in-domain data performs better and that deduplication of tokens should not be performed. In both approaches to producing training data, using perfect match sequences from the parallel data, ie. performing lexicon lookup, and using CSMT only where there is no perfect match, showed to produce best results.

Filtering the first CSMT hypothesis with an in-vocabulary filter proved to be more useful than filtering the top 5, regardless of the fact that many correct hypotheses can be found on those positions. High flectiveness of the language of interest and the danger of producing tokens with different endings covered in the IV filter is one possible explanation.

Finally, with our standardization approach we have shown that lemmatization errors produced on non-standard language are cut by half.

Regarding our future work, our primary goal is to extend our approach to more languages. We additionally plan on investigating a CSMT approach to standardization not limited to tokens, but applied on a wider context. By doing so we hope to deal with the 6.9% of tokens that are IV, but require standardization.

References

1. Aw, A., Zhang, M., Xiao, J., Su, J.: A Phrase-based Statistical Model for SMS Text Normalization. In: Proceedings of the COLING/ACL on Main Conference Poster Sessions, COLING-ACL 2006, pp. 33–40. Association for Computational Linguistics, Stroudsburg (2006)
2. Choudhury, M., Saraf, R., Jain, V., Mukherjee, A., Sarkar, S., Basu, A.: Investigation and Modeling of the Structure of Texting Language. Int. J. Doc. Anal. Recognit. 10(3), 157–174 (2007)
3. Kaufmann, M., Kalita, J.: Syntactic Normalization of Twitter Messages. In: Proceedings of the 8th International Conference on Natural Language Processing, ICON 2010 (2010)
4. Han, B., Cook, P., Baldwin, T.: Lexical Normalization for Social Media Text. ACM Trans. Intell. Syst. Technol. 4(1), 1–5 (2013)
5. Pennell, D., Liu, Y.: Toward text message normalization: Modeling abbreviation generation. In: 2011 IEEE International Conference on Acoustics, Speech and Signal Processing (ICASSP), pp. 5364–5367 (2011)
6. Pennell, D., Liu, Y.: A Character-Level Machine Translation Approach for Normalization of SMS Abbreviations. In: Proceedings of 5th International Joint Conference on Natural Language Processing, pp. 974–982. Asian Federation of Natural Language Processing, Chiang Mai (November 2011)
7. De Clercq, O.E., Desmet, B., Schulz, S., Lefever, E., Hoste, V.: Normalization of Dutch user-generated content. In: Proceedings of Recent Advances in Natural Language Processing, INCOMA, pp. 179–188 (2013)
8. Scherrer, Y., Erjavec, T.: Modernizing historical Slovene words with character-based SMT. In: BSNLP 2013 - 4th Biennial Workshop on Balto-Slavic Natural Language Processing, Sofia, Bulgarie, pp. 2013–2014 (July 2013)
9. Arhar, Š.: Učni korpus SSJ in leksikon besednih oblik za slovenino. Jezik in slovstvo 54(3-4), 43–56 (2009)
10. Rayson, P., Garside, R.: Comparing Corpora Using Frequency Profiling. In: Proceedings of the Workshop on Comparing Corpora, WCC 2000, vol. 9, pp. 1–6. Association for Computational Linguistics, Stroudsburg (2000)
11. Erjavec, T.: Automatic linguistic annotation of historical language: ToTrTaLe and XIX century Slovene. In: Proceedings of the 5th ACL-HLT Workshop on Language Technology for Cultural Heritage, Social Sciences, and Humanities, pp. 33–38. Association for Computational Linguistics, Portland (June 2011)

#impressme: The Language of Motivation in User Generated Content

Marc T. Tomlinson, David B. Bracewell, Wayne Krug, and David Hinote

Language Computer
2435 N. Central Expy, Suite 1200
Richardson, TX 75080
marc@languagecomputer.com

Abstract. An individual's ability to produce quality work is a function of their current motivation, their control over the results of their work, and the social influences of other individuals. All of these factors can be identified in the language that individuals use to discuss their work with their peers. Previous approaches to modeling motivation have relied on social-network and time-series analysis to predict the popularity of a contribution to user-generated content site. In contrast, we show how an individual's use of language can reflect their level of motivation and can be used to predict their future performance. We compare our results to an analysis of motivation based on utility theory. We show that an understanding of the language contained in comments on user generated content sites provides significant insight into an author's level of motivation and the potential quality of their future work.

1 Introduction

Creative inspiration is only part of the puzzle to the successful completion of an endeavor. Quality work requires the setting of goals, belief in your ability to succeed, and proper feedback from the community. Community members can provide critical support at times when we are feeling down or make us to step up to a new challenge by forcing the establishment of complex far-reaching goals. Unfortunately, community members can also reduce an individual's performance through disparaging comments or lulling them into a sense of well-being. Here we present an analysis of a system for determining the motivational implicatures present within a discourse and their effect on motivating an individual to produce quality user generated content.

We cast our approach to understanding the motivational content of a communication in terms of speech acts [1], which provide a theoretical framework to explore the motivational implicatures of an utterance. In this contribution we use the term motivational act to represent utterances by individuals that either reveal their motivation for an action or affect the motivation of another individual. We classify three main types of motivational acts. The first act that we look at are comments which indicate the value of an individual's contribution, or reward (i.e. comments like "great job" indicate positive social value for the individual's work). This act is further refined into separate categories for self-directed rewards, and reward statements directed at other individuals. The second factor looks for evidence that the individual has (or thinks they have) skill or control

A. Gelbukh (Ed.): CICLing 2014, Part II, LNCS 8404, pp. 176–187, 2014.
© Springer-Verlag Berlin Heidelberg 2014

to act within the environment ("I know I can do this"). Lastly, we identify comments that express goals or indicators of a goal orientation in an individual.

The motivational act expressed by each comment is classified using an approach based on distant supervision [2] and twitter. We first create a language model for each motivational act based on the words contained in tweets which have been marked with a hashtag which is used by individuals on twitter to identify tweets with a motivational message.

We then show how the language models trained on twitter can be used to capture the motivational meaning of a dialogue between a contributor and commenter on a user-generated-content site. The motivational acts can be used to predict the amount of effort the individual is likely to expend on future contributions. We use the community rated quality of an individual's subsequent contribution to the site, DeviantArt.com, as a proxy for the amount of effort the individual expended on their submission. Finally, we compare the quality of the classifiers and general framework to an approach based on a utility theory view of motivation which looks at changes in the quality of submissions by that individual over time.

2   Background and Related Work

Previous research has shown how social implicatures present within a group's discussions can provide insight into the quality of a groups contributions [3]. In the spirit of dialogue acts [4–6], social acts focus on the social implicature of the statement and thus more directly relate to the social intentions and goals of individuals. Researchers have recently begun to construct and annotate social acts. Bender et al. [7] create an annotated corpus of social acts relating to *authority claims* and *alignment moves* as well as a broader selection of social acts covering managerial influence, agreement, group affordance and others. These social acts can be used to infer the quality of the work (ratings of Wikipedia articles) which a group produces based on the social interactions between group members. In contrast to that work, we focus on how comments and expressions reveal or affect an individual's motivation and their correlation with the quality of future productions.

Researchers have also examined the quality of user generated comments on blog sites using shallower stylistic features and topical information. For instance Hsu et al. [8] and Khabiri et al. [9] analyzed the popular websites Slashdot and Digg to examine prediction of the probability that a given post will receive high marks from the community. These approaches use a variety of stylistic features, such as word counts, quotes, and hyperlink counts as well as information about the reputation of the individual creating the post, temporal features of the post, and structure features pertaining to the posts location on the page. This line of work does not attempt to address the issue of predicting the quality of future contributions by the same individual.

Alternatively, a considerable amount of work has looked at predicting the popularity of user generated content on the world-wide-web. Much of the work has considered people as one-dimensional products of their social network. For example, the popularity of a given picture posted on Flickr can be derived through an examination of the social network surrounding the poster. Previous work has found that approximately

50% of all favorite markings for a given post are generated by individuals connected to the post [10]. Similarly, Szabo and Huberman[11] predict the future popularity of a YouTube or Digg contribution based on a time-series analysis of its historical popularity. Their approach requires analyzing the first few hours, in the case of Digg, or 10 days, for YouTube, and shows that after that time the content follows a fairly predictable trajectory.

In addition to work in natural language processing and social network analysis, the inspiration for our approach comes from psychological theories of motivation. The predominant theory used for understanding an individual's motivation is based on prospect theory, an extension of utility theory [12]. Kehneman and Tversky discuss three concepts that affect how an individual values a reward. 1) Reference points - rewards are valued in how far they deviate above (positive reward, gain) or below (negative reward, loss) a given reference point; 2) Loss Aversion - avoiding a loss is treated as being more important than an equivalent gain (avoiding a loss of $10 is more important than gaining $10); 3) Diminishing sensitivity - the value of a change is not linear but decreases as the point gets further from the referent ($10 to $20 is a big jump, but $1,000,000 to $1,000,020 doesn't make much of a difference). These factors can be combined to create a model of the expected utility of an action for an individual and correspondingly, an individual's level of motivation to achieve the reward.

Prospect theory can be related to motivation and made more concrete through theories linking goal setting as the establishment of reference points [13]. These theories suggest that an individual's motivation is a function of the expected utility of their actions. Critically, the theory suggests that goals serve to establish reference points. For example, assuming that an individual has a goal to produce great art than the value of intermediate rewards (views, favorites, positive comments by community members) will be judged using the final goal as a reference point. As they get closer to their goal chances for small positive rewards will be more motivating because they are worth comparatively more, while setbacks will be devastating. Importantly, this framework suggests that individual's that meet their goals will be less motivated on future endeavors, because they have little utility. However, in ill-defined environments, such as user-generated content sites it is difficult to know how individuals value likes from the community and what an individual is using for a reference point. In the current work we utilize a model based on prospect theory as a comparison to one based on the motivational implicatures of the language used by individuals on user-generated content forums.

3   Motivational Language Uses

Several theories exist on how different expressions of motivational factors interact with an individual's future performance. Probably the most apparent factor is the setting of goals. Goals are expressions of intentions for a change of state which could require an action on the part of the individual. Examples of goals are, "I want to finish my paper". This goal expresses an intention for a specific action but requires making inferences about the probable rewards for the individual if they are successful. In contrast, a statement such as "I want to be famous" expresses a clear expectation for a reward resulting from some series of actions, but requires inference about the details of the future actions that the individual might precipitate to achieve their goal. Does the individual want to be a rock star or a serial killer? In contrast to explicit goals stated by an individual, goals can also be inferred by other people based on an analysis of the actions carried out by the individual. For example, if an individual repeatedly demonstrates their work to a community, one can infer that the individual likely has a goal to accomplish the end-product of the activity. The setting of goals for both action and inaction has been linked to many different motivational and long-term outcomes [14, 15].

An individual's motivation can also be signaled and inferred through their use of expressions which indicate a reward for an action. This can be derived from their communications and the communications of others. Rewards can come from an individual about their own work, "I really like my drawing", or from other community members, "your work is top-notch". These comments indicate a high value for the individual's action. In contrast, comments such as "you suck" indicate a negative reward for the individual's contribution, lowering its value and an individual's motivation for future work. However, some care needs to be taken in this as some people actually seek out and are motivated by negative comments [16].

The last motivational act is a statement indicating control (or lack thereof) over an action. Individual's that feel that they have control over the outcome of an action are more motivated to perform the action [17]. Individuals express their perceptions of their control over actions through statements such as "its really easy to x" (control) or "i feel i can't do anything right" (lack of control).

4   Detecting Motivational Acts through Distant Supervision

Goals, rewards, and control can be expressed in a myriad of different ways in text, sometimes very clearly "I want to do better", and sometimes only implied. This problem is analogous to that of classifying speech or dialogue acts. In this section we show how distant supervision can be used to create a language model which can identify motivational acts by a speaker.

One notorious problem with automatically recognizing the illocutionary force of an utterance is annotation of a supervised training set. Many utterances are ambiguous or signal multiple overlapping acts, this creates a notoriously difficult annotation task [18, 19]. To solve this problem we explore a novel approach for annotation through the use of a distant supervision framework [2]. Distant supervision involves the use of a small set of annotations that link to a larger knowledge base that contains noisy instances of those annotations. While the approach introduces noise into the system, the noise is mitigated by access to a very large collection of approximate annotations (in our case, millions of tweets containing these tags). Similar approaches have been used in sentiment and emotion analysis [20, 21]. In contrast, our approach learns the motivational implicature of an utterance from an individual's use of hashtags.

Hashtags are words or phrases that are often included in tweets to signify the topic of the tweet. Some hashtags have meanings that can be derived from the words making up the tag (e.g. #mygoal – is used to express goals of an individual), while others are related to Internet memes and require broader cultural knowledge (e.g. #fml used to express negative things happening in an individual's own life). When a user embeds a

Table 1. Example Hash Tags and Tweets for Goals, Control, and Rewards

Motivational Action	Sample Tags	Sample Tweets
Goal	#goalinlife, #mywish	"3 more days of studying"
Control	#dowhatisay, #kissmyfeet	"I defy the law of gravity"
Negative Reward Self	#fml, #crap	"I just locked the keys in my car"
Negative Reward Other	#worstdriverever, #awkward	"It does make me cringe"
Positive Reward Self	#whyismile, #victoryismine	"my cats make me smile"
Positive Reward Other	#ff, #thatsbadass	"Solar panels on the white house"

hashtag in a tweet, twitter provides a link to a page showing a collection of all of the tweets with that hashtag. This makes hashtags very popular for researchers trying to follow trending topics on twitter. To date, most of the research has focused on linking hashtags to topics, or sentiment. However, hashtags also provide annotation for non-semantic topics, such as goals, rewards, and many other social phenomena.

We considered models for identifying goals, perception of control, and rewards based trained from hashtags that were deemed to be relevant to those phenomena. Rewards were subdivided into four categories, positive self-orientation, positive other-orientation, negative self-orientation, and negative other-orientation. The hashtags were taken from lists of trending tags as well as generated through trial and error. An initial list of hashtags was generated by a single annotator who rated each tag on a scale of 1-5 as to the overall degree that tweets using that hashtag represented the concept of interest. This list was then refined by a second annotator.

Examples of hashtags exhibiting each of the characteristics are shown in Table 1. The goals represented by these hashtags are diverse but mostly mundane, examples for #mygoal range from "3 more days of studying #iwillsurvive #4.0 #mygoal" to "Looking for a bigger house By December I wane be out this house in a bigger house #mygoal". The English annotator identified 140 hashtags that were relevant to one of the dimensions (received a relevance score of 4 or 5 on a scale of 1-5).

We used a large collection of tweets (approximately 7.5 million). In our collection hashtags exhibiting control contained the largest number with approximately 315,000 tweets, while we only collected 110,000 tweets which were marked with a hashtag indicating positive rewards for the actions of other individuals, as shown in Table 1. For training and testing purposes we removed all URLs, hashtags, and @users from the tweets. We then discarded tweets that were less than two words long. This is very conservative, because we removed the classifier's ability to directly learn co-occurring hashtags, however we wanted to ensure that we would minimize deficient solutions and maximize the ability of the models to transfer from twitter to other genres of text, such as web forums.

Many of the most successful approaches to dialogue act classification on text have focused on integrating multi-layer models which examine both the utterance content and the surrounding utterances [22]. However, for this initial work we consider only the linguistic content of a single utterance.

Each motivational act detection model was trained to separate tweets that had been tagged with a hashtag that had been identified as signaling the particular motivational

Table 2. Training and testing sizes for N-gram classifiers with resultant accuracy and bias for labeling a tweet as a positive instance of the class. All test and train splits are 50/50 between positive and negative instances.

Motivational Action	# Hashtags	# Train	# Test	Accuracy
Goal	23	83,838	20,960	79.8
Control	18	153,136	38,286	70.2
Negative Reward Self	30	100,996	25,250	68.6
Negative Reward Other	47	157,582	39,396	69.6
Positive Reward Self	8	158,250	39,564	69.3
Positive Reward Other	5	103,948	25,988	78.9

act from a background model that represents tweets containing hashtags which signal one of the other motivational acts. Tweets containing hashtags from multiple acts were not trained or tested on.

We utilized a language model coupled with Naive-Bayes which considered n-grams between 2-4 words in length for each of the different motivational acts. The model compared the probability of all of the n-grams from 2-4 words long in the tweet given the motivational act to the probability of the n-grams given a background model. For evaluation purposes we considered the act to be present if the sequence of words was more likely given the act than in the background model, but for downstream usage, the model output the odds of the act compared to the background model.

The accuracy of the resultant classifiers (show in table 2 suggest that they are adequately capturing the differences between the categories, though there is some obfuscation of the true validity of the labels due to noisy use of the tags by individuals. Inspections of tweets with the labels suggest that many times the labels are used sarcastically. Anecdotally, we also examined a list of the top hashtags associated with instances labeled by our approach and found good generalization to novel hashtags. We looked at a list of the hashtags based on the average confidence of the labels being applied to the tweets containing those tags, we found many reasonable candidate tags. For example, tweets containing the hashtags #day1 and #day2 were among the most likely to be labeled as exhibiting a goal. This suggests that a model which spiders out proposing new hashtags based on co-occurrence and is able to then incorporate those instances would work very effectively and allow for fine tuning of the model.

The classifiers created above are capable of identifying the motivational signals found within an individual's language. In the next section we show how those classifiers can then be transferred to a novel domain and allow us to understand an individual's motivation by examining a complete discourse.

5   Modeling Motivation

In the previous section we showed how an individual utterance can be understood in terms of its motivational meaning. In this section we show one way to build a model

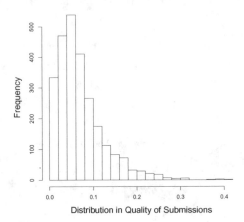

Fig. 1. Distribution of quality on DeviantArt. Quality is measured as the probability of an individual favoriting a contribution by an artist.

of the amount of effort an individual is willing to expend on a future endeavor by examining the motivational acts that the individual and other conversational participants use. For this effort we look at modeling the quality of an individual's contribution to the user generated content website DeviantArt.com.

5.1   Data

We collected and analyzed date from the user generated content site DeviantArt.com. DeviantArt is a community web-forum on which individuals post art-work that they created and receive comments on their posts. The art is very diverse including pictures, cartoons, and needlework, though as the name suggests many of the works are avant-garde. The website allows for artists to post their work and a forum for discussion between the artist and the community members about each contribution. Community members can identify their favorite pieces. The site is very well established and boasts over 2 million subscribers.

The data contained on DeviantArt and related sites is very valuable, because it provides easily inferable goals (respect from community members) as well as explicit statements that can improve these inferences, clear actions (production of art) by community members, and clear rewards for actions (community response, favorites, views). Most importantly, it has an easily inferred timeline of related actions which receive clear feedback at most states.

DeviantArt provides several different ways to assess the success of each piece of art. The website keeps track of the number of views that each piece of art receives, the number of comments it receives, and the number of times that individual's tag it as a favorite. A large percentage of the work examining user generated content looks at predicting the number of views that an item will receive. While previous work has reported that views are not well correlated with comments or favorites on Flickr [10], the views and favorites on DeviantArt are highly correlated ($\rho = .84, p < .001$, because of the distribution of these measures we are reporting the non-parametric rank-order correlation). Also in contrast to Flickr, comments and views, and comments and favorites are only weakly correlated ($\rho = .24$, $\rho = .22, p < .001$, respectively). This could be due to differences between the two communities, or differences in how browsing of new pieces is supported by the website.

Most previous work has looked at predicting the popularity of a piece, in contrast we break the trend by looking at a hybrid measure of the probability that an individual favorites a piece given that they viewed it. We feel that this measure provides a better characterization of the overall quality of a contribution. This is supported by an analysis of the correlation between the popularity of a piece and our measure. The probability that a piece is favorited given that it was viewed is the least correlated measure with the number of views; the number of comments on a piece is second-least correlated. For this paper we refer to our aggregated measure as the quality of a submission, but more research should be done to better characterize the differences in these measures.

To generate our data we sampled from the artists posting content on DeviantArt generating an initial list of 1103 artists. Each artist contributed, on average, 20 contributions for a total of 21,420 pieces of art. The mean number of views per contribution was 1545, while the median was 248. A preliminary analysis of the data showed that the variance in the quality of an artist's contributions was much higher for new artists than for experienced artists. The data suggest very interesting variations among individuals of different popularity and experience levels, but based on this preliminary analysis we restricted our analysis to those contributions where the artist had at least 1000 views and was at least the 20th post by the author. By restricting the data to those with 1000 views we also reduce the noise in the quality measure because of the size of denominator. This reduced our data set to 101 artists, with 2,059 contributions, and 127,622 comments on the artwork.

5.2   Using Comments to Infer Motivation

To understand how comments on a web forum can reflect an individual's level of motivation we created a model, the goals, reward, and control model (GRC), which determined the extent to which an individual's conversation with group members exhibited qualities of a motivated individual based on the motivational acts used within a conversation. We compared this model to a base-line model which used sentiment terms (Sentiment Model) instead of motivational acts and also to a second model which looked at predicting the motivation level of an individual based on the change in quality over time (Utility Model). Finally, we show how the GRC model can generate improved predictions through use of the utility model.

The GRC model uses an analysis of the artist's comments and the comments by other conversational participants to construct a model of the artist's motivation. Each of the comments were first identified as being made by the artist or by another individual. They were then labeled using the language models discussed in section 4 for each of the six motivational acts:Goal, Control, Negative Reward Self, Negative Reward Other, Positive Reward Self, Positive Reward other. Examples are shown in Table 3 For each

Table 3. Example Flickr Comments for Goals, Control, and Rewards

Motivational Action	Sample Comments
Goal	"Hoping to finish this in time for Easter."
Control	"Great details"
Negative Reward Self	"but it is very poor quality"
Negative Reward Other	"It looks a little bit flat with not much contrasts "
Positive Reward Self	"Thank you so much!"
Positive Reward Other	"Stunning photo! Congrats"

comment page the number times that language was used which exhibited each act were aggregated separately for the artist and community members. These twelve features were weighted and combined linearly, $\sum_{1..12} \beta_i F_i$.

Our base-line, sentiment model utilized an equivalent linear regression procedure except in place of the motivational acts we examined the presence or absence of positive and negative sentiment terms in each communication by the artist and the community. We used SentiWordNet [23] taking all terms with a positivity or negativity higher than .5.

Likewise, the utility model uses a linear combination of two factors to predict the quality of a future contribution. Our first factor is the expected utility of the next contribution, we set this as an exponential function of the quality of the current submission, Q_t^λ. We used a λ value of .3 to encode the diminishing sensitivity to higher rewards, though testing showed less than a 1 percent difference for realistic values of λ and a similar patterning of results. The second factor in the model accounts for goal related behavior (establishing the current piece as a goal) and adjusts the utility based on $Q_t - Q_{t-1}$ or $\Delta Q_{t-1,t}$. Individuals that accomplish their goals, $Q_t > Q_{t-1}$ are less motivated. For the results we split this model into one utilizing the first factor (quality) only and one showing the first and second factor combined.

The last model combines the GRC and the Utility Model to look at the interaction between the expected motivation of the individual based on their previous rewards and the way in which the individual and group is discussing their rewards, goals, and perceptions of control. This model replaces $\Delta Q_{t-1,t}$ from the utility model with the linguistic features, F_i.

5.3   Modeling Motivation and Results

For each of the four models we looked at two separate predictions. The first is how well we can predict the quality of the next submission by an artist, Q_{t+1}. Thus for the GRC Model we find the maximum likelihood estimate minimizing the error for the following equation

$$Q_{t+1} = \beta_0 + \sum_{1..12} \beta_i F_i.$$

using least squares and correlate our predicted value $\hat{Q}_{t+1}$ with the actual value Q_{t+1} on held out test data.

We utilized a 100-fold cross-validation procedure to find the correlation between the predicted quality of the next submission and the actual quality for each artist. The data for each artist was distributed randomly across the folds. The results are shown in the first column of Table 4. The GRC model achieves a correlation of .33, which is significantly greater than the baseline sentiment model. The .51 obtained by the utility model shows the correlation between the quality of a previous post and the next post, by adding in the difference in quality between the post at time t and $t-1$ we see another boost in performance to .53. Finally, the combined model achieves a correlation of .52, which is slightly better than the Quality based Utility model.

The above results suggest that the language individuals use reflects the quality of their future contributions. However, it is quite clear that the quality of the current submission is the best predictor of the quality of an individual's next submission. It could be that the comments in the post are only associated with the current quality and not the individuals future motivation, we want to measure changes in an individual's motivation over time.

One approach to measuring the change in motivation across time would be to consider if the quality of an individual's next submission is higher or lower than the quality of their current submission. However, this measure is subject to regression to the mean. Assuming that the quality of an individual's submission is normally distributed, if an individual produces an above average or below average submission, the next contribution is more likely to be below or above that contribution as a simple by-product of probability theory and not their level of motivation. Instead, we can look at the difference between the quality of a contribution at t+1 and its expected value based on the mean level of quality for the individual.

Our second prediction tests categorically, utilizing logistic regression, whether an individual's next post will be above or below average, defined as the moving average of the quality of their last 20 posts $\bar{Q}_t$. The results shown in the second column of Table 4 suggest that the model based on motivational acts performs at a level more similar to that based on utility theory. Additionally, the combined model performs better than either individual model. Critically, this second analysis shows that an individual's language usage reflects their level of motivation for subsequent contributions,

Table 4. Results for predicting the future quality of a contribution based on analysis of the users previous contributions and comments on those contributions. Differences of more than .01 are significant at $p < .05$ according to a paired t-test across testing folds.

Method	Q_{t+1} Correlation	$Q_{t+1} > \bar{Q}_t$ Classification Accuracy
Majority Class	NA	.54
Sentiment	.08	.55
GRC	.33	.58
Utility (Quality Only)	.51	.61
Utility & Difference	.53	.63
GRC + Utility	.52	.65

successfully predicting whether or not an individual will contribute content of unusually high or low quality compared to their average level of performance.

6   Conclusion

This paper serves as an initial demonstration of an approach for identifying an individual's motivation for future work. Our analysis was conducted by examining the motivational meaning of the language that the individual used in comments about their work and the language used by community members about the work. In addition, we showed that distant supervision of natural language classifiers can be used to identify, not just sentiment or semantics, but the language's effect on an individual's psychological state. For example, the language models can reveal their goals, perception of control over a situation, and the rewards that motivate them.

It is important to remember that our results show a link between language usage and future performance, but the results do not imply causality in a particular direction. The categorical prediction identifies individuals that will perform above average on their next submission, but the language could be a by-product of their motivation instead of the cause of their motivation. It will be up to future work to look at this in more detail and attempt to separate linguistic expressions that inform motivational states and those that change motivational states. In particular, to look at how the pattern of interactions between posters and community members can reveal the motivational level of the individual. Our focus in this contribution was in demonstrating a system for automatically detecting language uses that provides insight into an individual's level of motivation and that reflect the amount of effort an individual is likely to apply to their future performances.

Acknowledgment. This research was funded by the Intelligence Advanced Research Projects Activity (IARPA) through the Department of Defense US Army Research Laboratory (DoD / ARL). The U.S. Government is authorized to reproduce and distribute reprints for Governmental purposes notwithstanding any copyright annotation thereon. Disclaimer: The views and conclusions contained herein are those of the authors and should not be interpreted as necessarily representing the official policies or endorsements, either expressed or implied, of IARPA, DoD/ARL, or the U.S. Government.

References

1. Searle, J.R.: Speech Acts: An Essay in the Philosophy of Language. Cambridge University Press (1969)
2. Mintz, M., Bills, S., Snow, R., Jurafsky, D.: Distant supervision for relation extraction without labeled data. In: Proceedings of ACL-IJCNLP, vol. 2005 (2009)
3. Bracewell, D.B., Tomlinson, M.: The Social Actions of Successful Groups. In: IEEE Sixth International Conference on Semantic Computing (2012)
4. Core, M.G., Allen, J.F.: Coding Dialogs with the DAMSL Annotation Scheme. In: Traum, D. (ed.) AAAI Fall Symposium on Communicative Action in Humans and Machines, pp. 28–35. American Association for Artificial Intelligence (1997)

5. Stolcke, A., Shriberg, E., Bates, R., Coccaro, N., Jurafsky, D., Martin, R., Meteer, M., Ries, K., Taylor, P., Ess-Dykema, C.V.: Dialog Act Modeling for Conversational Speech. In: Applying Machine Learning to Discourse Processing, pp. 98–105. AAAI Press (1998)
6. Bunt, H.: The semantics of dialogue acts. In: International Conference on Computational Semantics, pp. 1–13 (2011)
7. Bender, E.M., Morgan, J.T., Oxley, M., Zachry, M., Hutchinson, B., Marin, A., Zhang, B., Ostendorf, M.: Annotating social acts: Authority claims and alignment moves in wikipedia talk pages. In: ACL HLT 2011, p. 48 (June 2011)
8. Hsu, C.-F., Khabiri, E., Caverlee, J.: Ranking Comments on the Social Web. In: 2009 International Conference on Computational Science and Engineering, pp. 90–97 (2009)
9. Khabiri, E., Hsu, C.F., Caverlee, J.: Analyzing and predicting community preference of socially generated metadata: A case study on comments in the digg community. In: AAAI Conference on Weblogs and Social Media (2009)
10. Cha, M., Mislove, A., Gummadi, K.P.: A measurement-driven analysis of information propagation in the flickr social network. In: Proceedings of the 18th International Conference on World Wide Web, WWW 2009, p. 721 (2009)
11. Szabo, G., Huberman, B.A.: Predicting the popularity of online content. Communications of the ACM 53(8), 80–88 (2010)
12. Kahneman, D., Tversky, A.: Kahneman & Tversky (1979) - Prospect Theory - An Analysis of Decision Under Risk.pdf. Econometrica 47(2), 263–291 (1979)
13. Heath, C., Larrick, R.P., Wu, G.: Goals as reference points. Cognitive Psychology 38(1), 79–109 (1999)
14. Albarracin, D., Hepler, J., Tannenbaum, M.: General Action and Inaction Goals: Their Behavioral, Cognitive, and Affective Origins and Influences. Current Directions in Psychological Science 20(2), 119–123 (2011)
15. Locke, E.A.: Toward a theory of task motivation and incentives. Organizational Behavior and Human Performance 3(2) (1968)
16. Finkelstein, S.R., Fishbach, A.: Tell Me What I Did Wrong: Experts Seek and Respond to Negative Feedback. Journal of Consumer Research 39(1), 22–38 (2012)
17. Ajzen, I.: The Theory of Planned Behavior. Organizational Behavior and Human Decision Processes 50, 179–211 (1991)
18. Fang, A., Bunt, H., Cao, J.: Collaborative Annotation of Dialogue Acts: Application of a New ISO Standard to the Switchboard Corpus. In: EACL 2012, pp. 61–68 (2012)
19. Bracewell, D.B., Tomlinson, M.T., Brunson, M., Plymale, J., Bracewell, J., Boerger, D.: Annotation of Adversarial and Collegial Social Actions in Discourse. In: 6th Linguistic Annotation Workshop, pp. 184–192 (July 2012)
20. Davidov, D., Tsur, O., Rappaport, A.: Enhanced Sentiment Learning Using Twitter Hashtags and Smileys. In: Coling, pp. 241–249 (August 2010)
21. Roberts, K., Roach, M., Johnson, J., Gurthrie, J., Harabagiu, S.M.: Empatweet: Annotating and detecting emotions on Twitter. In: Proceedings of LREC 2012, pp. 3806–3813 (2012)
22. Petukhova, V., Bunt, H.: Incremental dialogue act understanding. In: IWCS 2011 Proceedings of the Ninth International Conference on Computational Semantics, pp. 235–244 (2011)
23. Baccianella, S., Esuli, A., Sebastiani, F.: SentiWordNet 3.0: An Enhanced Lexical Resource for Sentiment Analysis and Opinion Mining. In: LREC, pp. 2200–2204 (2010)

Mining the Personal Interests of Microbloggers via Exploiting Wikipedia Knowledge

Miao Fan[*], Qiang Zhou, and Thomas Fang Zheng

CSLT, Division of Technical Innovation and Development,
Tsinghua National Laboratory for Information Science and Technology,
Department of Computer Science and Technology,
Tsinghua University, Beijing, 100084, China
fanmiao.cslt.thu@gmail.com

Abstract. This paper focuses on an emerging research topic about mining microbloggers' personalized interest tags from their own microblogs ever posted. It based on an intuition that microblogs indicate the daily interests and concerns of microblogs. Previous studies regarded the microblogs posted by one microblogger as a whole document and adopted traditional keyword extraction approaches to select high weighting nouns without considering the characteristics of microblogs. Given the less textual information of microblogs and the implicit interest expression of microbloggers, we suggest a new research framework on mining microbloggers' interests via exploiting the Wikipedia, a huge online word knowledge encyclopedia, to take up those challenges. Based on the semantic graph constructed via the Wikipedia, the proposed semantic spreading model (SSM) can discover and leverage the semantically related interest tags which do not occur in one's microblogs. According to SSM, An interest mining system have implemented and deployed on the biggest microblogging platform (Sina Weibo) in China. We have also specified a suite of new evaluation metrics to make up the shortage of evaluation functions in this research topic. Experiments conducted on a real-time dataset demonstrate that our approach outperforms the state-of-the-art methods to identify microbloggers' interests.

Keywords: Microblog, social tagging, interest, Wikipedia.

1 Introduction

As an emerging application in the Web 2.0 area, microblogging, which combines the short message service, social networking and broadcast medium, has attracted millions of online users to share their daily life and opinions [6]. Representative platforms such as Twitter and Sina Weibo allow users to freely post 140 English characters and Chinese words respectively in one microblog. The microblogs asynchronously posted by an individual user form a distinct microblog stream, which can be regarded as his/her daily life records. These records naturally indicate the owner's interests to some extent, and these interests may be the key factors for friends

---

[*] Corresponding author.

A. Gelbukh (Ed.): CICLing 2014, Part II, LNCS 8404, pp. 188–200, 2014.

discovering, microbloggers grouping or even personalized online advertising. Therefore, as a promising fundamental research direction, the studies on mining microbloggers' interests have emerged in recent years [11, 20].

Previous studies proposed several methods to identify microbloggers' interest tags.
They are, however, either adopting traditional keyword extraction approaches without
considering the characteristics of microblogs [20] or lack of reasonable evaluation
metrics and experimental results on real-time datasets [11]. Here, we list the defects
of previous approaches as follows,

- Due to the insufficiency of textual information in microblogs, traditional
  keyword extraction approaches, which are more suitable to deal with formal texts
  such as news articles and academic literatures, may not perform well on microblogs.

- Not all interest tags are expressed explicitly in the microblogs. Figure 1 shows
  a microblog of Professor Shaoping Ma working at Department of Computer
  Science and Technology in Tsinghua University, whose research interests are Information Retrieval and Search Engine. In this case, he reposted and commented
  another microblog which explicitly talked about several commercial search engines
  in China, such as Sogou (搜狗), Baidu (百度) and 360 (circled by red borders). But
  actually, we can conclude that the Search Engine is the most suitable tag to describe Professor Shaoping Ma's interest even though it did not occur in this microblog. Gupta et al [7] proposed that compared with keywords, tags are more abstract
  expressions to indicate one's interests, so that more semantic related interest tags
  which may not appear in a given individual's microblogs need to be mined.

- The short of human annotations on user interest tags also leads to the lack of
  evaluation metrics for this research topic, so that more reasonable evaluation metrics are required.

To take up these challenges, in this paper, we suggest a new and systematic research framework on mining the personalized interest tags of microbloggers via exploiting the Wikipedia knowledge. The idea was inspired by the intuition that the
Wikipedia contains the crowd-knowledge and rich lexical semantics.

Fig. 1. A microblog of Prof. Shaoping Ma   **Fig. 2.** A screenshot of Baidu's article in
Hudong Wiki

For example, Figure 2 show one screenshot of the articles of Baidu (百度) from Hudong Wiki . It is obvious that one of the potential interest tags of Prof. Shaoping Ma, Search Engine (搜索引擎), is included in both of the articles as a hyperlink title (circled by the red borders in Figure 2). Studies [2, 18] have proved that those hyperlinks in the Wiki articles are the widely accepted as semantic annotation resources. Therefore, we believe that they may help us to extend the semantics of one's microblogs, so that more semantically related interest tags can be identified.

2   Related Work

2.1   Most Related Research

Wu et al [20] firstly proposed the research about automatic personalized tag annotation for individual users based on the tweets they have ever posted on Twitter. They adopted the traditional keyword extraction approaches, TFIDF [15] and TextRank [12]. In their work, one user's tweets were all organized in a whole document and experimental results showed that those applied methods achieved comparable performances with keyword extraction on web pages.

Liu et al [11] pointed out the inefficiencies of adopting traditional keyword extraction methods without considering the characteristics of microblogs. They believed that there was a vocabulary gap between one's microblogs and his/her real intentions, so they proposed to bridge the gap with the help of word alignment models [1] in statistical machine translation. However, the performance of this method largely depended on the quality and the size of training data. What's more, in their work, they did not report the any experimental results on microblogs.

2.2   Other Research Topics on Microblogging

Due to the considerable amount of registered users on microblogging platforms, more and more researchers came to realize that microblogging is a challenging environment to conduct studies from different perspectives. Some of the studies explored several new research topics, such as hashtag retrieval [4] and real-time event detection [13, 14]. Some, on the other hand, still discussed traditional topics, i.e., recommendation [3], sentiment analysis [10], however, tried to propose new approaches given the characteristics of microblogs.

2.3   Applications of Wikipedia

Wikipedia is an open access online encyclopedia, in which knowledge from diverse categories is included. It thus has been used as an external semantic reinforcement to enhance the performance of many IR and NLP sub-tasks, such as sematic relatedness computing [5], query intent prediction [9], advertising keywords extraction from web pages [22], ontology construction [21], text clustering [8, 16] and classification [19].

3   Semantic Spreading Model

The semantic spreading model (SSM) will be described as follows. Section 3.1 will talk about the intuitive description on how to infer interest tags based on microblogs from the perspective of human beings. Then, several hypotheses that the computer could formulate are put forward. Based on these hypotheses, the SSM is proposed. Section 3.2 will discuss the core of SSM, the spreading activation algorithm. This algorithm runs on a semantic graph constructed by the Wikipedia and contains a key parameter, called semantic trigger parameter. How to estimate the value of the parameter will be discussed in Section 3.3. Section 3.4 will describe how to predict the interest tags of a microblogger, based on the spreading activation algorithm.

3.1   Intuition

The case of mining Prof. Shaoping Ma's interests in Section 1, to some extent, implies the process of how human beings predict one microblogger's interest tags based on his/her microblogs. Due to the insufficiency of textual information and the informal syntax in microblogs, it is more likely for people to judge one's interests on the lexical level. More specifically, when we judge one's interests from his/her microblogs, the microblogs will be checked one by one so that the keywords will be extracted as candidate tags. In most cases, these candidate tags are too specific to reflect one's interests or some of them indicate the same interest tag which does not even occur in the microblogs. To find these suitable interest tags, people prefer exploiting their own knowledge based on those candidates. Finally, all those tags are judged and weighted so that suitable interest tags will be selected.

According to the process above, we propose the following three hypotheses which could be formulated by computers.

H1: If a word can be inferred by multiple semantic related tags, it is more likely selected as the interest tag.

H2: If a tag can infer multiple words, the degrees of semantic relatedness between the tag and the words are different respectively.

H3: If a word is frequently mentioned or inferred by multiple microblogs from given microblogger, it more likely indicates his/her interests.

3.2   Speading Activation

Different from human beings, computers need to take advantage of external world knowledge base, such as Wikipedia, to infer more semantic related tags. Wikipedia, as an online encyclopedia, which is freely contributed by millions of users, can be exploited due to its rich semantic relationships.

Figure 3 shows the article of the WWW (万维网) in Hudong Wiki. Besides the normal text, some special hyperlinks, called outlinks, can be found in a wiki article. Studies on Wikipedia hyperlinks [2, 18] have proved that these outlinks (circled by red borders in Figure 4) reflect human's cognition on semantic relationship to some extent. Based on this sight, the Wikipedia can be reorganized as a semantic graph [17], in which each article denoted as a_j represents a node. $In(a_j)$ stands for the nodes that link to a_j and $Out(a_j)$ stands for the nodes that link from a_j.

According to H1, the score of a_j, denoted as $Score(a_j)$, is contributed by all the articles which link to a_j.

$$Score(a_j) = \sum_{a_i \in In(a_j)} \alpha_{i,j} * Score(a_i), \tag{1}$$

in which $\alpha_{i,j}$ is a parameter that shows the semantic trigger ability from a_i to a_j.

If we delve further, this "trigger" action will also take place another time at all the new nodes, such as a_j, just like a spreading process, so that more nodes will be "actived". Given this iteration process, formulate (1) can be revised as,

$$Score_p(a_j) = \sum_{a_i \in In(a_j)} \alpha_{i,j} * Score_{p-1}(a_i), \tag{2}$$

in which p stands for the iterations. Compared with other graph-based algorithms, spreading activation algorithm contains two terminal conditions,

$$(p \geq P) \text{ or } (\forall a_j, |Score_p(a_j) - Score_{p-1}(a_j)| \leq \delta). \tag{3}$$

P represents the times of upper bound that the algorithm iterates. Once $p \geq P$, the iteration stops. δ denotes the threshold. For each node, if the value of difference between two adjecent times is smaller than δ, the algorithm will stop either. Those contraints endow the spreading activation algorithm a characteristic that it does not need to execuate the iteration for each word on the whole graph until converge.

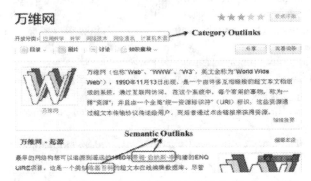

Fig. 3. The article of the World Wide Web in Hudong Wiki

3.3   Semantic Trigger Parameter

Section 3.2 mentioned the semantic trigger parameterα. According to **H2**, α stands for the degree of semantic relatedness between articles, which differs given diverse outlinks. More specifically, $\alpha_{i,j}$ represents the semantic trigger ability from a_i to a_j. We suppose that there are n articles in the whole graph, the semantic trigger metric,

$$A_{n*n} = \begin{pmatrix} \alpha_{1,1} & \cdots & \alpha_{1,n} \\ \vdots & \ddots & \vdots \\ \alpha_{n,1} & \cdots & \alpha_{n,n} \end{pmatrix}, \tag{4}$$

in which,

$$\alpha_{i,j} \begin{cases} = 1, & i = j \\ = 0, & a_j \notin Out(a_i) \\ > 0, & a_j \in Out(a_i) \end{cases}. \tag{5}$$

The key problem is how to estimate $\alpha_{i,j}$, when a_j is an outlink article of a_i. Figure 4 also shows another kind of hyperlinks (circled by green border). They declare the categories that an article belongs to. It is obvious that if an article belongs to diverse categories or higher hierarchy categories, the article contains more semantic information. As Gupta et al [7] proposed tags are more abstract expressions to indicate one's interests, articles with more semantic information are more likely to be chosen as interest tags. We define $\beta_{i,j}$ which represents whether article a_i belongs to category c_j.

$$\beta_{i,j} = \begin{cases} 1, & a_i \text{ belongs to } c_j \\ 0, & a_i \text{ does not belong to } c_j \end{cases}. \tag{6}$$

We suppose that there are m different categories in the whole graph. The article-category metric,

$$C_{n*m} = \begin{pmatrix} \beta_{1,1} & \cdots & \beta_{1,m} \\ \vdots & \ddots & \vdots \\ \beta_{n,1} & \cdots & \beta_{n,m} \end{pmatrix}, \tag{7}$$

in which $w_i = (\beta_{i,1}, \beta_{i,2}, \ldots, \beta_{i,m})$ represents the category vector of article a_i. When we estimate the semantic information of a_i, a straightforward way is to exploit the context category information of a_i. For each a_i, it contains not only w_i, but also w_k where $a_k \in Out(a_i)$. Therefore, the context category information of a_i is $w_i + \sum_{a_k \in Out(a_i)} w_k$. $\alpha_{i,j}$ represents the semantic information radio to some extent,

$$\alpha_{i,j} = \frac{(w_i + \sum_{a_k \in Out(a_i)} w_k) w_j^T}{(w_i + \sum_{a_k \in Out(a_i)} w_k) w_i^T}. \tag{8}$$

3.4 Micorblog Semantic Distribution

Section 3.2 and 3.3 demonstrate how to use spreading activation algorithm to predict the semantic tags from one article. Here, in this section, we will explain how to generate the interest tags of one microblogger based on the algorithm. After the semantic graph is constructed by the hyperlink structure of the Wikipedia, we need to provide the seeds from microblogs and initialize them with proper score. TFIDF is a suitable approach to extract seeds from microblogs belonging to user u. The TFIDF value of each seed is computed as,

$$Score(s_{i,u}) = \frac{N_{i,u}}{\sum_j N_{j,u}} Log\left(\frac{U}{U_i}\right), \tag{9}$$

where $N_{i,u}$ is the count of nouns in user u's microblogs. U is the total number of users in the dataset and U_i denotes the number of users whose microblogs contains seed i. Suppose that there are totally M tag-value pairs, each seed i as the initial article will generate a semantic distribution,

$$d_i = \{t_1: v_{i,1}, t_2: v_{i,2} ..., t_M: v_{i,M}\}. \tag{10}$$

According to **H3**, the microblog semantic distribution of user u is,

$$\sum d_i = \{t_1: \sum v_{i,1}, t_2: \sum v_{i,2}, ..., t_T: \sum v_{i,T}\}, \tag{11}$$

where T represents the total number of different tags. Then the top K tags with higher values are selected as interest tags.

4   Interest Mining System

Based on the semantic spreading model (SSM), we have implemented an interest mining system, called Social Card[1] and deployed it on the biggest microblogging platform (Sina Weibo) with more than 300 million registered users in China. In this section, we will describe the overall architecture of the Social Card System and take Prof. Shaoping Ma's Social Card as a case to study.

4.1   System Architecture

Social Card can automatically generate a name card with individual style for each microblogger. Once a microblogger signs permission and allows the system to extract his/her microblogs posted recently, Social Card will analyze them and automatically generate his/her personalized interest tags based on SSM. Those tags will be displayed by a special style tag cloud. A social card will be produced by combining this microblogger's personal information and his/her interest tag cloud.

Figure 4 shows the overall architecture of Social Card System. It contains two main parts, which are respectively called the personal information processing and the personalized interest tag mining.

The function of the part ahead is to access the personal information of a given microblogger and some available information is selected, such as his/her nickname, sex, birth date, Weibo homepage and even his/her icon.

As for the personalized interest tag mining subsystem, it will firstly gain a given microblogger's recent microblogs once the permission is passed via Sina Weibo API. Then, these microblogs are filtered as there is usually a lot of noise. Here, we list the type of noise and corresponding processing measures,

- **Reposted microblogs and comments:** A microblogger may repost or comment a microblog that he/she is interested in. However, in most circumstances, besides the original microblog and the comment from this microblogger, we can also gain other comments via API. In this system, we only retain the original microblogs and the corresponding comments from the target microblogger.

---

[1] http://app.weibo.com/detail/4Vmoqf?ref=appsearch

- **Emotion icons:** Even though we can see the emotion icons in many microblogs, we may ignore the fact these emotion icons are encoded as texts especially when we obtain one's microblogs via Sina Weibo API. Fortunately, they are encoded regularly and we can remove them with the help of regular expression.
- **Nicknames:** Numerous microblogs contain the nicknames of microbloggers. They are mostly mentioned with a prefix "@" and can be simply removed.

Next, we conduct word segmentation and part-of-speech tagging on the filtered microblogs and extract the nouns as seeds. The SSM initializes with these seeds and based on Hudong Wikipedia, one microblogger's interest tags are predicted. With the help of a special style tag cloud[2] given microblogger's interests are displayed. Finally, a social card is produced via adding the microblogger's personal information.

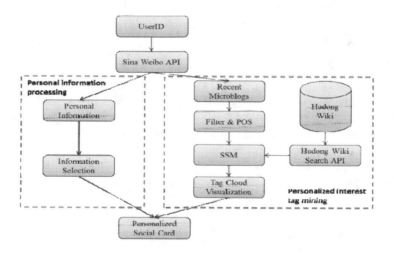

Fig. 4. The Architecture of Social Card System

4.2   Case Study

Figure 5 shows Prof. Shaoping Ma's social card generated by our system. His nickname on Sina Weibo is shown at the top-right part of this card. The left part displays the other personal information. On the right, a special tag cloud demonstrates his personalized interests which can be divided into two groups, research interests and hobbies respectively. From Prof. Shaoping Ma's social card, we can predict his research interests are about Search Engine (搜索引擎) and Information Retrieval (信息检索), as several tags with bigger font indicate those interests, such as searching (搜索), intelligent search (智能搜索) and artificial intelligence (人工智能). Other tags like camera (相机), Nikon (尼康) and photograph (照片) indicate that   his hobby is photography (摄影).

---

[2] http://www.wordle.net/

Fig. 5. Prof. Shaoping Ma's Social Card

5   Experiments

Besides our new approach, we have also re-implemented other state-of-the-art methods as comparisons, which will be discussed in Section 5.2. Experiments were conducted on those methods under the same real-time dataset (Section 5.1) which was
crawled from Sina Weibo API. Given the lack of evaluation metrics in this research
topic, we have specified several which are described in Section 5.3. In Section 5.4,
experiments were evaluated by those metrics and the results demonstrated that our
approach outperforms the other state-of-the-art methods.

5.1   Dataset

We prepared 1000 Sina Weibo users' recent 200 microblogs as our real-time dataset.
To construct a balance dataset, we randomly selected 500 famous users and 500 ordinary ones. Table 1 shows the several statistical properties of the dataset, in which N_u
denotes the number of users. N_w, N_n and N_v represent the number of words, the
number of nouns, the size of noun vocabulary, respectively.

Table 1. Statistical properties of the Sina Weibo dataset

N_u	N_w	N_n	N_v
1,000	7,582,318	1,095,121	25,767

5.2   Comparison Methods

To date, three other approaches have been proposed for microbloggers' interests mining, which are the TFIDF, TextRank and WTM.

Wu et al [20] firstly adopted the traditional keyword extraction approaches, TFIDF
and TextRank, respectively. For TFIDF, the microblogs posted by a user are regarded

as a whole document and value of each keyword is computed by E.q. (9). In Tex-tRank, each noun is regarded as a vertex V_i and the undirected edges $E(V_i)$ are add-ed between two nouns if they co-occur in at least one microblog. The weight of the edge that links V_i and V_j, w_{ji} is set to be the count of the co-occurrence. The value of a vertex $R(V_i)$ is determined by the other vertices $V_j \in E(V_i)$ that link to it. Start-ing with an arbitrarily assigned value, the value of V_i is updated iteratively until con-vergence, according to the following equation,

$$R(V_i) = (1-d) + \sum_{V_j \in E(V_i)} \frac{w_{ji}}{\sum_{V_k \in E(V_j)} w_{jk}} R(V_j). \qquad (12)$$

After Wu [20] and Liu [11] pointed out the inefficiencies of adopting traditional keyword extraction methods and proposed a novel approach called word trigger me-thod (WTM) based on the word alignment models (WAM) in statistical machine translation (SMT). The core idea of WTM is described as follows,

$$\Pr(p|d) = \sum_{w \in d} \Pr(p|w) \Pr(w|d), \qquad (13)$$

where the document d contains all the microblogs posted by the target user. The conditional probability of a candidate keyword p given document d depends on the probability of word w occurring in document d and the probability of candidate keyword p triggered by word w, in which $\Pr(p|w)$ is the translation probability obtained from the WAM. As WTM largely relies on the training data, we prepared 20,576 Hudong Wiki articles which relate to the nouns in the Sina Weibo dataset as the training data for WTM.

For the ranked lists of personalized tags of each microblogger generated by those methods, we selected the top 20 tags from each, dumped them into the evaluation pool and made sure that nobody knew which method each tag came from. We also invited five volunteers to score each tag (5-star) in the evaluation pool according to his/her microblogs. After averaging the score of each tag in the pool, a final ranked list of annotated personalized tags for each microblogger is generated as the gold standard. We denote this as the human evaluation (HE) approach.

5.3   Evaluation Metrics

Each of the five approaches (TFIDF, TextRank, WTM, SSM and HE) generated a set of rank-tag pairs respectively in which tags with higher values gained higher ranks. For example, $RT_y @ N$ represents the result set of the approach y,

$$\mathbf{RT}_y @ \mathbf{N} = \{(r,t) | t \in T_y@N, r = R_y(t), 1 \le r \le N\}, \qquad (14)$$

in which $R_y(t)$ stands for the rank of tag t, and $T_y @ N$ represents the top N tags in the approach y.

The Hit Rate (HR) shows the precision of interest tag mining approach y, com-pared with the gold standard set of human annotation T_{HE}. $|T_{HE}@ N \cap T_y@ N|$ denotes the count of tags that both exist in T_{HE} and T_y.

$$\mathbf{HR}_y@N = \frac{|T_{HE}@N \cap T_y@N|}{N}.$$ (15)

Compared with HR, the benefit of Mean Reciprocal Rank (MRR) is obvious as it can evaluate the approach y more precisely via considering the rank of tags further. To evaluate our result, we propose an Advanced Mean Reciprocal Rank (AMRR), which could evaluate the result with a rank list. Two approaches may gain equal value of HR but can be distinguished by AMRR as the rank of tags differs in these approaches. In AMRR, if a tag $t \in T_{HE}@N$ cannot be found in $T_y@N$, the value of $\frac{1}{R_y(t)}$ is set to zero.

$$\mathbf{AMRR}_y@N = \frac{1}{N}\sum_{t \in T_{HE}@N} \frac{1}{R_{HE}(t)} * \frac{1}{R_y(t)}.$$ (16)

Based on AMRR, the Ranking Improvement (IMP) can reveal the performance improvement between two difference approaches.

$$\mathbf{IMP}_{y,z}@N = \frac{1}{N}\sum_{t \in T_{HE}@N} \frac{1}{R_{HE}(t)} * \left(\frac{1}{R_y(t)} - \frac{1}{R_z(t)}\right).$$ (17)

5.4   Experimental Results

The parameters were all assigned the values that could help those state-of-the-art methods perform best, according to the literatures [11, 20]. In SSM, the upper bound of algorithm iterates P was set to 2, which provided the best results.

Table 2. HR @ N of TFIDF, TextRank, WTM and SSM (N = 5, 10, 20)

	TFIDF	TextRank	WTM	SSM
HR @ 5	0.3053	0.1053	0.3158	0.5062
HR @ 10	0.3263	0.1105	0.3053	0.5000
HR @ 20	0.3842	0.1842	0.3790	0.6079

Table 3 shows the average values of AMRR among TFIDF, TextRank, WTM and SSM, when top 5, 10, 20 tags are considered along with their ranks. Even considering the ranks of the tags, TextRank still shows the worst performance. TFIDF shows better performance on AMRR than WTM. On average, The AMRR value of SSM improves 0.0469, compared with the others.

Table 3. AMRR @ N of TFIDF, TextRank, WTM and SSM (N = 5, 10, 20)

	TFIDF	TextRank	WTM	SSM
AMRR @ 5	0.0734	0.0344	0.0658	0.1352
AMRR @ 10	0.0446	0.0201	0.0381	0.0757
AMRR @ 20	0.0250	0.0121	0.0216	0.0414

IMP is the comprehensive measure for evaluating the performance among these approaches. SSM outperforms the others especially on predicting high-weight tags. Overall, SSM improves 0.0364, 0.0619 and 0.0423 on average, compared with TFIDF, TextRank and WTM respectively.

6   Conclusion

In this paper, we have contributed a systematic research framework on mining microbloggers' interest tags via exploiting Wikipedia knowledge. More specifically, we have proposed a novel semantic spreading model (SSM) which exploits the semantic graph in Wikipedia. It can discover and leverage the potential interest tags, which do not occur in one's microblogs but semantically relate. To prove the feasibility of SSM, we also have implemented our idea into an interest mining system and deployed it on the biggest microblogging platform (Sina Weibo) in China. Due to the lack of evaluation metrics in this research topic, we have specified a suite of evaluation functions. Experiments on a real-time dataset demonstrate that our approach outperforms the state-of-the-art methods to identify microbloggers' interests.

Acknowledgements. This work is partly supported by National Program on Key Basic Research Project (973 Program) under Grant 2013CB329304 and National Science Foundation of China (NSFC) under Grant No. 61373075. Thanks to the comments from Yingnan Xiao and Chenchao Zhu.

References

1. Brown, P.F., Pietra, S.A.D., Pietra, V.J.D., Mercer, R.L.: The mathematics of statistical machine translation: parameter estimation. Computational Linguistics 19(2), 263–311 (1993)
2. Bu, F., Hao, Y., Zhu, X.: Semantic relationship discovery with Wikipedia structure. In: Proceedings of the 22nd International Joint Conference on Artificial Intelligence, pp. 1770–1775 (2011)
3. Chen, K., Chen, T., Zheng, G., Jin, O., Yao, E., Yu, Y.: Collaborative personalized tweet recommendation. In: Proceedings of the 35th Annual International Conference on Research and Development in Information Retrieval, pp. 661–670 (2012)
4. Efron, M.: Hashtag retrieval in a microblogging environment. In: Proceedings of the 33rd Annual International Conference on Research and Development in Information Retrieval, pp. 787–788 (2010)
5. Gabrilvich, E., Markovitch, S.: Computing Semantic Relatedness using Wikipedia-based Explicit Semantic Analysis. In: IJCAI 2007, pp. 1606–1610 (2007)
6. Kwak, H., Lee, C., Park, H., Moon, S.: What is Twitter, a social network or a news media? In: Proceedings of the 19th International Conference on World Wide Web, pp. 591–600 (2010)
7. Gupta, M., Li, R., Yin, Z., Han, J.: Survey on social tagging techniques. In: SIGKDD Explor., pp. 58–72 (2010)

8. Hu, J., Fang, L., Cao, Y., Zeng, H.-J., Li, H., Yang, Q., Chen, Z.: Enhancing text clustering by leveraging Wikipedia semantics. In: Proceedings of the 31st Annual International Conference on Research and Development in Information Retrieval, pp. 179–186 (2008)
9. Hu, J., Wang, G., Lochovsky, F., Sun, J., Chen, Z.: Understanding user's query intent with Wikipedia. In: Proceedings of the 18th World Wide Web Conference, pp. 471–478 (2009)
10. Jiang, L., Yu, M., Zhou. M., Liu, X., Zhao, T. : Target-dependent twitter sentiment classification. In: Proceedings of the 49th Annual Meeting of the Association for Computational Linguistics, pp. 151–160 (2011)
11. Liu, Z., Chen, X., Sun, M.: Mining the interests of Chinese microbloggers via keyword extraction. Front. Comput. Sci. 6(1), 76–87 (2012)
12. Mihalcea, R., Tarau, P.: Textrank: Bringing order into texts. In: Proceedings of the Conference on Empirical Methods in Natural Language Processing, Barcelona, Spain (2004)
13. Petrovic, S., Osborne, M., Lavrendo, V.: Streaming first story detection with application to Twitter. In: Proceedings of the North American Chapter of the ACL, pp. 181–189 (2010)
14. Sakaki, T., Okazaki, M., Matsuo, Y.: Earthquake shakes Twitter users: real-time event detection by social sensors. In: Proceedings of the 19th World Wide Web Conference, pp. 851–860 (2010)
15. Salton, G., Buckley, C.: Term-weighting approaches in automatic text retrieval. Information Processing and Management 24(5), 513–523 (1988)
16. Schonhofen, P.: Identifying document topics using the Wikipedia category network. In: Web Intell. Agent Syst., pp. 456–462 (2006)
17. Sowa, J.: Semantics of conceptual graphs. In: Proceedings of the Annual Meeting of the Association for Computational Linguistics, pp. 39–44 (1979)
18. Strube, M., Ponzetto, S.P.: Wikirelate! Computing semantic relatedness using Wikipedia. In: Proceedings of the 21st National Conference on Artificial Intelligence, Boston, MA (2006)
19. Wang, P., Hu, J., Zeng, H.-J., Chen, Z.: Using Wikipedia knowledge to improve text classification. Knowl. Inf. Syst. 19, 265–281 (2009)
20. Wu, W., Zhang, B., Ostendorf, M.: Automatic generation of personalized annotation tags for twitter users. In: Proceedings of the North American Chapter of the ACL, pp. 689–692 (2010)
21. Yu, J., Thom, J., Tam, A.: Ontology evaluation using Wikipedia categories for browsing. In: Proceedings of the 6th ACM Conference on Information and Knowledge Management, pp. 223–232 (2007)
22. Zhang, W., Wang, D., Xue, G.-R., Zha, H.: Advertising keywords recommendation for short-text Web pages using Wikipeda. ACM Trans. Intell. Syst. Technol. 3(2), Article 36, 25 pages (February 2012)

Website Community Mining from Query Logs with Two-Phase Clustering*

Lidong Bing, Wai Lam, Shoaib Jameel, and Chunliang Lu

Key Laboratory of High Confidence Software Technologies
Ministry of Education (CUHK Sub-Lab)
Department of Systems Engineering and Engineering Management
The Chinese University of Hong Kong, Shatin N.T., Hong Kong
{ldbing,wlam,msjameel,cllu}@se.cuhk.edu.hk

Abstract. A website community refers to a set of websites that concentrate on the same or similar topics. There are two major challenges in website community mining task. First, the websites in the same topic may not have direct links among them because of competition concerns. Second, one website may contain information about several topics. Accordingly, the website community mining method should be able to capture such phenomena and assigns such website into different communities. In this paper, we propose a method to automatically mine website communities by exploiting the query log data in Web search. Query log data can be regarded as a comprehensive summarization of the real Web. The queries that result in a particular website clicked can be regarded as the summarization of that website content. The websites in the same topic are indirectly connected by the queries that convey information need in this topic. This observation can help us overcome the first challenge. The proposed two-phase method can tackle the second challenge. In the first phase, we cluster the queries of the same host to obtain different content aspects of the host. In the second phase, we further cluster the obtained content aspects from different hosts. Because of the two-phase clustering, one host may appear in more than one website communities.

Keywords: Website Community, Query Logs, Tow-phase Clustering.

1 Introduction

The World Wide Web has been extensively developed since its first appearance two decades ago. Various applications on the Web have unprecedentedly changed humans' life. Web search provides us a fast and accurate access to the useful information on the Web. Online encyclopedia contains human knowledge in different areas and makes it accessible to every Web user. Online shopping saves us the time for searching items in the malls. Corresponding to various

---

* The work described in this paper is also supported by grants from the Research Grant Council of the Hong Kong Special Administrative Region, China (Project Code: CUHK413510).

A. Gelbukh (Ed.): CICLing 2014, Part II, LNCS 8404, pp. 201–212, 2014.
© Springer-Verlag Berlin Heidelberg 2014

applications on the Web, some important and challenging research topics have attracted a lot of attention from the research community. To facilitate a better Web search experience for the users, several directions are well studied such as search result ranking [20, 13], query log analysis [12, 23] and query reformulation [28, 4]; Online encyclopedias such as Wikipedia are employed to upgrade the performance of document classification and clustering methods [8, 17, 27] as well as to generate a hybrid ontology by merging them with the expert-edit ontology [22]; Different methods in Web information extraction (IE) are proposed to tackle the problem of extracting useful information from different types of Web pages, such as product description details extraction [29] and Web data record extraction [30].

Although the explosive growth and spread of the Web have resulted in a large scale information repository, it is still under-utilized due to the difficulty in automatically processing the information. One important task is to automatically organize and categorize a large amount of websites on the Web into different website communities. A website community refers to a set of websites that concentrate on the same or similar topics. Website community is different from online directories, such as Yahoo! Directory[1]. Online directories have predefined architecture, while the website community is not predefined and is adaptable according to the evolution of the real Web. There are two major challenges in website community mining task. First, the websites in the same topic may not have direct links among them because of competition concerns. One way to solve this problem is to calculate the content similarity of each pair of websites. This way is time consuming and cannot be generalized well. Second, a website may contain information about several topics. Accordingly, the website community mining method should be able to capture such phenomena and assigns such website into different communities.

In this paper, we propose a method to automatically mine website communities by exploiting the large scale Web query logs. Typically, each record in a query log data has user query, the clicked URL and other fields. Query log data can be regarded as a comprehensive summarization of the real Web. The queries that resulting a particular website clicked can be regarded as the summarization of the website content. Two websites in the digital video domain are clicked because of similar queries and they are thus connected by the queries indirectly. The strong summarization characteristic of query log data can help us overcome the first challenge mentioned above. The proposed two-phase method can tackle the second challenge. In the first phase, we cluster the queries of the same host to obtain different content aspects of the host. Then we construct sub-host vectors based on the obtained aspects of a host. In the second phase, we further cluster the sub-host vectors from different hosts and construct website communities. Because of the two-phase clustering, one host may appear in more than one website communities. The results of our website community mining can help us find the most important and popular website communities, the most popular hosts or URLs in these communities, as well as the most popular queries for

---

[1] http://dir.yahoo.com/

reaching these popular destinations. It also lets users know about the reliability of a website, because the number of users visiting a website can automatically tell the reliability of the information contained in the website.

2   Website Community Mining

2.1   Data Representation

In Web query log, a query q and a host h are the two basic items of an effective log record. q is the query issued by a search engine user and h is the website clicked by the user after browsing the list of the search results. After aggregating the click through results of different users for the same query q_j, the clicking information of query q_j can be denoted as a vector $\boldsymbol{q_j} : (h_{1j}, h_{2j}, \cdots, h_{ij}, \cdots)$, where h_{ij} is the weight of host h_i in this vector. Similarly, by aggregating the click through results of different users on the same host h_i, the clicking information of host h_i can be denoted as a vector $\boldsymbol{h_i} : (q_{i1}, q_{i2}, \cdots, q_{ij}, \cdots)$, where q_{ij} is the weight of query q_j in this vector.

Each of h_i and q_j can be regarded as a pseudo-document. The weight of each term in the pseudo-document can be calculated with the classical term frequency-inverse document frequency (TFIDF) method, which is a numerical value and reflects how important a term is to a document in a collection or corpus. We adopt the TFIDF method to calculate the weights in $\boldsymbol{h_i}$. Let f_{h_i,q_j} denote the raw frequency of query q_j resulting the host h_i clicked in the search query logs. The term frequency tf_{h_i,q_j} is calculated as:

$$tf_{h_i,q_j} = \begin{cases} 1 + \log f_{h_i,q_j} & if \ f_{h_i,q_j} > 0, \\ 0 & otherwise. \end{cases} \tag{1}$$

The inverse document frequency idf_{q_j} of q_j is a measure of whether q_j is common or rare among all hosts. It is calculated as:

$$idf_{q_j} = \log \frac{|H|}{|\{h \in H : q_j \in h\}|}, \tag{2}$$

where H is the collection of all hosts and $q_j \in h$ indicates the host h was clicked after at least one user issued query q_j. Finally, the weight q_{ij} of query q_j in the host vector $\boldsymbol{h_i}$ is calculated as:

$$q_{ij} = tf_{h_i,q_j} \times idf_{q_j}. \tag{3}$$

Similarly, the weight h_{ij} of host h_i in the query vector $\boldsymbol{q_j}$ is calculated as:

$$h_{ij} = tf_{q_j,h_i} \times idf_{h_i}, \tag{4}$$

where tf_{q_j,h_i} is the term frequency of host h_i clicked after the query q_j issued by users, and idf_{h_i} is the inverse document frequency of h_i.

2.2   Website Community Mining with Two-Phase Clustering

As we know, one site may contain several content topics. For example, Yahoo! portal contains information covering military, economy, etc. As a result, the queries related to Yahoo! are very diverse. In other words, the elements of vector h_i cover several topics. If h_i is directly fed into the community mining method, the host h_i will be grouped into a single community and the obtained community is not topic-cohesive. To tackle this problem, we propose a two-phase clustering method. In the first phase, we cluster the queries of the same host to obtain different content aspects of the host. Then we construct sub-host vectors based on the obtained aspects of a host. In the second phase, we perform clustering on sub-host vectors from different hosts and construct website communities. Therefore, the obtained website communities cover topics in a much finer granularity and a single host may appear in more than one website community according to its content aspects.

First Phase Clustering. In the first phase, we aim at mining different content topics covered by a particular host h_i. Let Q_{h_i} denote the set of queries that have non-zero weight in the vector h_i. We construct a matrix as shown below:

$$
\begin{pmatrix} q_1 \\ q_2 \\ \cdot \\ \cdot \\ q_{|Q_{h_i}|} \end{pmatrix} = \begin{pmatrix} h_{11} & h_{21} & \cdots\cdots & h_{|H|1} \\ h_{12} & h_{22} & \cdots\cdots & h_{|H|2} \\ & \cdots & & \cdots \\ & \cdots & & \cdots \\ h_{1|Q_{h_i}|} & h_{2|Q_{h_i}|} & \cdots\cdots & h_{|H||Q_{h_i}|} \end{pmatrix}.
\tag{5}
$$

Each row of the matrix is a query vector and each query results in at least one clicking on host h_i. Suppose q_i and q_j are issued for searching different topics. Although it is possible that some general websites such as Yahoo! will be clicked for both of them, the host sets clicked for them should be significantly different due to a large number of websites in the topics related to each of the queries. Based on this observation, we can get different topics in the host h_i by performing clustering on the query set Q_{h_i} with the feature vectors given in the above matrix. After this phase, the query set of a host is partitioned into several clusters, and each cluster represents a content aspect covered by the host h_i.

k-Means Algorithm. k-means algorithm is employed to conduct the clustering operation in this work. k-means is a method of cluster analysis and aims at partitioning a set of instances into k clusters. Each instance belongs to the cluster with the nearest mean. Given a set of instances $\{x_1, x_2, \cdots, x_n\}$, where each instance is a real vector, k-means clustering partitions the n instances into k clusters $(k \leq n)$ $C = \{C_1, C_2, \cdots, C_k\}$ so as to minimize the within-cluster sum of squares:

$$
\underset{C}{\operatorname{argmin}} \sum_{i=1}^{k} \sum_{x_j \in C_i} \|x_j - \mu_i\|,
\tag{6}
$$

```
Input:
    // A set of instance to be clustered
    I = {x₁, x₂, ⋯ , xₙ}
    k    // Number of clusters
Output:
    C = {C₁, C₂, ⋯ , Cₖ}    // Obtained clusters
Procedure k-means:
    // Without replacement sampling
    Sample an xᵢ as mean μⱼ of each Cⱼ
    For each xᵢ ∈ I
        Put xᵢ into Cⱼ = argmin_{C∈C} distance(xᵢ, μ)
    End
    While the mean of any cluster changes
        For each j ∈ {1..k}
            Recompute μⱼ = 1/|Cⱼ| Σ_{xᵢ∈Cⱼ} xᵢ
        End
        For each xᵢ ∈ I
            Put xᵢ into Cⱼ = argmin_{C∈C} distance(xᵢ, μ)
        End
    End
    Return C
```

Algorithm 1. Pseudocode for k-means.

where μ_i is the mean of instances in C_i.

The problem is computationally NP-hard and an efficient heuristic algorithm is employed to obtain a local optimum, which is described in Algorithm 1. The distance function $distance()$ is defined based on the commonly used cosine similarity, which is used to measure the similarity between two vectors and is calculated as follows:

$$cosine(x_i, x_j) = \frac{x_i \cdot x_j}{\|x_i\|\|x_j\|}. \tag{7}$$

The distance function is defined as:

$$distance(x_i, x_j) = 1 - cosine(x_i, x_j). \tag{8}$$

We intend to find at most 20 and at least 4 finer topics from a single host. Therefore, we can use the following formula to decide the number k in the k-means algorithm in the first phase:

$$k = \begin{cases} 4 & if \ |Q_{h_i}| = 20, \\ round(\frac{16}{1980} \times (|Q_{h_i}| - 20) + 4) & if \ 20 < |Q_{h_i}| \le 2000, \\ 20 & if \ |Q_{h_i}| > 2000. \end{cases} \tag{9}$$

Second Phase Clustering. Based on the clusters of queries obtained in the first phase, we compose sub-host vectors for each host h_i. Let C_l denote a cluster of queries. The sub-host vector of h_i constructed based on C_l can be denoted as

$h_i^l : (q_{i1}^l, q_{i2}^l, \cdots, q_{ij}^l, \cdots)$, where q_{ij}^l denotes the weight of q_j in the cluster C_l and is calculated as:

$$q_{ij}^l = \begin{cases} q_{ij} & if \ q_j \in C_l, \\ 0 & otherwise. \end{cases} \tag{10}$$

Then a topic matrix of a host is composed as below:

$$\begin{pmatrix} h_i^1 \\ h_i^2 \\ \cdot \\ \cdot \\ h_i^k \end{pmatrix} = \begin{pmatrix} q_{i1}^1 \ q_{i2}^1 \cdots\cdots q_{i|Q|}^1 \\ q_{i1}^2 \ q_{i2}^2 \cdots\cdots q_{i|Q|}^2 \\ \cdots \quad \cdots \\ \cdots \quad \cdots \\ q_{i1}^k \ q_{i2}^k \cdots\cdots q_{i|Q|}^k \end{pmatrix}, \tag{11}$$

where Q is the set of all queries in the query log data. Some cluster in C may contain very small number of queries and it is not the main topic of the host. We remove such small clusters i.e., the corresponding sub-host vectors, so that to avoid the possible noise. First we sort the clusters in a descending order of their size. Then we accumulate the queries from the top cluster to the bottom one. If the number of accumulated queries exceeds a fixed percentage threshold, the remaining clusters are pruned. Here we set the threshold to be 60%.

After the sub-host vectors are constructed for each host, we perform k-means clustering again on the entire set of sub-host vectors from different hosts to we obtain a set of clusters, i.e., website communities, composed of different sub-hosts. Our two-phase clustering method can capture more finer topics covered by the same host and it can cluster one host into different communities according to these finer topics. This provides more flexibility in the automatic community mining task.

3   Experiments

3.1   AOL Query Log Data

The query log data used in our experiments is the AOL data from [21] spanning three months from 1 March, 2006 to 31 May, 2006. The raw data is composed of queries and clicks recorded by a search engine. Each log record has 5 fields, namely, Anonymous User ID, Query, Query Time, Clicked URL, and URL Rank. The details of the fields are explained in Table 1. In the raw data, there are 19,442,629 lines of log records, 4,802,520 unique user ID, 1,416,831 unique URLs and 3,710,809 unique queries.

3.2   Data Preprocessing

One fragment of the raw data is given in Fig.1. It can be seen that the raw data contains a lot of noise records so that we first conduct data cleansing before the data is fed into our method. The cleansing operations include incomplete record removal, navigation query removal, and trivial host/query removal. As we can

Table 1. The details of the fields

	Meaning	Example
Anonymous User ID	Each search action has a unique Anonymous User ID, no user's information is revealed for privacy protection.	78253443
Query	The query issued by the user, it is case insensitive and the punctuation is removed.	lottery
Query Time	The time when the query was submitted for search.	2006-03-01 11:58:51
Clicked URL	If the user clicked on a search result, the domain portion (host) of the URL in the clicked result is stored.	www.calottery.com
URL Rank	If the user clicked on a search result, the rank of the clicked result is stored.	1

see in Fig.1, the Clicked URL field of some records is null, which indicates that the user did not click any result after he or she submitted the query. This kind of record is called incomplete record and should be removed. We also notice that quite a few queries are issued with the purpose of navigation, such as query "kbb.com" in Fig.1. These navigation queries are not helpful in semantic analysis and they are also removed. Finally, we also remove the trivial host and query. Trivial host means the host which has very few related queries in the entire log data. Similarly trivial query can be defined. We set the thresholds to be 20 and 10 for removing trivial hosts and trivial queries, respectively. This is an iterative cleansing process, because some trivial hosts are removed, some nontrivial queries may become trivial. After these cleansing operations, the remaining data set contains 3,031 hosts and 12,386 queries.

In Web search, different users have different searching habits. Even when they are searching for the same information, the queries issued may be different.

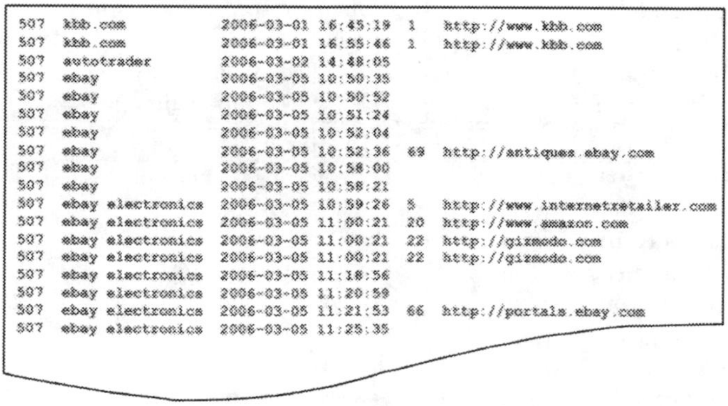

507	kbb.com	2006-03-01 16:45:19	1	http://www.kbb.com
507	kbb.com	2006-03-01 16:55:46	1	http://www.kbb.com
507	autotrader	2006-03-02 14:48:05		
507	ebay	2006-03-05 10:50:35		
507	ebay	2006-03-05 10:50:52		
507	ebay	2006-03-05 10:51:24		
507	ebay	2006-03-05 10:52:04		
507	ebay	2006-03-05 10:52:36	69	http://antiques.ebay.com
507	ebay	2006-03-05 10:58:00		
507	ebay	2006-03-05 10:58:21		
507	ebay electronics	2006-03-05 10:59:26	5	http://www.internetretailer.com
507	ebay electronics	2006-03-05 11:00:21	20	http://www.amazon.com
507	ebay electronics	2006-03-05 11:00:21	22	http://gizmodo.com
507	ebay electronics	2006-03-05 11:00:21	22	http://gizmodo.com
507	ebay electronics	2006-03-05 11:18:56		
507	ebay electronics	2006-03-05 11:20:59		
507	ebay electronics	2006-03-05 11:21:53	66	http://portals.ebay.com
507	ebay electronics	2006-03-05 11:25:35		

Fig. 1. A fragment of the AOL query log data

Therefore, we need to do query normalization to merge together the queries that have the same semantic meaning. Firstly, we remove the stop words in a query, such as "a", "of", "the", etc. Then, each remaining word in the query is lemmatized to its base form. Finally, because the search engines are keyword matching based, we can resort the words alphabetically in a query without affecting the searching results. For example, the query "store of movies dvd" will be converted into "store movies dvd" after stop words removing, then it is lemmatized into "store movie dvd", and finally it is reordered as "dvd movie store".

3.3 Results and Discussions

After the first phase, we obtained 7,875 sub-host vectors from the retained 3,031 hosts obtained after the cleansing. These sub-host vectors were fed into the

Table 2. Some example clusters, i.e., website communities

Cluster 1, 165 hosts	Cluster 2, 95 hosts
www.rankmytattoos.com	movies.yahoo.com
tattoo.about.com	movies.go.com
www.bullseyetattoos.com	www.the-numbers.com
www.vanishingtattoo.com	www.themovieinsider.com
www.tattoojohnny.com$^\beta$	www.hollywood.com
www.cheats.ign.com	movies.aol.com
www.cheatscodesguides.com	www.countingdown.com
www.tattoojohnny.com$^\alpha$	www.imdb.com
www.tattoonow.com	filmforce.ign.com
www.canismajor.com	movies.monstersandcritics.com
Cluster 3, 81 hosts	**Cluster 4, 95 hosts**
music.aol.com	www.internetautoguide.com
music.msn.com	www.carsearch.com
www.mp3.com	www.autobytel.com
www.allthelyrics.com	www.autotrader.com
www.rottentomatoes.com	www.modernracer.com
www.sing365.com	www.kbb.com
www.lyricsfreak.com	auto.consumerguide.com
www.lyricsdir.com	www.epinions.com
www.artistdirect.com	www.cars.com
www.lyricsdownload.com	www.automobilemag.com
Cluster 5, 48 hosts	**Cluster 6, 97 hosts**
www.123greetings.com	www.cars-on-line.com
www.bluemountain.com	www.oldride.com
www.hallmark.com	www.classicsandcustoms.com
yahoo.americangreetings.com	www.oldcars.com
www.dgreetings.com	www.antiquecar.com
www.1lovecards.com	www.classiccar.com
www.regards.com	www.vintagecars.about.com
www.egreetings.com	www.cars.com
www.americangreetings.com	www.oldcartrader.com
www.marlo.com	www.collectorcartraderonline.com

second phase to mine the website communities. We set the number of clusters k, i.e. the number of website communities, to be 100.

Six generated website communities are presented in Table 2. From the results, we can observe that the hosts belonging to the same topic are grouped in the same community. Cluster 1 is about "tattoos" and most of the websites in it are related to this type of information. Only one of the results in Cluster 1 has gone astray, namely, "cheats.ign.com". Cluster 2 includes a set of websites on movies and most of the sites are very popular ones such as "movies.yahoo.com", "hollywood.com", and "ibdb.com". Cluster 3 includes a set of websites on musics, such as "music.aol.com", "mp3.com", etc.

It can be seen that the same URL "tattoojohnny.com" has occurred twice in the Cluster 1. This is due to the two-phase clustering of our method that has been done. In the first phase of clustering, this host is divided into several sub-hosts, such as "tattoojohnny.com$^\beta$" and "tattoojohnny.com$^\alpha$", covering different content aspects of this host. The obtained sub-hosts from different hosts may have different topic granularity so that the sub-hosts from the same host may be clustered into the same community in the second phase. Therefore, our method is not sensitive on the difference among the hosts and it can automatically adjust the concerned granularity of topics.

The host "cars.com" appears in two clusters, namely, Cluster 4 and Cluster 6. From the hosts in Cluster 4, we may easily observe that this community focuses on providing some guidance in car trading (e.g. "internetautoguide.com" and "auto.consumerguide.com") as well as some mobile magazine (e.g. "automobilemag.com"). While the community obtained in Cluster 6 focuses on providing information about classic cars (e.g. "classiccar.com") and old car trading (e.g. "oldcartrader.com"). After checking the website of "cars.com" manually, we observe that this website provides information for both new cars and used

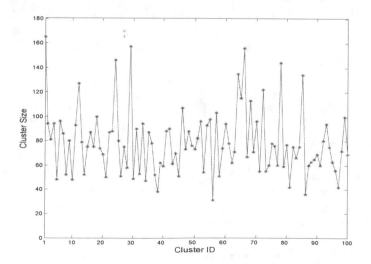

Fig. 2. Cluster size

cars. Because of this nature of the website, it is assigned into two communities by our two-phase mining method. Thus, different topics of the same site are well revealed in different communities.

The size of the mined communities is given in Fig. 2. The smallest community contains 30 hosts and the largest community contains 165 hosts. The diversity of the size is relatively small, which also suggests that the two-phase clustering strategy and small sub-host removal are effective to generate more reasonable clustering result.

4   Related Work

A number of papers have been published describing characteristics of query logs coming from some of the most popular search engines, including [1, 2, 9–11, 14, 16, 18, 19, 25]. Silverstein et al. [24] is the first to analyze a large query log of the AltaVista search engine containing about a billion queries submitted in a period of 42 days. Similarly to other works, their results showed that the majority of the users (in this case about 85%) only visit the first page of results. They also showed that 77% of the users' sessions end up just after the first query. As the authors stated, a smaller log could be influenced by ephemeral trends in querying (such as searches related to news just released, or to a new record released by a popular singer).

In [15] and [6], the authors analyzed a log made up of 7,175,648 queries issued to AltaVista during the summer of 2001. This second AltaVista log covers a time period almost three years after the first study was presented by Silverstein et al. and it is not as large as the first AltaVista log. Categorizing queries into topics is not a simple task. There are papers showing techniques for assigning labels to each query. Recent papers on the topic, including [3, 5, 7, 26], adopt a set of multiple classifiers subsequently refining the classification. Classification of the excite queries made in [25] shows that in no way is pornography a major topic of web queries, even though the top ranked query terms may indicate this. One sixth of the web queries have been classified as sex related. Web users look interested on a wide range of different topics, such as commerce, travel, and employment. Close to 10% of queries are about health and science.

5   Conclusions

In this paper, we proposed a method to automatically mine website communities by exploiting a large scale Web query logs. Query log data can be regarded as a comprehensive summarization of the real Web. The queries that lead to a particular website clicked are the summarization of the website content. The websites in the same topic are indirectly connected by the queries that convey information need in this topic. A two-phase method is proposed. In the first phase, we cluster the queries of the same host to obtain different content aspects of the host. In the second phase, we further cluster the obtained content aspects.

Because of the two-phase clustering, one host can appear in more than one website communities.

The results of our website community mining can help us find the most important and popular website communities, the most popular hosts or URLs in these communities, as well as the most popular queries for reaching to these popular destinations. It also lets users know about the reliability of a website. The amount of user visiting to a website can automatically tell the reliability of the information contained in the website.

References

1. Beitzel, S.M., Jensen, E.C., Chowdhury, A., Frieder, O., Grossman, D.: Temporal analysis of a very large topically categorized web query log. J. Am. Soc. Inf. Sci. Technol. 58(2), 166–178 (2007)
2. Beitzel, S.M., Jensen, E.C., Chowdhury, A., Grossman, D., Frieder, O.: Hourly analysis of a very large topically categorized web query log. In: Proceedings of the 27th Annual International ACM SIGIR Conference on Research and Development in Information Retrieval, pp. 321–328 (2004)
3. Beitzel, S.M., Jensen, E.C., Frieder, O., Lewis, D.D., Chowdhury, A., Kolcz, A.: Improving automatic query classification via semi-supervised learning. In: Proceedings of the Fifth IEEE International Conference on Data Mining, pp. 42–49 (2005)
4. Bing, L., Lam, W., Wong, T.L.: Using query log and social tagging to refine queries based on latent topics. In: Proceedings of the 20th ACM International Conference on Information and Knowledge Management, pp. 583–592 (2011)
5. Broder, A.Z., Fontoura, M., Gabrilovich, E., Joshi, A., Josifovski, V., Zhang, T.: Robust classification of rare queries using web knowledge. In: Proceedings of the 30th Annual International ACM SIGIR Conference on Research and Development in Information Retrieval, pp. 231–238 (2007)
6. Fagni, T., Perego, R., Silvestri, F., Orlando, S.: Boosting the performance of web search engines: Caching and prefetching query results by exploiting historical usage data. ACM Trans. Inf. Syst. 24(1), 51–78 (2006)
7. Gravano, L., Hatzivassiloglou, V., Lichtenstein, R.: Categorizing web queries according to geographical locality. In: Proceedings of the Twelfth International Conference on Information and Knowledge Management, pp. 325–333 (2003)
8. Hu, J., Fang, L., Cao, Y., Zeng, H.J., Li, H., Yang, Q., Chen, Z.: Enhancing text clustering by leveraging wikipedia semantics. In: Proceedings of the 31st Annual International ACM SIGIR Conference on Research and Development in Information Retrieval, pp. 179–186 (2008)
9. Jansen, B.J., Spink, A.: An analysis of web searching by european alltheweb. com users. Inf. Process. Manage. 41(2), 361–381 (2005)
10. Jansen, B.J., Spink, A.: How are we searching the world wide web?: a comparison of nine search engine transaction logs. Inf. Process. Manage. 42(1), 248–263 (2006)
11. Jansen, B.J., Spink, A., Koshman, S.: Web searcher interaction with the dogpile.com metasearch engine. J. Am. Soc. Inf. Sci. Technol. 58(5), 744–755 (2007)
12. Joachims, T.: Optimizing search engines using clickthrough data. In: Proceedings of the Eighth ACM SIGKDD International Conference on Knowledge Discovery and Data Mining, pp. 133–142 (2002)

13. Kleinberg, J.M.: Authoritative sources in a hyperlinked environment. Journal of the ACM 46(5), 604–632 (1999)
14. Koshman, S., Spink, A., Jansen, B.J.: Web searching on the vivisimo search engine. J. Am. Soc. Inf. Sci. Technol. 57(14), 1875–1887 (2006)
15. Lempel, R., Moran, S.: Predictive caching and prefetching of query results in search engines. In: Proceedings of the 12th International Conference on World Wide Web, pp. 19–28 (2003)
16. Mat-Hassan, M., Levene, M.: Associating search and navigation behavior through log analysis: Research articles. J. Am. Soc. Inf. Sci. Technol. 56(9), 913–934 (2005)
17. Ni, X., Sun, J.T., Hu, J., Chen, Z.: Cross lingual text classification by mining multilingual topics from wikipedia. In: Proceedings of the fourth ACM International Conference on Web Search and Data Mining, pp. 375–384 (2011)
18. Ozmutlu, H.C., Spink, A., Ozmutlu, S.: Analysis of large data logs: an application of poisson sampling on excite web queries. Inf. Process. Manage. 38(4), 473–490 (2002)
19. Ozmutlu, S., Spink, A., Ozmutlu, H.C.: A day in the life of web searching: an exploratory study. Inf. Process. Manage. 40(2), 319–345 (2004)
20. Page, L., Brin, S., Motwani, R., Winograd, T.: The pagerank citation ranking: Bringing order to the web. In: Proceedings of the 7th International World Wide Web Conference, pp. 161–172 (1998)
21. Pass, G., Chowdhury, A., Torgeson, C.: A picture of search. In: InfoScale 2006 (2006)
22. Ponzetto, S.P., Navigli, R.: Large-scale taxonomy mapping for restructuring and integrating wikipedia. In: Proceedings of the 21st International Jont Conference on Artifical Intelligence, pp. 2083–2088 (2009)
23. Radlinski, F., Joachims, T.: Query chains: learning to rank from implicit feedback. In: Proceedings of the Eleventh ACM SIGKDD International Conference on Knowledge Discovery in Data Mining, pp. 239–248 (2005)
24. Silverstein, C., Marais, H., Henzinger, M., Moricz, M.: Analysis of a very large web search engine query log. SIGIR Forum 33(1), 6–12 (1999)
25. Spink, A., Ozmutlu, H.C., Lorence, D.P.: Web searching for sexual information: an exploratory study. Inf. Process. Manage. 40(1), 113–123 (2004)
26. Vogel, D., Bickel, S., Haider, P., Schimpfky, R., Siemen, P., Bridges, S., Scheffer, T.: Classifying search engine queries using the web as background knowledge. SIGKDD Explor. Newsl. 7(2), 117–122 (2005)
27. Wang, P., Domeniconi, C.: Building semantic kernels for text classification using wikipedia. In: Proceedings of the 14th ACM SIGKDD International Conference on Knowledge Discovery and Data Mining, pp. 713–721 (2008)
28. Wang, X., Zhai, C.: Mining term association patterns from search logs for effective query reformulation. In: Proceedings of the 17th ACM Conference on Information and Knowledge Management, pp. 479–488 (2008)
29. Wong, T.L., Bing, L., Lam, W.: Normalizing web product attributes and discovering domain ontology with minimal effort. In: Proceedings of the Fourth ACM International Conference on Web Search and Data Mining, pp. 805–814 (2011)
30. Zhai, Y., Liu, B.: Web data extraction based on partial tree alignment. In: Proceedings of the 14th International Conference on World Wide Web, pp. 76–85 (2005)

Extracting Social Events Based on Timeline and User Reliability Analysis on Twitter

Bayar Tsolmon and Kyung-Soon Lee[*]

Division of Computer Science and Engineering, CAIIT, Chonbuk National University,
567 Baekje-daero, deokjin-gu, Jeonju-si, Jeollabuk-do
561-756 Republic of Korea
bayar_277@yahoo.com, selfsolee@chonbuk.ac.kr

Abstract. When some hot social issue or event occurs, it will significantly increase the number of comments and retweet on that day on twitter. Generally, an event can be extracted by its term frequency but it is hard to find an event that has a low term frequency. Because of this reason there can be a probability of missing important information. However, there is a kind of reliable user who is directly related to that event so that no matter how low the number of tweet is on that case. In this paper, we propose user reliability based event extraction method. The latent Dirichlet allocation(LDA) model is adapted with timeline analysis to extract high-frequency events. User behaviors are analyzed to classify reliable users who are directly related to the issue. Reliable low-frequency events can be detected based on reliable users. In order to verify the effectiveness of the proposed method, four social issues are selected and experimented on Korean twitter test set. The experimental results showed 97.2% in precision for the top 10 extracted events (P@10) on each day. This result shows that the proposed method is effective for extracting events in twitter corpus.

Keywords: Event extraction, timeline analysis, user behavior analysis, temporal LDA.

1 Introduction

With the rapidly increasing number of data on social networking services (SNS) lately, researches on event extraction attract more and more attention. Compared with the news and blog data, the SNS data are more widely used in the real-time event extraction. However, as the amount of SNS data increases, the noise also increases synchronously; thus a reliable event extraction method is being required.

Since the general event extraction methods [1, 2], depend on the term frequency, thus there has been a problem that some important information might be missed. Especially the frequency of a tweet's retweet that is irrelevant to the event such as rumor, spam or advertisement is high on the SNS environment. Besides, since posts are short and noisy, it is hard to extract reliable events easily via the frequency of

---

[*] Corresponding author.

A. Gelbukh (Ed.): CICLing 2014, Part II, LNCS 8404, pp. 213–223, 2014.
© Springer-Verlag Berlin Heidelberg 2014

words. Also, the events with low-frequency tweets cannot be extracted even though the users of the tweets are directly involved in the events.

In order to extract reliable low-frequency events as well as high-frequency events, we have observed the characteristics of users for event extraction as follows. On Twitter, there are some users who tend to write a lot about the issue. Such users are concerned about the issue for a long time and may publish valuable information on that issue. We assume that detecting active and reliable users for a certain issue allows to increase the effectiveness of the event extraction. We introduce to classify a Twitter user who is an important property for the event extraction based on user behavior analysis of two groups as socially well-known user and active user. If the socially well-known users mention about the particular event, it indicates that a significant social event happened. The active users are valuable users because they post important information every time an event occurs. Also, we have observed event features as follows. When any new event occurs on a certain day, there might be use of certain terms many times on that day as compared to use of those terms on any other normal day. Another aspect is that users write tweets to express their opinions on particular issues or events with positive or negative sentiments. We use a sentiment lexicon to consider the context of an event term which contains users' opinions.

In this paper, we propose an event extraction method based on timeline and user behavior analysis. Timeline and sentiment analysis is applied to the latent Dirichlet allocation (LDA) model [3] to cover multiple events which occur on the same day. Because, a sudden increase of topically similar posts usually indicates an event. To reflect the term distribution on timelines and user's sentiment on a specific event, event terms are extracted by chi-square measure and opinion scores. We have evaluated our method on a Korean Twitter data of four hot issues.

The rest of the paper is organized as follows: Section 2 presents related work; Section 3 describes our method of extracting events based on timeline and user behavior analysis. Section 4 shows the experimental results on a Korean tweet collection. We conclude the paper in Section 5.

2   Related Work

A rich set of studies has been conducted in various forms of social media. Tinati et al. [4] developed a model based upon the Twitter message exchange which enables us to analyze conversations around specific topics and identify key players in the conversation. A working implementation of the model helps categorize Twitter users by specific roles based on their dynamic communication behavior rather than an analysis of their static friendship network [5-7]. Xu et al. [8] performed a comprehensive review of the user posting behavior on the popular social media website, Twitter. Specifically, they assume that a user behavior is mainly influenced by three factors: breaking news, posts from social friends and user's intrinsic interest, and propose a mixture latent topic model to combine all these factors. Also, users who are categorized in an information sharing group tend to have a relatively large number of followers compare to follow. Information seekers have a large number of following compare to followers [9]. Kwak et al. [10] compared three different measures of influence-number of followers, page-rank, and the number of retweets-finding that the ranking of the most influential users differed depending on the measure.

Yang et al. [11] analyzed how the retweet behavior is influenced by factors: user, message, time, etc. Based on these important observations, they propose a semi-supervised framework of a factor graph model to predict users' retweet behaviors. Mendoza et al. [12] showed that the propagation of tweets that correspond to rumor differs from tweets that spread the news because rumors tend to be questioned more than news by the Twitter community. Sayyadi et al. [13] developed a new event detection algorithm creating a keyword graph and using community detection methods analogous to those used in social network analysis in order to discover events.

There are recent reports suggesting the use of sentiment analysis for detecting events. Pak et.al. [14] have shown to collect automatically a corpus for sentiment analysis and opinion mining purposes. They performed linguistic analysis of the collected corpus and explained discovered phenomena.

Topic models provide a principled way to discover hidden topics from large document collections. Standard topic models do not consider temporal information. A number of temporal topic models have been proposed to cover a topic changes over time [15]. Some of these models focus on the change of the topic distribution over time [16].

3   Event Extraction Based on Temporal LDA Model and Reliable Users

Generally when some event occurs, an event will be informed by authority users or through the mass media or normal users on Twitter. In addition, on each time of event occurrence there are some users who tend to write a lot about the issue. Such users are concerned about the issue for a long time and may publish valuable information on that issue. For example, the weather information from tweets of the official Twitter account of the National Weather Service is accurate and reliable. Also it could be quick and reliable to get the latest information for iPhone from tweets of the official Twitter account of Apple. Therefore, in order to extract accurate and reliable events, detecting reliable and active users about the issue is important. The most active and reliable users write tweets with references including news URLs, event related pictures and specific information.

Table 1. Tweet examples of an active user on earthquake issue

Date	Tweet example	# of retweet	Event
2010.11.16	4.5 scale earthquake in Sichuan, China http://j.mp/9mA5mR (China Daily news)	1	China earthquake
2011.02.23	Earthquake in New Zealand, South Korean 2 people missing	10	New Zeland earthquake
2011.03.11	7.0 scale earthquake in northeastern Japan, Tsunami warning...	84	Japan earthquake
2011.03.11	4.5 scale earthquake in Hawaii (AP)	24	Hawaii earthquake

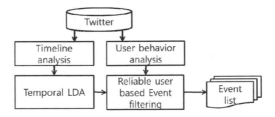

Fig. 1. The system structure of reliable user based event extraction

Since the existing event extraction methods depend on the number of retweets of tweets, there is a problem that some important tweets with fewer numbers of retweets might be missed. Table 1 shows tweet examples of an active user on "earthquake" issue by Twitter user id '*kbsnewstweet*' which is the official account of the Korean Broadcasting System (KBS). Here, even though the number of retweets is less, each tweet can be considered as an earthquake issue. We assume that reliable user information can be used   to extract reliable low-frequency events.

According to the our observation, we propose an event extraction method based on user behavior and timeline analysis as shown in Figure 1. Event terms are extracted based on timeline analysis and sentiment features; The temporal LDA is adapted to cluster high-frequency events on timeline (section 3.1); User behaviors are analyzed to detect active and reliable users (section 3.2); Events are filtered based on reliable users for the events by Temporal LDA (section 3.3).

3.1   Event Extraction Based on Temporal LDA Model

In this section, we propose event term extraction method based on temporal LDA topic model (T-LDA). The purpose of using the LDA model is to cover multiple events which occur on the same day. Generally, the basic event extraction methods are based on term frequency which has the disadvantage of not extracting those events. To figure out this problem, we have applied additional timeline weight on each term to the conventional LDA [3] topic model as shown in Figure 2. In the proposed model, the latent Dirichlet allocation(LDA) model is adapted with timeline analysis to extract high-frequency clustered events (shon in the Figure 3). Here, T represents time series, the parameter x indicates the additional weight of each term on time t. The weight x is calculated by *ChiOpScore* weighting by applying chi-square value and opinion score [17].

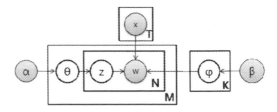

Fig. 2. Graphical representation of T-LDA model

When a new event occurs on a certain day, there might have a big occurrence of certain terms on that day as compared to occurence of those terms on any other normal day. The term significance of that day is measured by the chi-square statistic which measures the lack of independence between a term and the date. The users write tweets to express their opinions on particular issues or events with positive or negative sentiment words. A sentiment lexicon is used to detect the context of an event term which contains the user's opinion.

To reflect the term distribution on timelines and user's sentiments on a specific event, event terms are extracted by combining chi-square value and opinion score.

$$ChiOpScore(w, t_0) = \lambda \cdot ChiSquare(w, t_0) + (1 - \lambda) \cdot OpScore(w, t_0) \qquad (1)$$

where w is a bigram word, t_0 is a particular date, the parameter λ is set to 0.7 emprically. $OpScore$ is measured by combining the frequency of a term and the frequency of opinion words.

The $ChiOpScore$ value of timeline and sentiment anlaysis is applied to the standard LDA topic model to give an impact on the importance of event terms.

3.2   Reliable User Detection

We have classified a Twitter user who is an important property for event extraction. Users are classified according to tweeting behaviours into two groups: socially well-known users and active users. If the socially well-known users mention about the particular event, it indicates that a big social event happened. The highly active users are valuable ones because they post important information whenever a new event occurs.   We have defined these two groups as reliable users.

Detecting Socially Well-Known Users. Generally, socially well-known users on Twitter tend to have a lot number of tweets and retweets. To extract socially well-known users, we have adapted a HITS algorithm [18] by applying mentions, RTs (modified retweets) and retweets with an edge weight between user nodes [19].

$$AuthScore^{(T+1)}(p) = \sum_{q \to p} w_{qp} \times HubScore^T(q) \qquad (2)$$

$$HubScore^{(T+1)}(p) = \sum_{p \to q} w_{pq} \times AuthScore^T(q) \qquad (3)$$

The edge weight w_{qp} is as follows:

$$w_{qp} = \sum_{q \to p} FreqRT(q, p) + \sum_{q \to p} Mention(q, p) \qquad (4)$$

The top ranked 100 users with high **AuthScore** are selected as socially well-known users.

Detecting Active Users. Since there is a user who writes a lot of tweets about the issue and actively writes whenever the event related to the issue happens, these users regularly have relatively higher activity than other users. The following formula calculates the average weekly activity score [19].

$$Activity\ Score(u) = \frac{1}{W}\sum_{i=1}^{W} TweetFreq(u, d_i) \times RTFreq(u, d_i) \qquad (5)$$

where W shows the number of weeks; $TweetFreq$ shows the sum of tweets d that a user u wrote in the each ith week; $RTFreq$ represents the number of retweets d of the tweets written by a user u in the each ith week.

The top ranked 100 users with high **ActivityScore** are selected as active and reliable users. The events of the tweets by the detected reliable users can be extracted by the event filtering step.

3.3   Event Filtering Based on Reliable Users

There are some noise high frequency event terms which are not related to the event. In order to remove those noises, an event filtering approach is applied based on the reliable and highly active user's data. This process can give lower ranks to the highly frequent but unrelated event terms. On the other hand, it can give higher ranks to the low frequent but important event terms. The top events are selected for the reranked event terms.

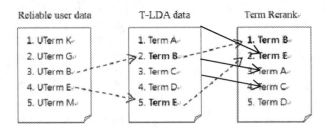

Fig. 3. Event filtering process

The event filtering procedure is illustrated in Figure 4. The procedure combines event terms based on T-LDA and reliable users.

Events are filtered based on the reliability user's tweets as following steps:

Step 1: The top 10 users are extracted according to the activity scores and authority scores. These 10 users are selected as reliable users.

Step 2: Event terms are selected by the tweets of the selected reliable users for the particular day.

Step 3: Priority tags are attached for the selected terms from both T-LDA and Step 2.

Step 4: Give higher ranks on the event terms with a tag, and give lower ranks to event terms with no tag.

When the term distribution on timelines becomes high, the probabilities of mentioning that term by reliable users increase. Even though the frequency of event term is less, at least one reliable user posted mentioned about that term, it could be extracted.

4   Experiments

4.1   Experimental Set-Up

We have evaluated the effectiveness of the proposed method on tweet collection. Four issues are chosen, and tweet documents for the issues are collected, spanning from November 1, 2010 to March 26, 2011 by Twitter API [20] (all issues and tweets are written in Korean). Table 2 shows the number of tweets related to each issue. Each record in this tweet data set contains the actual tweet body and the time when the tweet was published.

Table 2. Twitter data set

Issue	# of users	# of tweets
Park Ji-Sung	29,568	131,533
Kim Yu-Na	10,563	26.844
Earthquake	110,345	467,955
Cheonanham	19,473	84,195
Total	169,949	710,527

First, we exclude stop words about twitter data of collecting each issue, extracted noun and verb words using the Korean POS tagger. After preprocessing, we created a word pair of features in each tweet by using bi-gram with 3 window size.

4.2   Results on Reliable User Detection

The comparison methods for detecting reliable users are as in the following:

— Baseline: Twitter user's follower and following ratio
— AuthScore: Detecting socially well-known users (equation 2)
— ActivityScore: Detecting active and reliable users (equation 5)

The experimental results for the reliable user detection are shown in Table 3. The method based on Twitter user's follower and following ratio is used as a baseline for the extraction of reliable users. We have defined the top 10 users detected by each method as reliable users. In the result of the experiment, the proposed method showed better performance than the baseline. The method AuthScore and Activity Score achieved 77% and 92% respectively.

Table 3. The experimental results for the reliable user detection(P@10)

User classification	Baseline	AuthScore	ActivityScore
Park Ji-Sung	0.65	0.6	1.0
Kim Yu-Na	0.55	0.9	0.9
Earthquake	0.30	0.7	0.8
Cheonanham	0.65	0.9	1.0
Average	0.53	0.77	0.92

Table 4. Examples of the top reliable users

Issues	Top users ID	Description
Park Ji-Sung	theKFA	The Korean football association
Kim Yu-Na	FEVERSmedia	Fever skating fan forum
Earthquake	Kbsnewstweet	KBS news
Cheonanham	ROK_MND	Ministry of national defense

Since an experiment carried out by targeting users who are interested in the issue, a user who is socially popular about each issue, a user who has a reliable and direct correlation with an event could be extracted. Examples of the top extracted reliable users are shown in Table 4.

4.3   Results on Event Extraction

The comparison methods for event extraction are as follows:

— ChiOpScore: Event extraction based on timeline analysis and sentiment features
— T-LDA: Event extraction based on temporal LDA model
— Proposed method: Event filtering using reliable user tweet and T-LDA

Table 5. Event lists and answers for four issues

Issue	Event	Event List	Date	# of answer
Park Ji-Sung	E1	Park Ji-Sung Double goal	2010.11.07	5
	E2	Promote World Cup bid	2010.12.02	4
	E3	Season 6th Goal	2010.12.14	4
	E4	Park's facial injuries	2011.01.23	5
	E5	Ji-Sung retirement?	2011.01.31	7
	E6	Cha Beomgeun's confession	2011.02.01	2
Kim Yu-Na	E7	*Announcing* her new *programs*	2010.11.30	4
	E8	UNICEF *Goodwill Ambassador*	2010.12.02	3
	E9	Private training center	2010.12.27	6
	E10	Yu-Na's Devil Mask?	2011.01.26	3
	E11	Ultra - mini dress	2011.02.15	3
	E12	"Giselle" domestic training	2011.03.22	3
Earthquake	E13	Japan earthquake	2010.11.30	3
	E14	Jeju Island earthquake	2011.01.12	2
	E15	Signs of a volcanic eruption in Japan	2011.01.27	4
	E16	New Zealand earthquake	2011.02.22	6
	E17	Japan Tsunami Warning	2011.03.09	3
	E18	Japan Massive Earthquake	2011.03.11	1
Cheonanham	E19	Cheonanham torpedo attack	2010.11.04	3
	E20	'In-Depth 60 Minute' KBS broadcast	2010.11.17	2
	E21	Bombardment of Yeonpyeong Island	2010.11.23	3
	E22	'In-Depth 60 Minute' Scandal	2011.01.06	5
	E23	A Rumor about prepared for attack	2011.02.03	6
	E24	The 1st anniversary of Cheonanham	2011.03.21	4

Table 6. Summary of comparison results (P@10)

Issue No	Chi-OpScore	T-LDA	Proposed method	Issue No	ChiOpS-core	T-LDA	Proposed method
E1	5/5	5/5	5/5	E13	3/3	3/3	3/3
E2	4/4	4/4	4/4	E14	2/2	2/2	2/2
E3	4/4	4/4	4/4	E15	4/4	4/4	4/4
E4	5/5	5/5	5/5	E16	5/6	5/6	6/6
E5	6/7	6/7	7/7	E17	3/3	3/3	3/3
E6	2/2	2/2	2/2	E18	1/1	1/1	1/1
E7	4/4	4/4	4/4	E19	3/3	3/3	3/3
E8	3/3	3/3	3/3	E20	2/2	2/2	2/2
E9	5/6	5/6	6/6	E21	2/3	2/3	2/3
E10	3/3	3/3	3/3	E22	5/5	5/5	5/5
E11	2/3	2/3	2/3	E23	6/6	6/6	6/6
E12	3/3	3/3	3/3	E24	4/4	4/4	4/4
				Avg	94.3%	95.2%	97.2%

Table 5 shows 24 social events for six issues and the number of answers. All the events have taken place at a particular date. The answer sheet is judged by three human assessors for the top 10 event terms extracted by each method. We set a number of topics K to 20, α to 0.01 and β to 0.01 after some preliminary experiments. The T-LDA model was run for 500 iterations of Gibbs sampling. In our experiment the time T is spanning from November 1, 2010 to March 26, 2011, totaling 147 days.

Table 6 shows the result of the comparative experiment for 24 events. For each event, it shows the number of answers for the total answers. Here, '1/3' describes that one answer is detected by the proposed method among 3 answers. In the result of an experiment, T-LDA achieved 95.2%, and event extraction using reliable user showed 97.2%. From the result, a user who is reliable and highly active on the issue provided important information for each time the event related to issue occur. Our extraction method showed improvements in three events such as E5, E9 and E16 compared to the baseline.

The main limitation of the experiments with reliable user detection was to establish the gold standard. Firstly, the simplest approach for detecting reliable users is to apply a threshold criterion. But, determining threshold criteria for the various issue was not appropriate for our work. Secondly, we have selected the top 100 users detected by each method as reliable users. However the experimental results for those users was less effective than the baseline. For these reasons, we defined the top 10 users detected by each method as reliable users.

5   Conclusion

In this paper, we proposed the event extraction method based on timeline and user behavior analysis on Twitter. Event terms are extracted from the temporal LDA model, and reliable users are classified by modifying the HITS algorithm and user activity measurement. In order to give distinctive weight of an event term, term significance is filtered by reliable user data. The experimental results of the Korean tweet collection showed that the proposed method achieved 97.2% in average precision in the top 10 results. The result indicates that reliable users and the temporal LDA model allow to increase the effectiveness of the event extraction.

Discovering methods for better event expressions and less dependent on the number of tweets are future works.

Acknowledgements. This work was supported by Basic Science Research Program through the National Research Foundation of Korea (NRF) funded by the Ministry of Education, Science and Technology (NRF-2012R1A1A2044811).

References

1. Benson, E., Haghighi, A., Barzilay, R.: Event Discovery in Social Media Feeds. In: Proceedings of the 49th Annual Meeting of the Association for Computational Linguistics: Human Language Technologies, vol. 1, pp. 389–398. ACL (2011)
2. Popescu, A.M., Pennacchiotti, M.: Detecting Controversial Events from Twitter. In: Proceedings of the 19th ACM International Conference on Information and Knowledge Management, pp. 1873–1876. ACM (2010)
3. Blei, D.M., Ng, A.Y., Jordan, M.I.: Latent Dirichlet Allocation. The Journal of machine Learning Research 3, 993–1022 (2003)
4. Tinati, R., Carr, L., Hall, W., Bentwood, J.: Identifying Communicator Roles in Twitter. In: Proceedings of the 21st International Conference Companion on World Wide Web, pp. 1161–1168. ACM (2012)
5. Kardara, M., Papadakis, G., Papaoikonomou, T., Tserpes, K., Varvarigou, T.: Influence Patterns in Topic Communities of Social Media. In: Proceedings of the 2nd International Conference on Web Intelligence, Mining and Semantics, pp. 10–21. ACM (2012)
6. Weng, J., Lim, E.P., Jiang, J., He, Q.: Twitter Rank: Finding Topic-sensitive Influential Twitterers. In: Proceedings of the Third ACM International Conference on Web Search and Data Mining, pp. 261–270. ACM (2010)
7. Sun, B., Ng, V.T.: Identifying Influential Users by Their Postings in Social Network. In: Proceedings of the 3rd International Workshop on Modeling Social Media, pp. 1–8. ACM (2012)
8. Xu, Z., Zhang, Y., Wu, Y., Yang, Q.: Modeling User Posting Behavior on Social Media. In: Proceedings of the 35th International ACM SIGIR Conference on Research and Development in Information Retrieval, pp. 545–554. ACM (2012)
9. Java, A., Song, X., Finin, T., Tseng, B.: Why We Twitter: Understanding Microblogging Usage and Communities. In: Proceedings of WebKDD/SNA-KDD, pp. 556–565. ACM (2007)
10. Kwak, H., Lee, C., Park, H., Moon, S.: What is Twitter, a Social Network or a News Media? In: Proceedings of the 19th International Conference on World Wide Web, pp. 591–600. ACM (2010)
11. Yang, Z., Guo, J., Cai, K., Tang, J., Li, J., Zhang, L., Su, Z.: Understanding retweeting behaviors in social networks. In: Proceedings of the 19th ACM International Conference on Information and Knowledge Management, pp. 1633–1636. ACM (2010)

12. Mendoza, M., Poblete, B., Castillo, C.: Twitter Under Crisis: Can we trust what we RT? In: Proceedings of the First Workshop on Social Media Analytics, pp. 71–79. ACM (2010)
13. Sayyadi, H., Hurst, M., Maykov, A.: Event Detection and Tracking in Social Streams. In: Proceedings of ICWSM, pp. 311–314 (2009)
14. Pak, A., Paroubek, P.: Twitter as a Corpus for Sentiment Analysis and Opinion Mining. In: Proceedings of LREC, pp. 1320–1326 (2010)
15. Diao, Q., Jiang, J., Zhu, F., Lim, E.P.: Finding bursty topics from microblogs. In: Proceedings of the 50th Annual Meeting of the Association for Computational Linguistics: Long Papers, vol. 1, pp. 536–544. ACL (2012)
16. Blei, D.M., Lafferty, J.D.: Dynamic topic models. In: Proceedings of the 23rd International Conference on Machine Learning, pp. 113–120. ACM (2006)
17. Tsolmon, B., Kwon, A.-R., Lee, K.-S.: Extracting Social Events Based on Timeline and Sentiment Analysis in Twitter Corpus. In: Bouma, G., Ittoo, A., Métais, E., Wortmann, H. (eds.) NLDB 2012. LNCS, vol. 7337, pp. 265–270. Springer, Heidelberg (2012)
18. Kleinberg, J.M.: Authoritative Sources in a Hyperlinked Environment. Journal of the ACM 46(5), 604–632 (1999)
19. Tsolmon, B., Lee, K.S.: A Graph-based Reliable User Classification. In: Proceedings of the First International Conference on Advanced Data and Information Engineering, pp. 61–68. Springer, Singapore (2014)
20. Java library for Twitter API, http://twitter4j.org

Beam-Width Adaptation
for Hierarchical Phrase-Based Translation

Fei Su[1], Gang Chen[2], Xinyan Xiao[2], and Kaile Su[3]

[1] Department of Computer Science
Peking University
feisu@pku.edu.cn
[2] Youdao Inc.
Beijing, China
{chengang,xiaoxy}@rd.netease.com
[3] College of Mathematics, Physics and Information Engineering,
Zhejiang Normal University, Jinhua, China
kailepku@gmail.com

Abstract. In terms of translation quality, hierarchical phrase-based translation model (Hiero) has shown state-of-the-art performance in various translation tasks. However, the slow decoding speed of Hiero prevents it from effective deployment in online scenarios.

In this paper, we propose beam-width adaptation strategies to speed up Hiero decoding. We learn maximum entropy models to evaluate the quality of each span and then predict the optimal beam-width for it. The empirical studies on Chinese-to-English translation tasks show that, even in comparison with a competitive baseline which employs well designed cube pruning, our approaches still double the decoding speed without compromising translation quality. The approaches have already been applied to an online commercial translation system.

Keywords: Machine translation, Hierarchical Phrase-base model, Beam-Width Adaptation.

1 Introduction

The recent years have witnessed the proliferation of online translation services. In these applications, fast response and high efficiency are required. Hiero is proposed by Chiang in 2005[1]. It brings significant improvements in translation quality. However, we have encountered difficulties in efficiency while deploying Hiero as online translation service. Its decoding process is based on CKY algorithm, which treats every span (i.e. substring) of the input sentence as a CKY cell. For a sentence with n words, Hiero must deal with $(n+1)n/2$ spans (Figure 1). Let m denote the beam-width (maximum hypotheses quota in a span), the decoding time complexity is $O(mn^3)$. Our experiments on NIST data sets reveal that, with other circumstances being the same, the decoding time of Hiero is on average four times longer than that of the traditional phrase-based model [2].

Although the decoding of Hiero processes a huge number of spans, we observe that: 1. merely a very limited amount of spans contribute to the final 1-best results (i.e. those spans produce hypotheses which are involved in the composition

A. Gelbukh (Ed.): CICLing 2014, Part II, LNCS 8404, pp. 224–232, 2014.

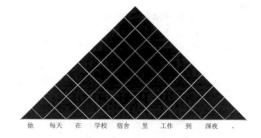

Fig. 1. Hiero decoding requires all spans be processed under a fixed beam-width

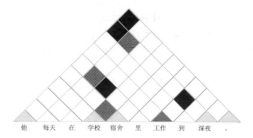

Fig. 2. Only a small amount of the spans contribute to the 1-best result in Hiero decoding. The gray level of a cell indicates the number of hypotheses needed in it.

of the 1-best results); 2. even in these fruitful spans, the adopted hypotheses are usually ranked top in the cube pruning step. To illustrate the observation, consider the decoding process for a Chinese sentence using our Hiero system in Figure 1, in which the cells are painted in different gray levels. A cell gets darker color if the related span requires more hypotheses (if the adopted hypothesis is ranked k^{th} in a span in cube pruning, then the span must contain at least k hypotheses). The white cell indicates that the span is useless. As we can see, a majority of the spans are unnecessarily processed.

To speed up Hiero, previous cube-pruning based methods [3–5] generate hypotheses for every span with a fixed beam-width. Inspired by cell closure work in context-free parsing [6, 7], we believe that beam-width should be adjusted according to a certain measurement for the "goodness" of a span.

Generally speaking, most of the spans in a sentence do not make sense grammatically (e.g. the span [*movie Les*] in the sentence "I watched the movie Les Misrables"). The hypotheses from them provide little help in decoding. Our statistics in Chinese-to-English(C2E) translation using Hiero show that: merely 5% of the spans contribute to the final translation. In those fruitful spans, 65% have adopted hypotheses ranked top 3, and 90% ranked top 20.

To improve the efficiency of Hiero decoding, we first evaluate the "quality" of a span. Based on that, we propose two strategies for beam-width adaptation. 1. Cell closure: low-quality spans are directly skipped in the chart parsing step; 2. Dynamic beam-width: instead of a fixed beam-width, a quality-determined beam-width is assigned to rest of the spans in the cube pruning step. The empirical studies on C2E translation show that our approaches have doubled the translation efficiency and achieve a satisfying performance.

The rest of the paper is organized as follows: in section 2 we discuss the establishment of span quality criterions. In section 3, we introduce two beam-width adaptation strategies. The details of the experiments are presented in section 4. The whole work is summarized in section 5.

2   Measuring Span Quality

Before the discussion of beam-width adaptation for a span, we firstly need to measure the "span quality" based on its lexical information. For a sentence s, the **quality** of a span which starts at ith word and ends at jth word is denoted as $Q(s, i, j)$.

At first, we consider $Q(s, i, j)$ as the probability of span contributing to the 1-best translation result. To learn that probability, we train a binary maximum entropy classifier which predicts *positive* if the span should be adopted in the 1-best result, and *negative* otherwise. we use the predicted probability as span quality:

$$Q_1(s, i, j) = P(positive|s, i, j) \tag{1}$$

To acquire the training data, we translate about 3 million Chinese sentences into English using Hiero. In the decoding step, for each span we record: 1. unigram and bigram lexical features; 2. whether it contributes to the 1-best translation result. Notice that the prediction result is apparently biased to *negative*, for the negative cases significantly outperform the positive cases in the training data.[1]

Besides Q_1, we also investigate other possible definitions for quality measurement. We believe that $Q(s, i, j)$ is mostly determined by whether the span has reasonable boundaries. Following [8] and [9], we train two maximum entropy classifiers C_{begin} and C_{end}. For each position of a sentence, C_{begin} (C_{end}) predicts *positive* if the position can be served as a begin (end) boundary of a span, and *negative* otherwise.

The training data of C_{begin} and C_{end} are also obtained during the decoding of the above 3 million sentences. For each position of the sentence, we record: 1. the boundary lexical features; 2. whether it is a begin (end) boundary of any adopted span. Under such circumstance, the problem of excess negative cases no longer exists.

It is quite obvious that the span will be inappropriate if either side of the boundary is inappropriate. Thus we present the second span quality definition using the probabilities predicted by C_{begin} and C_{end}:

$$Q_2(s, i, j) = min\{P_b(positive|s, i), P_e(positive|s, j)\} \tag{2}$$

---

[1] We follow [7] and weight positive cases λ times heavier than negative cases during training. In our experiments $\lambda = 20$.

Considering the limited precision of the two classifiers(about 72%~73%), a more conservative quality definition is proposed:

$$Q_3(s,i,j) = (P_b(positive|s,i) + P_e(positive|s,j))/2 \tag{3}$$

The above three introduced span quality definitions will be compared in the following experiments.

3 Beam-Width Adaptation Strategies

In this section, we introduce two strategies to speed up Hiero decoding. 1. Cell closure: the CKY cells of low-quality spans are closed in the chart parsing step; 2. Dynamic beam-width: instead of fixing beam-width, we adjust the beam-width of each span according to its quality in the cube pruning step.

3.1 Cell Closure

Considering $Q(s,i,j) \in [0,1]$, we have to fix a threshold (denoting τ), so that cells with span quality below τ will be closed, i.e., they are skipped in the chart parsing step and will not generate any hypotheses.

Notice that τ is selected as big as possible for more cells can be closed as long as the translation quality (BLEU) is preserved. We can see in the experimental section, cell closure strategy brings a slightly positive effect to translation quality.

After cell closure, the chart graph changes from Figure 1 to Figure 3. Notice that spans with length one or start with index zero must be reserved to guarantee the existence of at least one translation result.

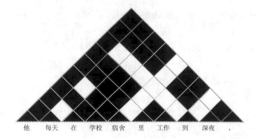

Fig. 3. Cell closure: low-quality spans (white) will not be processed in the chart parsing step

3.2 Dynamic Beam-Width

In section 1 we mentioned that about 65% of the fruitful spans have adopted hypotheses ranked top 3, and 90% ranked top 20, which means it is not economical for all spans to share the same beam-width in the cube pruning step.

In this strategy, for a span with $Q(s,i,j) > \tau$, its beam-width (denoted as $W(s,i,j)$) is adjusted dynamically according to its quality. The key issue here is to seek a suitable monotonic increasing function f:

$$W(s,i,j) = f(Q(s,i,j)) \qquad (4)$$

Notice that the curve of f is supposed to cross two points $(\tau, 0)$ and $(1, maxWidth)$ for the sake of consistency. $maxWidth$ denotes the maximum beam-width allowed in a span, which is a pre-fixed parameter in the traditional decoding.

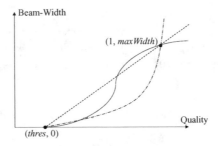

Fig. 4. Variety kinds of Width-quality curves

As Figure 4 shows, here are a variety of options for the function form. After the empirical studies on linear, quadratic and sigmoid functions, we find that the simple linear function yields the best performance:

$$W(s,i,j) = maxWidth * \frac{Q(s,i,j) - \tau}{1 - \tau} \qquad (5)$$

After the beam-width for a span is determined, the same number of hypotheses are generated using cube pruning algorithm [3]. After using dynamic beam-width strategy, the chart graph changes from Figure 3 to Figure 5.

Here we summarize the above two strategies. The beam-width is adjusted according to its span quality as follows:

$$W(s,i,j) = \begin{cases} 0 & Q(s,i,j) \leq \tau \\ f(Q(s,i,j)) & Q(s,i,j) > \tau \end{cases} \qquad (6)$$

4   Experiments

4.1   Basic Hiero Settings

We investigate the effectiveness of our approaches in C2E translation task. Our baseline Hiero system is implemented according to [3] using C++. Cube pruning is adopted with beam-width fixed at 100. GIZA++ is used for word alignment.

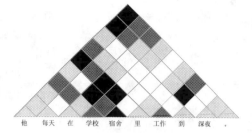

Fig. 5. Dynamic beam-width: the beam-width of each span is adjusted according to its quality

The translation parameters are tuned with MERT [10]. The development set is composed of about 1500 user queries from our online system. The training data consists of about 5M parallel sentence pairs extracted from web. NIST 05, 06 data sets and user request log of about 2000 sentences are used for evaluation purpose.

We first investigate how the constraint of beam-width affects translation speed and quality in our baseline system. Based on different value setting of maximum beam-width($maxWidth$), we record the average translation time for each sentence in our development set, and the total BLEU score. The result in Figure 6 shows that, (a) translation speed scales linearly as $maxWidth$ grows, while (b) after $maxWidth > 60$, a bigger beam width brings little improvements on translation quality. Therefor it is possible to improve translation efficiency by narrowing down beam-width without degrading translation quality. In the reminder of the section the details of the two beam-width adaptation strategies will be given.

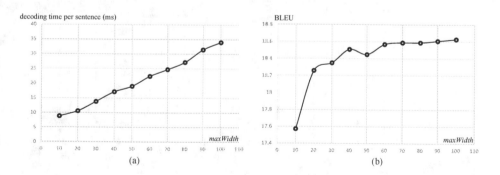

Fig. 6. The relationship between maximum beam-width and translation speed/quality

4.2   Cell Closure

Here we compare three span quality definitions Q_1, Q_2 and Q_3 in our development set. We plot in Figure 7 the variation of BLEU while cell closure is adopted with different value of τ. As we can see, comparing to Q_1 which is based on single

Table 1. System performance with beam-width adaptation strategies

Strategy	NIST 05		NIST 06		User Request Log	
	BLEU	Sec/Sent	BLEU	Sec/Sent	BLEU	Sec/Sent
Baseline	0.2911	0.280	0.2917	0.208	0.2047	0.186
Cell Closure	0.2912	0.171	**0.2924**	0.154	**0.2058**	0.139
Dynamic Beam-Width	0.2913	0.163	0.2905	0.120	0.2034	0.113
Closure & Dynamic	**0.2925**	**0.126**	0.2913	**0.101**	0.2045	**0.092**

maximum entropy model, Q_2 and Q_3 yield better performance. With $\tau = 0.3$ in Q_3, 40% of the spans can be pruned without degrading BLEU, in the meanwhile saving about 30% of translation time.

maxSpanLen constraint and glue rule [3] diminish the effect of the strategy. The spans longer than *maxSpanLen* will be directly processed by glue rule, and thus stay out of the chart parsing step. In our system settings (*maxSpanLen*=10) cell closure strategy avoids the generation of about 25% hypotheses, the ratio will be further raised with bigger *maxSpanLen*.

Based on the above results, we adopt Q_3 as the span quality measurement, and fix $\tau = 0.3$.

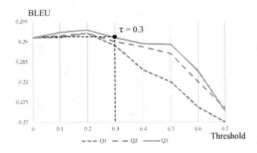

Fig. 7. The performance of three quality criterions in cell closure strategy

4.3   Dynamic Beam-Width

The experiment is conducted in our development set. Here we decide the best dynamic beam-width function f. Under the constraint of not degrading BLEU, linear, quadratic and sigmoid functions are compared. The best records of each function form are listed in Table 2. The experimental result shows that the simple linear function (5) yields the best performance: about 40% of hypotheses are pruned, in the mean while saving 40% of decoding time.

The performance of the two strategies on three evaluation sets is listed in Table 1. As we can see, span closure strategy not only improves translation efficiency, but also brings slightly positive inuence on BLEU. Comparing to the baseline with fixed beam-width cube pruning , the adoption of the two strategies have doubled the system translation speed. Our work is proven robust in all three data sets.

Table 2. The performance of various adjusting functions in dynamic beam-width strategy

Function form	Hypotheses derived (M)	Sec/Sent
Baseline	39.3	0.280
Linear	**23.4**	**0.160**
Quadratic (convex)	31.9	0.235
Quadratic (concave)	26.1	0.180
Sigmoid	29.7	0.201

5   Conclusion and Future Work

In this paper, we discuss speeding up Hiero decoding by suppressing the generation of hypotheses from inappropriate spans. We first evaluate the quality of each span, then prune the low-quality spans in the chart parsing step, and assign quality-determined beam-width for the rest of spans in the cube pruning step. The empirical studies on C2E translation show that our work has doubled the translation speed without compromising translation quality. The proposed approaches have already been adopted in our online system.

In the future, we intend to search for better span quality definition and dynamic beam-width function, and apply the approaches to MT in other language pairs. It is also interesting to apply our approaches to incremental decoding algorithm, and linguistic syntax based translation model.

References

1. Chiang, D.: A hierarchical phrase-based model for statistical machine translation. In: Proceedings of the 43rd Annual Meeting on Association for Computational Linguistics, pp. 263–270. Association for Computational Linguistics (2005)
2. Koehn, P., Och, F.J., Marcu, D.: Statistical phrase-based translation. In: Proceedings of the 2003 Conference of the North American Chapter of the Association for Computational Linguistics on Human Language Technology, vol. 1, pp. 48–54. Association for Computational Linguistics (2003)
3. Chiang, D.: Hierarchical phrase-based translation. Computational Linguistics 33, 201–228 (2007)
4. Hopkins, M., Langmead, G.: Cube pruning as heuristic search. In: Proceedings of the 2009 Conference on Empirical Methods in Natural Language Processing, vol. 1, pp. 62–71. Association for Computational Linguistics (2009)
5. Gesmundo, A., Henderson, J.: Faster cube pruning. In: IWSLT 2010: Proceedings of the 7th International Workshop on Spoken Language Translation (2010)
6. Roark, B., Hollingshead, K.: Linear complexity context-free parsing pipelines via chart constraints. In: Proceedings of Human Language Technologies: The 2009 Annual Conference of the North American Chapter of the Association for Computational Linguistics, pp. 647–655. Association for Computational Linguistics (2009)

7. Bodenstab, N., Dunlop, A., Hall, K., Roark, B.: Beam-width prediction for efficient context-free parsing. In: Proceedings of the 49th Annual Meeting of the Association for Computational Linguistics, Portland, Oregon. Association for Computational Linguistics (2011)
8. Roark, B., Hollingshead, K.: Classifying chart cells for quadratic complexity context-free inference. In: Proceedings of the 22nd International Conference on Computational Linguistics, vol. 1, pp. 745–751. Association for Computational Linguistics (2008)
9. Xiong, D., Zhang, M., Li, H.: Learning translation boundaries for phrase-based decoding. In: Human Language Technologies: The 2010 Annual Conference of the North American Chapter of the Association for Computational Linguistics, pp. 136–144. Association for Computational Linguistics (2010)
10. Zaidan, O.F.: Z-mert: A fully configurable open source tool for minimum error rate training of machine translation systems. The Prague Bulletin of Mathematical Linguistics 91, 79–88 (2009)

Training Phrase-Based SMT without Explicit Word Alignment

Cyrine Nasri, Kamel Smaili, and Chiraz Latiri

SMa^{r}T, LORIA, Campus scientifique,
BP 139, 54500 Vandoeuvre lès Nancy Cedex, France
{cyrine.nasri,smaili}@loria.fr,
chiraz.latiri@gnet.tn

Abstract. The machine translation systems usually build an initial word-to-word alignment, before training the phrase translation pairs. This approach requires a lot of matching between different single words of both considered languages. In this paper, we propose a new approach for phrase-based machine translation which does not require any word alignment. This method is based on inter-lingual triggers retrieved by Multivariate Mutual Information. This algorithm segments sentences into phrases and finds their alignments simultaneously. The main objective of this work is to build directly valid alignments between source and target phrases. The achieved results, in terms of performance are satisfactory and the obtained translation table is smaller than the reference one; this approach could be considered as an alternative to the classical methods.

Index Terms: Statistical Machine Translation, Inter-lingual triggers, Multivariate Mutual Information.

1   Introduction

The current best performing statistical machine translation systems are based on phrase-based models: the basic idea of phrase-based translation is to segment the given source sentence into phrases, then translate each phrase and finally compose the target sentence from these phrase translations.

It is important to point out that the current phrase-based models are not based on any deep linguistic concept.

Interestingly enough, the power of phrase-based translation is due to the quality of the phrase table. State-of-the-art statistical machine translation uses phrases as translation units to incorporate context into translation models, as described in [4], [14] and [15]. There are many ways to acquire such a table.

The mostly applied phrase pairs extraction method is the so-called Viterbi Extract [15]. In this approach, a source and a target phrase are considered to be translations of each other, if their words are only aligned within this phrase pair and not to the words outside. Collecting phrases and their corresponding translations extracted from all the sentences in the bilingual training corpus,

A. Gelbukh (Ed.): CICLing 2014, Part II, LNCS 8404, pp. 233–241, 2014.
© Springer-Verlag Berlin Heidelberg 2014

achieves a phrase table with a set of phrase pairs with scores indicating their translation accuracy.

The decoder based on log-linear model produces target sentences from left to right by covering the source phrases in a certain order [8]. The log-linear model uses several features such as relative frequencies of the phrase pairs, a word-based lexicon model, a target language model, a source phrase reordering model, as well as a word and phrase penalty model.

Currently, this is the most widely method used for producing phrases and decoding.

Other approaches have been investigated to obtain phrase pairs in less heuristic ways. Zhang in [16] presented an integrated phrase segmentation/alignment algorithm (ISA) for statistical machine translation, which segments and aligns phrases simultaneously. Without training a word alignment model, phrases are identified based on the similarities of mutual information values among word alignment points. Venugopal in [13] presented a technique that begins with an improved IBM models to create knowledge, that represents effectively local and global phrase contexts.

Another method proposed by Lavecchia in [8] retrieves valid linguistic phrases without using any alignments. This method identifies first the best part-of-speech phrases and then from these class phrases, they extracted the corresponding phrases which improve the perplexity of the source language. For instance, NOUN PRE NOUN is one of the retrieved part-of-speech phrases and from this pattern and the source corpus a phrase as *Table de Salon* is extracted. The obtained phrases are linguistically pertinent and consequently the derived phrases are also relevant. These obtained phrases are then used to rewrite the source training corpus in terms of phrases. The words of this phrase are gathered and used to rewrite the source training corpus.

In the following, we detail our method which is based on the inter-lingual triggers.

2   Inter-lingual Triggers

Inter-lingual triggers are inspired from triggers concept used in statistical language modeling [12]. A trigger is a set composed of words and its best correlated triggered words in terms of mutual information (MI). In [6], the authors proposed to determine correlations between words coming from two different languages. Each inter-lingual trigger is composed of a triggering source linguistic unit and its best correlated triggered target linguistic units. Based on this idea, they found among the set of triggered target units, potential translations of the triggering source words. Inter-lingual triggers are determined on a parallel corpus according to mutual information measure namely:

$$MI(a, b) = P(a, b) \log \frac{P(a, b)}{P(a)P(b)} \tag{1}$$

where a and b are respectively a source and a target words. $P(a, b)$ is the joint probabilities and $P(a)$ and $P(b)$ are marginal probabilities.

For each source unit a, the authors kept its k best target triggered units. This approach has been extended to take into account triggers of phrases [8]. The drawback of this method is that phrases are built in an iterative process starting from single words and joining others to them until the expected size of phrases is reached. In other words, at the end of the first iteration, sequences of two words are built, the following iteration produces phrase of three words and so on until the stop-criteria is reached. Then, once all the source phrases are built, their corresponding phrases in the target language are retrieved by using n-to-m inter-lingual trigger approach which means that a phrase of n words triggers a phrase of m words [8]. In order to avoid the propagation of errors due to the cascade of steps in the previous method, we propose a new approach based on multivariate mutual information which allows to retrieve source phrases given target ones.

3   Training Phrase with Multivariate Mutual Information

Multivariate Mutual Information (MMI) calculates the degree of correlation between n random variables. This concept is very interesting since we propose to take advantage of this principle by associating k words in the source language and r words in the target language with $n = k + r$.

$$MMI(A_1, A_2, ..., A_n) = P(A_1, A_2, ..., A_n) \log \frac{P(A_1, A_2, ..., A_n)}{P(A_1)P(A_2)...P(A_n)} \quad (2)$$

Our method allows creating inter-lingual triggers, their estimation is based on MMI. For instance for the trigger *petit déjeuner* $\rightarrow$ *breakfast*, we proceed as follows: $A_1 = petit$, $A_2 = déjeuner$, and $A_3 = breakfast$. $P(A_1, A_2, A_3)$ is the probability that the words *petit, déjeuner* and *breakfast* occur simultaneously. $P(A_1)$, $P(A_2)$, $P(A_3)$ are respectively the probabilities of *petit, déjeuner* and *breakfast*.

3.1   Selecting Phrases in Terms of Their Size

In this work, we started by identifying the longest phrase with their translations and then, the less longest and finally arriving to phrases of two words. This is motivated by the fact that we would like to appreciate the real contribution of each segment without the influence of its sub-segments. In fact, a long segment is linguistically more informative than a shorter one included into it. The algorithm, we proposed is based on retrieving phrases and their translations by using MMI as described in the following algorithm. For a fixed length of a trigger, the concatenation of its words constitute the source phrase. And all the words of its triggered sequence constitute the target phrase. It should be noted, that in this algorithm we suppose that the translation table is from English to French. Now we know that, in most cases for a French sentence, its equivalent in English is shorter. That is why, in the proposed algorithm, for each fixed length of phrase,

---

Algorithm 1. Oriented-Size Phrases Discovering (OSPD)

---

m : maximum size of a source phrase
n : maximum size of a target phrase
for i = m to 1 do
for j = n to 1 do
if i = j or i = j+1 then

1. Train triggers model $X_i \longrightarrow Y_j$ (X_i: source phrase composed of i words, Y_j : target phrase composed of j words)
2. Calculate $MMI(x_1, x_2, \ldots, x_i, y_1, y_2, ..., y_j)$ as shown in formula 2.
3. Include the retrieved phrases $Xi = x_1, x_2, \ldots x_i$ and their best translations $Y_j = y_1, y_2, \ldots y_j$ into the translation table

endif

Fig. 1. Algorithm 1: Oriented-Size Phrases Discovering (OSPD)

we look for a triggered phrase of the same length or longer by one word. More details are given in [9].

This method leads to remarkable triggers where the triggered words could be considered as potential translations of the trigger or very close in terms of meaning. For each source phrase, the 20 best triggers are kept. Table 1 illustrates some examples of obtained French-English triggers.

The experiments presented below have been conducted on the proceeding of the European Parliament[5]. We used French-English parallel corpus.

Table 2 shows the parallel corpus statistics used in our experiments. So far, we have only discussed how to collect a set of phrase pairs. More is needed to turn this set into a probabilistic phrase translation table. For that, we use the principle proposed in [8] to compute phrase translation probabilities. The translation table is obtained by assigning for each trigger a conditional probability calculated as follows:

$$\forall f, e_i \in Trig(f) \quad P(e_i|f) = \frac{MMI(e_i, f)}{\displaystyle\sum_{e_i \in Trig(f)} MMI(e_i, f)} \tag{3}$$

Table 1. Examples of retrieved phrases and their translations

French	English	MMI
autour de la table	around the table	0.0054
	around the	0.0022
	the table	0.0011
a été prise	was taken	0.001
	been taken	0.00037
	has been taken	0.00034
semaine denière	last week	0.016
	week	0.0095
	last	0.009

Table 2. An overview of the experimental material

Corpus	Sentences	English words	French words
Training	0,5M	15M	16,6M
Dev	1,4 K	14K	13,7K
Test	500	1153	1352

where $Trig(f)$ is the set of k English events triggered by the French event f.

In table 3, we present the results of OSPD method. S_1 corresponds to word-to-word translation and S8 corresponds to a translation using all the discovered phrases. The introduction of phrases of 8, 7 and 6 words improve the results by more than 2 points. This means that long phrases are suitable but not as relevant as the introduction of phrases of 5 words. These phrases bring more than 2.5 in terms of BLEU. But the best improvement is brought by sequence of words of 4,3, and 2 words: more than 4.5! Consequently, all these sequences of different sizes are necessary to improve the results. Phrases beyond of 8 words are not relevant. This method has a performance which is 1.1% less than the baseline method. This shows the feasibility to develop machine translation without any word alignment with acceptable results.

4 How to Improve This Method?

We would like to go further and to achieve results closer to those obtained by the baseline method which needs word alignment. For that, we analyze our method to identify its drawbacks.

4.1 Improving by Selecting the Best Phrases

One of the drawback of this method is that we keep for each phrase its 20 best translations. This could include bad translations which would corrupt the results. In fact, the examples given in table 4 are considered as bad translations but unfortunately yet they are in the translation table.

In [11] the authors argue that extracting only minimal phrases, i.e the smallest phrases pairs that map each entire sentence pairs, does not degrade performances. By using significance test, they remove unlikely phrase pairs which reduces the phrase table drastically and may even yield increases in performance [3].

We propose in the following pruning phrase table by incorporating best translation pairs in terms of MMI without the need of integrating sequences step by step in terms of their length. We will call this method **Best-Phrases Discovering (BP)**. To do that, we extract all the inter-lingual n-to-m-triggers as in [8] except, that instead of using classical mutual information, we use MMI. Then we sort all the discovered phrases in descending order according to their score. Only triggers that have a MMI greater than a fixed threshold are kept. The impact of this selection is positive in terms of BLEU as presented in the evaluation section.

Table 3. Evolution of BLEU in accordance to the length of phrases introduced in the translation table

Set	Selected Triggers	Score BLEU
S_1	1FR $\longrightarrow$ 1EN	34.16
S_2	S_1 + 8FR	36.32
S_3	S_2 + 7FR	36.36
S_4	S_3 + 6FR	36.8
S_5	S_4 + 5FR	39.58
S_6	S_5 + 4FR	41.12
S_7	S_6 + 3FR	43
S_8	S_7 + 2FR	**43.79**
	Baseline (Och method)	**44.3**

Table 4. Examples of bad English-French triggers

French phrase	English phrase	MMI $\times 10^{-5}$
les droits de l'homme	human rights situation	3.2
madame le président	president ladies and gentlemen	6
la semaine dernière	president last week	4.6

4.2   Lexical Weights

Lexical weights were proposed in [4] to validate the quality of alignments. Given a bilingual phrase to score, the objective consists in checking how well each source word translates into the target words it links to. When a source word links to multiple target words, the average of their translation probabilities is calculated. A source-to-target lexical weight is then the product of all scores. The same calculation is done from target to source, and the result is a pair of lexical weights between 0 and 1. Because in our method, we do not proceed to any alignment and because lexical weights is important in Moses, we decided to adapt the classical technique with one major change.

As, our method proposes phrases without any initial word-to-word alignments, we estimate a simple lexical translation probability distribution D based on word-to-word triggers [6]:

$$D(w_j|w_i) = \frac{MI(w_j|w_i)}{\sum MI(w_j|w_i)} \tag{4}$$

Where w_i is a word in a language i and w_j is a word in a language j.

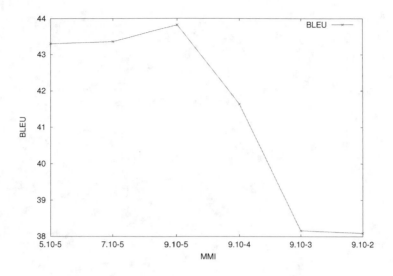

Fig. 2. Evolution of Bleu score in terms of MMI threshold on the DEV corpus

5   Evaluation

In the following, we present several experiments to evaluate the impact of the improvements we proposed in order to boost our initial method. The method is evaluated by comparing it to the baseline one with a refined alignments from Giza++ [10]. Default set of options is used: 2 translation probabilities, 2 lexical weights and length penalty.

In figure 2, we evaluate the impact of the method based on selecting the best phrases (BP). We can notice that the best performance is achieved for a MMI threshold equal to 9.10^{-5}. This curve shows also, if phrases are not selected judiciously, then the degradation of the performance is serious, more than 5 BLEU points are lost.

On the Test corpus, the best performance is achieved for a MMI threshold equal to 5.10^{-5} (figure 3). Obviously, the only threshold used is the one get on the development corpus which is close to the one get from the test corpus.

In this test, we improve slightly the results, the performance reaches a BLEU of 43.82. The results are still bellow the standard one (44.3). To improve our results we combined several translation tables OSPD, BP and the baseline one.

This table shows that by combining our two translation tables we are only 0.6% from the baseline method. This result confirms that it is possible to do almost as good as Och method without any word process alignment. By combining BP with the baseline method, we outperform the baseline result by 0.4%. This improvement reachs 0.6% when both translation tables (OSPD and BP) are combined with the baseline one. These last results illustrates that it is possible to improve the baseline translation table by using other phrases and for some of them, they get better scores than those proposed by the baseline one.

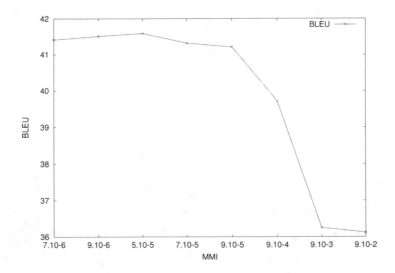

Fig. 3. Evolution of Bleu score in terms of MMI threshold on the Test corpus

Table 5. Experiments with different models

System	BLEU
Baseline	44.3
OSPD	43.79
BP	43.82
OSPD+BP	44.02
BP+Baseline	44.48
OSPD+BP+Baseline	44.57

Table 6. Size of the different translation tables

Baseline	OSPD	BP
33,3 M	51 M	22,9 M

Furthermore, our best method (BP) uses a smaller (12%) translation table than the baseline.

6   Conclusion

In conclusion, we proposed two methods which do not require any word align-ment. These two methods achieve results closer to the baseline method, around 1%, when they are used alone. When they are combined, we approach the base-line system and the difference is only about 0.6%. The size of the BP translation table is 12% smaller than the baseline one. Then one question may arise, could our community accept a new method of retrieving phrases which does not require any word alignment, getting results closer to the classical one and with a smaller and more cleaned translation table? Besides that, the obtained phrases have an added value since they can enrich those achieved by the baseline method since we have shown that they can improve the reference results.

References

[1] Abramson, N.: Information theory and coding. McGraw-Hill electronic sciences series. McGraw-Hill (1963)

[2] Hoang, H., Birch, A., Callison-burch, C., Zens, R., Aachen, R., Constantin, A., Federico, M., Bertoldi, N., Dyer, C., Cowan, B., Shen, W., Moran, C., Bojar, O.: Moses: Open source toolkit for statistical machine translation, 177–180 (2007)

[3] Johnson, H., Martin, J., Foster, G., Kuhn, R.: Improving Translation Quality by Discarding Most of the Phrase table. In: Proceedings of the 2007 Joint Conference on Empirical Methods in Natural Language Processing and Computational Natural Language Learning (EMNLP-CoNLL), pp. 967–975 (2007)

[4] Koehn, P., Franz, J.,Marcu, D.: Statistical phrase-based translation. In: Conference of the North American Chapter of the Association for Computational Linguistics on Human Language Technology, NAACL 2003, Edmonton, Canada, vol. 1, pp. 48–54 (2003)

[5] Koehn, P.: Europarl: A Parallel Corpus for Statistical Machine Translation. In: Conference Proceedings: the Tenth Machine Translation Summit, pp. 79–86 (2005)

[6] Lavecchia, C., Smaïli, K., Langlois, D., Haton, J.-P.: Using inter-lingual triggers for machine translation. In: INTERSPEECH, pp. 2829–2832 (2007)

[7] Lavecchia, C., Smaïli, K., Langlois, D., Haton, J.-P.: Using inter-lingual triggers for machine translation. In: INTERSPEECH, pp. 2829–2832 (2007)

[8] Lavecchia, C., Langlois, D., Smaïli, K.: Discovering phrases in machine translation by simulated annealing. In: Interspeech, pp. 2354–2357 (2008)

[9] Nasri, C., Smaïli, K., Latiri, C., Slimani, Y.: A new method for learning Phrase Based Machine Translation with Multivariate Mutual Information. In: NLP-KE 2012, HuangShan, China (2012)

[10] Och, F., Hermann, N.: A Systematic Comparison of Various Statistical Alignment Models. Computational Linguistics 29 (2003)

[11] Quirk, C., Menezes, A.: Do we need phrases? Challenging the conventional wisdom in Statistical Machine Translation. In: HLT-NAACL (2006)

[12] Tillmann, C., Ney, H., Lehrstuhl Fur Informatik Vi: Word Triggers and the EM Algorithm. In: Proceedings of the Workshop Computational Natural Language Learning (CoNLL), pp. 117–124 (1997)

[13] Venugopal, A., Vogel, S., Waibel, A.: Effective Phrase Translation Extraction from Alignment Models, pp. 319–326. ACL (2003)

[14] Zens, R., Och, F.J., Hermann, N.: Phrase-Based Statistical Machine Translation, pp. 18–32. Springer (2002)

[15] Zens, R., Ney, H.: Improvements in Phrase-Based Statistical Machine Translation. In: The Human Language Technology Conf, HLT-NAACL, pp. 257–264 (2004)

[16] Zhang, Y., Vogel, S., Waibel, A.: Integrated Phrase Segmentation and Alignment Algorithm for Statistical Machine Translation. In: Proceedings of International Conference on Natural Language Processing and Knowledge Engineering (NLP-KE 2003), Beijing, China (2003)

Role of Paraphrases in PB-SMT

Santanu Pal[1], Pintu Lohar[2], and Sudip Kumar Naskar[2]

[1] Universität Des Saarlandes, Saarbrücken, Germany
santanu.pal@uni-saarland.de
[2] Department of Computer Science & Engineering,
Jadavpur University, Kolkata, India
pintu.lohar@gmail.com, sudip.naskar@cse.jdvu.ac.in

Abstract. Statistical Machine Translation (SMT) delivers a convenient format for representing how translation process is modeled. The translations of words or phrases are generally computed based on their occurrence in some bilingual training corpus. However, SMT still suffers for out of vocabulary (OOV) words and less frequent words especially when only limited training data are available or training and test data are in different domains. In this paper, we propose a convenient way to handle OOV and rare words using paraphrasing technique. Initially we extract paraphrases from bilingual training corpus with the help of comparable corpora. The extracted paraphrases are analyzed by conditionally checking the association of their monolingual distribution. Bilingual aligned paraphrases are incorporated as additional training data into the PB-SMT system. Integration of paraphrases into PB-SMT system results in significant improvement.

Keywords: Paraphrasing, Statistical Machine Translation, Phrase Based, Comparable Corpora, Context.

1   Introduction

The quality of phrase-based statistical machine translation (PB-SMT) depends on word alignment and phrase translation estimation. The estimation of phrase translation can be achieved by using large amount of sentence aligned parallel training data. However, large volume of sentence aligned parallel data is not available for all language pairs. For those language pairs for which only limited training data are available, paraphrases can serve as indirect training data which can result in improved translation quality. A phrase or an idea can be represented or expressed in different ways in the same language by preserving the meaning of that phrase or idea by means of paraphrasing. Paraphrases can be used in various NLP applications such as Question Answering (QA), Summarization, Textual Entailment (TE), Machine Translation (MT), etc. In natural language generation the production of paraphrases allows for the creation of more varied and fluent text [1]. In MT, paraphrases can be used to alleviate the sparseness of training data [2], for handling OOV words, as well as to expand the reference translations in automatic MT evaluation [3-4]. Paraphrases can be collected form parallel corpora as

A. Gelbukh (Ed.): CICLing 2014, Part II, LNCS 8404, pp. 242–253, 2014.

well as from comparable corpora. Comparable corpora have recently received great attention in the field of NLP. Extraction of parallel fragments of texts, sentences and paraphrases from comparable corpora is particularly useful for any corpus based approaches to MT, especially for SMT [5-6] as the size of the parallel corpus plays a crucial role in the SMT performance. Out-of-vocabulary words or phrases also pose a major challenge in SMT. The OOV problem can be circumvented by paraphrasing the OOV words and phrases. Paraphrases of OOV words can be extracted from bilingual parallel corpora using another language as a pivot [7] and also taking into account contextual information. It can often be observed that SMT output contains untranslated words or phrases while similar type of phrases are available in the training data. This motivated us to extract paraphrases and use them in the state-of-the-art PB-SMT system which could take care of the problem of OOV.

In this work, first we extract paraphrases using bilingually calculated paraphrase probability from the baseline PB-SMT phrase table. Then we also collect paraphrases from comparable corpora. Successively the extracted paraphrases are examined and expanded according to the association of the monolingual distribution of the constituent words or contextual words. Finally, the bilingually aligned paraphrases are added to the parallel corpus as additional training data. Rarely occurring words such as named entities are handled differently. We extract the named entities (NEs) from the training data and align them through transliteration following the approach of [8].

The remainder of the paper is organized as follows. Section 2 discusses related works. The proposed system is described in Section 3. Section 4 presents the tools and resources used for the various experiments. Section 5 reports the results obtained together with some analysis. Section 6 concludes and provides avenues for further work.

2   Related Works

A significant number of works have been carried out on paraphrasing. The use of log-linear models for computing the paraphrase likelihood between pattern pairs and feature functions based on maximum likelihood estimation (MLE), lexical weighting (LW), and monolingual word alignment (MWA) has been presented in [9]. To rerank candidate paraphrases, [10] proposed bilingual paraphrase extraction using monolingual distributional similarity. [11] reports extraction of sense disambiguated paraphrases by pivoting through multiple languages. An unsupervised learning algorithm was presented in [12] for the identification of paraphrases from a corpus of multiple English translations of the same source text. A new and unique paraphrase resource was reported in [13], which contains meaning-preserving transformations between informal user-generated texts. Sentential paraphrases are extracted from a comparable corpus of temporally and topically related messages in Twitter which often express semantically identical information through distinct surface forms. A novel paraphrase fragment pair extraction method was proposed by [14] in which the authors used a monolingual comparable corpus containing different articles about the same topics or events. The procedure consisted of document pair extraction, sentence pair extraction and fragment pair extraction. A technique of clustering the phrases of a PB-SMT system using information in the phrase table itself was proposed in [15]. They incorporated phrase-cluster-derived probability estimates into a baseline log-linear feature combination that includes relative frequency and lexically-weighted conditional probability estimates. [16] describes a system that was built using baseline PB-SMT system by augmenting its phrase table with novel translation pairs generated by combining paraphrases with translation pairs learned directly from the bilingual training data. They investigated two methods for phrase table augmentation: source-side augmentation and target-side augmentation. A full-sentence paraphrasing technique was introduced by [17] where they demonstrated that the resulting paraphrases can be used to drastically reduce the number of human reference translations needed for parameter tuning without a significant decrease in translation quality. [18] reports deriving paraphrases from monolingual training data using distributional semantic similarity measures. Their method resulted in significant improvements in translation quality in a low-resource setting. Two methods - a shallow lexical substitution technique and a grammar-driven paraphrasing technique were proposed by [19]. They used these techniques to paraphrase a single reference, which, when used for parameter tuning, leads to superior translation performance over baselines that use only human-authored references. A novel approach to find translations of OOV words was proposed by [20]. They induced a lexicon by constructing a graph on source language monolingual text and employed a graph propagation technique in order to find translations for all the source language phrases. Experimental results showed that their graph propagation method significantly improves performance over two strong baselines under intrinsic and extrinsic evaluation metrics.

Most of the works belonging to paraphrasing for MT are carried out at the time of post-editing the MT output. However, the work presented here uses paraphrases as additional training material which could reduce the OOV problem and also might improve the word alignment as well. The first objective of the present work is to see whether paraphrases collected from comparable corpora can improve the overall MT quality. The second objective is to see whether contextual words added to the paraphrases and incorporated into the SMT model can bring any further improvement in the performance of the MT system. Finally, we add automatically aligned rare words such as NEs with this model.

We carried out the experiments on English–Bengali translation task. Bengali shows high morphological richness at lexical level. Besides, language resources in Bengali are not widely available. To the best of our knowledge this is the first attempt with paraphrasing for MT involving Bengali.

3   System Description

In the present work, paraphrases are identified both from the training corpus as well as from comparable corpora. The following two subsections detail them.

3.1   Paraphrase Extraction

We first identify paraphrases from the bilingual training corpus. The identification of paraphrases has been carried out by pivoting through phrases from the bilingual parallel corpus. We do not consider all phrases in the phrase table as potential candidates for paraphrasing; only linguistic phrases are considered. Initially we have extracted all the linguistic phrases (chunks) form the source side of the training data and then we find paraphrases for these linguistic chunks from the phrase table. The identification technique is similar to that of [7]. Let us consider the source phrase "*in the centre*" in an English-Bengali parallel corpus. The corresponding target phrase (কেন্দ্রে) is identified from the phrase table. Figure 1 illustrates the paraphrase extraction process. In this particular case, the source phrase "*in the centre*" has a target representation "কেন্দ্রে" found from the phrase table which in turn have another source representation "*in the heart*" in the phrase table; hence the two source phrases "*in the centre*" and "*in the heart*" can be considered as potential paraphrases conveying the same meaning.

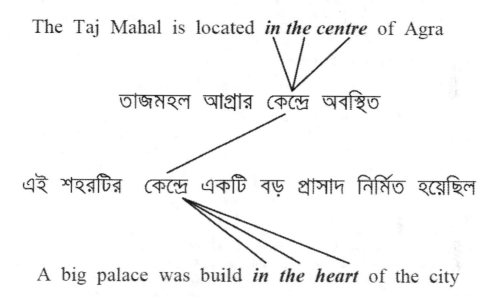

Fig. 1.

After extraction of potential linguistic paraphrase pairs, we compute the likelihood of them being paraphrases. For a potential linguistic paraphrase pair (e1, e2) we have defined a paraphrase probability p(e2|e1) in terms of the translation model probabilities p(f|e1), that the original English phrase e1 translates as a particular target language phrase f, and p(e2|f), that the candidate paraphrase e2 translates as the same foreign language phrase f. Since e1 can translate to multiple foreign language phrases, we sum over all such foreign language phrases. Thus the equation reduces to as follows:

$$\hat{e}2 \quad = \quad \begin{array}{c} \arg\max \ P(e_2|e_1) \\ e_2 \neq e_1 \end{array} \tag{1}$$

$$= \quad \begin{array}{c} \arg\max \\ e_2 \neq e_1 \end{array} \Sigma_f P(f|e_1)P(e_2|f) \tag{2}$$

We compute translation model probabilities using standard formulation from PB-SMT. So, the probability p(e|f) is calculated by counting how often the phrases e and f were aligned in the parallel corpus as follows:

$$p(elf) \quad = \quad \frac{count(e,f)}{\sum_f count(e,f)} \tag{3}$$

Using the equation (2) and (3) we calculate paraphrase probabilities from the phrase table. For example, let us consider that the phrase "in the centre" is to be paraphrased and we find two pivots "কেন্দ্রস্থল" and "কেন্দ্রে" through which the phrase "in the centre" is aligned with a candidate paraphrase "in the heart". Hence by equation (3) the corresponding paraphrase probability is calculated as:

p(in the heart|in the center) = [p(কেন্দ্রস্থল|in the center) x p(in the heart|কেন্দ্রস্থল)] + [p(কেন্দ্রে|in the center) x p(in the heart|কেন্দ্রে)].

3.2   Paraphrase Extraction Using Comparable Corpora

We collected comparable corpora from Wikipedia. Wikipedia[1] is an online collaborative encyclopedia available in a wide variety of languages. English Wikipedia is the largest in volume and contains millions of articles; there are many language editions with at least 100,000 articles. Wikipedia links articles on the same topic in different languages using "interwiki" linking facility. Wikipedia is an enormously useful resource for extracting parallel resources as the documents in different languages are already aligned. We first collect an English document from Wikipedia and then if there exists any inter-language link then we follows this link to find the same document in Bengali. Extracted English–Bengali document pairs from Wikipedia are already comparable since they talk about the same entity. Although each English–Bengali document pairs are comparable and they discuss about the same topic, most of the times they are not exact translation of each other; as a result parallel fragments of text are rarely found in these document pairs. The bigger the size of the fragment the less probable it is that its parallel version will be found in the target side. Nevertheless, there is always chance of getting parallel phrase, tokens or even sentences in comparable documents.

To collect comparable English–Bengali document pairs we designed a crawler. Based on an initial seed keyword list, the crawler first visits each English page of Wikipedia, saves the raw text (in HTML format), and then follows the cross-lingual link for each English page and collects the corresponding Bengali document. In this way, we collect English–Bengali comparable documents in the tourism domain. We retain only the textual information and all the other details are discarded. We extract overlapping English phrases from each document. The idea behind the concept of overlapping phrase is that according to linguistic theory, during translation, the intermediate constituents of the chunks do not usually take part in long distance reordering but only intra chunk reordering occurs. These chunks are combined together to make a longer phrase and phrases again make up a sentence. The entire process maintains the linguistic definition of a sentence. Breaking a sentence into N-grams would generate several phrases of length N but these phrases may not be linguistic phrases. For this reason, we avoided breaking the sentences into N-grams. For overlapping type of chunking, we set a value 'N', where N specifies the number of words inside a

---

[1] http://www.wikipedia.org/

linguistic phrase. For any sentence, we merge as many consecutive component chunks as possible by sliding an imaginary window over the sentence and extract overlapping phrases. Here, we condition the overlapping phrases on the number of words, i.e., the number of words in an overlapping phrase should never exceed N. The extracted overlapping phrases from each English document are then translated into Bengali through a baseline PB-SMT system. The baseline PB-SMT was trained on English–Bengali tourism domain corpus. The translated phrase is then searched for in the target comparable document. If it is found in the target document then the target phrase is directly fetched form the comparable target document and the source-target phrase pair are saved in a list. In this way, we extract a parallel phrase list. These parallel fragments of text, extracted from the comparable corpora are added with the tourism domain training corpus to enhance the performance of the baseline PB-SMT system. The parallel fragment extraction procedure is similar to the work of [6].

3.3   Paraphrase Expansion Using Context

Section 3.2 discusses extraction of linguistic paraphrases from the phrase table; however, the phrase table also contains some non-linguistic paraphrases. These paraphrases (i) may be incomplete phrase, or (ii) may contain some unnecessary words, or (iii) may need some word associated with it. For an incomplete phrase, we can directly collect the complete phrase from the parallel corpus itself. But for the other two problems, we analyzed them through stronger association measure. For problem (ii), the starting and ending surface word context of the paraphrases are analyzed. E.g., if a paraphrase contains the words $\{p_1, p_2, p_3 \ldots p_{n-1}, p_n\}$ then we compute the association of (p_1, p_2) and (p_{n-1}, p_n). If the association score is satisfactory then we store the paraphrase in a separate list; otherwise we discard these surface words. In case of problem (iii), we add context word at the starting and ending position of the paraphrases and analyze their association score in the same way as in (ii). Below we describe how we calculate the association score.

In this experiment, we have followed Point-wise Mutual Information (PMI) [21], Conditional probability, Log-likelihood Ratio (LLR) [22], Phi-coefficient and Co-occurrence measurement measures. Finally, a system combination model has been developed which gives a normalized weighted combination score to each of the association. A predefined cut-off score (above 70%) has been considered and the candidates having scores above the threshold value are considered having a proper association.

Point-Wise Mutual Information (PMI). It is an information-theoretic motivated measure for discovering interesting collocations [21]. Point-wise mutual information is mathematically defined as,

$$PMI(x, y) = log \frac{P(xy)}{p(x)p(y)} \tag{4}$$

where, $P(xy)$ = probability of the word x and y occurring together, $P(x)$ = probability of x occurring in the corpus and $P(y)$ = probability of y occurring in the corpus.

Conditional Probability. The probability of occurrence of a word x and given the context word y that is $p(x/y)$.

Log-Likelihood Ratio (LLR). Generally, it is the ratio between the probability of observing one component of a collocation given the other is present and the probability of observing the same component of a collocation in the absence of the other [22].

$$Log - Likelihood = -2 \sum_{i,j} f(i,j) \log \frac{f(i,j)}{f'(i,j)} \qquad (5)$$

Here the order of the words in the candidate collocation is irrelevant. We have adopted the probability using Baye's theorem by averaging the probability of w1 giving w2 and probability of w2 giving w1.

Phi-Coefficient. In statistics, the Phi coefficient Φ is a measure of association for two binary variables. The Phi coefficient is also related to the chi-square statistic as:

$$\emptyset = \sqrt{\frac{\chi^2}{n}} \qquad (6)$$

where n is the total number of observations and χ^2 is the chi-square distribution. Two binary variables are considered positively associated if most of the data falls along the diagonal cells. Here, the binary distinction denotes the positional information of the words. If we have a 2×2 table for two random variables x and y which denotes the presence of words w1 and w2 respectively, we have the following matrix:

	y=1	y=0	Total
x=1	n_{11}	n_{10}	n_{x1}
x=0	n_{01}	n_{00}	n_{x0}
Total	n_{y1}	n_{y0}	N

where, n_{11}=actual bigram <w1w2> count, n_{10}=frequency of bigram containing w1 but not w2, n_{01}=frequency of bigram containing w2 not w1, n_{00}=frequency of of bigrams neither containing w1 nor w2. n_{x1} and n_{x0} are the summation of their respective rows and n_{y1} and n_{y0} are the summation of their respective columns. Alternative words in place of absent w1 or w2 must be nouns. The phi coefficient that describes the association of x and y is

$$\emptyset = \frac{n_{11}n_{00} - n_{01}n_{10}}{\sqrt{n_{x1}n_{x0}n_{y1}n_{y0}}} \qquad (7)$$

Co-occurrence Measurement. We have used co-occurrence measurement by using the formula adopted by [23]

$$co(w1,w2) = \sum_{s \in S(w1,w2)} e^{-d(s,w1,w2)} \qquad (8)$$

where, co(w1,w2)=co-occurrence frequency between the words (after stemming), S(w1,w2)=set of all sentences where both w1 and w2 occurs, d(s,w1,w2)=distance between w1 and w2 in a sentence s in terms of number of words. For every adjacent occurrence of w1 and w2, co(w1,w2) increases by 1, but if in a sentence they are largely separated, it increases only marginally. This measurement is used further in calculating significant function.

4   Tools and Resources

A sentence-aligned English–Bengali parallel corpus contains 23,492 parallel sentences from the travel and tourism domain has been used in the present work. The corpus has been collected from the consortium-mode project "Development of English to Indian Languages Machine Translation (EILMT) System[2]". The Stanford Parser[3] and CRF chunker[4] [23] have been used for parsing and chunking in the source side of the parallel corpus, respectively.

The experiments were carried out using the standard log-linear PB-SMT model as our baseline system: GIZA++ implementation of IBM word alignment model 4, phrase-extraction heuristics described in [24], minimum-error-rate training [25] on a held-out development set, target language model trained using SRILM toolkit [26] with Kneser-Ney smoothing [27] and the Moses decoder [28] have been used in the present study.

5   Experiments and Results

We randomly identified 500 sentences each for the development set and the test set from the initial parallel corpus. The rest is considered as the training corpus. The training corpus was filtered with the maximum allowable sentence length of 100 words and sentence length ratio of 1:2 (either way). Finally the training corpus contained 22,492 sentences. In addition to the target side of the parallel corpus, a monolingual Bengali corpus containing 488,026 words from the tourism domain was used for the target language model. We experimented with different n-gram settings for the language model and the maximum phrase length and found that a 4-gram language model and a maximum phrase length of 7 produce the optimum baseline result on both the development and the test set. We carried out the rest of the experiments using these settings.

The experiments were carried out with various experimental settings: (i) Extracted paraphrases from the baseline PB-SMT phrase table are added directly as additional training data; (ii) Aligned paraphrases are extracted from comparable corpora and are treated as additional sentence level parallel data; (iii) Contextual information is added with the paraphrases; and (iv) rare words such as NEs are aligned and added with the

---

[2] The EILMT project is funded by the Department of Electronics and Information Technology (DEITY), Ministry of Communications and Information Technology (MCIT), Government of India.

[3] http://nlp.stanford.edu/software/lex-parser.shtml

[4] http://crfchunker.sourceforge.net/

parallel training data. Our hypothesis focuses mainly on the theme that prior alignment of bilingual paraphrases will result in improvement of the system performance in terms of translation quality.

In this work we created additional training material using these paraphrases; we replace each English paraphrase with all of its variants and thus create new training instances. For example, consider the English phrase *"throughout the year"* and its two paraphrases *"all year round"* and *"all around the year"*. Now we consider following sentences from our training data for each of these phrase and paraphrases.

(1a) Events, parties and festivals occur *throughout the year* and across the country.
(1b) Weather on all of the Hawaiian islands is very consistent, with only moderate changes in temperature *all year round.*
(1c) There is an intense agenda *all around the year* and the city itself is a collection of art and history.

In example1, the first sentence, the phrase *"throughout the year"* is replaced by its two paraphrases *"all year round"* and *"all around the year"* to create two additional sentences to be added to the existing training data. Similarly *"all year round"* and *"all around the year"* are replaced by the remaining two variants for the second and third sentence, respectively. In this way for these three training sentence, we can create six additional sentences from all combinations of replacement. Thus, effectively we have three types of replacements; (i) phrase-with-paraphrase, (ii) paraphrase-with-phrase and (iii) paraphrase-with-paraphrase. Following the approach described in Section 3 we were able to extract 1289 paraphrases from the baseline PB-SMT phrase table. The number of extracted paraphrases increases to 2846 when we use the comparable corpora.

Extrinsic evaluation was carried out on the MT quality using NIST [29] and BLEU [30] and the evaluation results are shown in Table 2. The baseline PB-SMT system (Exp 1 in Table 2) is the state-of-art PB-SMT system where GIZA++ with grow-diag-final-and has been used as the word alignment model.

Initially, 39,931 and 28,107 NEs were identified from the source and target sides of the parallel corpus respectively, of which 22,273 NEs are unique in English and 22,010 NEs in Bengali. A total of 14,023 NEs have been aligned through transliteration [8].

Table 2 shows that the performance improves substantially when we make use of the paraphrases to create additional parallel training examples. In experiment 2, we collected paraphrases from the baseline PB-SMT system which serves as additional sentences for the training data. Experiment 3 also follows the same procedure with the exception that in this case additional paraphrases are collected from comparable corpora. The performance also increases when we consider compositional contextual words with these paraphrases (experiment 4) which results in further improvements. When we add NE alignment with experiment 4, we obtain highest improvement.

Table 1. Evaluation results for different experimental setups

Experiments		Exp. No.	BLEU	NIST	OOV words
PB-SMT	Baseline	1	10.92	4.13	558
	Exp1 + Linguistic paraphrases with additional sentences	2	11.52	4.17	461
	Exp2 + paraphrases with comparable corpora + additional sentences	3	12.45	4.28	419
	Exp3 + non-linguistic paraphrase with Context word	4	12.82	4.31	396
	Exp4 + NE alignment	5	14.58	4.46	293

The evaluation result shows that the best system provided 3.66 absolute BLEU points (33.51% relative) improvement over the baseline system. The last column in table 2 shows the number of OOV words present in the resulting output of the corresponding experiment. As can be seen from Table 2, the percentage of OOV words present in the output reduces from 4.9% for the baseline system to 2.5% for the final system.

We also manually compared the translation outputs produced by our best system against the baseline outputs for a small subset of the test data. We found that our system results in more accurate lexical choices.

6   Conclusions and Future Works

In this paper we report on our work on the use of paraphrases in SMT. We extract paraphrases from the training corpus using bilingually calculated paraphrase probability from the baseline system. We also collect paraphrases from comparable corpora crawled from the web. The extracted paraphrases are examined and expanded according to the association of the monolingual distribution of the constituent words or contextual words. Using the bilingual aligned paraphrases as additional training data improves the system performance in terms of translation quality and results in fewer OOV words in the translation output.

As future work we would like to explore the paraphrase extraction technique using hybrid word alignments or hybrid MT method. We will also integrate the knowledge about paraphrase into the word alignment models as well as with in the MT workflow, this is another future direction for this work. We would also investigate into whether this approach can bring improvements of similar magnitude for larger training data.

Acknowledgements. The research leading to these results has received funding from the EU project EXPERT –the People Programme (Marie Curie Actions) of the European Union's Seventh Framework Programme FP7/2007-2013<tel:2007-2013>/ under REA grant agreement no. [317471].

References

1. Iordanskaja, L., Kittredge, R., Polguere, A.: Lexical Selection and Paraphrase in a Meaning-Text Generation Model. In: Paris, C.L., et al. (eds.) Natural Language Generation in Artificial Intelligence and Computational Linguistic, pp. 293–312. Kluwer Academic Publishers, Dordrecht (1991)
2. Callison-Burch, C., Koehn, P., Osborne, M.: Improved Statistical Machine Translation Using Paraphrases. In: The Main Conference on Human Language Technology Conference of the North American Chapter of the Association of Computational Linguistics, HLT-NAACL, pp. 17–24 (2006)
3. Denoual, E., Lepage, Y.: BLEU in characters: towards automatic MT evaluation in languages without word delimiters. In: The Second International Joint Conference on Natural Language Processing, pp. 81–86 (2005)
4. Kauchak, D., Barzilay, R.: Paraphrasing for automatic evaluation. In: The Main Conference on Human Language Technology Conference of the North American Chapter of the Association of Computational Linguistics (2006)
5. Heilman, M., Smith, N.A.: Tree edit models for recognizing textual entailments, paraphrases, and answers to questions. In: HLT 2010 Human Language Technologies: The 2010 Annual Conference of the North American Chapter of the Association for Computational Linguistics, pp. 1011–1019 (2010)
6. Gupta, R., Pal, S., Bandyopadhyay, S.: Improving MT System Using Extracted Parallel Fragments of Text from Comparable Corpora. In: 6th Workshop of Building and Using Comparable Corpora (BUCC). ACL, Sofia (2013)
7. Bannard, C., Callison-Burch, C.: Paraphrasing with Bilingual Parallel Corpora. In: ACL (2005)
8. Pal, S., Naskar, S.K., Pecina, P., Bandyopadhyay, S., Way, A.: Handling Named Entities and Compound Verbs in Phrase-Based Statistical Machine Translation. In: COLING 2010 Workshop on Multiword Expressions: from Theory to Applications (MWE 2010), Beijing, China, pp. 45–53 (2010)
9. Shiqi, Z., Haifeng, W., Ting, L., Sheng, L.: Extracting Paraphrase Patterns from Bilingual Parallel Corpora. Natural Language Engineering 15(4), 503–526 (2009)
10. Chan, T.P., Callison-Burch, C., Durme, B.V.: Reranking Bilingually Extracted Paraphrases Using Monolingual Distributional Similarity. In: GEometrical Models of Natural Language Semantics, GEMS (2011)
11. Aziz, W., Specia, L.: Multilingual WSD-like Constraints for Paraphrase Extraction. In: The Seventeenth Conference on Computational Natural Language Learning (CoNLL), Sofia, Bulgaria, pp. 202–211 (2013)
12. Barzilay, R., McKeown, K.R.: Extracting paraphrases from a parallel corpus. In: 39th Annual Meeting on Association for Computational Linguistics, pp. 50–57 (2001)
13. Xu, W., Ritter, A., Grishman, R.: Gathering and Generating Paraphrases from Twitter with Application to Normalization. In: ACL 2013 Workshop on Building and Using Comparable Corpora (2013)
14. Wang, R., Callison-Burch, C.: Paraphrase Fragment Extraction from Monolingual Comparable Corpora. In: Fourth Workshop on Building and Using Comparable Corpora, BUCC (2011)
15. Kuhn, R., Chen, C., Foster, G., Stratford, E.: Phrase Clustering for Smoothing TM Prob-abilities – or, How to Extract Paraphrases from Phrase Tables. In: COLING, Beijing, China (2010)

16. Fujita, A., Carpuat, M.: FUN-NRC: Paraphrase-augmented Phrase-based SMT Systems for NTCIR-10 PatentMT. In: The 10th NTCIR Conference, Tokyo, Japan, June 18-21 (2013)
17. Madnani, N., Ayan, N.F., Resnik, P., Dorr, B.J.: Using Paraphrases for Parameter Tuning in Statistical Machine Translation. In: The Second Workshop on Statistical Machine Translation, StatMT (2007)
18. Marton, Y., Callison-Burch, C., Resnik, P.: Improved Statistical Machine Translation Using Monolingually-Derived Paraphrases. In: The 2009 Conference on Empirical Methods in Natural Language Processing, EMNLP (2009)
19. Mehay, D.N., White, M.: Shallow and Deep Paraphrasing for Improved Machine Translation Parameter Optimization. In: The AMTA 2012 Workshop on Monolingual Machine Translation, MONOMT (2012)
20. Razmara, M., Siahbani, M., Haffari, G., Sarkar, A.: Graph Propagation for Paraphrasing Out-of-Vocabulary Words in Statistical Machine Translation. In: ACL (2013)
21. Church, K.W., Hanks, P.: Word association norms, mutual information, and lexicography. Journal on Computational Linguistics Archive 16(1), 22–29 (1990)
22. Dunning, T.: Accurate methods for the statistics of surprise and coincidence. Journal on Computational Linguistics - Special Issue on Using Large Corpora: I Archive 19(1), 61–74 (1993)
23. Phan, X.H.: Crfchunker: Crfenglish phrase chunker. In: PACLIC (2006)
24. Koehn, P., Och, F.J., Marcu, D.: Statistical phrase-based translation. In: HLT-NAACL, pp. 127–133 (2003)
25. Och, F.J.: Minimum Error Rate Training in Statistical Machine Translation. In: ACL (2003)
26. Stolcke, A.: SRILM - An Extensible Language Modeling Toolkit. In: International Conferance on Spoken Language Processing, vol. 2, pp. 901–904. Denver (2002)
27. Kneser, R., Ney, H.: Improved backing-off for M-gram language modeling. In: International Conference on Acoustics, Speech, and Signal Processing, ICASSP (1995)
28. Koehn, P., Hoang, H., Birch, A., Callison-Burch, C., Federico, M., Bertoldi, N., Cowan, B., Shen, W., Moran, C., Zens, R., Dyer, C., Bojar, O., Constantin, A., Herbst, E.: Moses: Open Source Toolkit for Statistical Machine Translation. In: ACL (2007)
29. Doddington, G.: Automatic evaluation of machine translation quality using n-gram cooccurrence statistics. In: Human Language Technology Conference, HLT, San Diego, CA, pp. 128–132 (2002)
30. Papineni, K., Roukos, S., Ward, T., Zhu., W.J.: BLEU: a method for automatic evaluation of machine translation. In: ACL (2002)

Inferring Paraphrases for a Highly Inflected
Language from a Monolingual Corpus

Kfir Bar and Nachum Dershowitz

School of Computer Science, Tel Aviv University, Ramat Aviv, Israel
{kfirbar,nachumd}@post.tau.ac.il

Abstract. We suggest a new technique for deriving paraphrases from a monolingual corpus, supported by a relatively small set of comparable documents. Two somewhat similar phrases that each occur in one of a pair of documents dealing with the same incident are taken as potential paraphrases, which are evaluated based on the contexts in which they appear in the larger monolingual corpus. We apply this technique to Arabic, a highly inflected language, for improving an Arabic-to-English statistical translation system. The paraphrases are provided to the translation system formatted as a word lattice, each assigned with a score reflecting its equivalence level. We experiment with the system on different configurations, resulting in encouraging results: our best system shows an increase of 1.73 (5.49%) in BLEU.

Keywords: Paraphrases, Arabic, Machine Translation.

1 Introduction

Paraphrases are pairs of text fragments, both in the same language, that have the same meaning in at least one context. Given a text, "paraphrasing" is the act of generating an alternate phrase that conveys the same meaning. Since the meaning of a text is determined only when its context is given, paraphrases are sometimes referred to as "dynamic translations". Paraphrases are also recognized as a bidirectional textual-entailment relation [13]. Identifying paraphrases is an important capability for many natural-language processing applications, including machine translation, as a possible workaround for the problem of limited coverage inherent in a corpus-based translation approach [11,29]. Other applications of paraphrasing include question answering [16,35,19] and automatic evaluation of summaries [40].

We usually distinguish between two levels of paraphrases: (1) *phrase* (sub-sentential) level refers to two variable-length text segments, each containing one or more words; and (2) *sentence* level, composed of two complete sentences.

We introduce a data-driven phrase-level paraphrasing technique and apply it to Arabic, a highly inflected language. The paraphrases are then employed to improve a phrase-based statistical translation system. The ideal setup for paraphrasing would probably be to have both a monolingual corpus and a bilingual parallel corpus as resources. However, since parallel corpora are not always available (for Arabic, there

A. Gelbukh (Ed.): CICLing 2014, Part II, LNCS 8404, pp. 254–270, 2014.

are only ones paired with English), we use monolingual documents as the primary resource for our paraphrasing algorithm. In addition, bilingual unaligned comparable documents (not translations) are used to suggest paraphrases. The paraphrase pairs are generated automatically by extracting similar phrases, similar on the lemma level, each of which occurs in one of a comparable pair of documents. In this work, a pair of comparable documents is composed of two news-related articles that appear to cover the same story. Like other, similar, works that utilize monolingual corpora for paraphrasing (e.g. [29]), we focus on the context in which the phrases occur in the text, where the context of a phrase is represented by some of its preceding and following words. We train a classifier to identify paraphrases through their context, supervised by an initial set of automatically annotated pairs.

This work makes the following contributions:

1. It proposes a new paraphrasing technique for monolingual corpora, especially as applied to a highly inflected language.
2. It compares the translation quality obtained by using different types of paraphrases.
3. It shows how to improve translation quality by tuning with translated paraphrase lattices.

Like most other Semitic languages, Arabic is highly inflected; therefore, data sparseness is much more noticeable than in English, and extracting paraphrases from a corpus turns out to be even more complicated. Arabic words are inflected for person, number and gender; prefixes and suffixes are added to indicate definiteness, conjunction, prepositions and possessive forms. Due to Arabic's rich morphology, we work on the lemma level. Consequently, our "paraphrases" sometimes include pairs with shared meaning, ignoring their inflection for number, gender, and person. The motivation is that such pairs often have similar English renderings.

We proceed as follows: Section 2 reviews some relevant previous work. Section 3 describes the details of our new monolingual paraphrasing technique. It is followed by a section explaining how paraphrases help in translation. In Section 5, we provide some experimental results, and finally we conclude in Section 6.

2   Related Work

There are several data-driven approaches for paraphrasing, which may be divided by the type of corpora they use. Some use monolingual corpora (e.g., [27,29]), some use parallel corpora, either monolingual (e.g., [4,31]) or bilingual (e.g., [1,40]), and others use comparable documents (e.g., [5,34,15,39,2]).

Barzilay and McKeown [4] extracted English paraphrases from monolingual parallel corpora. They marked a few identical aligned words as anchors and treated them as potential paraphrases. Following the co-training approach [8], they trained two classifiers, one to model the environment surrounding potential paraphrases and another to model the characteristics of paraphrases' words. In a previous work [2] we adapted that technique to derive Arabic paraphrases from comparable documents. Although we reported encouraging results, the quantity of paraphrases that particular technique

can produce is severely limited, and as a result we believe that it would be most difficult to employ this technique on a large enough scale to make a significant difference
for machine translation.

Bannard and Callison-Burch [1] used several bilingual parallel corpora of French
and Spanish paired with other languages, a technique known as "pivoting", to find
pairs of phrases in one language that translate similarly in one of the pivot languages.
Using this technique, Callison-Burch et al. [11] showed an improvement in the translation quality generated by a phrase-based statistical translation system. Callison-
Burch [10] and Zhao et al. [40] developed this approach further by adding syntactic
constraints to the extraction algorithms. To the best of our knowledge, there are no
available bilingual resources that pair Arabic with languages other than English and
that are aligned on the sentence level.

Marton et al. [29] used a relatively large monolingual corpus for deriving paraphrases in unsupervised settings to improve a phrase-based statistical translation system.
Generally speaking, potential paraphrases were found based on cosine similarity of
their distributional profile that captures the occurrences of the phrases' surrounding
words, modeled by log-likelihood [17]. They reported an improvement in translation
quality when the system is using relatively small bilingual corpora. The quality of
the resulted paraphrases is connected with the size of the monolingual corpus used by
the algorithm. A relatively large amount of monolingual data is needed for calculating
the statistics for the contextual words. Both Callison-Burch et al. [11] and Marton et
al. [29] derived paraphrases only for unseen input phrases, that is, phrases that do not
exist in the system's translation table. Jinhua et al. [24] considered paraphrases, derived by pivoting, also for input phrases that exist in the phrase table. They formatted
the input text, augmented with paraphrases, as a word lattice [18] and showed that it
outperformed a system that merely calculates paraphrases for unseen phrases. Inspired
by that, we restructure the input sentence as a lattice and augment it with paraphrases
of all the composing phrases, regardless of their presence in the phrase table. Furthermore, we show that tuning the system on such lattices helps improve the results.
Nakov and Ng [30] employed a similar lattice technique to help improve the results of
a Malay-to-English translation system by using Malay paraphrases of various sorts.
(Malay is another morphologically rich language, mainly based on a derivational
morphology, as opposed to the inflectional one in Arabic.)

Our experiments reconfirm the conclusion that paraphrasing aids translation (e.g.,
[11,29,24,30]), this time for Arabic.

3   Paraphrasing Technique

At its core, our paraphrasing technique takes inspiration from a number of the above
prior works. In a previous work [2] we extracted some morpho-syntactic features from
phrases' contextual words. We constructed pairs of phrases, each pair represented by a
single vector containing the weights of their features, as extracted from both phrases.
The pairs of phrases were extracted from comparable documents, simply by pairing
every phrase from one document with all phrases from its comparable partner.
We used a deterministic procedure to assign labels to some of the pairs indicating

whether a pair is composed of paraphrases or not and employed co-training, using those pairs as an initial training set, to label the unlabeled pairs. The labeling procedure considered pairs of similar phrases as paraphrases, and pairs of single words that were not identified as synonyms by a simple thesaurus as negative examples. The co-training learning algorithm focused on the context of both phrases and the words within the two phrases. The drawback is that newly discovered paraphrases were extracted merely from comparable documents; therefore, their number was relatively low and highly dependent on the quantity of comparable documents used. Obtaining a large amount of comparable data, as needed in that work, should considered much more challenging than obtaining plain monolingual texts, as in this work.

Since we are interested in improving machine translation, our system takes a given phrase and looks for its paraphrases. The input is paired with candidate phrases extracted from a relatively large monolingual corpus. Phrases that share a similar context with that of the input phrase are deemed paraphrases. To measure similarity of contexts, we first train a binary classifier using a relatively small set of annotated pairs, extracted from comparable documents using a technique like [2]. The limited-size corpus of bilingual comparable documents is, however, only used for training purposes. Going beyond our use of morpho-syntactic features (such as part of speech tags and base-phrase chunks), and inspired by the distributional similarity approach taken by [29], we define a new feature for capturing the semantic similarity of contexts, by representing words based on their frequency and co-occurrence with the phrase they are surrounding. In contrast to [29], who use co-occurrences of words and phrases as the model for finding paraphrases, we use it as part of a larger set of features considered by the context classifier.

3.1   Training a Context Classifier

In building a context classifier, we attempt to learn similarities between the contexts of two phrases that are deemed paraphrases. A context in our case is modeled by features extracted from the surrounding words of each of the two phrases. The number of words may vary for each individual feature. In particular, the classifier is trained to handle a binary classification problem: given pairs of phrases, decide which pairs are paraphrases and which are not. To supervise the training process, we deterministically generate a learning set of positive and negative examples, provided with their contexts. Those are collected from pairs of comparable documents, that is, different news articles that cover the same story. Obtaining comparable documents is currently done by a simple automated technique [39,2] from Arabic Gigaword (4th ed.) [33], a corpus of newswire stories published by several news agencies and grouped by publication date. The documents were pre-processed by AMIRA 2.0 [14], so that every word is assigned its lemma, full part-of-speech tag (excluding case and mood), base phrase chunks and named-entity recognition (NER) tags [6]. Pairing documents based on topic was done using the lemma-frequency vector of every document, taking those with cosine similarity above a threshold set heuristically to prefer precision to recall and considering only those articles that were published on the same day by different news agencies.

Given a pair of comparable documents, we begin by extracting all phrases (i.e., word sequences) of up to N words (here, N=6) from each document. We pair each phrase from one document with all the phrases from the other document, resulting in a relatively large set of pairs. Among those, we keep only those that we can label as positive or negative. A positive pair must comply with the following rules:

— Both phrases do not break a base phrase in the middle;
— both phrases contain at least one content word (non-functional, determined using the part-of-speech tag); and,
— both phrases match on the lemma level, word by word.

Since we work with words rather than senses, similar phrases do not always have the same meaning, given their local context. However, the fact that the phrases are taken from comparable documents suggests that they do share similar senses. Some positive examples are provided in Table 1.[1]

Table 1. Examples of positive phrase pairs

different number	*wqAl AlmSdr* ↔ *wqAlt AlmSAdr*
	"and the source said" ↔ "and the sources said"
different proclitic	*bt$kyl Hkwmp* ↔ *wt$kyl Hkwmp*
	"with establishment of a government" ↔
	"and establishment of a government"
exact match	*wzyr AlxArjyp* ↔ *wzyr AlxArjyp*
	"the minister for foreign affairs"

For negative examples, we select pairs of phrases that do not comply with the last rule and also not with one of the two others. This gives us enough confidence to believe that such phrase pairs are not paraphrases.

We use SVM [38] as the machinery for training the context classifier and employ a quadratic kernel, to enable the learning process to consider combinations of features. Technically, we used WEKA [23] as a framework combined with LibSVM [12]. The features that we use for training are described next.

3.2   Feature Extraction

We extract features from the contextual words surrounding each phrase. Given a pair of phrases, for each phrase we generate a vector that captures the *tf-idf* score of every lemma in context (the lemma frequency multiplied by the inverse document frequency). The context for this feature is delimited by 8 words before and after the phrase. The operational definition for a *document* for calculating the *idf* values is therefore a segment of 16 words. We collected a relatively large set of such documents from Arabic Gigaword (4th ed.).

---

[1] We are using Buckwalter transliteration for rendering Arabic script in ASCII [9].

The vectors are relatively sparse, each containing merely 16 non-zero values at most, while their dimension is much larger. Therefore, to increase the influence of the contextual lemmas that co-occur with their corresponding phrase more often than by chance, we calculate the pointwise mutual information (PMI) of every lemma appearing in the context of a specific phrase by collecting occurrences from chunks of about 10M words extracted from Arabic Gigaword. Finally, the *tf-idf* value of each lemma is multiplied by the relevant PMI value. Accordingly, every phrase P is represented by the following vector:

$$V_P = \{\langle l, \text{tf-idf}(l, P) \times \text{PMI}(l, P) \rangle \mid l \in \text{context}(P)\}$$

Working with lemmas is natural. The lemma groups together all the inflected perfective and imperfective forms of a verb, and all the inflected singular, dual and plural forms of a noun. Contextual words that are either not derived from a lemma or for which the morphological analyzer failed to find the lemma are considered with their surface form instead. This situation happens mostly (but not only) with named entities; hence, each named entity occurring in the context of P is replaced by a placeholder representing the entity type (e.g. person, organization, location).[2] Then, given a pair of phrases, we measure the cosine similarity of their vectors, and use it as a single numeric feature for classification. We checked the distribution of the context-similarity score over a sample of 1,735 phrase pairs corresponding to 867 positive and 868 negative pairs. Figure 1 shows the distribution, where the abscissa represents sub-intervals of the context-similarity value ranging from 0 to 1, that is, the first column represents the values between [0, 0.08), the second column represents the values between [0.08, 0.16), and so on. As expected, we note that a greater mass of the negative pairs is concentrated at the lower end of the scale, while the positive pairs move toward the right-hand side. We conclude that the context-similarity feature cannot be used deterministically for deciding positive or negative cases, however it can be combined with more relevant features and potentially help in classification. In addition to the cosine similarity of the two phrases, we use the part-of-speech and base-phrase tags of each contextual word, up to 6 words before and after each of the two phrases, taking into account their relative position in the sentence (cf. [2,4]).

3.3   Evaluation of the Context Classifier

Overall we extracted about 12,000 phrase pairs of various lengths. In order to prefer precision over recall, the number of negative examples was selected to be twice as many as the positive examples. We ran a 10-fold cross validation; the precision was 84.7 and recall was 79 (*F*-measure was 81.7). Essentially, our classifier prefers precision to recall. It means that it is capable of distinguishing between contexts of identical phrases, on the lemma level, and the context of non-paraphrases. However, since the paraphrases we are looking for are not part of this training set, and, in fact, are not necessarily composed of identical phrases, we cannot estimate the performance on real data, based on these results.

---

[2] Named entities were found by AMIRA 2.0 [14].

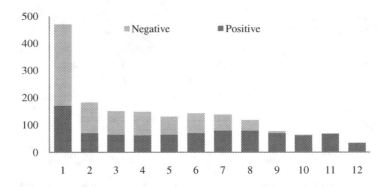

Fig. 1. The context-similarity distribution

3.4   Deriving Paraphrases from a Monolingual Corpus

To extract paraphrases for a given phrase, we use 10M words of Arabic Gigaword as a resource, corresponding to 2.7M indexed phrases. However, although the size of the corpus may affect the coverage, it does not affect the quality of the results, as we do not use it to capture any statistical information (PMI values are calculated based on a different part of the corpus).

We preprocess the corpus with AMIRA 2.0 and extract phrases of up to 6 words, where phrases may not break a base phrase in the middle and must contain at least one content (non-functional) word. Every phrase was indexed in a database, so that it can be searched by each of its lemmas.

Given a phrase P for paraphrasing, our algorithm begins by searching the database for potential candidate paraphrases, defined heuristically as phrases that have at least some percentage of words in common with P, matched on the lemma level. Then it pairs each candidate phrase with P, and decides whether they are paraphrases or not using the context classifier. Theoretically, every phrase may be considered a potential paraphrase of P; however, checking every phrase from the database, given P, is computationally infeasible. For now, we consider phrases that have at least 40% of their lemmas·in common, an ad-hoc threshold that was selected based on observations. A disadvantage of using this technique is that the extracted paraphrases are usually more structural. Considering better approaches, such as matching lemmas on the synonym level, is left for future investigation.

We consider the distribution provided by WEKA, which is calculated based on the distance of an instance from the separating hyperplane, to measure the quality of the returned paraphrases. In other words, we consider the distribution value as a confidence score. Moreover, to reflect the grammatical similarity of the phrases, that is, whether the paraphrase may actually replace P without being detrimental to the input-sentence structure, we calculate two language-model scores. They are calculated on the text containing the paraphrase of P within the original context of P. One score is a language-model log-probability of the sequence of words that lie to the left of the paraphrase, including the first word of the paraphrase itself, and the second is calculated for the last word of the paraphrase followed by the sequence of words to the right of the paraphrase. Both scores measure how likely it is to find the paraphrase in the same context as P. The language model is generated using a large monolingual corpus, on the lemma level, with SRILM [37].

4   Using Paraphrases in Translation

We experiment with an Arabic-to-English implementation of Moses [26], a statistical-machine-translation platform, aiming to improve its translation quality using different levels of paraphrases of fragments of the input sentence. Paraphrases can be derived either for any fragment of the input text, or only for unseen phrases (regardless of the system's ability to translate them using the translations of [some of] their fragments.) Even if a phrase is in the phrase table, there is a chance that the overall translation may be improved by translating one of its paraphrases, due to wrong alignments resulting in a bad translation. Despite the hypothetical benefit of considering paraphrases of phrases that exist in the phrase table, there is a risk that the system will prefer a translation of one of the paraphrases that was incorrectly identified to the translation of the original phrase. To deal with this, we assign scores to the paraphrases that reflect the quality of their equivalence, so that the system will judge them accordingly.

We follow [24] and format every input sentence along with its paraphrases as a word lattice [18], that is, a directed acyclic graph, with every node uniquely labeled and every edge containing a token and a weight. A lattice is mainly used when parts of the input sentence are ambiguous and, instead of selecting merely one interpretation in the usual way, the lattice encodes multiple interpretations, each encoded with a plausibility weight.

Given a tokenized input Arabic sentence of N tokens for translation, we begin by initiating a lattice that captures the transition of the individual tokens linearly. (We use the D3 tokenization scheme [22].) We create a lattice of $N+1$ nodes and N edges, each representing a token. Every edge is assigned the value 1 (the maximum value in our case), keeping the lattice faithful to the input text. Then we add bypasses to the lattice, reflecting the paraphrases found by our paraphrasing algorithm. Paraphrases are generated for all phrases of the input sentence that are composed of up to 6 words and do not break a base phrase. To control the complexity we allow every phrase to have at most 3 paraphrases, each assigned with a confidence score higher than a threshold. The number 3 was determined mainly based on observations; the threshold is learned based on experiments, which we show in the next section. We refer to this structure as a *paraphrase lattice*. Figure 2 shows a paraphrase lattice representing the input sentence $a\ b\ c\ d$ augmented with one paraphrase of 4 tokens $x\ y\ z\ w$, replacing the phrase $b\ c$.

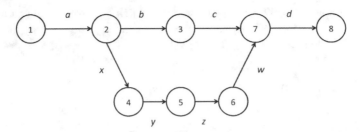

Fig. 2. Paraphrase lattice

We assign weights to every edge in the lattice to reflect the chance that a specific paraphrase represents its corresponding input phrase. Those weights are considered by the decoder as part of the log-linear model of the translation system. In particular, Moses introduces an additional feature function, referred to as *InputFeature*, which represents the input type; the weight of that feature function, as combined in the log-linear model, allows the decoder to consider different paths of the input lattice while keeping in mind other factors, such as the translation and language models. The weight of InputFeature may be either set manually or tuned automatically with, for example, Minimum Error Rate Training (MERT) [31] on translated lattices. We take both approaches: (1) tuning the translation system on plain translated segments and setting the weight of the InputFeature function manually; and (2) tuning the translation system on translated lattices, and as a result, adapting the weights of all feature functions (including InputFeature) automatically.

Moses allows one to assign edges with several weights, whereas the log-linear model considers each individually. In this paper, we use three weights: (1) the context-classifier confidence score; (2) the left language-model score; and (3) the right language-model score. The weight of a single outbound edge is always 1. When there are several edges departing from the same node (where a paraphrase path begins), we normalize the score by dividing each component by the total sum of all values of the same component on the other sibling edges.

5   Experimental Approach and Results

We use Arabic paraphrases in translation and automatically measure their effect on the translation quality. Our baseline is an Arabic-to-English Moses instance, using different sizes of bilingual corpora, focusing on the newswire domain. We employ an English 5-gram language model, generated from a monolingual corpus of about 30 million words, and tune the system with MERT using bilingual texts containing 130K Arabic words. The Arabic text is tokenized following D3 [22] using MADA 3.1 [21,36] and the English text is tokenized using the default Moses tokenizer.

We test all the systems on the same evaluation set, the newswire part of the 2009 NIST OpenMT Evaluation set [20], containing 586 sentences, corresponding to 20,671 D3 tokens.

We begin by investigating how the threshold on the confidence score of the context classifier affects the overall translation performance. Clearly, as we decrease the threshold, the number of generated paraphrases grows larger; however, the quality of the paraphrases is likely to decrease (recall that we limit every phrase with maximum number of 3 paraphrases). The values of the confidence score that we observe are mainly concentrated in the range [0.91, 1]. Therefore, we use several threshold values within that range. Figure 3 shows the BLEU [32] scores calculated for a system that uses a bilingual corpus containing one million Arabic words, running under different threshold values. It is clear from the results that, when using a relatively high threshold, the translation quality gets better, while the number of generated paraphrases decreases. At 0.97 we see a steep drop in the number of qualified paraphrases, and as a result the BLEU score slightly declines. Overall, the results are encouraging, as we may learn from this that the confidence score affects the translation results as expected: With low thresholds, we get a relatively large number of paraphrases that do not appropriately reflect the meaning of its generating phrase, hence may be detrimental to the final result. Accordingly, we use 0.96 in the following experiments.

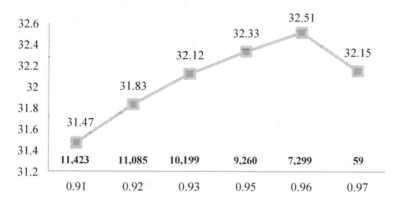

Fig. 3. BLEU scores of a system running on paraphrase lattice with different thresholds on the confidence score as returned by the context classifier. The abscissa represents the threshold values and the ordinate is the corresponding BLEU score. The numbers in boldface indicate the overall number of paraphrases generated using each threshold value.

To evaluate the contribution of paraphrasing to the translation results, we setup several baseline frameworks. Instead of deriving paraphrases from a corpus, we consider verbal and nominal synonyms, extracted with the help of a thesaurus. Since Arabic WordNet [7] is limited, we generated an Arabic thesaurus copying the simplistic technique of [3], namely, we look at the list of lemmas provided by SAMA 3.1 [28] and extract pairs of lemmas that share at least one English gloss in common. Those pairs are deemed synonyms. Overall we extracted about 120K pairs corresponding to about 20K lemmas.

As for paraphrases, synonyms of input words are added to the lattice. Since synonyms are single words, we allow any number of synonyms to be generated for a single input word. As we do not have a confidence score under this setting, we assign equal weight to all synonyms, including the word itself. The language-model scores are calculated in the same way as for paraphrases. Note that synonyms are provided on the lemma level; hence, they must be inflected to reflect the form of the original input word. For example, given an input verb *EvrwA*, "they discovered", derived from the lemma *Eavar-u_1*, the thesaurus returns the synonym *ka$af-i_1*. To generate the required form *k$fwA*, we employ Almor [21], an Arabic morphological generator, and provide it with the morphological features of the original word, as extracted by MADA.

Given a verb, our paraphrasing algorithm very often identifies different inflected forms of the same verb as paraphrases. For the most part, such forms can be generated deterministically, regardless of the context in which they occur. Therefore, we build another baseline lattice, *Morph Gen*, which contains all the inflected forms generated by Almor that have the same English translation as the original form. For example, for a verb given in its perfective-form/singular/3rd-person, we generate only the form that is inflected for the opposite gender. This is because in English, singular 3rd person, past-tense verbs are combined with *s* or *es*. We manually crafted a few more such rules.

We experiment with different sizes of bilingual corpora and the BLEU scores are presented in Table 2. The best improvement over the baseline is 0.91 in BLEU score, observed when using a lattice containing paraphrases and all synonyms, running with a bilingual corpus of 500K Arabic words. When we increase the size of the bilingual corpus, the improvement is eroded, although it persists. This observation complies with the observations made in similar works [11,29]. The system that uses paraphrases outperforms all other baselines, including those that use synonyms. Moreover, we observe that synonyms moderately improve the final translations over the baseline when using a relatively small bilingual corpus. Synonyms also help further improve the results when combined with paraphrases. Using different inflected forms, represented by Morph Gen, was found to be counterproductive. In fact, the results obtained from using the Morph-Gen lattice are consistently a little worse than the baseline. This teaches us that the improvement obtained by using paraphrases was not due to verbs that get paraphrased simply as different inflected forms, (although such cases do exist). We used paired bootstrap resampling [25] for calculating statistical significance ($p < 0.05$) over the baseline. The improvements we get by paraphrases, syn/paraphrases on all corpus sizes are all statistically significant. The improvements we get by the synonym lattices are not significant, however.

Table 3 shows the total number of phrases for which the system generated at least one paraphrase/synonym, corresponding to the portion of them that do not appear in the bilingual corpora we use. Generally speaking, we learn that among the phrases that got paraphrased by the system, there are more phrases that appear in the bilingual corpora the translation system uses as a resource. It implies that the system benefits from paraphrases of phrases that could be translated merely using the bilingual texts in the usual way.

Table 2. Evaluation results for different size (in millions of Arabic words) bilingual corpora on different lattices. Improvements over the baseline are in boldface.

Corpus size →	0.5M	1M	1.5M	4.5M
Baseline	31.48	32.18	32.75	34.20
Verb Synonyms	31.34	32.06	32.38	34.11
Noun Synonyms	**31.60**	**32.20**	32.26	33.97
All Synonyms	**31.50**	**32.31**	32.30	34.07
Morph Gen	31.44	32.00	32.40	34.02
Paraphrases	**32.28**	**32.51**	**33.19**	**34.21**
Syn/paraphrases	**32.39**	**32.72**	**33.28**	34.12

Table 3. The number of unseen phrases that were paraphrased by the system. The *Total Generated* column represents the total number of phrases for which the system created at least one paraphrase. Every other column represents the number of unseen phrases corresponding to the size (in millions of Arabic words) of the bilingual corpus used by the translation system.

	Total Generated	Unseen in 0.5M	Unseen in 1M	Unseen in 1.5M	Unseen in 4.5M
Verb Synonyms	466	34	32	32	29
Noun Synonyms	649	25	22	18	14
All Synonyms	1,115	59	54	50	43
Paraphrases	7,299	331	217	211	193
Syn/paraphrases	8,414	390	225	219	199

Table 4 compares the way some seen/unseen phrases got translated by the system with and without paraphrasing.

So far, all our experiments were executed on a system that was merely tuned on the original sentences, formatted as word lattices, but including neither paraphrases nor synonyms. The weight of the InputFeature function, which affects the preferences of the decoder, was assigned arbitrarily to be 0.1. As a next step, we repeat the same experiments, minus the less productive ones, this time with a system that was tuned with MERT on the same development set, formatted as paraphrase lattices and augmented with paraphrases. The results are presented in Table 5. The best statistically significant improvement of +1.73 (5.49%) BLEU points over the baseline is obtained by the system that uses 500K Arabic words, tuned with MERT on the paraphrase lattices. For the most part, tuning the parameters for paraphrases helps improve the translations. But we see a slight drop for the larger corpus, suggesting that the weights assigned to other features were slightly miscalculated. We are in the process of trying to alleviate this.

Table 4. Examples of paraphrases and their translations. The Arabic text is tokenized according to the D3 scheme [22]. Columns from left to right: (1) the original Arabic phrase; (2) the way it was translated by the baseline system (<unseen> means that the phrase was not translated as a whole); (3) the paraphrase that was used by a system with paraphrasing capabilities; (4) the way the paraphrase was translated; and (5) our comments.

Original phrase	Baseline translation	Paraphrase	Paraphrase Translation	Comments
dblwmAsywn bwlndywn	<unseen>	*dblwmAs bwlndy*	Polish diplomat	Different number
Al+ qyAdp Al+ Eskryp Al+ jnwbyp	<unseen>	*Al+ qyAdp Al+ Eskryp fy Al+ jnwb*	the military headquarter of the south	Different phrasing
w+ y}n Al+ Tfl	the Child and *y}n*	*lmA*A bkY Al+ Tfl*	why the child cried	Synonyms helped to improve translation
nATq b+ Asm Al+ xArjyp	a spokesman of the foreign affairs	*Al+ mtHdv b+ Asm wzArp Al+ xArjyp*	the spokesman of the ministry for foreign affairs	
wADAf >n AlEskryyn	The military *wADAf*	*ADAf >n Aljy$*	he added that the army	Wrong tokenization (*wADAf*) was fixed through paraphrasing
Al+ A$hr Al+ Axyrp	the last months	*Al+ AyAm Al+ Axyrp*	the last days	Wrong paraphrasing resulting in wrong translation
nzE Al+ slAH Al+ nwwy	The elimination of weapon for mass destruction	*Ant$Ar Al+ AslHp Al+ nwwyp*	the spreading of weapon for mass destruction	Antonyms are identified wrongly as paraphrases

Table 5. Results of some of the experiments from Table 2, repeated after tuning the system with paraphrases. (TOPL = tuned on paraphrase lattice) indicates that the system was tuned on paraphrase lattices.

Corpus size →	0.5M	1M	1.5M	4.5M
Baseline	31.48	32.18	32.75	34.20
All Synonyms	31.50	32.31	32.30	34.07
All Synonyms (TOPL)	**31.89**	**32.47**	**32.45**	33.73
Paraphrases	32.28	32.51	33.19	34.21
Paraphrases (TOPL)	**33.01**	**33.11**	**33.46**	34.19
Syn/paraphrases	32.39	32.72	33.28	34.12
Syn/paraphrases (TOPL)	**33.21**	**33.43**	**33.68**	34.10

6   Conclusions

We have demonstrated the potential of using Arabic paraphrases and synonyms to improve the results of a statistical translation system. We presented a new technique for paraphrasing from a monolingual corpus, supported by a context classifier that was trained using examples from a relatively small set of comparable documents. As a result, the resulting algorithm does not require large quantities of text to calculate word statistics. Arabic is highly inflected; therefore, working on the lemma level was natural. Although some of the derived paraphrases were in fact different inflected forms of their corresponding original phrases, we found that this was not the salient reason for improvement. We configured our algorithm to prefer precision over recall by merely considering phrases that have some lemmas in common with the subject phrase. Improving this technique should improve the results even more. We may conclude that the translation system benefits from using MERT on paraphrase lattices to adjust the weight of the InputFeature function, resulting in better final translations. Our immediate intent is to apply this paraphrasing technique to additional languages with complex morphology.

References

1. Bannard, C., Callison-Burch, C.: Paraphrasing with Bilingual Parallel Corpora. In: The 43rd Meeting of the Association for Computational Linguistics (ACL), pp. 597–604. Ann Arbor, MI (2005)
2. Bar, K., Dershowitz, N.: Deriving Paraphrases for Highly Inflected Languages from Comparable Documents. In: The 24th International Conference on Computational Linguistics (COLING), Mumbai, India (2012)

3. Bar, K., Dershowitz, N.: Using Semantic Equivalents for Arabic-to-English Example-Based Translation. In: Soudi, A., Farghaly, A., Neumann, G., Zbib, R. (eds.) Challenges for Arabic Machine Translation, pp. 49–72. John Benjamins Publishing Company (2012)
4. Barzilay, R., McKeown, K.: Extracting Paraphrases from a Parallel Corpus. In: The 43rd Meeting of the Association for Computational Linguistics (ACL), Toulouse, France, pp. 50–57 (2001)
5. Barzilay, R., Lee, L.: Learning to Paraphrase: An Unsupervised Approach Using Multiple-Sequence Alignment. In: The North American Chapter of the Association for Computational Linguistics (HLT-NAACL), Edmonton, Canada, pp. 16–23 (2003)
6. Benajiba, Y., Diab, M., Rosso, P.: Arabic Named Entity Recognition: An SVM-based Approach. In: The Arab International Conference on Information Technology (ACIT), Hammamet, Tunisia (2008)
7. Black, W., Elkateb, S., Vossen, P.: Introducing the Arabic WordNet Project. In: The 3rd Global Wordnet Conference (GWC), Jeju, South Korea, pp. 295–299 (2006)
8. Blum, A., Mitchell, T.: Combining Labeled and Unlabeled Data with Co-training. In: The 11th Annual Conference on Computational Learning Theory (COLT), Madison, WI, pp. 92–100 (1998)
9. Buckwalter, T.: Buckwalter Arabic Morphological Analyzer Version 1.0. LDC Catalog number LDC2002L49 (2002)
10. Callison-Burch, A.: Syntactic Constraints on Paraphrases Extracted from Parallel Corpora. In: The Conference for Empirical Methods in Natural Language Processing (EMNLP), Honolulu, Hawaii, pp. 196–205 (2008)
11. Callison-Burch, C., Koehn, P., Osborne, M.: Improved Statistical Machine Translation using Paraphrases. In: The North American Association for Computational Linguistics (NAACL), New York City, NY (2006)
12. Chang, C.C., Lin, C.J.: LIBSVM: A Library for Support Vector Machines. ACM Transactions on Intelligent Systems and Technology 2, 27:1–27:27 (2011)
13. Dagan, I., Glickman, O.: Probabilistic Textual Entailment: Generic Applied Modeling of Language Variability. In: PASCAL Workshop on Learning Methods for Text Understanding and Mining, Grenoble, France, pp. 26–29 (2004)
14. Diab, M.: Second Generation Tools (AMIRA 2.0): Fast and Robust Tokenization, POS Tagging, and Base-Phrase Chunking. In: MEDAR 2nd International Conference on Arabic Language Resources and Tools, Cairo, Egypt (2009)
15. Dolan, W.B., Quirk, C., Brockett, C.: Unsupervised Construction of Large Paraphrase Corpora: Exploiting Massively Parallel News Sources. In: The 20th International Conference on Computational Linguistics (COLING), Geneva, Switzerland (2004)
16. Duboue, P.A., Chu-Carroll, J.: Answering the Question you wish They had Asked: The impact of Paraphrasing for Question Answering. In: Human Language Technology-North American Association for Computational Linguistics (HLT-NAACL), New York City, NY, pp. 33–36 (2006)
17. Dunning, T.: Accurate Methods for the Statistics of Surprise and Coincidence. Computational Linguistics 19(1), 61–74 (1993)
18. Dyer, C., Muresan, S., Resnik, P.: Generalizing Word Lattice Translation. In: The 46th Annual Meeting of the Association for Computational Linguistics: Human Language Technologies (ACL-HLT), Columbus, Ohio, pp. 1012–1020 (2008)
19. Fader, A., Zettlemoyer, L., Etzioni, O.: Paraphrase-Driven Learning for Open Question Answering. In: The 51st Meeting of the Association for Computational Linguistics (ACL), Sofia, Bulgaria (2013)

20. Garofolo, J.: NIST OpenMTEval2009,
    http://www.itl.nist.gov/iad/mig/tests/mt/2009/
21. Habash, N., Rambow, O.: Arabic Tokenization, Morphological Analysis, and Part-of-speech Tagging in One Fell Swoop. In: The Conference of American Association for Computational Linguistics, pp. 578–580. Ann Arbor, MI (2005)
22. Habash, N., Rambow, O., Roth, R.: MADA+TOKAN: AToolkit for Arabic Tokenization, Diacritization, Morphological Disambiguation, POS Tagging, Stemming and Lemmatization. In: The Second International Conference on Arabic Language Resources and Tools, The MEDAR Consortium, Cairo, Egypt (2009)
23. Hall, M., Frank, E., Holmes, G., Pfahringer, B., Reutemann, P., Witten, I.H.: The WEKA Data Mining Software: An Update. SIGKDD Explorations 11, 10–18 (2009)
24. Jinhua, D., Jiang, J., Way, A.: Facilitating Translation Using Source Language Paraphrase Lattices. In: The Conference on Empirical Methods in Natural Language Processing (EMNLP), Stroudsburg, PA, pp. 420–429 (2010)
25. Koehn, P.: Statistical Significance Tests For Machine Translation Evaluation. In: The Conference on Empirical Methods in Natural Language Processing (EMNLP), Barcelona, Spain, pp. 388–395 (2004)
26. Koehn, P., Hoang, H., Birch, A., Callison-Burch, C., Federico, M., Bertoldi, N., Cowan, B., Shen, W., Moran, C., Zens, R., Dyer, C., Bojar, O., Constantin, A., Herbst, E.: Moses: Open Source Toolkit for Statistical Machine Translation. In: The 45th Meeting of the Association for Computational Linguistics (ACL), Prague, Czech Republic (2007)
27. Lin, D., Pantel, P.: Discovery of Inference Rules for Question Answering. Natural Language Engineering 7(4), 343–360 (2001)
28. Maamouri, M., Graff, D., Bouziri, B., Krouna, S., Bies, A., Kulick, S.: Standard Arabic morphological analyzer (SAMA), Version 3.1. Linguistic Data Consortium, Philadelphia, PA (2010)
29. Marton, Y., Callison-Burch, C., Resnik, P.: Improved Statistical Machine Translation Using Monolingually-Derived Paraphrases. In: The Conference on Empirical Methods in Natural Language Processing (EMNLP), Singapore (2009)
30. Nakov, P., Ng, H.T.: Translating from Morphologically Complex Languages: A Paraphrase-Based Approach. In: The Meeting of the Association for Computational Linguistics (ACL), Portland, OR (2011)
31. Och, F.J.: Minimum Error Rate Training in Statistical Machine Translation. In: The 41st Annual Meeting of the Association for Computational Linguistics (ACL), Sapporo, Japan, pp. 160–167 (2003)
32. Papineni, K., Roukos, S., Ward, T., Zhu, W.J.: Bleu: A Method for Automatic Evaluation of Machine Translation. In: The 40th Annual Meeting of the Association for Computational Linguistics (ACL), Philadelphia, PA, pp. 311–318 (2002)
33. Parker, R., Graff, D., Chen, K., Kong, J., Maeda, K.: Arabic Gigaword, 4th edn. Linguistic Data Consortium, LDC2009T30, Philadelphia, PA (2011) ISBN 1-58563-532-4
34. Quirk, C., Brockett, C., Dolan, W.: Monolingual Machine Translation for Paraphrase Generation. In: The 2004 Conference of Empirical Methods in Natural Language Processing (EMNLP), Barcelona, Spain, pp. 142–149 (2004)
35. Riezler, S., Vasserman, A., Tsochantaridis, I., Mittal, V., Liu, Y.: Statistical Machine Translation for Query Expansion in Answer Retrieval. In: The 45th Annual Meeting of the Association for Computational Linguistics (ACL), Prague, Czech Republic, pp. 464–471 (2007)

36. Roth, R., Rambow, O., Habash, N., Diab, M., Rudin, C.: Arabic Morphological Tagging, Diacritization, and Lemmatization Using Lexeme Models and Feature Ranking. In: The 46th Annual Meeting of the Association for Computational Linguistics (ACL), Columbus, Ohio, pp. 117–120 (2008)
37. Stolcke, A.: SRILM – An Extensible Language Modeling Toolkit. In: The International Conference on Spoken Language Processing, pp. 901–904. Denver, CO (2002)
38. Vapnik, V., Cortes, C.: Support Vector Networks. Machine Learning 20, 273–297 (1995)
39. Wang, R., Callison-Burch, C.: Paraphrase Fragment Extraction from Monolingual Comparable Corpora. In: The Fourth Workshop on Building and Using Comparable Corpora (BUCC), Istanbul, Turkey (2011)
40. Zhao, S., Wang, H., Liu, T., Li, S.: Pivot Approach for Extracting Paraphrase Patterns from Bilingual Corpora. In: The 46th Annual Meeting of the Association for Computational Linguistics (ACL), Columbus, OH, pp. 780–788 (2008)

Improving Egyptian-to-English SMT by Mapping Egyptian into MSA

Nadir Durrani[1,*], Yaser Al-Onaizan[2], and Abraham Ittycheriah[2]

[1] University of Edinburgh
dnadir@inf.ed.ac.uk
[2] IBM T.J. Watson Research Center
{onaizan,abei}@ibm.com

Abstract. One of the aims of DARPA BOLT project is to translate the Egyptian blog data into English. While the parallel data for MSA[1]-English is abundantly available, sparsely exists for Egyptian-English and Egyptian-MSA. A notable drop in the translation quality is observed when translating Egyptian to English in comparison with translating from MSA to English. One of the reasons for this drop is the high OOV rate, where as another is the dialectal differences between training and test data. This work is focused on improving Egyptian-to-English translation by bridging the gap between Egyptian and MSA. First we try to reduce the OOV rate by proposing MSA candidates for the unknown Egyptian words through different methods such as spelling correction, suggesting synonyms based on context etc. Secondly we apply convolution model using English as a pivot to map Egyptian words into MSA. We then evaluate our edits by running decoder built on MSA-to-English data. Our spelling-based correction shows an improvement of 1.7 BLEU points over the baseline system, that translates unedited Egyptian into English.

1 Introduction

The use of dialectal Arabic has been previously only limited to the speech whereas written texts were produced using MSA. With the rapidly increasing availability of the colloquial text, due to influx of social media in the Arabic-speaking countries in recent times, there has been interest in translating forums and blogs. DARPA GALE project aimed at translating news wire and parliamentary proceedings. The focus in BOLT project has shifted towards translating blog data and different dialects of Arabic, more specifically Egyptian which is considered to be the most widely used dialect after MSA. The new focus of translating dialect and blog data presents numerous challenges. An immediate bottle-neck is the lack of NLP resources for various dialects. Secondly the blog data is user-generated therefore noisy and lack standardization in orthography [1].

While the parallel data for MSA-English is abundantly available, sparsely exists for Egyptian and other dialects of Arabic. A notable drop in the quality of translation is observed when translating Egyptian blog data using translation models built on top of

---

* Work done during an internship at IBM T.J. Watson Research Center.
[1] MSA = Modern Standard Arabic.

A. Gelbukh (Ed.): CICLing 2014, Part II, LNCS 8404, pp. 271–282, 2014.
© Springer-Verlag Berlin Heidelberg 2014

MSA-English parallel data. Table 1 shows results, in terms of BLEU [2], when decoding MSA (Gale-dev10 set) and Egyptian (tahyyes dev set) using two different decoders TRL [3] and DTM [4]. Some of this drop can be attributed to the OOV rate which is high as 3.66% when translating Egyptian. While the rest occurs because of the dialectal differences and data mismatch

The comparison of BLEU scores in Table 1, however, may not be fair because it is obtained from running decoders on different dev sets. In order to ensure that these numbers reflect a true comparison, we did a pilot study. We randomly took a set of 100 Egyptian sentences and got them translated to MSA through a human translator. We then decoded both MSA and EGY sentences into English which used models trained on MSA-En parallel data. Both TER [5] and BLEU scores (Table 2) for the Egyptian sample were worse by more than 3 points as compared to that of the output of MSA. The high OOV rate (3.68%) in the English output of Egyptian sample, also confirms the result in Table 1.

Table 1. Baseline Comparison for MSA and EGY (Egyptian)

	TRL	DTM	OOV Rate
MSA	22.83	28.81	0.64%
EGY	19.49	22.29	3.66%

Table 2. Translating 100 Sentences of MSA and EGY into English

TRL	TER	BLEU	OOV Rate
MSA	58.52	18.77	1.07%
EGY	61.75	15.22	3.68%

Our focus in this paper is to improve Egyptian-to-English translation by bridging the gap between Egyptian and MSA. We try to learn dialectal differences between Egyptian and MSA and map the Egyptian words, that our system does not know how to translate or translate well, to their corresponding MSA words.

The paper is organized as follows. In Section 2 we provide a review of previous work In Section 3, we study the patterns of OOV words in the Egyptian data, dialectal differences between Egyptian and MSA, and discuss methods to propose MSA candidates for these. In Section 4, we describe a model to rank the candidates in a stack-based decoding framework. In Section 5, we present results on the OOV handling. In Section 6 we try to exploit a very small Egyptian-English corpus, using English as a pivot to map Egyptian to MSA using the well-known convolution model . In Section 7 we conclude the paper.

2   Previous Work

A plentiful amount of research has been spent in developing resources and natural language processing of MSA [6]. However, research on different dialects of Arabic is relatively sparse [7,8]. In this section we discuss previous work on machine translation of dialects. A hybrid machine translation that uses both rule-based and statistical methods to transform an Egyptian sentence into a diacritized MSA sentence was proposed in [9]. The input sentence is first tokenized and pos-tagged through a statistical model. A rule-based model, built on top of Egyptian-MSA lexicon, is then used to transfer the source into diacritized MSA. [10] also improved dialectal translation in hybrid machine translation by normalizing dialectal Arabic on character and morpheme level using a dialect-specific morphological analyzer. By applying their processing to the training and test corpora, they observed an improvement in the translation quality by approximately 2% on web text in terms of BLEU score. Like this paper, [11] focus on translating OOV words in the dialect Arabic. They propose paraphrases of the source language words. The candidates are obtained by applying morphological analysis on the input and mapping the affixes of OOV words into their MSA counterparts. The transformation is only applied to the affixes and not to the stems. The resulting candidates are then fed into MSA-to-English SMT as an input lattice. Their methodology gives an improvement of 0.56 BLEU points on a test set having an OOV Rate of 1.51%. In a recent effort, [1] built a 1.5 M words Dialectal (Levantine and Egyptian)-to-English data using Amazon's Mechanical Turk crowdsourcing service. In their study, they showed that a system built on small amount of dialectal data (1.5 Million words) improves the translation quality by more than 6 BLEU points than their MSA counterpart built from an enormous amount of data (150 Million Words). One of the interesting finds in their paper is that adding Dialect-English parallel data to the training (MSA-English parallel data) proves much more benifical than using Dialect-MSA parallel data to first transform dialect into MSA and then translating from MSA to English. The former is better than latter by a difference of roughly 2 BLEU points. [12] used character level transformation including morphological, phonological and spelling changes to narrow down the gap between Egyptian and MSA, subsequently improving Egyptian-to-English machine translation.

Other notable contributions towards building NLP resources for dialect discussed as follows. [7] built parser for spoken dialect using MSA tree-bank. [13] minned the web to extract a Dialect-to-MSA lexicon. A statistical morphological segmenter for Iraqi and Levantine speech transcripts was built by [14]. Their supervised algorithm for morpheme segmentation reduces the OOV rate.

3   Patterns of OOV and Methods for Candidate Suggestions

In this section, we study the patterns of errors in the Egyptian dev set and propose methods to give MSA candidates for these. After analyzing some data, we classified the error patterns into five sets i) substring repetition, ii) compounding errors iii) spelling differences, iv) dialect specific errors and v) true OOV words. We briefly discuss each of these categories.

3.1   Substring Repetition

Most noticeable but rather small number of errors in the blog data appeared due repetition of a character/substring in a known word. See Figure 1 for reference. In the first word, the susbstring ﻻ repeats, in third, fourth and fifth words characters و , ف and

Table 3. Dialectal Rules to Convert Egyptian to MSA

Egyptian	MSA
bXXX	XXX
XXXh	XXXp
XXXp	XXXh
XXXy	XXX +y
XXXNa	XXX +Na
H |h (y|n|t|A) XXX	s (y|n|t|A) XXX

ي [2] repeat. For a third consecutive appearance of a character or substring in a string, one can be sure that it is spurious and can be safely deleted. For all such instances of OOV words we remove the spurious characters one at a time and hypothesize the ones found in the MSA vocabulary.

باسلاااااااااالالام يا سلام عشروووووووووووون الففففف جنييييه

ياسلام يا سلام عشرون الف جنيه

Ref: how nice how nice twenty thousand pounds

Fig. 1. Character and Substring Repetition

3.2   Compounding Errors

A small number of errors were caused due to multiple words conjoined into a single token. For example in an OOV word بالحشدوالهتافات, بالحشد (crowd) and الهتافات (chants) are compounded into a single token through a joining morpheme و (and). We propose MSA candidates for such OOV words by splitting them into their right components.

3.3   Spelling Differences

From the error analysis of several hundred Egyptian sentences we noticed some dialectal specific corrections that can be applied to transform an Egyptian word into its MSA counterpart. For example in the Egyptian word بالصهاينه (Zionist), character ب can be dropped from the beginning and the ending character ه can be changed to character ة to form an MSA word الصهاينة (Zionist). After looking at a sample of hundered MSA-Egyptian word pairs, a list of rules (given in Table 3) is extracted. We apply these transformation rules to the error word and hypothesize the ones found in the MSA vocabulary.

To do some further analysis, we extracted a list of 5000 most frequent Egyptian words, with context, from a 2 Million word monolingual corpus. We then got these translated into MSA through human translator. Of the 5000 words approximately 70%

---

[2] The shapes of Arabic characters ف and ي change to ة and ‌ respectively, in context because of the cursive nature of Arabic script.

were translated to themselves (source word). Of the remaining 30% words that the human choose to translate differently, 33% can be transformed to MSA by applying single-edit distance to the original word. Another 16.4% can be converted to MSA by applying two edits. This provides a strong motivation to use the spelling correction mechanism as one of the techniques to propose MSA candidates. In our spelling correction module we apply all possible single edits (deletions, substitution and insertions) to the unknown word to get the candidate strings and hypothesize the ones that are found in the MSA vocabulary.

3.4   Dialect-Based Errors and True OOVs

The fourth class of OOV words contain purely dialectal Egyptian words to which applying spelling correction does not yield an MSA word. For example Egyptian word مستنية (waiting) has an alternative منتظرة in MSA. Finally a portion of errors constitute name entities like سكسكة (Sekska). We call this fifth category of unknown words as True OOVs.

Context-based Synonym Suggestions. In order to specifically target the last two classes of errors we use a technique similar to proximity based synonym acquisition [15,16]. The idea is to propose a synonym for an unknown word based on the language model context. The intuition is that synonyms of a word are likely to share the same context. For example consider a sentence "Barking like a tyke". Assume that "tyke" is unknown to our translation model. Based on the context "Barking like a" we might be able to produce candidates like "dog", "doggy", "bitch" etc. We do not use a fixed radius of words to propose candidates but also take advantage of language model back off with smoothing when proposing candidates.

4   Model

4.1   Egyptian-to-MSA Decoder

In the last section we discussed a bunch of methods to propose candidates for the unknown Egyptian words. Now we devise a model to score these candidates. Say we observe an Egyptian sentence $Z = Z_1, \ldots U_i, \ldots U_j, \ldots Z_n$ having unknown words U_i and U_j. For a known word Z_i we simply hypothesize the source word itself. For an unknown word U_j, we propose a list of MSA candidates $\{A_1 \ldots A_m\}$ using one or several methods discussed in the last section. Then we search for the best viterbi path $A = Z_1, \ldots A_i, \ldots A_j, \ldots Z_n$ according to:

$$p(A|Z) = \operatorname*{argmax} \prod_i^n p(A_i|Z_{i-k+1} \ldots Z_{i-1}) \tag{1}$$

where k is the order used for monolingual language model. We train a 5-gram language model. We use a stack-based search with a beam-search algorithm similar to that used in Pharoah [17] to select the best viterbi path. A large monolingual language model built on MSA data is used to score the candidates. The decoder decodes monotonically covering one Egyptian word at a time. For each Egyptian sentence we get a 1-best MSA sentence which is then decoded using MSA-to-English decoder.

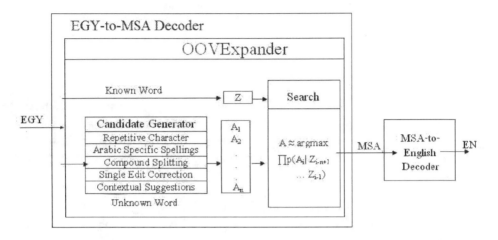

Fig. 2. Model for Egyptian-to-English Translation

4.2   Overall Model

The overall model for Egyptian-to-English translation is shown in Figure 2. Egyptian is first converted into MSA through an Egyptian-to-MSA decoder (as discussed in last section). The 1-best output is than passed to an MSA-to-English decoder. For the baseline system we bypass Egyptian-to-MSA decoding and translate Egyptian text using MSA-to-English decoder.

The decoding approach in [11] is superior to ours from the prospective that all MSA candidates of an Egyptian word are directly hypothesized into an MSA-to-English SMT system. In comparison our system proposes candidates and selects 1-best of these in a preprocessing step. The decision is based on just one monolingual language model feature. Because the language model is built on the large monolingual MSA data and the test set is Egyptian dialect, there is a domain mismatch and often time context does not prove to be useful to select the best candidates. In case of a tie, the decoder randomly picks one of the candidates. This is not the case in their system where n-best candidates are directly hypothesized in the MSA-to-English SMT, best English translation is chosen based on all features in the MSA-English MT model. Our approach, however, differs from the prospective that we propose MSA candidates for stems and not the affixes, whereas their work is only limited to affixes.

5   Results

In this section we discuss results obtained from running different proposed methods for OOV handling. For development purpose we test our edits to the baseline Egyptian input using TRL decoder because it is much efficient than the DTM decoder.

Table 4. OOV Handling Results – Applied To = Number of Egyptian words processing is applied to – Freq5 = Words having frequency $1 \leq 5$

System	TER	BLEU	Applied To
Baseline	60.40	19.49	0
A: Repeats	60.39	19.51	21
Dialect Rules	60.34	19.65	145
B: Spelling	60.14	19.95	510
C: Compounding	61.31	19.83	603
Synonyms	61.04	19.62	666
A + B	60.14	19.96	531
A + B + C	60.34	20.01	629
A + B + Freq5	59.96	20.01	863
A + B + C + Freq5	60.16	20.06	961

Our best component result (See Table 4) is obtained by the spelling correction module which hypothesize candidates that are at a single edit distance of the unknown word. Dialectal rules show small improvements because these are applied to a small number of words. Using compound splitting as a method of proposing candidates results in drop in TER score. Applying compounding to all OOV words hurts performance because most OOV words contain smaller components which have different meanings. Using synonym acquisition as a method of proposing MSA candidates also did not improve the results. Figure 3 shows examples of good and bad candidate proposed using this method. In Figure 3 (a) for the unknown word هيوفقك (help), and given the history وانشاء الله ربنا (God will), the language model propose all the candidate words that are actions, people expect from God such as يوفقك (help), يحميه (protect), يرزقك (bless) etc. The language model score selects the right candidate in this case. An example of bad synonym is shown in Figure 3(b) where a list of proper names are proposed given the context عزيزتي (My dear), the language model selects another proper name كاتيا (Katia) rather than the correct candidate نوارة (Nawarah).

Table 4 also show different system combination that we attempted. In system $A + B + C$ and $A + B + C + Freq5$, compounding is applied only to those OOVs for which candidates are not proposed through other methods. In the last two rows we also propose candidates for the low frequency words appearing one to five times in the translation table. The results improve slightly.

Table 5 show gains obtained on dev and test set by running DTM decoder instead of TRL to test the preprocessed Egyptian input. Our front-end handling of OOV dialectal Egyptian words show an improvement of 1 BLEU point on dev and 1.7 BLEU points on the test set.

(a)

وانشاء الله ربنا هيوفقك ويوفق

وانشاء الله ربنا يوفقك ويوفق

Ref: And Allah willing, God will help you succeed

(b)

عزيزتي نواره

عزيزتي كاتبا

Ref: My dear Nawarah

Fig. 3. Examples of Good (a) and Bad (b) Synonym Suggestions

Table 5. OOV Handling Results – Using DTM decoder

System	Dev		Test	
DTM	TER	BLEU	TER	BLEU
Baseline	56.88	23.87	59.86	22.00
A + B + Freq5	56.36	24.77	58.91	23.64
A + B + C + Freq5	56.51	24.85	58.95	23.72

6   Mapping Egyptian to MSA through Pivoting

In this part of the paper we shift our focus towards mapping all Egyptian words, and not just the OOVs, into MSA. We use the well-known Convolution Model, previously used for an Arabic information retrieval task [18]. The idea is to pivot English as an informant between Egyptian and MSA. The model for mapping an Egyptian word Z to an MSA word A is given as:

$$p_c(A|Z) = \sum_i^n p(A|e_i)p(e_i|Z) \tag{2}$$

We use Model-1 to estimate the probability distributions $p(e|Z)$ and $p(A|e)$. The probability distribution $p(e|Z)$ is estimated from the 8.5K parallel sentences of Egyptian-English and $p(A|e)$ is estimated using 300K sentences of MSA-English. In order to overcome the sparse $p(e|Z)$ distribution we interpolate it with the Model-1 distribution $p_a(e|Z)$ built from the MSA-English corpus as:

$$p(e|Z) = \lambda p_z(e|Z) + (1 - \lambda)p_a(e|Z)$$

Equation 2 can then be rewritten as:

$$p_c(A|Z) = \sum_i^n p(A|e_i)[\lambda p_z(e_i|Z) + (1 - \lambda)p_a(e_i|Z)]$$

We select the top-10 MSA candidates according to the convolution model and search for the MSA sentences that maximizes the viterbi probability. The overall search (Equation 1, Section 4) is now redefined as:

$$p(A|Z) = \text{argmax} \prod_{i}^{n} p(A_i|Context)p_c(A_i|Z_i)$$

The results are shown in Table 6. Applying model to all Egyptian words in the dev set we notice a significant drop in the translation quality of Egyptian-to-English translation (See Z_1 in Table 6). Applying model to all Egyptian words in the dev set, hurts the translation quality because we are editing some words that our MSA-to-English system knows how to translate. In order to verify this hypothesis we tried to apply the model only to those Egyptian words that are frequent in the Egyptian-English data and less frequent in MSA-English data. $Z_x A_y$ in Table 6 means that candidates are proposed for an Egyptian word that appears at least x times in the Egyptian-English corpus and at most y times in the MSA-English corpus. However, results did not improve than the baseline system for any value of x and y. As the values for x and y are tightened i.e. model is only applied to only the words that are less frequent in the rich MSA-English corpus, we end up proposing candidates for only less than 450 words. If we try to apply model to only the Egyptian words that are seen at least 5 times in the Egyptian-English corpus (to avoid noisy alignments), the model is being applied to only 196 words and the results start converging towards the baseline.

Table 6. Convolution Model – Proposed = Number of Egyptian words model is applied to and an MSA word different than Egyptian is selected in search

System	TER	BLEU	Proposed
Baseline	60.40	19.49	
Z_1	62.99	17.43	3361
$Z_1 A_{15}$	60.82	19.18	450
$Z_5 A_{20}$	60.59	19.33	196
$Z_{10} A_{500}$	60.66	19.29	362
$Z_{50} A_{100}$	60.53	19.41	62
UNK Word	60.48	19.50	80

The accuracy of the model in Table 6 is judged by the BLEU score of the MSA-to-English system. In order to scrutinize the results, we evaluated the accuracy of the convolution model in a more direct fashion. We used the 5000 most frequent Egyptian words list (also mentioned in Section 3.3). From this list we removed the Egyptian words that were translated to themselves. A remaining list of 1473 words is then used as a test set to evaluate the accuracy of the convolution model. The results are shown in Table 7. The 1-best and 10-best accuracies of the model are ~7% and ~22% respectively.

Table 7. Convolution Model Accuracy

	1-Best	10-Best
1473 Words	~7%	~22 %
5000 Words	~20 %	~50 %

In an another attempt to analyze the results we used the 100 sample sentences that were translated to MSA by human (Recall Section 1). In our results here we also measure the BLEU score at the intermediate step taking the human translated MSA as reference. The results of this controlled experiment is shown in Table 8. We see a slight improvement in both Egyptian-to-MSA and MSA-to-En systems, when applying the model to $Z_3 A_{20}$ i.e. Egyptian words that occured at least 3 times in the Egyptian-English corpus and at most 20 times in the MSA-English corpus. But the model is applied to only 50 words in this case. The BLEU score of 33.07 and TER score of 68.30 in the Egyptian-to-MSA baseline suggests that humans changed the Egyptian text significantly. However, when we try to apply the convolution model to all Egyptian words Z_1 we see a significant drop in the BLEU score.

Table 8. Convolution Model – EGY-to-MSA = BLEU score for the edited Egyptian taking human translated MSA as reference, MSA-to-En = BLEU score for the edited Egyptian after decoding into English, Proposed = Number of Egyptian words model is applied to and an MSA word different than Egyptian is selected in search

System	EGY-to-MSA	MSA-to-En	Proposed
Baseline	33.07	15.22	
Z_1	24.88	12.74	808
$Z_1 A_{20}$	32.23	15.14	123
$Z_3 A_{20}$	33.27	15.34	50
$Z_5 A_{25}$	33.23	15.29	42

We tried to analyze the output and found that most of the errors occur because of the sparse and noisy $p_z(e|Z)$ distribution built on the 8.5K Egyptian-to-English parallel data. The English informants proposed by the Egyptian word Z are incorrect. See Table 9 for examples.

Table 9. Convolution Model – Example Candidates Suggestion According to $p_z(e|Z)$

Word	Correct	Suggested	Output
دستّوريه	Constitutional	Top	أعلي
ثوره	Revolution	Qadafi	القذافي
ربنا	Lord	Unbelievers	الكفار

7   Conclusion

In this paper we showed that the quality of Egyptian-to-English SMT can be improved by trying to map Egyptian to MSA, for which we have richer, more reliable translations. We proposed several methods to bridge the gap between Egyptian and MSA. We removed repetitions, applied Egyptian specific mappings, tried spelling correction, used compound splitting and suggested synonyms based on context etc. We also applied convolution model using English as a pivot to map Egyptian words into MSA. Our spelling-based correction showed improvement of 1.7 BLEU points over the baseline system, that translates unedited Egyptian into English.

Acknowledgments. We would like to thank the anonymous reviewers for their helpful feedback and suggestions. Nadir Durrani received funding from the European Union Seventh Framework Programme (FP7/2007-2013) under grant agreement n° 287658. This publication only reflects the authors views.

References

1. Zbib, R., Malchiodi, E., Devlin, J., Stallard, D., Matsoukas, S., Schwartz, R., Makhoul, J., Zaidan, O.F., Callison-Burch, C.: Machine translation of arabic dialects. In: The 2012 Conference of the North American Chapter of the Association for Computational Linguistics, Montreal. Association for Computational Linguistics (2012)
2. Papineni, K., Roukos, S., Ward, T., Zhu, W.J.: Bleu: a method for automatic evaluation of machine translation. In: Proceedings of the 40th Annual Meeting on Association for Computational Linguistics, ACL 2002, Morristown, NJ, USA, pp. 311–318. Association for Computational Linguistics (2002)
3. Tillmann, C., Ney, H.: Word reordering and a dynamic programming beam search algorithm for statistical machine translation. Computational Linguistics 29, 97–133 (2003)
4. Ittycheriah, A., Roukos, S.: Direct translation model 2. In: Proceedings of the North American Chapter of the Association for Computational Linguistics and Human Language Technologies Conference (NAACL-HLT), Rochester, NY (2007)
5. Snover, M., Dorr, B., Schwartz, R., Micciulla, L., Makhoul, J.: A study of translation edit rate with targeted human annotation. In: Proceedings of Association for Machine Translation in the Americas, pp. 223–231 (2006)
6. Habash, N.: Introduction to Arabic Natural Language Processing. In: Synthesis Lectures on Human Language Technologies. Morgan & Claypool Publishers (2010)
7. Chiang, D., Diab, M.T., Habash, N., Rambow, O., Shareef, S.: Parsing arabic dialects. In: EACL (2006)
8. Habash, N., Rambow, O.: Magead: A morphological analyzer and generator for the arabic dialects. In: Proceedings of the 21st International Conference on Computational Linguistics and 44th Annual Meeting of the Association for Computational Linguistics, Sydney, Australia, pp. 681–688. Association for Computational Linguistics (2006)
9. Abo Bakr, H.M., Shaalan, K., Ziedan, I.: A hybrid approach for converting written egyptian colloquial dialect into diacritized arabic. In: Proceedings of the 6th International Conference on Informatics and Systems, INFOS2008, Cairo, Egypt (2008)
10. Sawaf, H.: Arabic dialect handling in hybrid machine translation. In: Proceedings of the Conference of the Association for Machine Translation in the Americas (AMTA), Denver, Colorado. Association for Machine Translation in the Americas (2010)

11. Salloum, W., Habash, N.: Dialectal to standard arabic paraphrasing to improve arabic-english statistical machine translation. In: Proceedings of the First Workshop on Algorithms and Resources for Modelling of Dialects and Language Varieties, Edinburgh, Scotland, pp. 10–21. Association for Computational Linguistics (2011)

12. Sajjad, H., Darwish, K., Belinkov, Y.: Translating dialectal arabic to english. In: Proceedings of the 51st Annual Meeting of the Association for Computational Linguistics, Sofia, Bulgaria. Short Papers, vol. 2, pp. 1–6. Association for Computational Linguistics (2013)

13. Al-Sabbagh, R., Girju, R.: Mining the web for the induction of a dialectical arabic lexicon. In: Chair, N.C.C., Choukri, K., Maegaard, B., Mariani, J., Odijk, J., Piperidis, S., Rosner, M., Tapias, D. (eds.) Proceedings of the Seventh International Conference on Language Resources and Evaluation (LREC 2010), Valletta, Malta, European Language Resources Association, ELRA (2010)

14. Riesa, J., Yarowsky, D.: Minimally supervised morphological segmentation with applications to machine translation. In: Chair, N.C.C., Choukri, K., Maegaard, B., Mariani, J., Odijk, J., Piperidis, S., Rosner, M., Tapias, D. (eds.) Proceedings of the 7th Conf. of the Association for Machine Translation in the Americas (AMTA 2006), Cambridge, MA. Association for Machine Translation in the Americas, AMTA (2006)

15. Baroni, M., Bisi, S.: Using cooccurrence statistics and the web to discover synonyms in technical language. In: Proceedings of LREC 2004, pp. 1725–1728 (2004)

16. Hagiwara, M., Ogawa, Y., Toyama, K.: Selection of effective contextual information for automatic synonym acquisition. In: Proceedings of the 21st International Conference on Computational Linguistics and 44th Annual Meeting of the Association for Computational Linguistics, Sydney, Australia, pp. 353–360. Association for Computational Linguistics (2006)

17. Koehn, P.: Pharaoh: A beam search decoder for phrase-based statistical machine translation models. In: Frederking, R.E., Taylor, K.B. (eds.) AMTA 2004. LNCS (LNAI), vol. 3265, pp. 115–124. Springer, Heidelberg (2004)

18. Franz, M., McCarley, J.S.: Arabic information retrieval at ibm. In: TREC (2002)

Bilingually Learning Word Senses for Translation

João Casteleiro, Gabriel Pereira Lopes, and Joaquim Silva

Universidade Nova de Lisboa,
Faculdade de Ciências e Tecnologia,
Departamento de Informática,
2829-516 Caparica, Portugal
casteleiroalves@gmail.com, {gpl,jfs}@fct.unl.pt

Abstract. All words in every natural language are ambiguous, specially when translation is at stake. In translation tasks, there is the need for finding out adequate translations for such words in the contexts where they occur. In this article, a bilingual strategy to cluster words according to their meanings is described. A publicly available parallel corpora sentence aligned is used. Word senses are discriminated by their translations and by the words occurring in a window, both in the source and target language parallel sentences. This strategy is language independent and uses a correlation algorithm for filtering out irrelevant features. Clusters obtained were evaluated in terms of *F-measure* (getting an average rating of 94%) and their homogeneity and completeness was determined using *V-Measure* (getting an average rating of 83%). Learned clusters are then used to train a support vector machine to tag ambiguous words with their translations in the contexts where they occur. This task was also evaluated in terms of *F-measure* and confronted with a baseline.

1 Introduction

Word sense ambiguity is present in all words in all natural languages and is a fundamental characteristic of human language. This characteristic has consequences in translation as it is necessary to find the right sense and the correct translation for each single and multi-word. A system for automatic translation from English into Portuguese should know how to translate the word *sentence* either as *frase* (a structurally independent grammatical unit of one or more words), or as *sentença* (the judgement formally pronounced upon a person convicted in criminal proceedings). As the efficiency and effectiveness of a translation system depends on the meaning of the text being processed, disambiguation will always be beneficial and even necessary.

There are two main approaches to tackle the issue of Word Sense Disambiguation, supervised and unsupervised learning. The former refers to the machine learning task of inferring a function from labelled training data. The latter works with unlabelled information, where we do not know the classification of the data in training sample. In Natural Language Processing, since the production of labelled training data is expensive, people will often prefer to learn from unlabelled data, but will try to "feed" their algorithms with some knowledge

A. Gelbukh (Ed.): CICLing 2014, Part II, LNCS 8404, pp. 283–295, 2014.

sources (EuroWordNet's)[1]. One way to work around the knowledge acquisition bottleneck present in both approaches is use an hybrid solution. We propose the use of a validated knowledge source (validated lexicon) automatically extracted from unlabelled bilingual parallel corpora, without any semantic and grammatical tag. This way we can guarantee a source of reliable and valid knowledge, with limited information.

In this paper we present an approach of ambiguous words classification, with the final purpose of assisting automatic and human translators on translation processes. Our ultimate goal is to try to increase the accuracy of machine translation when faced with expressions more complex, more ambiguous and less frequent than general. To achieve our target we propose a semi-supervised strategy to classify words, based on clusters of senses strongly correlated by associated features, automatically extracted[1] from a sentence aligned parallel corpora.

2   Related Work

Word Sense Disambiguation (WSD) is a Natural Language Processing problem that began to be studied as earlier as 1950, according to Gale, Church and Yarowsky[2]. First approaches to word sense disambiguation always passed by building specific rules applied to monolingual dictionaries, which made the solution rigid and without effect on a large scale. In the sixties, Bar-Hillel placed as a central point the difficulty of sense disambiguation in paper [3], where he states that he saw no means by which the sense of the word *pen* in the sentence *The box was in the pen* could be determined automatically.

In last decades, several studies have been presented in the field. Li and Church in [4] state that should not be necessary to look at the entire corpus to know if two words are strongly associated or not, thus, they proposed an algorithm for efficiently computing word associations. In [5], Mihalcea and Moldovan introduced a strategy that takes advantage of the sentence context. The words are paired and an attempt is made to disambiguate one word within the context of other word. Gamallo, Agustini and Lopes[6] presented in 2005 an unsupervised strategy to partially acquire syntactic-semantic requirements of nouns, verbs and adjectives from partially parsed monolingual text corpora. The goal is to identify clusters of similar positions by identifying the words that define their requirements extensionally.

In the past years, two main approaches have been studied that relied on supervised and knowledge-based methods. Ponzeto and Navigli[7] presented a large-scale method for automatic enrichment of a computational lexicon with encyclopedic relational knowledge. Zhi Zhong and Hwee Tou Ng in [8], show that application of these solutions in the field go beyond translation systems, whereas Moraliyski and Dias [9], presented a new methodology for synonym detection based on the combination of global and local distributional similarities of pairs of words.

The validity of using parallel corpora for bilingual word sense disambiguation has been assessed by several studies. Bansal et all[10] proposed in 2012 an unsupervised method for clustering translations of words through point-wise mutual information, based on a monolingual and parallel corpora. Apidianaki and He[11] presents a clustering algorithm for cross-lingual sense induction that generates bilingual semantic inventories from parallel corpora. In [12], authors described a statistical technique for assign senses to words based on the context in which they appear. Incorporating the method in a machine translation system, they have achieved to significantly reduce error rate. In 2003, Diab [13] addresses the problem of WSD from a multilingual perspective, expanding the notion of context to encompass multilingual evidence. Tufis et all in [14], present a method that exploits word clustering based on automatic extraction of translation equivalents, being supported by available aligned wordnets.

We propose the extension and changes of several works in the field, with the goal of increasing the accuracy of word sense disambiguation systems when faced with expressions more complex, ambiguous and less frequent than general. The system presented here, is different from the mentioned above ([10],[11],[12],[13], [14]), once, as far as we know, a such specific, expressive and validated bilingual lexicon, automatically extracted, was never used in a language independent strategy to build sense clusters strongly correlated, through a covariance-based filter.

3   Approach Outline

With the goal of classifying ambiguous words based on clusters of information, according to their features similarity, the developed research involves executing a set of personalized and typical steps to achieve it, which include: selection of word senses through a lexicon strongly validated; representation of contexts, by extracting features whenever any ambiguous word occurs in a parallel corpora; features correlation, based on a correlation measure; automatically learn word clusters; use of *Support Vector Machine* classifier (SVM) to fit ambiguous expressions in the corresponding cluster.

We will describe each step below.

3.1   Selection of Word Senses

Selection of word senses consists in seeking possible meanings for a specific word, which can be made by consulting a source of reliable information. In this approach we will use the *ISTRION EN-PT* lexicon, that is a bilingual and strongly validated data source, resultant from the project *ISTRION*[2]. This lexicon contain approximately 850.000 automatically extracted[1] and manually validated entries for the pair of languages English-Portuguese. To study the different contexts and meanings of potentially ambiguous words, we must identify them.

---

[2] http://citi.di.fct.unl.pt/project/project.php?id=97

In *ISTRION* lexicon, words occur alone (eg. plant) and in multi-words (eg. aluminium plant, aquatic plant, ...), with their translations. We will extract all the possible meanings reflected in the translation. Thus, for each ambiguous word in the source language (eg. English) we will extract the different senses existent in the target language (eg. Portuguese). In table 1 we can observe all the possible meanings extracted in the target language for the ambiguous word *plant*.

At the end of word sense selection process, we will obtain a data-set (pairs of ambiguous words and senses) with all possible meanings contained in the lexicon, for 10 ambiguous words selected (discriminated in table 7).

Table 1. Example of the sense selection for the ambiguous word **plant**

Ambiguous Word	Sense
Plant	Instalação (facility)
Plant	Central (industrial facility)
Plant	Plantas (green plant)
Plant	Instalações (facilities)
Plant	Vegetais (vegetables)
Plant	Fábrica (factory)
Plant	De as plantas (vegetables)
Plant	Vegetal (vegetable)
Plant	De plantas (vegetable)

For the moment, we have chosen to focus our approach on English and Portuguese languages, however, the research is language independent, as it could have been done with any language pair.

3.2   Features Extraction

The task of disambiguation aims at determining which of the senses of an ambiguous word is invoked in a particular use of that word. This is done by looking at the context of the pairs of words at stake. A monolingual context can be either a window including both N words to the left of the target word and N words to its right. In a bilingual perspective, as we propose, with parallel corpora sentence aligned, we get different contexts if we just look to one language translations or if we look at both languages. Local context features with bilingual words evidence starts from the assumption that incorporating knowledge from more than one language into the feature vector will be more informative than only using monolingual features[15]. Thus, in the sentence pair retrieval from bilingual parallel corpora "produced in plant facilities \t produzido nas instalações da fábrica", the context of the target pair of words "plant (fábrica)" could be [produced; facilities; produzido; instalações]. Here we stripped "in, da, nas", whereas in phrase "the potted plant is wilting \t o vaso de plantas está a murchar" the context for the pair of words "plant (plantas)" is [potted; wilting; vaso; murchar], again with no stop words.

In our approach, after some evidence tests, we have chosen to set a window of 3 words to left and right of the target word in the source language (eg. English) and 1 word to left and right of the target word in the target language (eg. Portuguese). We opt to use a sentence aligned parallel corpora (composed by Europarl[3] and DGT[4]), from which we extract all different features for the ambiguous target pair "(Ambiguous Word) \t (Sense N)" that "fall" within the setted window, stripping the stop words and noting the frequency and position. Thus, assuming that the source language is English and the target language is Portuguese, for a distance of three non functional words at left from the ambiguous word, we add the tag "enL3_" and increment the occurrence frequency. In the same way, for a distance of one word at left from the sense present in the ambiguous pair, we add the tag "ptL1_" and increment the occurrence. In table 2 and 3 we have represented the feature extraction process for a sentence of a bilingual (English-Portuguese) parallel corpora. In table 2, all non-functional words distant apart 3 words from the ambiguous word were extracted from the source language. In table 3 we extract only the direct non-functional "neighbours" of the ambiguous word in the target language. All extracted words were tagged with information about the position and language.

Table 2. Extracted features from English language, labelled with distance for ambiguous pair *plant(central)*

-	enL3_	-	-	enL2_	-	-	enL1_	-	-	enR1_	-	enR2_
a	contract	for	the	construction	of	the	power	plant	in	question	was	signed

Table 3. Extracted features from Portuguese language, labelled with distance for ambiguous pair *plant(central)*

-	-	-	-	-	-	ptL1_	-	-	ptR1_	
foi	assinado	um	contrato	para	a	construção	da	central	termoeléctrica	

In order to distribute different weights for different features that are at different distance values from the target word, we have chosen to apply penalties to more distant words. By other words, as the feature deviates from ambiguous word, it will have less weight in the final decision. Thus, for words that are two non functional words apart from the target word, the square root is applied to the frequency value. For words that are three non functional words apart from the target word, the cubic root is applied to the frequency value (table 4).

3.3 Features Correlation

At the end of feature extraction task we obtained 10 matrices, each one with N lines, related to all different pairs "(Ambiguous Word) \t (Sense N)" extracted

---

[3] http://www.statmt.org/europarl/

[4] http://ipsc.jrc.ec.europa.eu/?id=197

Table 4. Features extraction for ambiguous pair *plant (central)*

Ambiguous Pair	Features	Frequency	Final Weight
plant(central)	enL3_contract	2	$\sqrt[3]{2}$
	enL2_construction	3	$\sqrt{3}$
	enL1_power	51	51
	enR1_question	1	1
	enR2_signed	2	$\sqrt{2}$
	ptL1_construção	2	2
	ptR1_termoeléctrica	8	8
	...	...	...

and M columns, concerning the total features extracted for those pairs. Due to the huge amount of features and our purpose of clustering, we opted for transforming the initial matrix into a new and more compact matrix with information about the correlation of each sense with all others (table 5), for this we used a covariance-based correlation measure (Equation 1).

A covariance (Equation 2) is a measure of how much two random variables change together. If the values of one variable mainly correspond to the values of the other variable, the variables tend to show similar behaviour and the covariance is positive. In the opposite case, covariance is negative. Note that this process is computationally heavy. The system needs to compute all relations between all translations of a word and the corresponding feature weights. If the number of features is very large, the processing time increases proportionally.

$$Corr(x,y) = \frac{Cov(x,y)}{\sqrt{Cov(x,x)} + \sqrt{Cov(y,y)}} \quad (1)$$

$$Cov(x,y) = \frac{1}{m-1} \cdot \sum_{f=f1}^{fm} [(w(x,f) - w(x,.)).(w(y,f) - w(y,.))] \quad (2)$$

where *"f"* represents a feature and *"w"* its weight.

$$w(x,.) = \frac{1}{m} \cdot \sum_{f=f1}^{fm} w(x,f) \quad (3)$$

Table 5. "**plant**" sense correlation

Senses	Feature1	Feature2	...	Feature N	
Instalação (facilities)	w(Feature1)	w(Feature2)	...	w(FeatureN)	**Corr(Instalação,Central)**
Central (facilities)	w(Feature1)	w(Feature2)	...	w(FeatureN)	
Planta (green plant)	w(Feature1)	w(Feature2)	...	w(FeatureN)	...
...	...	...	...	...	...

*equation 3 represents the average weight of the relation between sense "**x**" with all features.*

$$w(x, f) = Weight\ of\ sense\ "\mathbf{x}"\ with\ feature\ "\mathbf{f}".\qquad(4)$$

At the end of correlation process we will obtain a correlation matrix with N x N values where the diagonal is always 1, once it represents the correlation between a meaning with itself. The other values range -1 and 1, as the correlation between words is stronger or not.

3.4   Clusters Construction

At this stage we already have the knowledge to automatically learn sense clusters. Our goal is to join similar senses of the same ambiguous word in the same cluster, based on features correlation. Through the analysis of correlation matrix we easily induce sense relations. In order to streamline the task of creating clusters, we opt to use *WEKA* tool[16] with *X-means*[17] algorithm, that is *K-Means* extended by an improved-structure, with a solved *K-Means* shortcoming. With *X-means* the user does not need to supply the number of clusters K. Instead, the algorithm returns the best solution for the correlation matrix presented as input (table 6): two clusters for "plant", where it may be translated by "instalação","central","instalações" or "fábrica" (cluster 0), or as "plantas","vegetais", "vegetal", "de as plantas", "de plantas" (cluster 1).

Table 6. Result of the clustering process for ambiguous word *plant*

Cluster	Sense
Cluster 0	Instalação (Facilities)
Cluster 0	Central (Facilities)
Cluster 1	Plantas (*Green* plant)
Cluster 0	Instalações (Facilities)
Cluster 1	Vegetais (*Green* plant)
Cluster 0	Fábrica (Facilities)
Cluster 1	De as plantas (*Green* plant)
Cluster 1	Vegetal (*Green* plant)
Cluster 1	De plantas (*Green* plant)

4   Experiments and Results

4.1   Clusters Evaluation

In order to determine the consistency of the obtained clusters, all of these will be evaluated with *V-measure* and *F-measure*. Regarding *V-measure*, we introduce two criteria presented in [18], homogeneity and completeness. A clustering process is considered homogeneously well-formed if all of its clusters contain only data points which are members of a single class. Comparatively, a clustering result satisfies completeness if all data points that are members of a given class are elements of the same cluster. So, increasing the homogeneity of a clustering solution often results in decreasing its completeness. These two criteria run roughly in opposition. V-measure is an entropy-based measure which reflects how successfully the homogeneity and completeness have been satisfied. This measure will be used to evaluate the resulting clusters.

To analyse the mathematical equations presented in (5), (6), (7), (8), assume that N data points exists, together with a set of classes, $C = \{Ci|i=1,...,n\}$ and a set of clusters, $K=\{Ki|1,...,m\}$. $A=\{a_{ij}\}$ is the contingency table produced by the clustering algorithm, where a_{ij} is the number of data points that are members of class Ci and elements of cluster Kj.

Homogeneity is defined as:

$$h = \begin{cases} 1 & if \quad H(C,K) = 0 \\ 1 - \frac{H(C|K)}{H(C)} & else \end{cases}$$

where

$$H(C|K) = -\sum_{k=1}^{|K|}\sum_{c=1}^{|C|} \frac{a_{ck}}{N} log \frac{a_{ck}}{\sum_{c=1}^{|C|} a_{ck}} \tag{5}$$

$$H(C) = -\sum_{c=1}^{|C|} \frac{\sum_{k=1}^{|K|} a_{ck}}{n} log \frac{\sum_{k=1}^{|K|} a_{ck}}{n} \tag{6}$$

The value of homogeneity varies between 1 and 0. In the perfectly homogeneous case, the value $H(K|C)$ is 0 and consequently the homogeneity is 1. In an imperfect situation homogeneity is 0, that happens when the class distribution within each cluster is equal to the overall class distribution.

Completeness is defined as:

$$c = \begin{cases} 1 & if \quad H(K,C) = 0 \\ 1 - \frac{H(K|C)}{H(K} & else \end{cases}$$

where

$$H(K|C) = -\sum_{C=1}^{|C|}\sum_{K=1}^{|K|} \frac{a_{ck}}{N} log \frac{a_{ck}}{\sum_{K=1}^{|K|} a_{ck}} \tag{7}$$

$$H(K) = -\sum_{K=1}^{|K|} \frac{\sum_{C=1}^{|C|} a_{ck}}{n} log \frac{\sum_{C=1}^{|C|} a_{ck}}{n} \tag{8}$$

Similarly to determination of homogeneity, the results of completeness also vary between 1 and 0. In the perfectly complete case, H(K|C) = 0, and the completeness is 1. In the worst case scenario, each class is represented by every cluster with distribution equal to the distribution of cluster sizes, H(K|C) is maximal and equals to H(K), consequently the completeness is 0. *V-measure* is then a measure for evaluating clusters, which studies the harmonic relationship between homogeneity and completeness (9).

$$V\beta = \frac{2 * h * c}{h + c} \qquad (9)$$

In order not to base the approach only on a single evaluation measure of clusters, we also used the known F-measure evaluation[19]. As *V-measure*, this one also varies between 0 and 1, being the former the worst scenario and the latter the optimal.

Table 7. Sense Clusters Evaluation

Cluster	*V-Measure*	*F-Measure*
Plant	1.0	1.0
Sentence	0.77	0.90
Motion	1.0	1.0
Tank	0.43	0.8
Train	0.81	0.90
Cold	0.77	0.88
Chair	1.0	1.0
Right	0.52	0.88
Interest	1.0	1.0
General	1.0	1.0

Analysing the results of sense clusters as a whole (table 7), we can easily understand that almost all clusters are well formed. If we just look to the *F-measure* evaluation, the results varies between 0.88 and 1, revealing a high value of precision and recall, since this evaluation measure is based on these two parameters. If we look to *V-measure*, despite almost all clusters have obtained high evaluation scores, two of them were below the desired *(Tank, Right)*. After an analyses of these two clusters, we noticed the fact that homogeneity of clusters hasn't been compromised, ie, formed clusters have only data points which are members of a single class. The same isn't true for completeness, revealing that not all considered data is part of clusters. This happens due to the absence in parallel corpora of the ambiguous pairs that were initially extracted from *ISTRION* lexicon, which leads to a poor features extraction task and consequently a poor characterization of the ambiguous meanings.

4.2   Classifier

The purpose of the classification stage is to determine how the disambiguation system behaves when faced with a set of potentially ambiguous sentences (containing ambiguous words). For instance, in the expression *"No sentence is completely irrelevant to the general topic of this paragraph"*, our goal is to classify the ambiguous words *"sentence"* as a string of words satisfying the grammatical rules of a language and *"general"* as the main or major parts of something rather than the details. For this, we need to classify each ambiguous word in the sentence according to the corresponding sense cluster. To accomplish the classification process, for each one of the 10 ambiguous words studied, 15 potentially ambiguous sentences were extracted from a corpora that was not used in the learning stage, totalling 150 expressions.

The classification task usually involves the use of two separate training and testing sets. The training phase is closely related with the stage of clustering achieved previously, since we used the acquired knowledge from clusters to train the system. Each cluster is encoded by the presence or absence of all features that belong to all clusters related with a particular ambiguous word, and a "target value" (i.e. the class labels) corresponding to the sense cluster. In what concerns to the testing phase, our goal is to encode each testing sentence with ambiguous words, extracted from a corpora which was not used in the training phase, and confront it with the training set. To encode each expression, we analyse the presence or absence in the sentences of all the features used in the training set. This manner we ensure that both training and testing sets are normalized in the same way, and can be confronted against each other.

The classifier used was a support vector machine (LIBSVM) [20] with the *Radial Basis Function (RBF)* as kernel type. This kernel non-linearly maps samples into a higher dimensional space, so it can handle the case when the relation between class labels and attributes is non-linear.

4.3   Classifier Results

The results obtained with the application of *SVM* classifier are evaluated using *F-measure*. For each one of the 150 ambiguous sentences classified, given the

Table 8. Classification Task Evaluation

Expression with ambiguous word	F-Measure (SVM Classifier)	F-Measure (Baseline)
Plant	0.71	0.33
Sentence	0.76	0.42
Motion	0.90	0.37
Tank	0.90	0.33
Train	0.90	0.42
Cold	0.93	0.42
Chair	0.70	0.28
Right	0.52	0.27
Interest	1.00	0.33
General	0.80	0.33

results obtained we determined the precision, recall and then *F-measure* per ambiguous word. In order to have a baseline for comparison, the same experiments described above were performed using a simple decision method: the most probable sense of each ambiguous word was used.

Looking at the results shown in table 8, and establishing a comparison with the used baseline, it is evident that the use of a reliable knowledge source (composed by ISTRION lexicon and parallel corpora) to build cohesive sense clusters is an asset in the train of *SVM* classifier, leading the disambiguation system to present very optimistic results.

5   Conclusions and Future Work

We presented a bilingual Word Sense Disambiguation approach that learns word sense clusters and then uses learned contextual information for classifying expressions according to the sense of ambiguous words occurring there. In contrast to other monolingual and multilingual approaches to WSD, the system presented offers a balance regarding the knowledge information sources used. First, it does not require grammatically and semantically very rich lexical resources; second, it does not work without a minimum reliable knowledge. Another difference lies in the features used to characterize ambiguous senses. We use information present in more than one language, allowing an increase in correlation characteristics, and leading to a construction of highly homogeneous clusters. The experimental results clearly demonstrate that the sense disambiguation system presented guarantees assistance to translation processes. Analysing more specifically the results, mainly with regard to sense clusters obtained, with valid values of homogeneity and completeness, we can state that those learned clusters are highly reliable sources of information, supporting the whole process of disambiguation and learning. The results obtained in the classification process with *F-measure* evaluation, and using the baseline to compare them, allow us to be optimistic about the role that this approach can have on translation systems.

In future research we would like to improve the system in several ways. We really believe that the use of different sets of features (e.g., a larger window in target side) can further enhance the correlation values obtained, and consequently the quality of the formed clusters. We also think that the encoding of the input expressions in the classification stage can be improved, namely by using positional features rather than global features. Another increment concerns the use of validated multilingual information to build sense clusters (more than 2 languages). It is also our intention, to test our approach with existing datasets, used on cross-lingual WSD tasks of SemEval[21]. Finally, we intend to incorporate the approach in a translation system, enhancing the accuracy of the same when faced with expressions more complex, more ambiguous and less frequent than usual.

References

1. Aires, J., Lopes, G.P., Gomes, L.: Phrase translation extraction from aligned parallel corpora using suffix arrays and related structures. In: Lopes, L.S., Lau, N., Mariano, P., Rocha, L.M. (eds.) EPIA 2009. LNCS (LNAI), vol. 5816, pp. 587–597. Springer, Heidelberg (2009)
2. Gale, W.A., Church, K.W., Yarowsky, D.: A method for disambiguating word senses in a large corpus. Computers and the Humanities 26, 415–439 (1992)
3. Bar-Hillel, Y.: The present status of automatic translation of languages. Advances in Computers 1, 91–163 (1960)
4. Li, P., Church, K.W.: A sketch algorithm for estimating two-way and multi-way associations. Computational Linguistics 33, 305–354 (2007)
5. Mihalcea, R., Moldovan, D.I.: A method for word sense disambiguation of unrestricted text. In: Proceedings of the 37th Annual Meeting of the Association for Computational Linguistics on Computational Linguistics, pp. 152–158. Association for Computational Linguistics (1999)
6. Gamallo, P., Agustini, A., Lopes, G.P.: Clustering syntactic positions with similar semantic requirements. Computational Linguistics 31, 107–146 (2005)
7. Ponzetto, S.P., Navigli, R.: Knowledge-rich word sense disambiguation rivaling supervised systems. In: Proceedings of the 48th Annual Meeting of the Association for Computational Linguistics, pp. 1522–1531. Association for Computational Linguistics (2010)
8. Zhong, Z., Ng, H.T.: Word sense disambiguation improves information retrieval. In: Proceedings of the 50th Annual Meeting of the Association for Computational Linguistics. Long Papers, vol. 1, pp. 273–282. Association for Computational Linguistics (2012)
9. Moraliyski, R., Dias, G.: One sense per discourse for synonymy extraction. In: International Conference on Recent Advances in Natural Language Processing, RANLP 2007, vol. 2, pp. 383–387 (2008)
10. Bansal, M., DeNero, J., Lin, D.: Unsupervised translation sense clustering. In: Proceedings of the 2012 Conference of the North American Chapter of the Association for Computational Linguistics: Human Language Technologies, pp. 773–782. Association for Computational Linguistics (2012)
11. Apidianaki, M., He, Y., et al.: An algorithm for cross-lingual sense-clustering tested in a mt evaluation setting. In: Proceedings of the International Workshop on Spoken Language Translation, pp. 219–226 (2010)
12. Brown, P.F., Pietra, S.A.D., Pietra, V.J.D., Mercer, R.L.: Word-sense disambiguation using statistical methods. In: Proceedings of the 29th Annual Meeting on Association for Computational Linguistics, pp. 264–270. Association for Computational Linguistics (1991)
13. Diab, M.T., et al.: Word sense disambiguation within a multilingual framework (2003)
14. Tufiş, D., Ion, R., Ide, N.: Fine-grained word sense disambiguation based on parallel corpora, word alignment, word clustering and aligned wordnets. In: Proceedings of the 20th International Conference on Computational Linguistics, p. 1312. Association for Computational Linguistics (2004)
15. Lefever, E., Hoste, V., De Cock, M.: Five languages are better than one: an attempt to bypass the data acquisition bottleneck for wsd. In: Gelbukh, A. (ed.) CICLing 2013, Part I. LNCS, vol. 7816, pp. 343–354. Springer, Heidelberg (2013)

16. Hall, M., Frank, E., Holmes, G., Pfahringer, B., Reutemann, P., Witten, I.H.: The weka data mining software: an update. ACM SIGKDD Explorations Newsletter 11, 10–18 (2009)
17. Pelleg, D., Moore, A.W., et al.: X-means: Extending k-means with efficient estimation of the number of clusters. In: ICML, pp. 727–734 (2000)
18. Rosenberg, A., Hirschberg, J.: V-measure: A conditional entropy-based external cluster evaluation measure. In: EMNLP-CoNLL, vol. 7, pp. 410–420 (2007)
19. Rijsbergen, V. (ed.): Information Retrieval, 2nd edn. Information Retrieval Group. University of Glasgow (1979)
20. Chang, C.C., Lin, C.J.: Libsvm: a library for support vector machines. ACM Transactions on Intelligent Systems and Technology (TIST) 2, 27 (2011)
21. Lefever, E., Hoste, V.: Semeval-2010 task 3: Cross-lingual word sense disambiguation. In: Proceedings of the 5th International Workshop on Semantic Evaluation, pp. 15–20. Association for Computational Linguistics(2010)

Iterative Bilingual Lexicon Extraction from Comparable Corpora with Topical and Contextual Knowledge

Chenhui Chu, Toshiaki Nakazawa, and Sadao Kurohashi

Graduate School of Informatics, Kyoto University, Kyoto, Japan
{chu,nakazawa}@nlp.ist.i.kyoto-u.ac.jp, kuro@i.kyoto-u.ac.jp

Abstract. In the literature, two main categories of methods have been proposed for bilingual lexicon extraction from comparable corpora, namely topic model and context based methods. In this paper, we present a bilingual lexicon extraction system that is based on a novel combination of these two methods in an iterative process. Our system does not rely on any prior knowledge and the performance can be iteratively improved. To the best of our knowledge, this is the first study that iteratively exploits both topical and contextual knowledge for bilingual lexicon extraction. Experiments conduct on Chinese–English and Japanese–English Wikipedia data show that our proposed method performs significantly better than a state–of–the–art method that only uses topical knowledge.

1 Introduction

Bilingual lexicons are important for many bilingual natural language processing (NLP) tasks, such as statistical machine translation (SMT) [1, 2] and dictionary based cross–language information retrieval (CLIR) [3]. Since manual construction of bilingual lexicons is expensive and time–consuming, automatic construction is desirable. Mining bilingual lexicons from parallel corpora is a possible method. However, it is only feasible for a few language pairs and domains, because parallel corpora remain a scarce resource. As comparable corpora are far more widely available than parallel corpora, extracting bilingual lexicons from comparable corpora is an attractive research field.

In the literature, two main categories of methods have been proposed for bilingual lexicon extraction from comparable corpora, namely topic model based method (TMBM) [4] and context based method (CBM) [5]. Both methods are based on the Distributional Hypothesis [6], stating that words with similar meaning have similar distributions across languages. TMBM measures the similarity of two words on cross–lingual topical distributions, while CBM measures the similarity on contextual distributions across languages.

In this paper, we present a bilingual lexicon extraction system that is based on a novel combination of TMBM and CBM. The motivation is that a combination of these two methods can exploit both topical and contextual knowledge to measure the distributional similarity of two words, making bilingual lexicon

A. Gelbukh (Ed.): CICLing 2014, Part II, LNCS 8404, pp. 296–309, 2014.

extraction more reliable and accurate than only using one knowledge source. The key points for the combination are as follows:

- TMBM can extract bilingual lexicons from comparable corpora without any prior knowledge. The extracted lexicons are semantically related and provide comprehensible and useful contextual information in the target language for the source word [4]. Therefore, it is effective to use the lexicons extracted by TMBM as a seed dictionary, which is required for CBM.
- The lexicons extracted by CBM can be combined with the lexicons extracted by TMBM to further improve the accuracy.
- The combined lexicons again can be used as the seed dictionary for CBM. Therefore the accuracy of the lexicons can be iteratively improved.

Our system not only maintains the advantage of TMBM that does not require any prior knowledge, but also can iteratively improve the accuracy of bilingual lexicon extraction through combination CBM. To the best of our knowledge, this is the first study that iteratively exploits both topical and contextual knowledge for bilingual lexicon extraction. Experimental results on Chinese–English and Japanese–English Wikipedia data show that our proposed method performs significantly better than the method only using topical knowledge [4].

2   Related Work

2.1   Topic Model Based Methods (TMBM)

TMBM uses the Distributional Hypothesis on topics, stating that two words are potential translation candidates if they are often present in the same cross–lingual topics and not observed in other cross–lingual topics [4]. It trains a Bilingual Latent Dirichlet Allocation (BiLDA) topic model on document–aligned comparable corpora, and identifies word translations relying on word–topic distributions from the trained topic model. This method is attractive because it does not require any prior knowledge.

Vulić et al. [4] first propose this method. Later, Vulić and Moens [7] extend this method to detect highly confident word translations by a symmetrization process and the one-to-one constraints, and demonstrate a way to build a high quality seed dictionary using both BiLDA and cognates. Liu et al. [8] develop this method by converting document–aligned comparable corpora into a parallel topic–aligned corpus using BiLDA topic models, and identify word translations with the help of word alignment. Richardson et al. [9] exploit this method in the task of transliteration.

Our study differs from previous studies in using a novel combination of TMBM and CBM.

2.2   Context Based Methods (CBM)

From the pioneering work of [10, 11], various studies have been conducted on CBM for extracting bilingual lexicons from comparable corpora. CBM is based

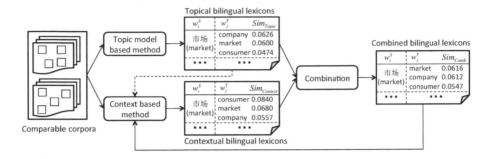

Fig. 1. Bilingual lexicon extraction system

on the Distributional Hypothesis on context, stating that words with similar meaning appear in similar contexts across languages. It usually consists of three steps: context vector modeling, vector similarity calculation and translation identification that treats a candidate with higher similarity score as a more confident translation. Previous studies use different definitions of context, such as window–based context [11, 5, 12–15], sentence–based context [16] and syntax–based context [17, 18] etc. Previous studies also use different measures to compute the similarity between the vectors, such as cosine similarity [16, 17, 14, 15], Euclidean distance [11, 18], city–block metric [5] and Spearman rank order [12] etc.

Basically, CBM requires a seed dictionary to project the source vector onto the vector space of the target language, which is one of the main concerns of this study. In previous studies, a seed dictionary is usually manually created [5, 17], and sometimes complemented by bilingual lexicons extracted from a parallel corpus [16, 15] or the Web [14]. In addition, some studies try to create a seed dictionary using cognates [12, 13], however this cannot be applied to distant language pairs that do not share cognates, such as Chinese–English and Japanese–English. There are also some studies that do not require a seed dictionary [10, 11, 18]. However, these studies show lower accuracy compared to the conventional methods using a seed dictionary.

Our study differs from previous studies in using a seed dictionary automatically acquired without any prior knowledge, which is learned from comparable corpora in an unsupervised way.

3   Proposed Method

The overview of our proposed bilingual lexicon extraction system is presented in Figure 1. We first apply TMBM to obtain bilingual lexicons from comparable corpora, which we call topical bilingual lexicons. The topical bilingual lexicons contain a list of translation candidates for a source word w_i^S, where a target word w_j^T in the list has a topical similarity score $Sim_{Topic}(w_i^S, w_j^T)$. Then using the topical bilingual lexicons as an initial seed dictionary, we apply CBM to obtain

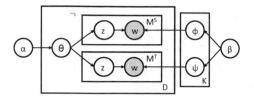

Fig. 2. The BiLDA topic model

bilingual lexicons, which we call contextual bilingual lexicons. The contextual bilingual lexicons also contain a list of translation candidates for a source word, where each candidate has a contextual similarity score $Sim_{Context}(w_i^S, w_j^T)$. We then combine the topical bilingual lexicons with the contextual bilingual lexicons to obtain combined bilingual lexicons. The combination is done by calculating a combined similarity score $Sim_{Comb}(w_i^S, w_j^T)$ using the $Sim_{Topic}(w_i^S, w_j^T)$ and $Sim_{Context}(w_i^S, w_j^T)$ scores. After combination, the quality of the lexicons can be higher. Therefore, we iteratively use the combined bilingual lexicons as the seed dictionary for CBM and conduct combination, to improve the contextual bilingual lexicons and further improve the combined bilingual lexicons.

Our system not only maintains the advantage of TMBM that does not require any prior knowledge, but also can iteratively improve the accuracy by a novel combination with CBM. Details of TMBM, CBM and combination method will be described in Section 3.1, 3.2 and 3.3 respectively.

3.1 Topic Model Based Method (TMBM)

In this section, we describe TMBM to calculate the topical similarity score $Sim_{Topic}(w_i^S, w_j^T)$.

We train a BiLDA topic model presented in [19], which is an extension of the standard LDA model [20]. Figure 2 shows the plate model for BiLDA, with D document pairs, K topics and hyper–parameters α, β. Topics for each document are sampled from a single variable θ, which contains the topic distribution and is language–independent. Words of the two languages are sampled from θ in conjugation with the word–topic distributions ϕ (for source language S) and ψ (for target language T).

Once the BiLDA topic model is trained and the associated word–topic distributions are obtained for both source and target corpora, we can calculate the similarity of word–topic distributions to identify word translations. For similarity calculation, we use the *TI+Cue* measure presented in [4], which shows the best performance for identifying word translations in their study. *TI+Cue* measure is a linear combination of the *TI* and *Cue* measures, defined as follows:

$$Sim_{TI+Cue}(w_i^S, w_j^T) = \lambda Sim_{TI}(w_i^S, w_j^T) + (1 - \lambda)Sim_{Cue}(w_i^S, w_j^T) \quad (1)$$

TI and *Cue* measures interpret and exploit the word–topic distributions in different ways, thus combining the two leads to better results.

The *TI* measure is the similarity calculated from source and target word vectors constructed over a shared space of cross–lingual topics. Each dimension of the vectors is a *TF–ITF* (term frequency – inverse topic frequency) score. *TF–ITF* score is computed in a word–topic space, which is similar to *TF–IDF* (term frequency – inverse document frequency) score that is computed in a word–document space. *TF* measures the importance of a word w_i within a particular topic z_k, while *ITF* of a word w_i measures the importance of w_i across all topics. Let $n_k^{(w_i)}$ be the number of times the word w_i is associated with the topic z_k, W denotes the vocabulary and K denotes the number of topics, then

$$TF_{i,k} = \frac{n_k^{(w_i)}}{\sum_{w_j \in W} n_k^{(w_j)}}, ITF_i = log\frac{K}{1 + |k : n_k^{(w_i)} > 0|} \tag{2}$$

TF–ITF score is the product of $TF_{i,k}$ and ITF_i. Then, the *TI* measure is obtained by calculating the cosine similarity of the K dimensional source and target vectors. Let S^i be the source vector for a source word w_i^S, T^j be the target vector for a target word w_j^T, then cosine similarity is defined as follows:

$$Cos(w_i^S, w_j^T) = \frac{\sum_{k=1}^{K} S_k^i \times T_k^j}{\sqrt{\sum_{k=1}^{K} (S_k^i)^2} \times \sqrt{\sum_{k=1}^{K} (T_k^j)^2}} \tag{3}$$

The *Cue* measure is the probability $P(w_j^T | w_i^S)$, where w_j^T and w_i^S are linked via the shared topic space, defined as:

$$P(w_j^T | w_i^S) = \sum_{k=1}^{K} \psi_{k,j} \frac{\phi_{k,i}}{Norm_\phi} \tag{4}$$

where $Norm_\phi$ denotes the normalization factor given by $Norm_\phi = \sum_{k=1}^{K} \phi_{k,i}$ for a word w_i.

3.2   Context Based Method (CBM)

In this section, we describe CBM to calculate the contextual similarity score $Sim_{Context}(w_i^S, w_j^T)$.

We use window–based context, and leave the comparison of using different definitions of context as future work. Given a word, we count all its immediate context words, with a window size of 4 (2 preceding words and 2 following words). We build a context by collecting the counts in a bag of words fashion, namely we do not distinguish the positions that the context words appear in. The number of dimensions of the constructed vector is equal to the vocabulary size. We further reweight each component in the vector by multiplying by the *IDF* score following [17], which is defined as follows:

$$IDF(t, D) = log\frac{|D|}{1 + |\{d \in D : t \in d\}|} \tag{5}$$

where $|D|$ is the total number of documents in the corpus, and $|\{d \in D : t \in d\}|$ denotes number of documents where the term t appears. We model the source and target vectors using the method described above, and project the source vector onto the vector space of the target language using a seed dictionary. The similarity of the vectors is computed using cosine similarity (Equation 3).

As initial, we use the topical bilingual lexicons extracted in Section 3.1 as seed dictionary. Note that the topical bilingual lexicons are noisy especially for the rare words [7]. However, since they provide comprehensible and useful contextual information in the target language for the source word [4], it is effective to use the lexicons as a seed dictionary for CBM.

Once contextual bilingual lexicons are extracted, we combine them with the topical bilingual lexicons. After combination, the quality of the lexicons will be improved. Therefore, we further use the combined lexicons as seed dictionary for CBM, which will produce better contextual bilingual lexicons. Again, we combine the better contextual bilingual lexicons to the topical bilingual lexicons. By repeating these steps, both the contextual bilingual lexicons and the combined bilingual lexicons will be iteratively improved.

Applying CBM and combination one time is defined as one iteration. At iteration 1, the topical bilingual lexicons are used as seed dictionary for CBM. From the second iteration, the combined lexicons are used as seed dictionary. In all iterations, we produce a seed dictionary for all the source words in the vocabulary, and use the Top 1 candidate to project the source context vector to the target language. We stop the iteration when the predefined number of iterations have been done.

3.3   Combination

TMBM measures the distributional similarity of two words on cross–lingual topics, while CBM measures the distributional similarity on contexts across languages. A combination of these two methods can exploit both topical and contextual knowledge to measure the distributional similarity, making bilingual lexicon extraction more reliable and accurate. Here we use a linear combination for the two methods to calculate a combined similarity score, defined as follows:

$$Sim_{Comb}(w_i^S, w_j^T) = \gamma Sim_{Topic}(w_i^S, w_j^T) + (1 - \gamma)Sim_{Context}(w_i^S, w_j^T) \qquad (6)$$

To reduce computational complexity, we only keep the Top–N translation candidates for a source word during all the steps in our system. We first produce a Top–N candidate list for a source word using TMBM. Then we apply CBM to calculate the similarity only for the candidates in the list. Finally, we conduct combination. Therefore, the combination process is a kind of re–ranking of the candidates produced by TMBM. Note that both $Sim_{Topic}(w_i^S, w_j^T)$ and $Sim_{Context}(w_i^S, w_j^T)$ are normalized before combination, where the normalization is given by:

$$Sim_{Norm}(w_i^S, w_j^T) = \frac{Sim(w_i^S, w_j^T)}{\sum_{n=1}^{N} Sim(w_i^S, w_n^T)} \qquad (7)$$

where N is the number of translation candidates for a source word.

4   Experiments

We evaluated our proposed method on Chinese–English and Japanese–English Wikipedia data. For people who want to reproduce the results reported in the this paper, we released a software that contains all the required code and data at `http://www.CICLing.org/2014/data/24`.

Note that Wikipedia is a special type of comparable corpora, because document alignment is established via interlanguage links. For many other types of comparable corpora, it is necessary to perform document alignment as an initial step. Many methods have been proposed for document alignment in the literature, such as IR–based [21, 22], feature–based [23] and topic–based [24] methods. After document alignment, our proposed method can be applied to any type of comparable corpora.

4.1   Experimental Data

We created the experimental data according to the following steps. We downloaded Chinese[1] (20120921), Japanese[2] (20120916) and English[3] (20121001) Wikipedia database dumps. We used an open–source Python script[4] to extract and clean the text from the dumps. Since the Chinese dump is a mixture of Traditional and Simplified Chinese, we converted all Traditional Chinese to Simplified Chinese using a conversion table published by Wikipedia[5]. We aligned the articles on the same topic in Chinese–English and Japanese–English Wikipedia via the interlanguage links. From the aligned articles, we selected 10,000 Chinese–English and Japanese–English pairs as our training corpora.

We preprocessed the Chinese and Japanese corpora using a tool proposed by Chu et al. [25] and JUMAN [26] respectively for segmentation and Part–of–Speech (POS) tagging. The English corpora were POS tagged using Lookahead POS Tagger [27]. To reduce data sparsity, we kept only lemmatized noun forms. The vocabularies of the Chinese–English data contain 112,682 Chinese and 179,058 English nouns. The vocabularies of the Japanese–English data contain 47,911[6] Japanese and 188,480 English nouns.

---

[1] `http://dumps.wikimedia.org/zhwiki`

[2] `http://dumps.wikimedia.org/jawiki`

[3] `http://dumps.wikimedia.org/enwiki`

[4] `http://code.google.com/p/recommend-2011/`
`source/browse/Ass4/WikiExtractor.py`

[5] `http://svn.wikimedia.org/svnroot/mediawiki/branches/`
`REL1_12/phase3/includes/ZhConversion.php`

[6] The vocabulary size of Japanese is smaller than that of Chinese and English, because we kept only common, sahen and proper nouns, place, person and organization names among all sub POS tags of noun in JUMAN.

4.2   Experimental Settings

For BiLDA topic model training, we used the implementation PolyLDA++ by Richardson et al. [9][7]. We set the hyper–parameters $\alpha = 50/K, \beta = 0.01$ following [4], where K denotes the number of topics. We trained the BiLDA topic model using Gibbs sampling with $1,000$ iterations. For the combined *TI+Cue* method, we used the toolkit BLETM obtained from Vulić et al. [4][8], where we set the linear interpolation parameter $\lambda = 0.1$ following their study. For our proposed method, we empirically set the linear interpolation parameter $\gamma = 0.8$, and conducted 20 iterations.

4.3   Evaluation Criterion

We manually created Chinese–English and Japanese–English test sets for the most 1,000 frequent source words in the experimental data with the help of Google Translate[9]. Following [4], we evaluated the accuracy using the following two metrics:

- Precision@1: Percentage of words where the Top 1 word from the list of translation candidates is the correct one.
- Mean Reciprocal Rank (MRR) [28]: Let w be a source word, $rank_w$ denotes the rank of its correct translation within the list of translation candidates, V denotes the set of words used for evaluation. Then MRR is defined as:

$$MRR = \frac{1}{|V|} \sum_{w \in V} \frac{1}{rank_w} \tag{8}$$

Note that we only used the Top 20 candidates from the ranked list for calculating MRR.

4.4   Results

The results for the Chinese–English and Japanese–English test sets are shown in Figure 3, where "Topic" denotes the lexicons extracted only using TMBM described in Section 3.1, "Context" denotes the lexicons extracted only using CBM method described in Section 3.2, "Combination" denotes the lexicons after applying the combination method described in Section 3.3, "K" denotes the number of topics and "N" denotes the number of translation candidates for a word we compared in our experiments.

In general, we can see that our proposed method can significantly improve the accuracy in both Precision@1 and MRR metrics compared to "Topic". "Context" outperforms "Topic", which verifies the effectiveness of using the lexicons extracted by TMBM as seed dictionary for CBM. "Combination" performs better

---

[7] https://bitbucket.org/trickytoforget/polylda
[8] http://people.cs.kuleuven.be/~ivan.vulic/
software/BLETMv1.0wExamples.zip
[9] http://translate.google.com

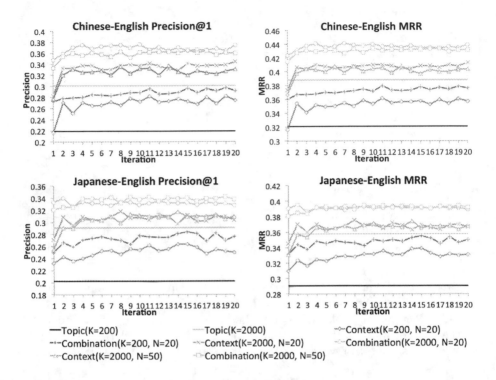

Fig. 3. Results for Chinese–English and Japanese–English on the test sets

than both "Topic" and "Context", which verifies the effectiveness of using both topical and contextual knowledge for bilingual lexicon extraction. Moreover, iteration can further improve the accuracy, especially in the first few iterations. Detailed analysis for the results will be given in Section 5.

5   Discussion

5.1   Why Are Our "Topic" Scores Lower Than [4]?

The "Topic" scores are lower than the ones in [4], which are over 0.6 when $K = 2000$. The main reason is that the experimental data we used is much more sparse. Our vocabulary size is from tens of thousands to hundreds of thousands (see Section 4.1), while in [4] it is only several thousands (7,160 Italian and 9,166 English nouns). Moreover, the number of document pairs we used for training is less than [4], which is 10,000 compared to 18,898 pairs.

Another reason is the evaluation method. It may underestimate simply because of the incompleteness of our test set (e.g. our system successfully finds the correct translation "vehicle" for the Chinese word "车", but our test set only contains "car" as the correct translation.).

5.2 How Does the Proposed Method Perform on Different Language Pairs?

Our proposed method is language–independent, which is also indicated by the experimental results on two different language pairs of Chinese–English and Japanese–English. In Figure 3, we can see that although the "Topic" scores and the absolute values of improvement by our proposed method on Chinese–English and Japanese–English are different because of the different characteristics of the data, the improvement curves are similar.

5.3 How Many Iterations Are Required?

In our experiments, we conducted 20 iterations. The accuracy improves significantly in the first few iterations, and after that the performance becomes stable (see Figure 3). We suspect the reason is that there is an upper bound for our proposed method. After several iterations, the performance nearly reaches that upper bound, making it difficult to be further improved, thus the performance becomes stable. The iteration number at which the performance becomes stable depends on the particular experimental settings. Therefore, we may conclude that several iterations are enough to achieve a significant improvement and the performance at each respective iteration depends heavily on the experimental settings.

5.4 How Does the Number of Topics Affect the Performance?

According to [4], the number of topics can significantly affect the performance of the "Topic" system. In our experiments, we compared 2,000 topics that show the best performance in [4], to a small number of topics 200. Similar to [4], using 2,000 topics is significantly better than 200 topics for the "Topic" lexicons.

For the affect on the improvement by our proposed method, the improvements over "Topic" are smaller on 2,000 topics than the ones on 200 topics for both "Context" and "Combination". We suspect the reason is that the absolute values of improvement on the seed dictionary cannot lead to the same level of improvement for CBM. At iteration 1, the improvement of the "Topic" scores cannot fully reflect on the "Context" scores. Thus, the "Context" scores are lower than the "Topic" scores for 2,000 topics, while they are similar to or higher than the "Topic" scores for 200 topics (see Figure 3). The performance at iteration 1 impacts the overall improvement performance for the future iterations.

5.5 How Does the Number of Candidates Affect the Performance?

In our experiments, we measured the difference using 20 and 50 translation candidates for each word. The results show that using more candidates slightly decreases the performance (see Figure 3). Although using more candidates may increase the percentage of words where the correct translation is contained within the Top N word list of translation candidates (Precision@N), it also leads to

Table 1. Improved examples of "研究↔research" (left) and "施↔facility" (right)

Candidate	Sim_{Topic}	$Sim_{Context}$	Sim_{Comb}	Candidate	Sim_{Topic}	$Sim_{Context}$	Sim_{Comb}
research	0.0530	**0.2176**	**0.0859**	facility	0.0561	0.1127	**0.0674**
scientist	0.0525	0.1163	0.0653	center	0.0525	**0.1135**	0.0647
science	**0.0558**	0.0761	0.0599	building	0.0568	0.0933	0.0641
theory	0.0509	0.0879	0.0583	landmark	**0.0571**	0.0578	0.0572
journal	0.0501	0.0793	0.0559	plan	0.0460	0.1007	0.0570

more noisy pairs. According to our investigation on Precision@N of the two settings, the difference is quite small. For Chinese–English: Precision@20=0.5620, Precision@50=0.5780, while for Japanese–English: Precision@20=0.4930, Precision@50=0.5030. Therefore, we suspect the decrease is because the negative effect outweighs the positive. Furthermore, using more candidates will increase the computational complexity. Therefore, we believe a small number of candidates such as 20 is appropriate for our proposed method.

5.6 What Kind of Lexicons Are Improved?

Although TMBM has the advantage of finding topic related translations, it lacks of the ability to distinguish candidates that have highly similar word–topic distributions to the source word. This weakness can be solved with CBM.

Table 1 (left) shows an improved example of the Chinese word "研究↔research". All the candidates identified by "Topic" are strongly related to the topic of academia. The differences among the Sim_{Topic} scores are quite small, because of the high similarities of the word–topic distributions between these candidates and the source word, and "Topic" fails to find the correct translation. However, the differences in contextual similarities between the candidates and the source word are quite explicit. With the help of $Sim_{Context}$ scores, our proposed method finds the correct translation. Based on our investigation on the improved lexicons, most improvements belong to this type, where the Sim_{Topic} scores are similar, while the $Sim_{Context}$ scores are easy to distinguish.

Table 1 (right) shows an improved example of the Japanese word "施↔facility". The Sim_{Topic} scores are similar to the ones in the example on the left side of Table 1 that are not quite distinguishable, and "Topic" fails to find the correct translation. The difference is that CBM also fails to find the correct translation, and the Top 2 $Sim_{Context}$ scores are quite similar. The combination of the two methods successfully finds the correct translation, although this could be by chance. Based on our investigation, a small number of improvements belong to this type, where both Sim_{Topic} and $Sim_{Context}$ scores are not distinguishable.

5.7 What Kind of Errors Are Made?

As described in Section 5.5, for nearly half of the words in the test sets, the correct translation is not included in the Top N candidate list produced by TMBM.

We investigated these words and found several types of errors. The majority of errors are caused by unsuccessful identification despite topic alignment being correct (e.g. Japanese word "手↔player" is translated as "team"). Some errors are caused by unsuccessful topic alignment between the source and target words (e.g. Japanese word "置↔establishment" is translated as "kumagaya" which is a Japanese city name). There are also errors caused by words that do not clearly fit into one topic (e.g. Chinese word "爵士↔jazz/sir" may belong to either a musical or social topic). The remaining errors are due to English compound nouns. There are several pairs that contain English compound nouns in our test sets (e.g. "香港↔Hong Kong" in Chinese–English, and "ソ↔soviet union" in Japanese–English). Currently, our system cannot deal with compound nouns, and we leave it as future work for this study.

There are still some errors for words with their correct translation included in the Top N candidate list produced by TMBM, although our proposed method significantly improves the accuracy. Based on our investigation, most errors happen in the case that either the "Topic" or "Context" gives a significantly lower score to the correct translation than the scores given to the incorrect translations, while the other gives the highest or almost highest score to the correct translation. In this case, a simple linear combination of the two scores is not discriminative enough, and incorporating both scores as features in a machine learning way may be more effective.

6   Conclusion and Future Work

In this paper, we presented a bilingual lexicon extraction system exploiting both topical and contextual knowledge. Our system is based on a novel combination of TMBM and CBM, which does not rely on any prior knowledge and can be iteratively improved. Experiments conducted on Chinese–English and Japanese–English Wikipedia data verified the effectiveness of our system for bilingual lexicon extraction from comparable corpora.

As future work, firstly, we plan to compare different definitions of context for CBM. Secondly, we plan to conduct experiments on other comparable corpora rather than Wikipedia, where document alignment is required beforehand. Finally, we plan to extend our system to handle compound nouns, rare words and polysemy.

Acknowledgments. The first author is supported by Hattori International Scholarship Foundation[10]. We also thank the anonymous reviewers for their valuable comments.

References

1. Brown, P.F., Della Pietra, S.A., Della Pietra, V.J., Mercer, R.L.: The mathematics of statistical machine translation: Parameter estimation. Association for Computational Linguistics 19, 263–312 (1993)

---

[10] http://www.hattori-zaidan.or.jp

2. Koehn, P., Hoang, H., Birch, A., Callison-Burch, C., Federico, M., Bertoldi, N., Cowan, B., Shen, W., Moran, C., Zens, R., Dyer, C., Bojar, O., Constantin, A., Herbst, E.: Moses: Open source toolkit for statistical machine translation. In: Proceedings of the 45th Annual Meeting of the Association for Computational Linguistics Companion Volume Proceedings of the Demo and Poster Sessions, Prague, Czech Republic, pp. 177–180. Association for Computational Linguistics (2007)

3. Pirkola, A., Hedlund, T., Keskustalo, H., Järvelin, K.: Dictionary-based cross-language information retrieval: Problems, methods, and research findings. Information Retrieval 4, 209–230 (2001)

4. Vulić, I., De Smet, W., Moens, M.F.: Identifying word translations from comparable corpora using latent topic models. In: Proceedings of the 49th Annual Meeting of the Association for Computational Linguistics: Human Language Technologies, Portland, Oregon, USA, pp. 479–484. Association for Computational Linguistics (2011)

5. Rapp, R.: Automatic identification of word translations from unrelated english and german corpora. In: Proceedings of the 37th Annual Meeting of the Association for Computational Linguistics, College Park, Maryland, USA, pp. 519–526. Association for Computational Linguistics (1999)

6. Harris, Z.S.: Distributional structure. Word 10, 146–162 (1954)

7. Vulić, I., Moens, M.F.: Detecting highly confident word translations from comparable corpora without any prior knowledge. In: Proceedings of the 13th Conference of the European Chapter of the Association for Computational Linguistics, Avignon, France, pp. 449–459. Association for Computational Linguistics (2012)

8. Liu, X., Duh, K., Matsumoto, Y.: Topic models + word alignment = a flexible framework for extracting bilingual dictionary from comparable corpus. In: Proceedings of the Seventeenth Conference on Computational Natural Language Learning, Sofia, Bulgaria, pp. 212–221. Association for Computational Linguistics (2013)

9. Richardson, J., Nakazawa, T., Kurohashi, S.: Robust transliteration mining from comparable corpora with bilingual topic models. In: Proceedings of the Sixth International Joint Conference on Natural Language Processing, Nagoya, Japan, pp. 261–269. Asian Federation of Natural Language Processing (2013)

10. Rapp, R.: Identifying word translations in non-parallel texts. In: Proceedings of the 33rd Annual Meeting of the Association for Computational Linguistics, Cambridge, Massachusetts, USA, pp. 320–322. Association for Computational Linguistics (1995)

11. Fung, P.: Compiling bilingual lexicon entries from a non-parallel english-chinese corpus. In: Proceedings of the 3rd Annual Workshop on Very Large Corpora, pp. 173–183 (1995)

12. Koehn, P., Knight, K.: Learning a translation lexicon from monolingual corpora. In: Proceedings of the ACL 2002 Workshop on Unsupervised Lexical Acquisition, Philadelphia, Pennsylvania, USA, pp. 9–16. Association for Computational Linguistics (2002)

13. Haghighi, A., Liang, P., Berg-Kirkpatrick, T., Klein, D.: Learning bilingual lexicons from monolingual corpora. In: Proceedings of ACL 2008, HLT, Columbus, Ohio, pp. 771–779. Association for Computational Linguistics (2008)

14. Prochasson, E., Fung, P.: Rare word translation extraction from aligned comparable documents. In: Proceedings of the 49th Annual Meeting of the Association for Computational Linguistics: Human Language Technologies, Portland, Oregon, USA, pp. 1327–1335. Association for Computational Linguistics (2011)

15. Tamura, A., Watanabe, T., Sumita, E.: Bilingual lexicon extraction from comparable corpora using label propagation. In: Proceedings of the 2012 Joint Conference on Empirical Methods in Natural Language Processing and Computational Natural Language Learning, Jeju Island, Korea, pp. 24–36. Association for Computational Linguistics (2012)

16. Fung, P., Yee, L.Y.: An ir approach for translating new words from nonparallel, comparable texts. In: Proceedings of the 36th Annual Meeting of the Association for Computational Linguistics and 17th International Conference on Computational Linguistics, Montreal, Quebec, Canada, vol. 1, pp. 414–420. Association for Computational Linguistics (1998)
17. Garera, N., Callison-Burch, C., Yarowsky, D.: Improving translation lexicon induction from monolingual corpora via dependency contexts and part-of-speech equivalences. In: Proceedings of the Thirteenth Conference on Computational Natural Language Learning (CoNLL 2009), Boulder, Colorado, pp. 129–137. Association for Computational Linguistics (2009)
18. Yu, K., Tsujii, J.: Extracting bilingual dictionary from comparable corpora with dependency heterogeneity. In: Proceedings of Human Language Technologies: The 2009 Annual Conference of the North American Chapter of the Association for Computational Linguistics, Boulder, Colorado. Companion Volume: Short Papers, pp. 121–124. Association for Computational Linguistics (2009)
19. Mimno, D., Wallach, H.M., Naradowsky, J., Smith, D.A., McCallum, A.: Polylingual topic models. In: Proceedings of the 2009 Conference on Empirical Methods in Natural Language Processing, Singapore, pp. 880–889. Association for Computational Linguistics (2009)
20. Blei, D.M., Ng, A.Y., Jordan, M.I.: Latent dirichlet allocation. Journal of Machine Learning Research 3, 993–1022 (2003)
21. Utiyama, M., Isahara, H.: Reliable measures for aligning japanese-english news articles and sentences. In: Proceedings of the 41st Annual Meeting of the Association for Computational Linguistics, Sapporo, Japan, pp. 72–79. Association for Computational Linguistics (2003)
22. Munteanu, D.S., Marcu, D.: Improving machine translation performance by exploiting non-parallel corpora. Computational Linguistics 31, 477–504 (2005)
23. Vu, T., Aw, A.T., Zhang, M.: Feature-based method for document alignment in comparable news corpora. In: Proceedings of the 12th Conference of the European Chapter of the ACL (EACL 2009), Athens, Greece, pp. 843–851. Association for Computational Linguistics (2009)
24. Zhu, Z., Li, M., Chen, L., Yang, Z.: Building comparable corpora based on bilingual lda model. In: Proceedings of the 51st Annual Meeting of the Association for Computational Linguistics, Sofia, Bulgaria. Short Papers, vol. 2, pp. 278–282. Association for Computational Linguistics (2013)
25. Chu, C., Nakazawa, T., Kawahara, D., Kurohashi, S.: Exploiting shared Chinese characters in Chinese word segmentation optimization for Chinese-Japanese machine translation. In: Proceedings of the 16th Annual Conference of the European Association for Machine Translation (EAMT 2012), Trento, Italy, pp. 35–42 (2012)
26. Kurohashi, S., Nakamura, T., Matsumoto, Y., Nagao, M.: Improvements of Japanese morphological analyzer JUMAN. In: Proceedings of the International Workshop on Sharable Natural Language, pp. 22–28 (1994)
27. Tsuruoka, Y., Miyao, Y., Kazama, J.: Learning with lookahead: Can history-based models rival globally optimized models? In: Proceedings of the Fifteenth Conference on Computational Natural Language Learning, Portland, Oregon, USA, pp. 238–246. Association for Computational Linguistics (2011)
28. Voorhees, E.M.: The TREC-8 question answering track report. In: Proceedings of the Eighth TExt Retrieval Conference (TREC-8), pp. 77–82 (1999)

Improving Bilingual Lexicon Extraction from Comparable Corpora Using Window-Based and Syntax-Based Models

Amir Hazem and Emmanuel Morin

Laboratoire d'Informatique de Nantes-Atlantique (LINA)
Université de Nantes, 44322 Nantes Cedex 3, France
{Amir.Hazem,Emmanuel.Morin}@univ-nantes.fr

Abstract. This paper proposes two strategies for combining a window-based and a syntax-based context representation for the task of bilingual lexicon extraction from comparable corpora. The first strategy involves combining the scores assigned to translations by both models and using them for ranking and selection; the second strategy involves a combination of the context features provided by the two models prior to applying the lexicon extraction method. The reported results show that the combination of the two context representations significantly improves the performance of bilingual lexicon extraction compared to using each of the representations individually.

Keywords: Comparable corpora, Bilingual lexicon extraction, Context representation, Dependency relations.

1 Introduction

The identification of word translations from comparable corpora has become one of the most investigated applications. The main motivation for exploiting comparable corpora[1] rather than parallel corpora[2] is their availability and easier acquisition thanks to the web. Previous work has shown that translational word pairs can be automatically derived from the statistical distribution of words in bilingual comparable corpora [1,2], assuming that there is a correlation between the co-occurrences of words which are translations of each other. The main assumption underlying lexicon extraction from comparable corpora is that the associations of words and their contexts are preserved. A well-known approach that makes use of this assumption is the *standard approach*. It was introduced in [1,2] and later re-used and improved by many researchers [3–6] among others. The principle is to characterize each word of the source and the target language by its context vector. Each entry w_j of the context vector of a word w_i corresponds to the association between w_i and w_j. To translate a word w_i, its context

---

[1] Comparable corpora are collections of documents of different languages, generally written at the same period and treating the same topics.

[2] Parallel corpora are collections of documents which are translation of each other.

A. Gelbukh (Ed.): CICLing 2014, Part II, LNCS 8404, pp. 310–323, 2014.

vector is first translated using a seed lexicon (bilingual dictionary), then the similarity between the translated context vector and all the context vectors of the words of the target language is measured, and finally the translation candidates are ranked according to their similarity scores.

A central point of the *standard approach* is certainly the representation of the context words. The main question that remains is how to choose the words that constitute the context of a given word? Many studies have been conducted in this direction [6–8]. Some researchers consider the context of a word w_i as a window of n words surrounding w_i [2, 8, 9]. We refer to this strategy as the window-based model. One noteworthy problem of this model is the size of the selected window. On the one hand, if the window size is too small, some relevant words can be missed. On the other hand, if the window size is too large, more noise can be introduced by taking into account non relevant words. Rapp (1999) and Morin *et al.* (2007) for instance, used a 7 window size (3 words before and 3 words after a given word w_i) while Prochasson and Morin (2009) used 25 words, etc. Other researchers used more refined context representations. Otero (2008) for instance, proposed to extract syntactic dependencies for each word by using robust parser. This technique alleviates the problem of window size since no window is used at all. The dependency relations are extracted following some syntactic rules defined by specific patterns [11]. We refer to this strategy as a syntax-based model. Otero (2008) has compared the classical window-based and the syntax-based models and has shown that using dependency relations provides better results than the windowing approach on large comparable corpora (about 10 million words). A first approach proposed in [12] combines the window-based model at sentence level with three dependency models which extract the information regarding predecessors, successors and siblings. The combined probabilistic model has shown a significant improvement on an English-Japanese comparable corpora. Our approaches differ from the work of Andrade *et al.* (2011) in at least two points: (i) while they used three dependency relations (predecessor, successor and sibling) to build their models, we experienced seven dependency relations as defined in [10]. (ii) We exploited these relations in addition to the window-based model in order to improve the performance of the *standard approach*. Our contribution is mainly a re-interpretation of the context representation of the *standard approach* while [12] proposed to combine four statistical models to identify likely translation candidates based on pivot words which are positively associated.

It has been stated that the *standard approach* is more efficient when using large comparable corpora [2,10]. The more data we have, the better is the context representation and the better will be the performance of bilingual lexicon extraction methods. In this paper, we focus on the task of bilingual terminology extraction from specialized comparable corpora. Domain specific comparable corpora are often of a small size. This particularity renders the *standard approach* more sensitive to context representation and often leads to lower performance. Due to the lack of information conveyed by specialized comparable corpora, we believe that more attention should be paid to context characterization and propose to push the context representation study a step further. In this paper, we first revisit the *standard approach* using the two main context representation strategies, then we introduce two new approaches that exploit both: the window-based and the syntax-based models together in order to build more discriminant context vectors. We show that our approaches significantly outperform the state of the art on four English-French specialized comparable corpora.

The remainder of this paper is organized as follows: Section 2 introduces the *standard approach* traditionally dedicated to the task of bilingual lexicon extraction from comparable corpora. Section 3 describes the window-based and syntax-based context representations. Our approach is then presented in Section 4. Section 5 describes the different linguistic resources. Section 6 evaluates the contribution of our approaches through different experiments. Finally, Section 7 presents our conclusion and some perspectives.

2   Standard Approach

The *standard approach* has been used in many works [1,2,6,8,10,13,14], among others. The main difference between these works resides in some specificities of the task such as: context representation and co-occurrence estimation, context vector similarity measures, corpus characteristics, etc. A representation of the *standard approach* is given in Figure 1.

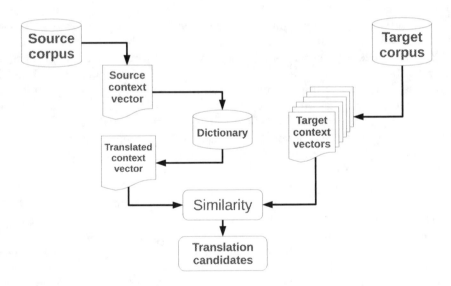

Fig. 1. Illustration of the *standard approach*

The *standard approach* relies on the assumption that a word and its translation tend to appear in the same lexical contexts. Hence, the extraction of translation pairs can be obtained by measuring their context similarity. First, the context vector v_i of the source word w_i is built. It contains all the words of the source language that occur with w_i. These words are weighted according to association measures such as the mutual information [15], the log-likelihood [16] or the discounted odds-ratio [6] for instance. The entries of the context vector v_i are then translated using a bilingual dictionary. Only the first most frequent translations found in the dictionary for a word are used. Words with no entry in the dictionary are discarded. Finally, a similarity measure is used to score each word, w_t, in the target language with respect to the translated context vector. Usual measures of vector similarity include the cosine similarity [17] or the weighted Jaccard index (WJ) [18] for instance. The candidate translations of the word w_i are the target words ranked following the similarity score.

3   Context Representation

In this section we present the two main word context representation strategies used in the *standard approach*.

3.1   Window-Based Model

The window-based model, also known as the bag-of-words model consists in building a set of words surrounding a given word w_i taken from the entire corpus. For each occurrence of w_i, n words based on the window size are selected. Let us consider the following sentence for instance:

...The *risk* of *recurrence* or **death** for this *patient occurring* within five years after the diagnosis is high...

If we choose a window of size 5 (2 words before and 2 words after w_i), and after applying some preprocessing such as removing function words and lemmatization, we obtain the following normalized sentence:

... *risk recurrence* **death** *patient occur* ...

According to this example, the context of the word **death** is: {*risk, recurrence, patient, occur*}. To build its final context vector, we repeat the described process for all the occurrences of **death** in the corpus.

3.2   Syntax-Based Model

This technique consists in the identification of syntactic dependencies between words [19]. Hence, context vectors of words will contain syntactic information [10,20]. A dependency relation is defined as a binary asymmetric relation between a head or a parent word and a modifier or a dependent word [20]. The dependency relations form a tree that inter-connects all the words of a given sentence. A word can have several modifiers but each word can only modify one word at a time [19]. The root of the tree (also called the head) does not modify any word of a sentence. Let us consider the example proposed in [19]:

I have a brown dog

A list of tuples is used to represent a dependency tree: (word, category, head, relationship)

- word: represents a node of the tree
- category: constitutes the lexical category of the word
- head: specifies which token is modified by the given word
- relationship: represents a label given to the dependency relation (subj for subject, spec for specifier...)

The < sign means predecessor and > means successor. The dependency tree for the previous example is represented in Table 1.

Table 1. Representation of a dependency tree for the above example

Modifier	Category	Head	Type
I	N	< *have*	subj
have	V	-	-
a	Det	< *dog*	spec
brown	Adj	< *dog*	adjn
dog	N	> *have*	comp

More details concerning syntactic dependencies are given in [21] and more specifically, for the tasks of words disambiguation and attachment resolution. According to Otero (2008) there are 3 elementary notions:

- Lexical words (Mary, ran, fast, nice, etc.)
- Syntactic dependencies (subject, direct object, prepositional relation between two nouns, prepositional relation between verbs and nouns, etc.)
- Lexico-syntactic model (words and their lexical category are represented in terms of dependency (Noun + subj + Verb))

Lexical words represent a set of properties {Nouns, Verbs, Adjectives, Adverbs ... }, while syntactic dependencies and lexico-syntactic models are defined as operations on the set of properties. Hence, a dependency relation takes as input two sets of properties, and gives as output a more restrictive set that consists of the intersection of the two input sets. Table 2 summarizes the dependency relations as defined in [10].

If we take the word **recurrence** for instance, there is a Lmod dependency relation with the adjective **local**. In the process of building the dependency relations context vector of **recurrence**, we count the number of times that **local** occurs as a left modifier (Lmod) of **recurrence** in the corpus. This will constitute one entry of the context vector of **recurrence**. This process is repeated for all the words of the corpus.

Table 2. Syntactic dependency relations

Relation	Type	Example
Lmod	Left modifier if (Adj-Noun) relation	local-recurrence
Rmod	Right modifier if (Noun-Adj) relation	number-insufficient
modN	Noun modifier if (Noun-Noun) relation	breast-cancer
Lobj	Left object if (Noun-Verb) relation	study-demonstrate
Robj	Right object if (Verb-Noun) relation	have-effect
PRPN	Prepositional if (Noun-PRP-Noun) relation	malignancy-in-woman
PRPV	Indirect object if (Verb-PRP-Noun) relation	occur-in-portion

4   Combination Approach

In the previous section we have presented the two main context representation strategies, namely the window-based and the syntax-based models. If we consider the information conveyed by the windowing approach as a rough quantitative model in which words within the same window are of equal importance, we can notice the inconsistency of this proceeding. Words within the same window are not necessarily of the same importance vis-à-vis a word to characterize. This can motivate the choice of a more fine-grained modeling by using dependency relations [10,20] instead of the traditional bag-of-words model. That said, dependency relation modeling needs a large amount of data to be efficient. In the case of specialized comparable corpora which are often of a small size, syntax representation may not guarantee good performance also because of parsing errors. In addition, the window-based approach captures syntagmatic relations when using small windows and paradigmatic relations when using large windows [22], which is more difficult to obtain with the syntax model. This motivates our proposition to take into account strengths and weaknesses of both context representations in order to improve the quality of bilingual lexicon extraction from comparable corpora. We present in the following subsections two context combination strategies that we refer to as the *post combination* and the *prior combination* approaches.

4.1   Post Combination Approach

In information retrieval tasks, combining different ranked lists returned by multiple search engines in response to a given query, is often used in such a way as to optimize the performance of the combined ranking [23]. Since the *standard approach* using the window-based or the syntax-based strategies results in two distinct rankings, this provides an appropriate framework for exploiting information conveyed by the rankings. The *post combination approach* consists in taking as input the scores of the ranked lists returned by each approach, and then combining the two lists using a simple arithmetic combination of scores. Scores in different rankings are compatible since they are based on the same similarity measure (i.e. on the same scale). It should be noted that other combinations of scores and ranks were assessed, but the arithmetic combination gave the best results. Using scores as fusion criteria, we compute the similarity score of a candidate by summing its scores from each list returned by each method:

$$s_{comb}(w) = \lambda \times s_{window}(w) + (1 - \lambda) \times s_{syntax}(w) \tag{1}$$

where $s_{comb}(w)$ is the final score of the word w. $s_{window}(w)$ is the score returned by the *standard approach* using the window-based context representation, and $s_{syntax}(w)$ is the score returned by the *standard approach* using the syntax-based context representation. The confidence given to each context representation approach is fixed by the parameter λ^3 ($\lambda \in [0,1]$). Experiments on a held-out corpus showed that the best performance is obtained for a λ that varies between 0.55 and 0.65, giving more importance to the *standard approach* based on the windowing model.

4.2   Prior Combination Approach

The context vector aims at capturing a set of contextual information of a given word w. In the case of the window-based model, this information is defined as the counts of words that occur with w within a window of size n. In the case of the syntax-based model, this information is the counts of words that occur and share dependency relations with w. In the *prior combination approach*, we consider the context vector of w as a descriptor that contains multivariate information: (i) one global co-occurrence information returned by the window-based model, and (ii) one more specific co-occurrence information returned by the syntax-based model. Since these two representations are not exclusive, we propose to merge them into the same context vector in order to capture both window and syntax relations of w. An example to illustrate our method is given in Table 3.

The context combination method considers, in one single vector, both window-based and syntax-based models. If we take for instance the word **regional** represented in the example, we can see that according to the windowing model, **regional** occurs 13 times with the word **recurrence** while the syntax model found that **regional** occurs twice as its left modifier (Lmod). The prior combination approach takes into account the two types of information, and considers that **regional** occurs 13 times with **recurrence**, including twice as its left modifier. Another advantage is that if the window-based or the syntax-based model misses a type of information, as we can see for the word **oestrogen** for instance (see Table 3), the combination takes it into account thanks to the windowing approach in this case.

---

^{3} λ is a single value used for the whole corpus.

Table 3. Example of the context representations of the word **recurrence** and its co-occurrence counts, according to the window-based and the syntax-based approaches, as well as their prior combination

Window-based	Syntax-based	Prior-combination
$regional_{13}$	$regional_{Lmod_2}$	$regional_{13}$
		$regional_{Lmod_2}$
$local_5$	$local_{Lmod_1}$	$local_5$
		$local_{Lmod_1}$
$oestrogen_1$	-	$oestrogen_1$
$rate_{65}$	$rate_{modN_{29}}$	$rate_{65}$
	$rate_{PRPV_3}$	$rate_{modN_{29}}$
		$rate_{PRPV_3}$

5 Linguistic Resources

In this section, we present the material used in our experiments namely: (i) the comparable corpora, (ii) the bilingual seed lexicon, (iii) the terminology reference list and (iv) the dependency relation tool.

5.1 Comparable Corpora

We built four English-French specialized comparable corpora. Details for each corpus are given hereafter:

Diabetes corpus The French part of the diabetes corpus has been first crawled from the web using several keywords such as: *diabète* (*diabetes*), *obésité* (*obesity*) and *alimentation* (*food*). Then, it has been filtered manually by keeping medical domain documents only. This results in a French part of 250,000 words (4,983 distinct words). The English part has been extracted from the medical website PubMed[4] using keywords such as: *diabetes, obesity* and *food*. This results in an English part of 255,000 words (5,362 distinct words).

Wind energy corpus For the wind energy corpus, the documents have been crawled from the web using the *Babook* crawler [24] based on several keywords such as: *wind, rotor, blade* for the English part and *vent, rotor, pale* for the French part. This results in an English-French corpus of 320,000-332,000 words (5,580-6,025 distinct words).

---

[4] www.ncbi.nlm.nih.gov/pubmed/

Volcano corpus The volcano corpus has been mainly built manually based on electronic academic books, glossaries, scientific magazines, etc. Some crawled documents from the web have been added. This results in an English-French corpus of 400,000-423,000 words (about 8,623-9,142 distinct words).

Breast cancer corpus For the breast cancer corpus, documents from the medical domain within the sub-domain of breast cancer have been extracted from the Elsevier website[5]. The selected documents were those containing in the title or the keywords, the term 'cancer du sein' in French and 'breast cancer' in English. This results in an English-French corpus of 500,000-523,000 words (6,081-7,376 distinct words).

All the documents of the four bilingual corpora have been normalized using tokenization, part-of-speech tagging and lemmatisation.

5.2   Bilingual Seed Lexicon

We used the English-French ELRA-M0033 [6] dictionary as a bilingual seed lexicon. This resource contains about 200,000 entries mainly from the general language.

5.3   Evaluation Lists

To build the terminology evaluation lists, we selected the English-French pairs of single word terms (SWTs) which occur more than four times in each part of the comparable corpus[7]. Only nouns were selected, this results in 244 English/French SWTs for the diabetes corpus, 150 English/French SWTs for the wind energy corpus, 158 English/French SWTs for the volcano corpus and 321 English/French SWTs for the breast cancer corpus. It should be noted that the terminology evaluation lists required for the task of bilingual lexicon extraction from comparable corpora are often of a small size due to the specificity of specialized domains [3, 9, 25], etc.

5.4   Dependency Relation Parser

To extract the dependency relations for each word of the comparable corpora, we used the parser provided by Otero (2008)[8]. Let us consider for instance, the following sentence taken from the diabetes corpus:

The parser produces a left modifier relation (Lmod) between *accurate* and *assessment*, a right object relation (Robj) between *facilitate* and *assessment*, a left modifier relation (Lmod) between *family* and *structure* and a prepositional relation (PRPN) between *assessment* and *structure*.

---

[5] www.elsevier.com

[6] www.elra.info

[7] Distributional methods can not be applied to rare words.

[8] http://gramatica.usc.es/pln/tools/deppattern.html

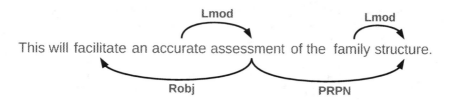

Fig. 2. Illustration of the dependency relation parsing of a sentence taken from the diabetes corpus

6   Experiments and Results

In this section, we first present the experimental setup used in our evaluation and then, we present different experiments using the window-based, the syntax-based and the proposed approaches.

6.1   Experimental Setup

The *standard approach* requires three parameters to be set, namely: (i) the context representation (window-based or syntax-based), (ii) the association measure, and (iii) the similarity measure. We chose to apply the main association and similarity measures for both context representations. We present the results using the point-wise mutual information [15] associated with the Cosine similarity, denoted by PMI-Cos, the discounted odds ratio [6] also associated with the Cosine similarity, denoted by DOR-Cos and the log-likelihood [16] associated with the Jaccard measure [18], denoted by LL-Jacc. Other combinations were assessed but the ones chosen have shown the best performance. For the window-based representation, we used a window of seven words. For a complete study, Laroche and Langlais (2010) carried out different experiments regarding the influence of these parameters on the quality of bilingual lexicon extraction. For the translation step of the *standard approach*, whenever the bilingual dictionary provides several translations for a word, all the entries are considered but weighted according to their frequency in the target language [26]. The results of the different approaches were evaluated using the mean average precision (MAP) [27]:

$$MAP = \frac{1}{|W|} \sum_{i=1}^{|W|} \frac{1}{Rank_i} \qquad (2)$$

where $|W|$ corresponds to the size of the evaluation list, and $Rank_i$ corresponds to the ranking of a correct translation candidate i.

6.2   Results

In order to evaluate the influence of the representation of the context on the bilingual terminology extraction task, we carried out experiments on four specialized

Table 4. Results (MAP%) of the *standard approach* using different context representations on four specialized comparable corpora (the improvements indicate a significance at the 0.01 level using the Student t-test †)

	Diabetes	Volcano	Wind energy	Breast cancer	
window	15.5	21.7	15.6	22.6	PMI-Cos
syntax	10.4	23.3	16.5	18.9	
prior_comb	5.5	18.3	13.8	17.9	
post_comb	26.6†	39.6†	26.4†	29.4†	
window	16.5	30.3	20.2	24.8	DOR-Cos
syntax	10.0	18.4	13.7	14.6	
prior_comb	23.7†	44.8†	29.6†	32.8†	
post_comb	23.6†	45.9†	31.3†	32.0†	
window	20.6	46.8	24.2	27.9	LL-Jacc
syntax	17.8	30.2	23.0	19.2	
prior_comb	27.0†	50.6†	32.2†	34.2†	
post_comb	27.8†	51.0†	31.0†	32.8†	

comparable corpora using three combination measures of the *standard approach* (PMI-Cos, DOR-Cos and LL-Jacc). Table 4 shows a comparison between the *standard approach* using the window-based representation, denoted by *window*, the *standard approach* using the syntax-based representation, denoted by *syntax* and our two proposed approaches denoted by *prior_comb* for the prior combination approach and *post_comb* for the post combination approach.

As a preliminary remark, we can notice that the *window* approach outperforms the *syntax* approach in almost all the configurations. The lower performance of the *standard approach* based on dependency relations (*syntax*) is certainly due to the lack of data conveyed by specialized comparable corpora in addition to some parsing errors. We can also note that the results differ (sometimes noticeably) according to the bilingual comparable corpora. This is not surprising since comparable corpora are different in terms of quality and evaluation lists.

Concerning the *prior_comb* approach, we can notice that except for the PMI-Cos configuration, there are significant improvements comparing to the *window* and *syntax* approaches. The lower results obtained using the PMI-Cos parameters may be due to the fact that the point-wise mutual information tends to overestimate low frequencies and underestimate high frequencies. As the *prior_comb* approach introduces many dependency relations with low frequencies, this can increase the overestimation effect of the PMI measure. For the *post_comb* approach, we notice that it significantly outperforms the *window* and *syntax* approaches in all configuration. Also, we can see that the *post_comb* approach is not sensitive to the PMI-Cos configuration. This can be explained by the fact that no modification on the context vectors entries is needed since this approach is based on the outputs combination of the *window* and *syntax* approaches.

The significant improvements obtained by the two proposed approaches suggest that combining window-based and syntax-based representations is suitable for the task of bilingual terminology extraction from comparable corpora. That said, if the difference of results between the two combination strategies is not substantial, the *post_comb* approach needs to run twice the *standard approach* (one run per contextual representation). Aiming at enriching the context vector with more features in the future, running the *standard approach* once for each feature can rapidly become fastidious if the number of features increases. From this point of view, we believe that the *prior_comb* approach is more appropriate for adding more features and can be beneficial for future work on the enrichment of context vectors. In addition, the *prior_comb* approach can be extended (or adapted) by adding other dependency relations depending on the parser used. For future work, we also want to contrast different parsers to evaluate the *prior_comb* approach. If the *prior* combination of the window-based and the syntax-based models has shown significant improvements for the task of bilingual terminology extraction from comparable corpora, our model can be used in other tasks such as word sense disambiguation and synonym extraction for instance. This is also another direction for future work.

7 Conclusion

In this paper, we have presented two context representation strategies based on bag-of-words and dependency relation models, for the task of bilingual terminology extraction from specialized comparable corpora. The proposed approaches obtained significant improvements when compared to the traditional window-based and syntax-based approaches on four specialized comparable corpora. The results obtained lend support to the hypothesis that combining the two representations is more suitable for small corpora. We believe that our prior context representation model can be relevant in NLP applications that use context representation such as word sense disambiguation or synonym extraction for instance. It is our hope that this work will encourage further exploration on the enrichment of context vectors with multiple information.

Acknowledgments. The research leading to these results has received funding from the French National Research Agency under grant ANR-12-CORD-0020 (CRISTAL project).

References

1. Fung, P.: A statistical view on bilingual lexicon extraction: From parallel corpora to non-parallel corpora. In: Farwell, D., Gerber, L., Hovy, E. (eds.) AMTA 1998. LNCS (LNAI), vol. 1529, pp. 1–17. Springer, Heidelberg (1998)

2. Rapp, R.: Automatic identification of word translations from unrelated english and german corpora. In: Proceedings of the 37th Annual Meeting of the Association for Computational Linguistics (ACL 1999), College Park, MD, USA, pp. 519–526 (1999)
3. Chiao, Y.C., Zweigenbaum, P.: Looking for candidate translational equivalents in specialized, comparable corpora. In: Proceedings of the 19th International Conference on Computational Linguistics (COLING 2002), Tapei, Taiwan, pp. 1208–1212 (2002)
4. Prochasson, E., Morin, E.: Anchor points for bilingual extraction from small specialized comparable corpora. TAL 50(1), 283–304 (2009)
5. Yu, K., Tsujii, J.: Extracting bilingual dictionary from comparable corpora with dependency heterogeneity. In: Proceedings of Human Language Technologies: The 2009 Annual Conference of the North American Chapter of the Association for Computational Linguistics, NAACL-Short 2009, Boulder, Colorado, Companion Volume: Short Papers, pp. 121–124 (2009)
6. Laroche, A., Langlais, P.: Revisiting context-based projection methods for term-translation spotting in comparable corpora. In: Proceedings of the 23rd International Conference on Computational Linguistics (COLING 2010), Beijing, China, pp. 617–625 (2010)
7. Gaussier, E., Renders, J.M., Matveeva, I., Goutte, C., Déjean, H.: A geometric view on bilingual lexicon extraction from comparable corpora. In: Proceedings of the 42nd Annual Meeting of the Association for Computational Linguistics (ACL 2004), Barcelona, Spain, pp. 526–533 (July 2004)
8. Morin, E., Daille, B., Takeuchi, K., Kageura, K.: Bilingual Terminology Mining – Using Brain, not brawn comparable corpora. In: Proceedings of the 45th Annual Meeting of the Association for Computational Linguistics (ACL 2007), Prague, Czech Republic, pp. 664–671 (2007)
9. Déjean, H., Sadat, F., Gaussier, E.: An approach based on multilingual thesauri and model combination for bilingual lexicon extraction. In: Proceedings of the 19th International Conference on Computational Linguistics (COLING 2002), Taipei, Taiwan, pp. 218–224 (2002)
10. Otero, P.G.: Evaluating two different methods for the task of extracting bilingual lexicons from comparable corpora. In: Proceedings of LREC 2008 Workshop on Comparable Corpora (LREC 2008), Marrakech, Marroco, pp. 19–26 (2008)
11. Otero, P.G.: Learning bilingual lexicons from comparable english and spanish corpora. In: Proceedings of Machine Translation Summit XI, pp. 191–198 (2007)
12. Andrade, D., Matsuzaki, T., Tsujii, J.: Effective use of dependency structure for bilingual lexicon creation. In: Gelbukh, A. (ed.) CICLing 2011, Part II. LNCS, vol. 6609, pp. 80–92. Springer, Heidelberg (2011)
13. Ismail, A., Manandhar, S.: Bilingual lexicon extraction from comparable corpora using indomain terms. In: Proceedings of the 23rd International Conference on Computational Linguistics (COLING 2010), Beijing, China, pp. 481–489 (2010)
14. Bouamor, D., Semmar, N., Zweigenbaum, P.: Context vector disambiguation for bilingual lexicon extraction from comparable corpora. In: Proceedings of the 51st Annual Meeting of the Association for Computational Linguistics (ACL 2013), Sofia, Bulgaria, pp. 759–764 (2013)
15. Fano, R.M.: Transmission of Information: A Statistical Theory of Communications. MIT Press, Cambridge (1961)
16. Dunning, T.: Accurate methods for the statistics of surprise and coincidence. Computational Linguistics 19(1), 61–74 (1993)

17. Salton, G., Lesk, M.E.: Computer evaluation of indexing and text processing. Journal of the Association for Computational Machinery 15(1), 8–36 (1968)
18. Grefenstette, G.: Explorations in Automatic Thesaurus Discovery. Kluwer Academic Publisher, Boston (1994)
19. Lin, D.: Dependency-based evaluation of minipar. In: Proceedings of the Workshop on the Evaluation of Parsing Systems, First International Conference on Language Resources and Evaluation (LREC 1998), Granada, Spain (1998)
20. Garera, N., Callison-Burch, C., Yarowsky, D.: Improving translation lexicon induction from monolingual corpora via dependency contexts and part-of-speech equivalences. In: Proceedings of the 13th Conference on Computational Natural Language Learning (CoNLL 2009), Boulder, Colorado, pp. 129–137 (2009)
21. Otero, P.G.: The meaning of syntactic dependencies. Linguistik Online (2008)
22. Grefenstette, G.: Corpus-derived first, second and third-order word affinities. In: Proceedings of the 6th Congress of the European Association for Lexicography (EURALEX 1994), Amsterdam, The Netherlands, pp. 279–290 (1994)
23. Aslam, J.A., Montague, M.: Models for Metasearch. In: Proceedings of the 24th Annual International ACM SIGIR Conference on Research and Development in Information Retrieval (SIGIR 2001), New Orleans, Louisiana, USA, pp. 276–284 (2001)
24. Groc, C.D.: Babouk: Focused web crawling for corpus compilation and automatic terminology extraction. In: Proceedings of the IEEE-WICACM International Conferences on Web Intelligence, Lyon, France, pp. 497–498 (2011)
25. Daille, B., Morin, E.: French-english terminology extraction from comparable corpora. In: Dale, R., Wong, K.-F., Su, J., Kwong, O.Y. (eds.) IJCNLP 2005. LNCS (LNAI), vol. 3651, pp. 707–718. Springer, Heidelberg (2005)
26. Hazem, A., Morin, E.: Ica for bilingual lexicon extraction from comparable corpora. In: Proceedings of the 5th Workshop on Building and Using Comparable Corpora (BUCC 2012), Istanbul, Turkey (2012)
27. Manning, D.C., Raghavan, P., Schütze, H.: Introduction to information retrieval. Cambridge University Press (2008)

An IR-Based Strategy for Supporting Chinese-Portuguese Translation Services in Off-line Mode

Jordi Centelles[1], Marta Ruiz Costa-jussà[1], Rafael E. Banchs[1], and Alexander Gelbukh[2]

[1] Institute for Infocomm Research, Singapore
[2] Centro de Investigación en Computación, Instituto Politécnico Nacional, Mexico
{visjcs,vismrc,rembanchs}@i2r.a-star.edu.sg, gelbukh@gelbukh.com

Abstract. This paper describes an Information Retrieval engine that is used to support our Chinese-Portuguese machine translation services when no internet connection is available. Our mobile translation app, which is deployed on a portable device, relies by default on a server-based machine translation service, which is not accessible when no internet connection is available. For providing translation support under this condition, we have developed a contextualized off-line search engine that allows the users to continue using the app.

Keywords: machine translation, translation aid, Chinese, Portuguese.

1 Introduction

Machine translation applications have gained a lot of popularity in recent years. Currently, statistical approaches to machine translation are dominating the market, as they allow for automatically learning translation tables from parallel corpora (Brown et al 1993, Koehn et al 2003). The main problem for this approaches is the high amount of resources they consume regarding to memory and computational power. Due to this, most translation applications operate under a client-server architecture in which the client only provides a dummy interface while all the computations are carried out on a remote server. The main limitation of this scheme is that the client required internet connection to be available.

In this work, we present a search-based strategy for supporting machine translation services when internet connection is not available. More specifically, our proposed strategy, which is based on Information Retrieval technologies, is designed to support our Chinese-Portuguese translation service that has been deployed at the client side as a mobile app. The proposed strategy, allows for the mobile app to continued operating, with limited capabilities, on off-line mode when no internet connection is available. The off-line mode also includes contextualization strategies that allow improving the system performance based on user preferences, location and time.

The rest of the paper is structure as follows. In section 2, we describe the original Chinese-Portuguese on-line translation service. In section 3, we present the proposed off-line mode strategy and its contextualization capabilities. Finally, in section 4, we present our conclusion and proposed future directions of research.

A. Gelbukh (Ed.): CICLing 2014, Part II, LNCS 8404, pp. 324–330, 2014.
© Springer-Verlag Berlin Heidelberg 2014

2   Chinese-Portuguese On-line Translation Services

In this section we describe the original Chinese-Portuguese on-line translation service. First, we present a brief overview on the Chinese-Portuguese machine translation engine (the server side), and then we present a detailed description of the mobile app that connects to this service (the client side).

2.1   Chinese-Portuguese Translation System

In order to build our machine translation system, we have used a standard phrase-based statistical machine translation based on Moses (Koehn et al., 2007). This well-known approach splits the source sentence to translate in segments and it assigns to each segment a bilingual phrase from a phrase-table. Bilingual phrases are translation units that contain source words and target words. These bilingual phrases have different scores associated to them (including conditional, posterior and lexical probabilities). Among the list of bilingual phrases, the decoder is in charge of selecting the ones that maximize the linear combination of feature functions. Such strategy is known as the log-linear model (Och and Ney, 2002). The two main feature functions are the translation model and the target language model. Additional models include phrase and word penalty and reordering.

Our system is a corpus-based approach where the key for translation quality is regarding the quality and quantity of the corpus used for training. Generally speaking, translation between distant language pairs follows pivot approaches through English (or other major-resourced language) because of the lack of parallel data to train the direct approach. The main advantage of our system is that we are using the direct approach and at the same time, we rely on a rather large corpus which has been properly preprocessed.

Regarding data preprocessing we have done the following:

- For Chinese, we have segmented the data using the Stanford Segmenter tool (Tseng et al., 2005).
- For Portuguese, we have true cased the data and tokenized it with Moses tools.

Moses was used with the standard configuration. Different training domain corpus where concatenated to a single training corpus. We have corpora from different domains available. In particular we have used the following ones:

- TAUS. Data provided by this organization include translation memories of technical content.
- In-house. This corresponds to a small corpus in the transportation and hospitality domains

Statistics for the training corpus are presented in Table 1. Just to give an idea of the quality of our translation system we report the automatic and human evaluation

Table 1. Corpus details

Dataset	Parameter	Chinese	Portuguese
TAUS Train	Number of sentences	5 M	
	Running words	57 M	62 M
	Vocabulary	648 K	200 K
TAUS Dev	Number of sentences	808	
	Running words	11 K	12 K
	Vocabulary	3.0 K	3.4 K
TAUS Test	Number of sentences	721	
	Running words	9.9 K	10.9 K
	Vocabulary	2.8 K	3.3 K
In-house	Number of sentences	729	
	Running words	4.1 K	4.7 K
	Vocabulary	737	890

Table 2. Translation results

Translation direction	Domain / Dataset	Quality (BLEU)
Chinese-to-Portuguese	TAUS	37.97
	In-house	4.49
Portuguese-to-Chinese	TAUS	39.58
	In-house	6.48

results for Chinese-Portuguese. For fine-tuning the translation engines, we have used the TAUS development dataset and, then, we have tested with the TAUS and In-house test. Results are shown in terms of the standard metric BLEU in Table 2.

2.2   Chinese-Portuguese Translation App

The Android app for the Chinese-Portuguese translation client was programmed with the Android development tools (ADT). It is a plug-in for the Eclipse IDE that provides the necessary environment for building an app.

The Android-based app is depicted in Figure 1. For the communication between the Android app and the server we use the HTTPClient interface. Among other things, it allows a client to send data to the server via, for instance, the POST method, as used on the website case.

In addition to the base translation system, the app also incorporates Automatic Speech Recognition (ASR), Optical Character Recognition technologies as input methods (OCR), Image retrieval and Language detection (Centelles et al., 2013).

Also, the system uses a database to store the translation performed by the system and keep track of the most used translations. To create the databases we used the popular open source database management system: MySQL.

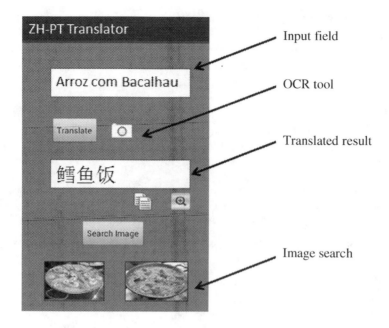

Fig. 1. Android-based Chinese-Portuguese translation client application

3   Off-line Search-Based Translation System

In this section we describe our proposed search-based off-line strategy to support the Chinese-Portuguese translation service. First, we describe our search engine implementation for translation, and then, we present the developed contextualization strategy for improving the performance of the system.

3.1   Search Engine for Translation

In most information retrieval applications the user provides a query aiming at recovering documents that are relevant to the query. The translation task can be seen as conceptually similar, in the sense that the user provides a source sentence to be translated (a query) aiming at obtaining a meaningful translation for it.

In our proposed approach to translating by means of information retrieval we construct two composed indexes, one in each language, in which pointers to each other are also included. This index construction is performed in three steps:

- Common translation collection: we collect the most commonly Chinese and Portuguese sentences and their respective translations from the translation service. This bilingual data collection is updated on a monthly basis according to the activity of the on-line registered users.

- Bilingual dictionary match: form the collected bilingual sentence pairs, a bilingual dictionary is used to identify Chinese and Portuguese term translations simultaneously occurring in the sentence pairs, which are replaced by entry codes in the dictionary. The entries of the used bilingual dictionary correspond with nouns and adjectives that are commonly observed in the translated pairs.
- A Chinese index in constructed by using the processed Chinese sentences and, in the same way, a Portuguese index is constructed by using the processed Portuguese sentences. The two indexes include pointers to each other so each Portuguese sentence points to its corresponding Chinese translation and each Chinese sentence points to its corresponding Portuguese translation.

These indexes are implemented by using the bag-of-words approach, for which the TF-IDF weighting scheme is used (Salton and Buckley 1988). For searching across the indexes, cosine similarity metric is used for ranking the retrieved outputs. Given a user input in the source language, the retrieval process is implemented in two steps:

- Dictionary match: the input sentence is evaluated for occurrences of terms from the bilingual dictionary. In case a term is detected, it is replaced by its corresponding entry code.
- Source search: two searches are performed over the source language index, the first one involves the original sentence provided by the user, and the second one involves the processed sentence (if terms have been found on it). The retrieved sentence with highest cosine similarity score is then selected.

Finally, the translation is constructed by using the corresponding sentence pair from the target language index:

- Sentence extraction: the target sentence corresponding to the selected source sentence is extracted from the target index if the obtained cosine similarity is high enough (current threshold value is 0.85).
- Sentence post edition: if the selected target sentence includes one or more dictionary entry codes on it, they are replaced by their corresponding dictionary forms before providing the final translation to the user.

Figure 2, illustrates the index construction, search, and translation generation processes used for the off-line translation system implementation.

3.2   Contextualized Translation Services

Finally, in this section we describe our contextualization strategy for improving the quality of the off-line translation service.

For providing the system with contextualization capabilities, each requested translation and its corresponding result from the online service are logged in the system along with the following types of metadata:

- User information: unique identification number for the user requesting the translation.

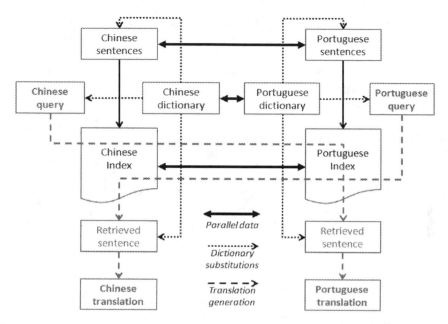

Fig. 2. Proposed approach for off-line translation by means of an information retrieval strategy over the collection of the most commonly requested translation pairs

- Location information: spatial coordinates as provided by the GPS service of the mobile device at the moment the translation was requested.
- Time information: time stamp for the specific hour and day at which the translation was requested.
- Semantic information: a semantic categorization of the specific topic the requested translation belongs to.

These four types of metadata are used to train a personalized predictive model able to estimate which are the most probable translations the current user might be requesting in the next 24 hours, based on the current context (user-location-time) and previous translation history.

This model is updated every time the system is using the online mode, and the corresponding translation indexes and dictionaries are refreshed based on the model predictions. In this way, when going off-line, a personalized and contextualized translation service is locally available for the user.

4   Conclusions and Future Work

In this work, we have described an Information Retrieval engine that is used to support our Chinese-Portuguese machine translation services when no internet connection is available. Our mobile translation app, which is deployed on a portable device, relies by default on a server-based machine translation service, which is not accessible when no internet connection is available. For providing translation support under this condition, we have developed a contextualized off-line search engine that allows the users to continue using the app.

As future work we plan to improve our off-line solution by incorporating predictive suggestions, so the system can suggest source sentences to the user by using partial inputs as queries for searching across the source index. We also want to improve the contextualization capabilities by including user dependent models for spatial and time localization.

Acknowledgements. This work is supported by the Seventh Framework Program of the European Commission through the International Outgoing Fellowship Marie Curie Action (IMTraP-2011-29951). The authors also want to thank the Institute for Infocomm Research for its support and permission to publish this research.

References

1. Brown, P.F., Della Pietra, S.A., Della Pietra, V.J., Mercer, R.L.: The mathematics of statistical machine translation: parameter estimation. Computational Linguistics 19(2), 263–311 (1993)
2. Centelles, J., Costa-Jussà, M.R., Banchs, R.E.: CHISPA on the GO A mobile Chinese-Spanish translation service for travelers in trouble. Accepted for Publication at the 14th Conference of the European Chapter of the Association for Computational Linguistics (EACL 2014) Demo Track (2014)
3. Koehn, P., Och, F.J., Marcu, D.: Statistical phrase-based translation. In: Proc. of HLT/NAACL 2003, pp. 127–133 (2003)
4. Koehn, P., Hoang, H., Birch, A., Callison-Burch, C., Federico, M., Bertoldi, N., Cowan, B., Shen, W., Moran, C., Zens, R., Dyer, C., Bojar, O., Constantin, A., Herbst, E.: Moses: Open source toolkit for statistical machine translation. In: Proceedings of the 45th Annual Meeting of the Association for Computational Linguistics (ACL 2007), Prague, Czech Republic, pp. 177–180 (June 2007)
5. Tseng, H., Chang, P., Andrew, G., Jurafsky, D., Manning, C.: A conditional random field word segmenter. In: Fourth SIGHAN Workshop on Chinese Language Processing (2005)
6. Salton, G., Buckley, C.: Term-weighting approaches in automatic text retrieval. Information Processing and Management 24(5), 513–523 (1988)

Cross Lingual Snippet Generation
Using Snippet Translation System

Pintu Lohar[1], Pinaki Bhaskar[1], Santanu Pal[2], and Sivaji Bandyopadhyay[1]

[1] Department of Computer Science and Engineering, Jadavpur University,
Kolkata - 700032, India
[2] Universität Des Saarlandes, Saarbrücken, Germany
{pintu.lohar,pinaki.bhaskar}@gmail.com,
santanu.pal@uni-saarland.de, sivaji_cse_ju@yahoo.com

Abstract. Multi Lingual Snippet Generation (MLSG) systems provide the users with snippets in multiple languages. But collecting and managing documents in multiple languages in an efficient way is a difficult task and thereby makes this process more complicated. Fortunately, this requirement can be fulfilled in another way by translating the snippets from one language to another with the help of Machine Translation (MT) systems. The resulting system is called Cross Lingual Snippet Generation (CLSG) system. This paper presents the development of a CLSG system by Snippet Translation when documents are available only in one language. We consider the English-Bengali language pair for snippet translation in one direction (English to Bengali). In this work, a major concentration is given towards translating snippets with simpler but excluding deeper MT concepts. In experimental results, an average BLEU score of 14.26 and NIST score of 4.93 are obtained.

Keywords: Snippet Translation, Cross Lingual Snippet Generation, Statistical Machine Translation, Snippet Generation.

1   Introduction

A snippet is defined as a small piece of something. In the case of Information Retrieval or any Search Engine, Snippet is a one or two line query-biased summary of the retrieved document. In multilingual lingual snippet generation, snippets are generated from a set of texts or sentences in multiple languages, which are present in a same document.

The consortia of Cross Lingual Information Access (CLIA) formed in the year of 2006 with 10 consortia members of IIT Bombay, IIT Kgp, IIIT Hyderabad, CADC Pune, CDAC Noida, Jadavpur University, AU-KBC, AU-CEG, ISI Kolkata and Utkal University. The objective is to develop a CLIA system, which can cross search in three different languages: One IL (Indian Language), Hindi and English. So in the CLIA system, if a query is given in any of the six Indian languages (e.g. Hindi, Marathi, Bengali, Punjabi, Tamil and Telugu) then the system will search for the documents in that specific Indian language and in Hindi as well as in English. In the CLIA

A. Gelbukh (Ed.): CICLing 2014, Part II, LNCS 8404, pp. 331–342, 2014.

system there are two cross lingual search available; one is IL-Hindi and another is IL-English. Hence we had to develop a Snippet Generation and a Snippet Translation module which can generate snippet from documents in any of these seven languages i.e. English and six Indian languages and then translate the snippets from English-to-IL and Hindi-IL. But the system can be trained for any language pair in any direction for translation using bilingual parallel corpora for the corresponding language pairs.

The present work is motivated by the fact that, many users might want to obtain snippets in the languages of their choice while the snippets are available only in one language. This can be done with the help of a snippet translation system as mentioned earlier. In this paper, a snippet translation system has been proposed based on baseline phrase-based SMT system (PB-SMT) [1].

The rest of this paper is organized as follows. Section 2 summarizes some related works. Section 3 discusses about the system architecture of the overall system. Section 4 describes Snippet Generation module followed by the Snippet Translation module in section 5. Section 6 deals with the experiments. Section 7 shows the evaluation. Finally, we conclude and outline future work in Section 8.

2   Related Work

Most of the research works related to this task or field is either on development of MT systems or summarization systems. Little and less number of research works have been done on snippet generation and translation. Currently, most successful summarization systems follow the extractive summarization framework. These systems first rank all the sentences in the original document set and then select the most salient sentences to compose summaries for a good coverage of the concepts. For the purpose of creating more concise and fluent summaries, some intensive post-processing approaches are also appended on the extracted sentences. E.g., redundancy removal [2] and sentence compression [3] approaches are used to make the summary more concise. Sentence re-ordering approaches [4] are used to make summary more fluent.

A lot of research work has been done in the domain of both query dependent and independent summarization, like MEAD [5], NeATS [6] and XDoX [7]. A document graph based query focused multi-document summarization system has been described by [8], [9] and [10]. In [11], they investigated into the utility of document summarization in the context of IR, more specifically in the application of so-called query-biased summaries. In [12], they explored the algorithms and data structures required as part of a search engine to allow efficient generation of query-biased snippets. They began by proposing and analyzing a document compression method that reduces snippet generation time by 58% over a baseline using the zlib compression library. A system, eXtract [13], addressed this important yet open problem. They identified that a good XML result snippet should be a self-contained meaningful information unit of a small size that effectively summarizes this query result and differentiates it from others, according to which users can quickly assess the relevance of the query result.

There is significant number of works done in Cross Lingual Information Retrieval (CLIR) and Cross Lingual Information Access (CLIA) as well. In [14], they focused on query translation, disambiguation of multiple translation candidates and query expansion with various combinations. A word alignment table can be learnt by an SMT system [15] trained on aligned parallel sentences, to map a query in source language into an equivalent query in the language of the target document collection. In [16], they explored how a combined probability model of term translation and retrieval can reduce the effect of translation ambiguity.

SMT systems have undergone many research works since the past few years. PB-SMT models [1] outperformed the word-based models. An alignment template approach for PB-SMT [17] allows for general many-to-many relations between words. A PB-SMT model that uses hierarchical phrases is also presented in [18]. Few related works in English-Bengali SMT systems are also noticeable. The PB-SMT system can be improved [19] by single tokenizing the MWEs on both sides of the parallel corpus. The addition of transliteration module [20] to handle out-of-vocabulary (OOV) words for low-density languages like Bengali is also suggested.

In the present work, the same system for snippet generation [21], [22] have been followed. We have used PB-SMT system for translating the generated English snippet to user's language i.e. query language (Bengali).

3   The System Architecture

3.1   CLIA System

In this section the overview of the system framework of the current CLIA system has been described. The CLIA system has been developed on the basic architecture of Nutch[1], which use the architecture of Lucene[2]. Nutch is an open source search engine, which supports only the monolingual search in English, etc. The architecture of Nutch has been used in CLIA. Various new or modified features of CLIA system have been added or modified into Nutch architecture. The main feature of CLIA is the cross lingual search, which needs the query translation and snippet translation.

Higher-level system architecture of CLIA system has been shown in the figure 1. The major modules in the output processing of the CLIA system are the Snippet Generation and Snippet Translation modules, which generate and translate the snippets of all the retrieved documents.

3.2   Snippet Translation Module

This is the final module of the system and the main focus of this work. The PB-SMT based Snippet translation system framework has been shown in the figure 2. One noticeable observation from the figure 2 is that in the preprocessing step, a snippet is split into n sentences and the baseline system generates m translated sentences, where $m = n \; or \; m \neq n$. Later is true when the system generates different number of output sentences than the actual number. Snippet translation system is discussed in section 5.

---

[1] http://nutch.apache.org/
[2] http://lucene.apache.org/

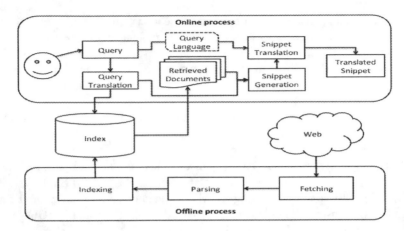

Fig. 1. System Architecture of CLIA

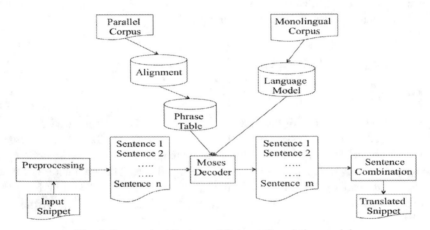

Fig. 2. System Architecture of Snippet Translation module

4   Snippet Generation

4.1   Key Term Extraction

Key Term Extraction module has three sub modules like Query Term extraction, Title Words Extraction and Meta Keywords Extraction. First, the Multiword Word Expressions (MWE) and Named Entities (NE) are identified and tagged in the given query. All the stop words are removed from the untagged query words. Then, the query is translated into the desired language between English or Hindi. Query Term Extraction module extracts all the query terms with their Boolean relations (AND or OR). Title words are also extracted and used as the keywords of the document in this system. If the meta keywords are available, all the meta keywords are extracted to use as more keywords.

4.2   Sentence Extraction

The document text is parsed and the parsed text is used to generate the snippet. This module will take the parsed text of the documents as input, filter the input parsed text and extract all the sentences from the parsed text. The text in the query language are identified and extracted from the document using the Unicode character list using Wikipedia [23]. Filtered parsed text has been parsed to identify and extract all sentences in the documents. As the sentence marker '.' (dot) is not only use as a sentence marker, it has other use also like point and in abbreviation like Mr., Prof., U.S.A. etc. A possible list of abbreviation has to create to minimize the ambiguity.

4.3   Top Sentence Identification

All the extracted sentences are now searched for the keywords i.e. query terms, title words and meta keywords. Extracted sentences are given some weight according to search and ranked on the basis of the calculated weight.

Weight Assigning. There are basic three components in the sentence weight like query term dependent score, title word dependent score and meta keyword dependent score. Term dependent scores (Q_s) of the sentence s are calculated using equation 1.

$$Q_s = \sum_{q=1}^{n_q} F_q \left(20 + \left(n_q - q + 1\right) \left(\sum_p \left(1 - \frac{f_p^q - 1}{N_s} \right) \right) \times t \right) \tag{1}$$

where, q is the no. of the term, n_q is the total no. of term, f_p^q is the position of the word which was matched with the term q in the sentence s, N_s is the total no. of words in sentence s, t is the priority of the term and

$$F_q = \begin{array}{l} 0; \; if \; term \, q \, is \, not \, found \\ 1; \; if \; term \, q \, is \, found \end{array} \tag{2}$$

At the end of the equation 1, the calculated term dependent score is multiplied by t to set the priority as equation 3.

$$t = \begin{array}{l} 3; \;\; if \; term \; is \; query \; term \\ 2; \;\; if \; term \; is \; title \\ 1; \;\; if \; ter \; is \; meta \; keyword \end{array} \tag{3}$$

So, the final weight (W_s) of each sentence (s) is calculated by simply adding all the three scores like mentioned in the equation 4.

$$W_s = Q_s + T_s + K_s \tag{4}$$

Sentence Ranking. Sentences are sorted in descending order of their weight. Now, top three ranked sentences are taken for the Snippet Generation. If all these three sentences are small enough to fit into the maximum length of a snippet without trimming, then after this module the system goes directly to the Snippet Generation module to generate the snippet. Otherwise it goes through the Snippet Unit Selection module.

4.4   Snippet Unit Selection

Snippet unit is basically a phrase or clause of a sentence. The snippet units are extracted from top three sentences in this module using the syntactic information available in the sentences. The sentences are split into snippet units according to brackets, semi colon (';'), coma (',') etc. Then, same Weight assigning module is used to calculate the weights of snippet units too. The snippet units are sorted in descending order of their weight in the same way of Sentence Ranking module.

4.5   Snippet Generation

As [8] using equation 5, the module selects the ranked snippet units subject to maximum length of the snippet has been reached.

$$\sum_i l_i S_i < L \tag{5}$$

where, l_i is the length of snippet unit i, S_i is a binary variable representing the selection of snippet unit i and L (=100 words) is the maximum length of the snippet. Now, the selected snippet units are reordered according to their order of appearance in the text and then returned as a generated snippet.

5   Snippet Translation

5.1   Phrase-Based SMT

Phrase-Based SMT systems are increasingly popular SMT systems. The motivation behind using PB-SMT is that the words may not be the best candidates for smallest unit of translation. Sometimes one word in source language translates into multiple words in target language or vice-versa. Word-based models often break down in these cases. A bilingually translated sentence aligned parallel corpus is a prerequisite for PB-SMT systems. The more the data, the better the output would be. The basic concept is to apply statistical analysis to the training data to eventually generate the phrase table, which provides the key elements (phrases) to build the output translation while translating the text in source language to the text in target language. A reordering table may also be generated to deal with the difference in word order between the source and the target language. The phrase table contains a large list of candidate phrases for potential translations. As these phrases are statistically generated, they need not be linguistically motivated phrases. These are simply sequence of words found in source and target sentences in the training data. The source sentence phrases may overlap with each other and they can also have several translations. In this case a subset of the entries in the table should be selected to make a translation of the sentence. The members of the selected subset must then be arranged in a specific order to produce a decent translated output. For this purpose the structure of the target language is statistically analyzed using language models. Several language model toolkits such as SRILM[3], IRSTLM[4] etc. are used in current SMT systems.

---

[3] http://www.speech.sri.com/projects/srilm/

According to [24], Phrase-Based Statistical MT (PB-SMT) Model can be described by the following equation:

$$e_{best} = argmax_e \prod_{i=1}^{I} \varphi(\bar{f_i}|\bar{e_i})d(start_i - end_{i-1} - 1) \prod_{i=1}^{|e|} p_{LM}(e_i|e_1 \dots e_{i-1}) \quad (6)$$

Here, the source and target sentences are divided into I phrases and each source phrase $\bar{f_i}$ is translated into a target phrase $\bar{e_i}$. Since we mathematically inverted the translation direction in the noisy channel, the phrase translation probability $\varphi(\bar{f_i}|\bar{e_i})$ is a translation from target to source. $start_i$ is the position of the first word of the source input phrase that translate to the ith target phrase and end_i is the position of the last word of that source phrase. Reordering distance is computed as $start_i - end_{i-1} - 1$. A language model p_{LM}, given the history of certain number of previous words is also integrated in the model to test the fluency of the target output. Finally e_{best} is generated that is the highest probable candidate among all possible combinations. Hence, the PB-SMT model ensures that, (i) the source phrases match the target words (.); (ii) phrases are reordered appropriately (d); and (iii) the output is fluent in target language (p_{LM}).

5.2 Snippet Translation Using PB-SMT

Snippets generally do not contain grammatically well-formed sentences. Most of the times, they are comprised of only informative phrases. These phrases may not be grammatically linked with each other. PB-SMT systems are well suited for translating sentences but they can also be fruitful when applied to translate the snippets.

Let us consider two example snippets from the outputs of the Snippet Generation system:

1) *Bangalore is the third busiest airport in India, with over 10 million passengers a ... Airlines will also provide wheelchair assistance... Bangalore Airport Hotels:...*
2) *New Delhi Hotels - The Lalit hotels one of the best five star luxury hotels in Delhi... Delhi hotels, the Lalit offers world class services in 5 star hotels in Delhi and...*
3) *Programme for Pushkar fair 2012, Pushkar fair 2012 programme details. Dates, timings of competitions, attractions, activities and more. Ceremonies and...*

In the above examples, the segments of each snippet are separated by either multiple dots (...) or a sentence boundary marker. In snippet-1, the first segment is an almost perfect sentence except its last part. The second one is a perfect sentence and the third one is just a phrase. On the other hand, in snippet-2, the first segment starts with a phrase and its remaining part does not contain any verb. The second one is also an almost correct sentence but it lacks continuation. Finally, snippet-3 contains only phrases. For these reasons, translating snippets is more difficult than translating normal sentences. Though the snippets may not be the container of well-formed sentences, most of the times their meaning can be perceived by the readers.

---

[4] http://sourceforge.net/projects/irstlm/

6   Experiments

6.1   Tools and Resources

As the training set we used sentence-aligned English-Bengali parallel corpus containing 36,668 parallel sentences, a bilingual parallel dictionary of 53,857 words, a bilingual parallel corpus of 55,548 Named-Entities from the travel and tourism domain. The corpus has been collected from the consortium-mode project "Development of Cross Lingual Information Access (CLIA) System[5]". The Stanford Parser[6] has been used for parsing the source English side of the parallel corpus.

In the present work; GIZA++ word alignment[7], SRILM language modeling toolkit and the Moses toolkit[8] have been used as the baseline system. The bengali sentences in the parallel corpus are used as the monolingual corpus to build the language models. Table 1 shows the parallel corpus statistics for English-Bengali language pair.

Table 1. Corpus Statistics for English-Bengali language pair

Parallel Sentences	Parallel Named-Entities	Parallel Dictionary Words	Total Parallel Entries
36,668	55,548	53,857	146,073

6.2   Experimental Setup

Preprocessing. English follows the Subject-Verb-Object (SVO) word order, while Bengali generally follows Subject-Object-Verb (SOV) word order. This major difference leads to higher reordering costs for translation between these two languages. Hence, we reorder the source side of the training data (English sentences) in SOV order using Stanford Parser. The generated snippets are preprocessed as: (a) First, cleaned, tokenized and lowercased and the multiple dots (...) are considered as a segment boundary marker.

Thus, after preprocessing, snippet-1 in section 5.2 can be split into following three segments. (i) *bangalore is the third busiest airport in india , with over 10 million passengers a* (ii) *airlines will also provide wheelchair assistance .and* (iii) *bangalore airport hotels :*. These segments are combined together to form the output.

Translation Model. We trained the system with 3-gram, 4-gram and 5-gram language models and a maximum phrase length of 7. The alignment heuristic is grow-diag-final. The inclusion of NEs and dictionary words in the training set maximizes

---

[5] The CLIA project is funded by the Department of Electronics and Information Technology (DEITY), Ministry of Communications and Information Technology (MCIT), Government of India.
[6] http://nlp.stanford.edu/software/lex-parser.shtml
[7] https://code.google.com/p/giza-pp/
[8] https://github.com/moses-smt/mosesdecoder

the availability of training data. When a word is not translated, we consider it as an NE and forward as an input to the transliteration module.

Transliteration Model. The transliteration model is built using the technique described by [25]. According to this technique, a word, represented as a sequence of letters of the source language s = $s_1^J = s_1 \ldots s_j \ldots s_J$ needs to be transcribed as a sequence of letters in the target language, represented as t = $t_1^I = t_1 \ldots t_i \ldots t_I$. The problem of finding the best target language letter sequence among the transliterated candidates can be represented as:

$$t_{best} = arg \max_t \{Pr(t|s)\} \tag{7}$$

Reformulating the above equation using Bayes Rule:

$$t_{best} = arg \max_t p(t|s)p(s) \tag{8}$$

We built the transliteration model using 3-gram language model, a maximum phrase length of 3.

7 Evaluation

We manually developed two reference sets with 129 translated snippets per reference set. Since the evaluation part of the whole system is under development, so far we have developed these reference translations for English-Bengali language pair only. This is a challanging task to create reference translations because snippets are not good reprentatives of meaning preservation. We tried to make these reference sets as different as possible but at the same time as close to each other in meaning. Depending upon the completeness of meaning preservation in snippets we developed the corresponding complete/incomplete translations. Table-2 illustrates this with 2 examples alongwith the outputs of the baseline system with 3-gram language model. As evaluation metrics, two widely used metrics BLUE [26] and NIST [27] have been used. Table-3 shows the evaluation results. Note that, both the BLEU and NIST scores increase as the order of language model increases. On an average, we achieved BLEU and NIST score of 14.263 and 4.9314 respectively. Even higher order n-gram models are yet to be explored.

Finally, analysing the outputs in Table 2 following observations can be made:

1. Due to few translation errors in the system, few words are: (a)missing translation and (b)translated but wrongly placed. For example, in snippet-2 the word "Mandu" is not translated in bengali. The word "Mandavgarh" is in the second sentence in input but translated as "মান্দাভগার্ই" in the first sentence of the output. Moreover, the translated word "পরজীবী" in output-1 is an unwanted word. It has no word in input to be translated from. These errors reduces the evaluation score.
2. The mismatch between multiple dots(...) in reference sets and bengali sentence boundary marker '।' ("daadi") in outputs reduces the scores significantly.

3. In many cases few words in translated snippets are not same but equivalent or similar in meaning with the words in reference sets. Moreover, few words are inflected and does not match to the words in any of the reference sets. For example, in snippet-1, the word 'বিশ্ববিদ্যালয়ের', present in both the reference sets is an inflected version of the translated word 'বিশ্ববিদ্যালয়' in the output. This kind of dissimilarity also results in reduced score.

Table 2. Translated references and outputs

SL No.	1	2
English Snippet	The excavated ruins of Nalanda University in Bihar have a unique place in history. Being the first international residential university, which ...	Mandu is a historical city situated near Indore city of Madhya Pradesh. Know more about Mandavgarh in India.
Translated Reference-1	ইতিহাসে বিহারের নালন্দা বিশ্ববিদ্যালয়ের নিখাত ধ্বংসাবশেষের এক অনন্য স্থান আছে । প্রথম অন্তর্জাতীয় আবাসিক বিশ্ববিদ্যালয় হিসেবে , যা ...	মান্ডু হল মধ্য প্রদেশে ইন্দোর শহরের কাছাকাছি অবস্থিত একটি ঐতিহাসিক শহর । ভারতে মান্দাবগড় সম্পর্কে আরও জানুন ।
Translated Reference-2	বিহারের নালন্দা বিশ্ববিদ্যালয়ের নিখাত ধ্বংসাবশেষের একটি অনন্য স্থান রয়েছে ইতিহাসে । প্রথম আর্ন্তজাতিক আবাসিক বিশ্ববিদ্যালয় হিসেবে , যেটা ...	মধ্য প্রদেশে ইন্দোর শহরের কাছাকাছি অবস্থিত মান্ডু একটি ঐতিহাসিক শহর । ভারতে মান্দাবগড় সম্পর্কে আরও জানুন ।
Output Snippets	বিহারের নালন্দা বিশ্ববিদ্যালয় অবস্থিত এই থননকার্যে পাওয়া ইতিহাসে এক অনন্য অনুবৃতি প্রথম অন্তর্জাতীয় আবাসিক বিশ্ববিদ্যালয় পরজীবী আছে , যা ।	মান্দাভগাই একটি ঐতিহাসিক শহর এগোনো । ইন্দোর শহর সম্পর্কে আরও জানতে । হল ভারতে অবস্থিত ।

Table 3. Evaluation Results

Language Model	BLEU	NIST	Average BLEU	Average NIST
3-gram	14.23	4.7331		
4-gram	14.24	5.0267	**14.263**	**4.9314**
5-gram	**14.32**	**5.0345**		

8   Conclusions and Future Work

In this paper we showed how a cross lingual snippet generation system could be implemented with the help of snippet translation system. In the presence of documents in multiple languages, MLSG systems are the best way for snippet generation without sacrificing meaning preservation that is often caused by translation errors in CLSG systems using snippet translation. But if the documents are present in only one language we can translate the generated snippets into the language of available choices provided that a parallely-aligned translated corpus must be present for the source-target language pair. We used a baseline PB-SMT system without further exploring deeper aspects of SMT systems. However, baseline PB-SMT systems do not produce satisfactory outputs for low-resource languages thereby reduces the meaning relevance. The performance can be further improved with availability of larger corpora.

In future, we would like to investigate on extracting parallel texts from bilingual comparable corpora to increase the size of parallel corpus. Additionally, both the automatic and manual evaluations for all Indian languages under the CLIA project need to be done to validate the performance of the whole system. We would also like to explore the application of advanced SMT techniques for further improvement in our system. Moreover, another benefit can be gained by merging the results from MLSG and CLSG systems when much of the documents are available in English, but less number of documents is available in under resourced languages.

Acknowledgements. The work has been carried out with support from Department of Electronics and Information Technology (DEITY), Ministry of Communications and Information Technology (MCIT), Government of India funded research project "Development of Cross Lingual Information Access (CLIA) System", Phase II.

References

1. Koehn, P., Och, F.J., Marcu, D.: Statistical phrase-based translation. In: Conference of the North American Chapter of the Association for Computational Linguistics on Human Language Technology, vol. 1, pp. 48–54 (2003)
2. Carbonell, J., Goldstein, J.: The Use of MMR, Diversity-based Reranking for Reordering Documents and Producing Summaries. In: ACM SIGIR, pp. 335–336 (1998)
3. Knight, K., Marcu, D.: Statistics-based summarization - step one: Sentence compression. In: The American Association for Artificial Intelligence Conference (AAAI), pp. 703–710 (2000)
4. Barzilay, R., Elhadad, N., McKeown, K.R.: Inferring strategies for sentence ordering in multidocument news summarization. J. Artificial Intelligence Research. 17, 35–55 (2002)
5. Radev, D.R., Jing, H., Styś, M., Tam, D.: Centroid - based summarization of multiple documents. J. Information Processing and Management. 40, 919–938 (2004)
6. Lin, C.Y., Hovy, E.H.: From Single to Multidocument Summarization: A Prototype System and its Evaluation. In: ACL, pp. 457–464 (2002)
7. Hardy, H., Shimizu, N., Strzalkowski, T., Ting, L., Wise, G.B., Zhang, X.: Cross-document summarization by concept classification. In: SIGIR, pp. 65–69 (2002)
8. Bhaskar, P., Bandyopadhyay, S.: A Query Focused Multi Document Automatic Summarization. In: The 24th Pacific Asia Conference on Language, Information and Computation (PACLIC 24). Tohoku University, Sendai (2010)
9. Bhaskar, P., Bandyopadhyay, S.: A Query Focused Automatic Multi Document Summarizer. In: The International Conference on Natural Language Processing (ICON), IIT, Kharagpur, India (2010)
10. Bhaskar, P.: Query Focused Language Independent Multi-document Summarization and Information Retrieval for English and Bengali. Jian, A. (ed.). LAMBERT Academic Publishing, Saarbrücken (2013) ISBN 978-3-8484-0089-8
11. Tombros, A., Sanderson, M.: Advantages of Query Biased Summaries in Information Retrieval. In: SIGIR (1998)
12. Turpin, A., Tsegay, Y., Hawking, D., Williams, H.E.: Fast Generation of Result Snippets in Web Search. In: SIGIR (2007)

13. Huang, Y., Liu, Z., Chen, Y.: Query Biased Snippet Generation in XML Search. In: SIGMOD, Vancouver, BC, Canada (2008)
14. Reddy, M.V., Hanumanthappa, M., Kumar, M.: Cross Lingual Information Retrieval Using Search Engine and Data Mining. ACEEE International Journal on Information Technology (2011)
15. Jagarlamudi, J., Kumaran, A.: Cross-lingual Information Retrieval for Indian Languages. In: Peters, C., Jijkoun, V., Mandl, T., Müller, H., Oard, D.W., Peñas, A., Petras, V., Santos, D. (eds.) CLEF 2007. LNCS, vol. 5152, pp. 80–87. Springer, Heidelberg (2008)
16. Xu, J., Weischedel, R.: Cross-lingual information retrieval using hidden Markov models. In: Joint SIGDAT Conference on Empirical Methods in Natural Language Processing and Very Large Corpora: Held in Conjunction with the 38th Annual Meeting of the Association for Computational Linguistics, vol. 13 (2000)
17. Och, F.J., Ney, H.: The Alignment Template Approach to Statistical Machine Translation. In: ACL (2004)
18. Chiang, D.: A Hierarchical Phrase-Based Model for Statistical Machine Translation. In: 43rd Annual Meeting on Association for Computational Linguistics (2005)
19. Pal, S., Naskar, S.K., Bandyopadhyay, S.: MWE Alignment in Phrase Based Statistical Machine Translation. In: The XIV Machine Translation Summit, pp. 61–68 (2013)
20. Islam, M.Z., Tiedemann, J., Eisele, A.: English to Bangla Phrase – Based Machine Translation. In: The 14th Annual Conference of The European Association for Machine Translation, Saint-Raphaël, France, pp. 27–28 (2010)
21. Bhaskar, P., Bandyopadhyay, S.: Cross Lingual Query Dependent Snippet Generation. International Journal of Computer Science and Information Technologies (IJCSIT) 3(4), 4603–4609 (2012) ISSN: 0975-9646
22. Bhaskar, P., Bandyopadhyay, S.: Language Independent Query Focused Snippet Generation. In: Catarci, T., Forner, P., Hiemstra, D., Peñas, A., Santucci, G. (eds.) CLEF 2012. LNCS, vol. 7488, pp. 138–140. Springer, Heidelberg (2012)
23. List of Unicode characters on Wikipedia, http://en.wikipedia.org/wiki/List_of_Unicode_characters
24. Koehn, P.: Statistical machine translation. Cambridge University Press (2010)
25. Rama, T., Gali, K.: Modeling machine transliteration as a phrase based statistical machine translation problem. In: Named Entities Workshop: Shared Task on Transliteration, pp. 124–127 (2009)
26. Papineni, K., Roukos, S., Ward, T., Zhu, W.J.: BLEU: a method for automatic evaluation of machine translation. In: ACL (2002)
27. Doddington, G.: Automatic evaluation of machine translation quality using n-gram cooccurrence statistics. In: Human Language Technology Conference (HLT), San Diego, CA, pp. 128–132 (2002)

A Novel Machine Translation Method
for Learning Chinese as a Foreign Language

Tiansi Dong and Armin B. Cremers

Bonn-Aachen International Center for Information Technology
Dahlmannstraße 2, 53113 Bonn, Germany
{dongt,abc}@bit.uni-bonn.de

Abstract. It is not easy for western people to learn Chinese. Native
German speakers find it difficult to understand how Chinese sentences
convey meanings without using cases. Statistical machine translation
tools may deliver correct German-Chinese translations, but would not
explain educational matters. This article reviews some interdisciplinary
research on bilingualism, and expounds on how translation is carried out
through *cross-linguistic cue switching* processes. Machine translation approaches are revisited from the perspective of cue switching concluding
that: the word order cue is explicitly simulated in all machine translation approaches, and the case cue being implicitly simulated in statistical
machine translation approaches can be explicitly simulated in rule-based
and example-based machine translation approaches. A convergent result
of machine translation research is to advocate an explicit deep-linguistic
representation. Here, a novel machine translation method is motivated
by blending existing machine translation methods from the viewpoint of
cue-switching, and is firstly aimed as an educational tool. This approach
takes a limited amount of German-Chinese translations in textbooks
as examples, whose cues can be manually obtained, and for which we
have developed MultiNet-like deep linguistic representations and cross-
linguistic cue-switching processes. Based on this corpus, our present tool
is aimed at helping native German speakers to learn Chinese more efficiently, and shall later be expanded to a more comprehensive machine
translation system.

1 Introduction

More and more people are interested in learning Chinese, and more and more
universities provide courses for learning Chinese as a foreign language. Besides
the picture-like characters, western people wonder how complex meanings can
be conveyed by short Chinese sentences. It may indeed be hard even for language
specialists to correctly identify structures of Chinese sentences. Thus, it is little
surprising that, e.g., most German students find it difficult to translate German
sentences into Chinese. Existing machine translation softwares provide limited
help. For example, while Google translation software provides German to Chinese
translation, students cannot learn much from the statistical approach to their

A. Gelbukh (Ed.): CICLing 2014, Part II, LNCS 8404, pp. 343–354, 2014.

translation. They desire a computer-aided learning tool, which can show step by step how a German sentence should be transformed into a Chinese sentence.

It would seem easier to learn a second language which has a similar structure with the mother language. For example, native Spanish speakers are at an advantage when learning Portuguese, whereas for German speakers, it would be much easier to learn English than Chinese. The explanation from psychology is that each language uses specific cues to connect linguistic forms and meanings. It needs more effort to learn a new language with a different set of cues. German language mainly uses the case cue, while Chinese language has no case cue at all. It is a demanding task even for native Chinese teachers to systematically explain how meanings are carried by a sequence of Chinese characters. We propose a computational approach to such systematical explanation: Each explanation is formally represented as a piece of linguistic knowledge and some operations; all these representations are integrated and implemented as a software for language learning. Cues shall be explicitly represented in the software, so that it would be suitable for learning.

The rest of the article is structured as follow: Section 2 reviews the aggregated language model in psychology and the mental difference between early-learned and late-learned L2 bilingual speakers; section 3 reviews research work on cue in psychology and revisited machine translation approaches from the perspective of cue-switching; section 4 proposes a machine translation framework by integrating the cue-switching perspective and existing machine translation approaches; section 5 outlines a software tool for learning and teaching Chinese as a foreign language at Bonn University; section 6 summarizes the article and lists some on-going work.

2   The Aggregated Language Model

Do bilinguals have two separate language systems in mind, each supporting one language, or one system for both languages? Sufficient researches in psycholinguistics support an aggregated language model, that is, later learned languages are superimposed on the earlier learned language. This idea can be traced back to Freud [16]. For example, researchers found that lexemes in different languages are simultaneously activated providing non-selective lexical access in both languages, cf. e.g., [11], [10]. This explains why bilinguals often mix lexemes in different languages. The aggregation can reach such a degree that some Spanish-English bilinguals may pronounce neither in Spanish nor in English, as observed in [39]. A general agreement in psycholinguistics is on the separation between meanings and forms. Meanings (concepts) are shared by both languages, cf. e.g.,[20], [8], [7], [6], while forms are specific to each language, e.g., [20],[33], [17]. A hierarchical model of meanings and forms for lexeme acquisition is illustrated in Figure 1. For non-proficient late-learned L2 speakers, there will be no direct link between L2 and C, and for bilinguals who learned two languages from birth, there will be no direct link between L1 and L2, i.e., [30], [22].

For the perspective of human memory system, mother tongues are used as skills by the *procedural* memory system, while late-learned second languages are

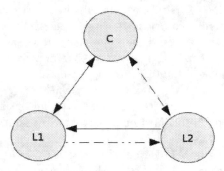

Fig. 1. The *Revised Hierarchical Model* by Kroll and Stewart (1994). 'C' is the conceptual representation; 'L1' is the lexical representation of the mother language; 'L2' is the lexical representation of the second language.

used as knowledge by the *declarative* memory system, e.g., [1], [34], [42], [35]. The declarative memory system has a rapid learning ability and can acquire new knowledge fast, while the procedural memory system acquires knowledge slowly and gradually, e.g., [29]. Late-learned-L2 bilinguals can master their second language at a high-level of proficiency, and result in a neuro-cognitive pattern similar to that of mother tongue speakers, [36, p.153], for the performance of a skill is gradually improved through practicing.

3   Cue and Its Usage in Machine Translation

The aggregated linguistic model, as well as the memory systems, can be used to explain learning any two languages. However, why is it harder for German students to learn Chinese than English? This functional difference can be explained in terms of *cue*.

3.1   Translation as Cross-linguistic Cue Switching

A cue is a piece of information which can be used to determine the relationship between meanings and forms, [25, 24]. Each language uses a particular set of cues. German relies on both agreement cues and animacy cues, English relies overwhelmingly on word order, e.g., [25], Chinese relies on cues of passive marker 被/by, animacy, word order, object marker 把/hold, indefinite marker ——/one, cf.e.g., [24]. When a sentence in language A is read by a bilingual speaker, cues of language A are used to transform the sentence into its meaning; then cues of language B are used to transform the meaning into a sentence of language B.

Cross-linguistic cue switching may not be trivial. The Guugu Yimithirr people, living in North Queensland, Australia, only use absolute orientations, such as "I left it on the southern edge of the western table in your house", cf. [23, p.100]. English speakers would use relative orientations in terms of "left", "right", "front", and "back". Translation of spatial descriptions in the two languages demands a total perspective switching in orientation reference framework. The problem is that the preferred perspective and the granularity differ, as noted in [31, p. 430], and is sometimes even incompatible without extra information. The precondition for cross-linguistic cue switching is that their semantic representations must be compatible.

Researchers tried to find universal acquisition order of linguistic cues. For example, does it hold that the acquisition of agent-object relations (animacy cue) is before the acquisition of word order cues? Does it hold that the acquisition of word order cues come before the acquisition of grammatical morphology? Researchers found that SVO word order was the first cue for sentence comprehension for English children, cf. [3], and that Hungarian and Polish children acquired case inflections in sentence comprehension and production before word orders, cf. [26], [41]. These findings deny both possible universal cue acquisition orders.

The cue-switching from language A to language B is a different functional path from the cue-switching from language B to language A. This can be inferred from some bilingual patients. Lady E.M. in [2] spoke Venetian as her mother language (L1) and Italian as her second language (L2). After a stroke, she was not able to speak L1, but could surprisingly speak L2 proficiently. Her language ability was slowly recovered in a surprising manner: She could speak L2 in proficiency; word translation from L1 into L2 was more accurate and quick than vice versa. Lady A.D. in [28] spoke French as L1 and Arabic as L2. Four days after a traffic accident, she could only speak L2. Her language abilities in L1 and L2 were recovered slowly and unstable – sometimes she spoke L1 better than L2, sometimes vice versa. The gentleman in [28] spoke both French (L1) and English (L2) fluently. After an operation, his English was firstly recovered. On the next week, he could speak French, but failed to speak English. On the fourth week, his L1 and L2 were recovered.

3.2   Cues in Statistical Machine Translation Techniques

The statistical machine translation (SMT) is the most popular and the most widely-studied method. The idea dates back to Weaver[40](1949), and was reintroduced in early 90s, cf. [5]. The central idea can be stated as follows: Given a sentence f in French, we look for the sentence e in English which maximizes $\mathbf{P}(e|f)$, written as $argmax_e\, \mathbf{P}(e|f)$. That is, the sentence e in English is *most likely* the translation of the sentence f in French. The term '*most likely*' can be approached by the Bayes rule: $argmax_e\, \mathbf{P}(e|f) = argmax_e\, \mathbf{P}(e) * \mathbf{P}(f|e)$. $\mathbf{P}(e)$ is the 'Language Model'. $\mathbf{P}(f|e)$ is the 'Translation Model'.

$\mathbf{P}(e)$ is the probability that e occurs in English. The bigram approach can compute $\mathbf{P}(e)$ through $\mathbf{b}(y|x)$, which is the probability that word y follows word x. $\mathbf{b}(y|x)=number\text{-}of\text{-}occurrences(xy)/number\text{-}of\text{-}occurrences(x)$, eg. [21]. $\mathbf{b}(am|I)$ is the probability that am follows I; $\mathbf{b}(is|I)$ is the probability that is follows I. Given a qualified corpus, $\mathbf{b}(am|I)$ should be greater than $\mathbf{b}(is|I)$. In this way, we can say that the agreement cue and the word order cue are statistically encoded. Using trigram $\mathbf{b}(z|xy)$ or N-gram $\mathbf{b}(w_n|w_1 \ldots w_{n-1})$, more cues can be encoded. Unfortunately, within the statistical language model it is not clear which cue supports a high probability. In the translation model, $\mathbf{P}(f|e)$, transitions of word order cues between two languages are explicitly modeled through several probabilities, such as alignment probability, position probability, distortion probability.

3.3   Cues in Example-Based Machine Translation Techniques

Example-based techniques decompose the input into several phrases with the assumption that each phrase can be directly translated into phrases in the target language by analogy, and the output in the target language is composed using these translated fragments, [27], [4].

This technique was inspired by a naive observation of the second language acquisition of human adults, and views translation as a process of analogy, as described in [27, pp.173]: "Let us reflect about the mechanism of human translation of elementary sentences at the beginning of foreign language learning. A student memorizes the elementary English sentences with the corresponding Japanese sentences. The first stage is completely a drill of memorizing lots of similar sentences and words in English, and the corresponding Japanese. Here we have no translation theory at all to give to the student. He has to get the translation mechanism through his own instinct". The original technique did not represent cue information. However, it is quite easy to add cue representation into example-based techniques – just thinking of a similar scenario: "at the beginning of foreign language learning, students not only memorize the elementary English sentences with the corresponding Japanese sentences, but also know two different ways in expressing meanings in the two languages...".

3.4   Cues in Rule-Based Machine Translation Techniques

The classical approach to machine translation is the rule-based approach. The main principle of the approach is to find a structural link between the input in the source language and the output in the target language, while preserving the meaning. Structural links normally refer to relations between grammatical rules. It is easy for grammatical rules to encode word order cues, case cues, even anamicy cues. The problem of this approach is that rules automatically carry word order cues, and for languages which do not use word order cues, this becomes an unnecessary constraint.

3.5   Cues in Inter-lingual Machine Translation Techniques

The *inter-lingual* translation techniques use a canonical representation independent of the source and the target languages, which somehow serves as the meaning of the source sentence. Translation is carried out by transferring the source language into the canonical representation, and further into the target language. It is natural to use cues in this approach. The critique of this approach, however, is that such canonical meaning representation does not exist even for formal languages, [43].

4   Blending Cue-Switching Translation with Existing Techniques

The ultimate quality of natural language software is simply evaluated by the quality of meaning production. In order to improve the translation reliability, researchers have attempted to combine different approaches, for instance, an example-based translation approach with a rule-based translation approach, e.g., [37, 32], or a rule-based translation approach with a statistical translation approach, e.g., [9], or adding formal domain knowledge into translation system, e.g., [15]. In one of the largest research projects on NLP in Germany, Verbmobil, cf. [38], researchers tried almost all possible hybrid methods and also developed new methods within established techniques. However, the translation quality is still not as good as expected, cf. [37]. Each of the existing machine translation techniques encodes or can encode, explicitly or implicitly, linguistic cues. Simple combination might not help much, in part because cues are language specific and often implicitly modeled. Here, we opt for a hybrid method by blending these techniques from the perspective of cue-switching.

The scenario for example-based machine translation in [27] is that students start learning a new language by memorizing elementary new phrases, sentences, and the corresponding mother language expression, without knowing the ways that meanings are expressed in the new language. Our scenario is that students also memorize how meanings are conveyed in both languages. That is, cues are taught explicitly with elementary examples. For Chinese students to learn German, they will get to know how meanings are expressed by different cases. For German students to learn Chinese, they will remember how meanings are expressed by word orders or other devices. In computational simulation, we use MultiNet-like meaning representation formalism for elementary German and Chinese phrases and sentences, [18], [13], and explicitly represent cues between these elementary forms and the corresponding meaning representations. Thus, we have a primitive translation system by blending example-based translation technique and cue-switching. As we use explicit meaning representation for both languages, we also blend example-based translation with inter-lingual translation. The meaning representation of the sentence *Ich fällte einen Baum mit einer Axt*(I cut down a tree with an axe) is illustrated in Figure 2.

We promote this elementary approach by adding rules into the system, so that translation can be carried out for new inputs. In rule-based translation approaches, rules are often used to capture word order information among parts of speeches. However, rules can also represent the animacy cue, categorical relations, and also case cues. For example, to improve the example in Figure 2, we can add a rule for animacy cue: the actor of *fällen* is of animacy; we add a rule for case cue: the affected role of *fällen* is accusative, the associated object of the affected role is dative. Instead of expressing these rules explicitly in logical forms, we integrate them implicitly into meaning representations, supported by a knowledge-base of ontologies. Three ontological relations are: *is_a*, *generalized_by*, and *has_feature_of*. For example, "I *is_a* PERSON", "PERSON *has_feature_of* ANIMACY". Roles of *fällen* are categorized by ontologies. For example, its actor

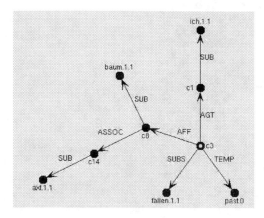

Fig. 2. Meaning representation of the sentence *Ich fällte einen Baum mit einer Axt* using the MultiNet formalism

role is categorized by ANIMACY. Only ontologies within this category can be the actor. For the German language which conveys meanings through animacy and case cues, there would be little traditional rules on word orders. When a long German sentence is translated into a very short Chinese sentence, their MultiNet-like meaning representations are not the same. The meaning representation of Chinese is normally located at a more abstract layer. In this case, we introduce rules for transitions between these meaning representations. The traditional rule-based translation approach is thus blended into the primitive system, and forms an extended system.

The elementary translation system can be pictured as the first year student learning a new language, who learns lots of simple expressions and understands different ways of meaning expressions in the new language. The extended system can be viewed as the second year student, who learns rules and structures of the language, and can analyze new sentences. A third year student can master the new language through a large amount of readings. His language proficiency improves along with the world knowledge thus acquired. Statistical approaches can be further blended with the extended system, to augment the knowledge-base and semantic parsing. There is some work in the literature, for example, the MultiNet representation can be statistically acquired, [19], currently with the correctness of around 70% for sentences of German Wikipedia.

5   A Computer-Aided Tool for German Speakers to Learn Chinese

To help learning and teaching Chinese as a second language, we are developing a computer-aided tool for the Department of Sinology, University of Bonn. Chinese is notorious for lacking of inflection information. This introduces difficulties not only to German students, but also to native Chinese teachers. We first developed

Fig. 3. A snapshot of the cue input interface. "Mir" is singular and the dative form of "ICH"(I), which is "PERSON". The receptor role of "GEFALLEN"(like) should be "PERSON" in the dative form. If its object role and its receptor role are filled, a complete meaning is delivered.

a software to collect all paired German-Chinese words, phrases, and sentences which either appear in their textbooks, or are taught in classes. This sets up the knowledge-base to compare translation performances between students and our software system. Based on these pair corpora, we developed MultiNet-like meaning representations and cues to acquire these meaning representations. We developed software tools for manually inputting cues carried by words, phrases and sentences, as illustrated in Figure 3.

Meaning representations of German expressions are automatically acquired using this cue knowledge. For the above example: "Das Auto gefällt mir gut" (I like the car very much), our algorithm first identifies basic meanings. For simplicity, we use "AUTO", "GEFALLEN", "ICH", "GUT" as basic meaings for "auto", "gefallen", "mir", and "gut", respectively. The algorithm then searches meaning frames among these basic meanings. Here, "GEFALLEN" is the meaning frame, which needs an object role and a receptor role to complete a meaning. According to the input knowledge, the receptor role shall be in the category of "PERSON" and in the dative form. The algorithm searches for suitable basic meanings for each roles. Here, "ICH" for the receptor role, and "AUTO" for the object role. The real procedure for simple German sentences is presented in Algorithm 1.

Meaning acquisition from Chinese sentences is more difficult, in part because there is no space between Chinese characters. We need to first group Characters into words, and this already relies on the overall understanding of the meanings. Some results on approaching a meaning-based segmentation method within a "slow intelligent" framework are presented in [12].

6   Summary and On-going Work

In this paper, we outlined a novel translation method and its first application in second-language education. Translation is carried out in three phases: firstly, a

---

Algorithm 1. Automatic acquisition of meanings based on cue knowledge

---

input : a phrase or a sentence
output: a meaning representation
TableOfBasicClasses ← ∅;
Result ← ∅;
foreach *word in the input* **do**
> ListOfBasicClasses ← turn *word* into a list of basic class, each containing cue information, one of its basic meaning, corresponding meaning type;
> Append ListOfBasicClasses to TableOfBasicClasses;

CurrentMaximumEvaluation ← 0;
foreach *reading in* TableOfBasicClasses **do**
> identify meaning frames in this reading;
> ListOfSegmenations ← making partitions of the reading, according to different grouping strategies;
> **foreach** *segmentation in* ListOfSegmenations **do**
>> fill basic classes into roles of meaning frames;
>> Evaluation ←evaluate the quality of meaning frames;
>> **if** Evaluation ≥CurrentMaximumEvaluation **then**
>>> Result ← current filled meaning frame;

return Result;

---

sentence in the source language is transformed into its meaning representation, secondly, the style of this meaning representation is transformed into the style in the target language, and lastly, a sentence in the target language is produced. Our current work is put early developed systems together, cf. [13], [14]. The integrated system can be used as a learning tool, and later turned into a more comprehensive machine translation system. To this end we have shown here that cue switching can blend very well with other methods of machine translations.

Our integrated approach requires a language knowledge base, which can be acquired in second language eduction. We are developing a software tool for the knowledge input and interactive use in Chinese learning. The insights gained in this step will be fed into the design of a general machine translation method.

Acknowledgement. We thank Jörg Zimmermann for commenting on an earlier version.

References

[1] Addridge, J.W., Berridge, K.C.: Coding of serial order by neostriatal neurons: A "natural action" approach to movement sequence. Journal of Neuroscience 18, 2777–2787 (1998)

[2] Aglioti, S., Beltramello, A., Girardi, F., Fabbro, F.: Neurolinguistic and follow-up study of an unusual pattern of recovery from bilingual subcortical aphasia. Brain 119, 1551–1564 (1996)

[3] Bates, E., MacWhinney, B.: Competition, Variation, and Language Learning. In: MacWhinney, B. (ed.) Mechanisms of Language Acquisition, pp. 157–193. Lawrence Erlbaum, Hillsdale (1987)

[4] Block, H.U.: Example-based Incremental Synchronous Interpretation. In: Wahlster, W. (ed.) Verbmobil: Foundations of Speech-to-Speech Translation, pp. 411–417. Springer (2000)

[5] Brown, P.F., Cocke, J., Della Pietra, S.A.: A Statistical Approach to Machine Translation. Computational Linguistics 16(2), 79–85 (1990)

[6] Caramazza, A., Brones, I.: Semantic classification by bilinguals. Canadian Journal of Psychology 34(1), 77–81 (1980)

[7] Chen, H.-C.: Lexical processing in a non-native language: Effects of language proficiency and learning strategy. Memory and Cognition 18(3), 279–288 (1990)

[8] Chen, H.-C., Ng, M.-L.: Semantic facilitation and translation priming effects in Chinese-English bilinguals. Memory and Cognition 17, 454–462 (1989)

[9] Chen, Y., Jellinghaus, M., Eisele, A., Zhang, Y., Hunsicker, S., Theison, S., Federmann, C., Uszkoreit, H.: Combining multi-engine translations with Moses. In: Proceedings of the Fourth Workshop on Statistical Machine Translation, StatMT 2009, pp. 42–46. Association for Computational Linguistics, Stroudsburg (2009)

[10] Colomé, À.: Lexical activation in bilinguals' speech production: language specific or language independent? Journal of Memory and Language 45(4), 721–736 (2001)

[11] Costa, A., Caramazza, A.: Is lexical selection in bilinguals language-specific? further evidence from Spanish-English and English-Spanish bilinguals. Bilingualism Language and Cognition 2(3), 231–244 (1999)

[12] Dong, T., Cui, P.: Slow Intelligent Segmentation of Chinese Sentences using Conceptual Interval. In: The 19th International Conference on Distributed Multimedia Systems, Brighton, UK, pp. 51–54 (2013)

[13] Dong, T., Glöckner, I.: Word Expert Translation from German into Chinese in the Slow Intelligence Framework. In: The 17th International Conference on Distributed Multimedia Systems, Florence, Italy, pp. 16–21 (2011)

[14] Dong, T., Glöckner, I.: Relating Slow Intelligence Research to Bilingualism. In: The 18th International Conference on Distributed Multimedia Systems, Miami, USA, pp. 3–8 (2012)

[15] Emele, M.C., Dorna, M., Lüdeling, A., Zinsmeister, H., Rohrer, C.: Semantic-Based Transfer. In: Wahlster, W. (ed.) Verbmobil: Foundations of Speech-to-Speech Translation, pp. 359–376. Springer (2000)

[16] Freud, S.: Zur Auffassung der Aphasien. Eine kritische Studie. Franz, Leipzig und Wien (1891); Translated by E. Stengel. International University Press, New York (1953)

[17] Gerard, L., Scarborough, D.L.: Language-specific lexical access of homographs by bilinguals. Journal of Experimental Psychology: Learning, Memory, and Cognition 15(2), 305–315 (1989)

[18] Helbig, H.: Knowledge Representation and the Semantics of Natural Language. Springer, Heidelberg (2006)

[19] Helbig, H., Hartrumpf, S.: Word class functions for syntactic-semantic analysis. In: Proceedings of the 2nd International Conference on Recent Advances in Natural Language Processing, Tzigov Chark, Bulgaria, pp. 312–317 (1997)

[20] Kirsner, L., Smith, M.C., Lockhart, R.S., King, M.L., Jain, M.: The bilingual lexicon: Language-specific units in an integrated network. Journal of Verbal Learning and Verbal Behavior 23(4), 519–539 (1984)

[21] Knight, K.: A statistical MT tutorial workbook. Prepared in Connection with the JHU Summer Workshop (1999)
[22] Kroll, J.F., Stewart, E.: Category Interference in Translation and Picture Naming: Evidence for Asymmetric Connections between Bilingual Memory Representations. Journal of Memory and Language 33, 149–174 (1994)
[23] Levinson, S.C.: Language and Cognition: The Cognitive Consequences of Spatial Description in Guugu Yimithirr. Journal of Linguistic Anthropology 7(1), 98–131 (1997)
[24] Li, P., Bates, E., MacWhinney, B.: Processing a Language without Inflections: A Reaction Time Study of Sentence Interpretation in Chinese. Journal of Memory and Language 32, 169–192 (1993)
[25] MacWhinney, B., Bates, E., Kliegl, R.: Cue validity and sentence interpretation in English, German, and Italian. Journal of Verbal Learning and Verbal Behavior 23, 127–150 (1984)
[26] MacWhinney, B., Pleh, C., Bates, E.: The development of sentence interpretation in Hungarian. Cognitive Psychology 17, 178–209 (1985)
[27] Nagao, M.: A framework of a mechanical translation between Japanese and English by analogy principle. In: Elithorn, A., Banerji, R.R. (eds.) Artificial and Human Intelligence. Elsevier Science Publishers (1984)
[28] Paradis, M., Goldblum, M.C., Abidi, R.: Alternate antagonism with paradoxical translation behavior in two bilingual aphasic patients. Brain and Language 15, 55–69 (1982)
[29] Poldrack, R.A., Packard, M.G.: Competition between memory systems: Converging evidence from animal and human studies. Neuropsychologia 41, 245–251 (2003)
[30] Potter, M.C., So, K.-F., van Eckardt, B., Feldman, L.B.: Lexical and conceptual representation in beginning and proficient bilinguals. Journal of Verbal Learning and Verbal Behavior 23(1), 23–38 (1984)
[31] Rosch, E., Mervis, C.B., Gray, W., Johnson, D., Boyes-Braem, P.: Basic objects in natural categories. Cognitive Psychology 8, 382–439 (1976)
[32] Sánchez-Martínez, F., Forcada, M.L., Way, A.: Hybrid rule-based Example-Based MT: Feeding Apertium with Sub-sentential Translation Units. In: Forcada, M.L., Way, A. (eds.) Proceedings of the 3rd Workshop on Example-Based Machine Translation, Dublin, Ireland, pp. 11–18 (2009)
[33] Scarborough, D.L., Gerard, L., Cortese, C.: Independence of lexical access in bilingual word recognition. Journal of Verbal Learning and Verbal Behavior 23(1), 84–99 (1984)
[34] Schacter, D.L., Tulving, E. (eds.): Memory systems. MIT Press, Cambridge (1994)
[35] Squire, L.R., Knowlton, B.J.: The medial temporal lobe, the hippocampus, and the memory systems of the brain. In: Gazzaniga, M.S. (ed.) The New Cognitive Neurosciences, pp. 765–780. MIT Press, Cambridge (2000)
[36] Ullman, M.T.: A Cognitive Neuroscience Perspective on Second Language Acquisition: The Declarative/Procedural Model. In: Sanz, C. (ed.) Mind and Context in Adult SLA: Methods, Theory, and Practice, pp. 141–178. Georgetwon University Press, Washington (2005)
[37] Uszkoreit, H., Xu, F., Liu, W.: Challenges and Solutions of Multilingual and Translingual Information Service Systems. In: Jacko, J.A. (ed.) HCI, Part IV, HCII 2007. LNCS, vol. 4553, pp. 132–141. Springer, Heidelberg (2007)
[38] Wahlster, W. (ed.): Verbmobil: Foundations of Speech-to-Speech Translation. Springer (2000)
[39] Walters, J.: Bilingualism: The Sociopragmatic-Psycholinguistic Interface. Lawrence Erlbaum Associates, Mahwar (2005)

[40] Weaver, W.: Translation. In: Locke, W.N., Booth, A.D. (eds.) Machine Translation of Languages: Fourteen Essays, pp. 15–22. MIT Press, Cambridge (1949)

[41] Weist, R., Konieczna, E.: Affix processing strategies and linguistic systems. Journal of Child Language 12, 27–36 (1985)

[42] Willingham, D.B.: A Neuropsychological Theory of Motor Skill Learning. Psychological Review 105, 558 (1998)

[43] Woods, W.A.: Semantics and quantification in natural language question answering. In: Bobrow, D.G., Collins, A. (eds.) Representation and Understanding, pp. 35–82. Academic Press, New York (1975)

A New Relevance Feedback Algorithm
Based on Vector Space Basis Change

Rabeb Mbarek[1], Mohamed Tmar[1], and Hawete Hattab[2]

[1] Multimedia Information Systems and Advanced Computing Laboratory,
High Institute of Computer Science and Multimedia, University of Sfax, Sfax, Tunisia
rabeb.hattab@gmail.com, mohamedtmar@yahoo.fr
http://www.miracl.rnu.tn
[2] Umm Al-Qura University
hattab.hawete@yahoo.fr

Abstract. The idea of Relevance Feedback is to take the results that
are initially returned from a given query and to use information about
whether or not those results are relevant to perform a new query. The
most commonly used Relevance Feedback methods aim to rewrite the
user query. In the Vector Space Model, Relevance Feedback is usually
undertaken by re-weighting the query terms without any modification
in the vector space basis. With respect to the initial vector space basis
(index terms), relevant and irrelevant documents share some terms (at
least the terms of the query which selected these documents). In this paper we propose a new Relevance Feedback method based on vector space
basis change without any modification on the query term weights. The
aim of our method is to build a basis which optimally separates relevant
and irrelevant documents. That is, this vector space basis gives a better
representation of the documents such that the relevant documents are
gathered and the irrelevant documents are kept away from the relevant
ones.

Keywords: Vector space basis, Relevance Feedback, Query.

1 Introduction

The Vector Space Model (VSM) [13], is an algebraic model for representing text
documents (and any objects, in general) as vectors of identifiers, such as, for example, index terms. For the most vector space based Information Retrieval (IR)
and Relevance Feedback (RF) models, the set of index terms (original vector
space basis) generates documents and queries. Document coordinates depend
on the well-known $tf * idf$ weighting method, and query coordinates are computed using a similar weighting method and eventually re-estimated using a RF
approach, such as the Rocchio's algorithm [15]. Although several term weighting and RF methods have been proposed, only a few approaches consider that
changing the original vector space basis is an issue of investigation. The vector
space basis change consists of using a transition matrix[1]. By changing the vector

---

[1] The algebraic operator responsible for change of basis.

A. Gelbukh (Ed.): CICLing 2014, Part II, LNCS 8404, pp. 355–366, 2014.

space basis, each vector coordinate changes depending on this matrix. According to [9], if we change the basis, then the inner product changes and so the Cosine function behavior changes[2].

With respect to the original vector space basis, relevant and irrelevant documents share some terms (at least the terms of the query which selected these documents). To avoid this problem, it suffices to generate each document by phrases. And so, this representation can optimally separates relevant and irrelevant documents. To model this approach, it suffices to remark that each phrase is a combination of index terms. Let us define the following matrix: each column is generated by a phrase, that is each column contains the combination coefficients of this phrase with respect to index terms. This matrix is the transition matrix from the original basis (index terms) to a basis composed by phrases.

The best framework that could make our assumption into application is RF: the user shows relevant and irrelevant documents in an initial ranking and instead of reformulating the query, we change the basis in which it is written (as well as the documents). In other words, a vector space based Information Retrieval System (IRS) provides different results if it just rewrites the documents onto different vector space bases. Since the original vector space basis is not necessary the best one, from the infinite set of vector space bases, there exists a one that optimally separates relevant and irrelevant documents. In this paper we attempt to keep the IRS discover it. That is, it suffices to find a vector space basis which gives a better representation of the documents such that the relevant documents are gathered and the irrelevant documents are kept away from the relevant ones. In this paper, we propose a RF method based on this assumption.

This paper is organized as follows. Section 2 presents the related work. Sections 3 and 4 describe our RF approach based on the vector space basis change. Section 5 shows the evaluation results obtained from user study expriment. The conclusion and future works are presented in section 6.

2   Related Work

The VSM is here adopted as model of reference. A recent reconsideration of the geometry underlying IR, and indirectly of the VSM, was done in [13]. It proved very effective in retrieving documents of different languages, subjects, size, and media, thanks to a number of proposed and tested weighing schemes and applications which have made it a sound framework.

2.1   Relevance Feedback

The RF has been used in several IR models: the VSM [15,7], the probabilistic model [14,4], the language model [5], and the bayesian network retrieval model [2]. Most of the proposed approaches consist in adding new terms to the initial query and re-weighting original terms [17]. RF is covered in several books (e.g.,

---

[2] By the same, Dice, Jaccard and Overlap functions behavior changes.

Manning et al. [12]) and surveys Ruthven and Lalmas [16]. A dedicated track (i.e., the RF track) was run at TREC in 2008 and 2009.

Rocchio's algorithm is a classic algorithm for incorporating RF into the VSM [15]. The basic idea of this method is to add an average weight of each term within the set of relevant documents to the original (or initial) query vector, and to subtract an average weight within the set of irrelevant ones from this vector.

Pseudo-Relevance Feedback (PRF) assumes that the top ranked k documents of the initial retrieval are relevant and extracts expansion terms from them. Croft and Harper [4] first suggested this technique for estimating the probabilities within the probabilistic model. Due to the sensitivity to the quality of top k documents, PRF is not robust to the quality of the initial retrieval. Several approaches have been proposed to improve the robustness of PRF. The main strategies used could be categorized as follows:

- To refine the feedback document set so that instead of using all the top k documents, we could choose a subset of it Sakai et al. [18].
- Instead of using all the terms obtained through feedback for query refinement, only use a subset of important terms to avoid introducing query drift Cao et al. [3].
- Dynamically decide when to apply PRF instead of using it for all queries Amati et al. [1].
- Varying the importance of each feedback document Tao and Zhai [20].
- Using a large external collection like Wikipedia or the web as a source of expansion terms besides those obtained through PRF, Xu et al. [21], Zhou et al. [22].
- Lv and Zhai [8] proposed a positional relevance model where the terms in the document which are nearer to the query terms are assigned more weight.
- Recently, Zhou et al. [23] proposed a novel approach to PRF inspired by collaborative filtering.

2.2 Vector Space Basis Change

The Latent Semantic Indexing (LSI) [6] foundation is based on the assumption that there are many semantic relations between terms (synonymy, polysemy...), whereas capturing these relations by using semantic resources such as ontologies is complex, an alternative statistical solution could be taken into account by Singular Value Decomposition (SVD). This method results on a new vector space basis, with a lower dimension than the original one (all index terms), and in which each component is a linear combination of the indexing terms.

In [9,10], a context is modeled by a vector space basis and its evolution is modeled by linear transformations (transition matrices) from one basis to another. The basic idea is that, a vector space basis models a document or query descriptors. The semantics of a document or query descriptors depend on context. A basis can be derived from a context. Therefore, a basis of a vector space is the construct to model context. Also, change of context can be modeled by linear transformations from one base to another. In particular, RF is an example of context change [9].

Recently, Mbarek and Tmar [11] have studied the problem of separateness of relevant and irrelevant documents. Let g_R be the centroïd of relevant documents. The authors built a vector space basis which minimizes the quotient of first, the sum of squared distances between each relevant document and g_R and second, the sum of squared distances between each irrelevant document and g_R. To solve this optimization problem they used a necessarily test which is: the first partial derivative of the quotient is equal to zero. We remark that the calcul on the quotient is difficult and so applying the second partial derivative test (a sufficient condition) is very difficult.

In this paper, we will study the concavity of the function difference between first, the sum of squared distances between each relevant document and g_R and second, the sum of squared distances between each irrelevant document and g_R. This study allows us to find the minimums and the maximums of this function.

3   Vector Space Basis Properties

The RF consists of building a query, which is in most cases a representation that allows the relevant documents to appear in the top of the ranking list as more as possible. From a certain point of view, it can be viewed as a representation that keeps the relevant documents gathered to their centroïd and the irrelevant ones far from it. Each document d_i is represented in a vector space by $d_i = (w_{i1}, w_{i2}, ...w_{iN})^T$ where w_{ij} is the weight of term t_j in document d_i and N is the number of index terms[3]. As for us our approach is independent of the term weighting method.

The Euclidian distance between documents d_i and d_j is given by:

$$dist(d_i, d_j) = \sqrt{\sum_{k=1}^{N} (w_{ik} - w_{jk})^2}$$
$$= \sqrt{(w_{i1} - w_{j1}...w_{iN} - w_{jN}).(w_{i1} - w_{j1}...w_{iN} - w_{jN})^T}$$
$$= \sqrt{(d_i - d_j)^T.(d_i - d_j)}$$

By changing the basis using a transition matrix M, the distance between 2 vectors d_i^* and d_j^* which are respectively d_i and d_j rewritten in the new basis is given by:

$$dist(d_i^*, d_j^*) = dist(M.d_i, M.d_j)$$
$$= \sqrt{(M.d_i - M.d_j)^T.(M.d_i - M.d_j)}$$
$$= \sqrt{(d_i - d_j)^T.M^T M.(d_i - d_j)}$$

The vector space basis which optimally separates relevant and irrelevant documents is represented by a matrix M^* called the optimal transition matrix. M^* puts the relevant documents gathered to their centroïd g_R and the irrelevant documents far from it.

---

[3] x^T is the transpose of x.

g_R is done by:

$$g_R = \frac{1}{|R|} \sum_{d \in R} d$$

Where R is the set of relevant documents.

By the same, using a transition matrix M, we obtain:

$$M.g_R = M.(\frac{1}{|R|} \sum_{d \in R} d) = \frac{1}{|R|} \sum_{d \in R} M.d$$

The optimal matrix M^* should minimize the sum of squared distances between each relevant document and g_R i.e:

$$M^* = \arg\min_{M \in M_n(\mathbb{R})} \sum_{d \in R} dist^2(M.d, M.g_R) \tag{1}$$

$$= \arg\min_{M \in M_n(\mathbb{R})} \sum_{d \in R} (Md - Mg_R)^T.(Md - Mg_R)$$

$$= \arg\min_{M \in M_n(\mathbb{R})} \sum_{d \in R} (d - g_R)^T.M^T M.(d - g_R)$$

By the same, the optimal matrix M^* should maximize the sum of squared distances of each irrelevant document and g_R, which leads on the following:

$$M^* = \arg\max_{M \in M_n(\mathbb{R})} \sum_{d \in S} dist^2(M.d, M.g_R) \tag{2}$$

$$= \arg\max_{M \in M_n(\mathbb{R})} \sum_{d \in S} (Md - Mg_R)^T.(Md - Mg_R)$$

$$= \arg\max_{M \in M_n(\mathbb{R})} \sum_{d \in S} (d - g_R)^T.M^T M.(d - g_R)$$

where S is the set of irrelevant documents.

Equations 1 and 2 result on the following single equation:

$$M^* = \arg\max_{M \in M_n(\mathbb{R})} [\sum_{d \in S} (d - g_R)^T.M^T M.(d - g_R) - \sum_{d \in R} (d - g_R)^T.M^T M.(d - g_R)]$$

We put

$$\tag{3}$$

$$f(M) = \sum_{d \in S} (d - g_R)^T.M^T M.(d - g_R) - \sum_{d \in R} (d - g_R)^T.M^T M.(d - g_R) \tag{4}$$

4   Optimal Transition Matrix and Relevance Feedback

4.1   Optimal Transition Matrix Identification

In this section we attempt to solve the Equation 3 which leads to the the optimal transition matrix we look for.

If $M = (m_{ij})_{1 \leq i,j \leq n}$ is a solution of Equation 3, then $\forall \quad 1 \leq i \leq j \leq n$

$$\frac{\partial[\sum\limits_{d \in S} (d - g_R)^T.M^T M.(d - g_R) - \sum\limits_{d \in R} (d - g_R)^T.M^T M.(d - g_R)]}{\partial m_{ij}} = 0 \quad (5)$$

And so
$$\sum_{d \in S} (d - g_R)^T.(\frac{\partial M}{\partial m_{ij}})^T M.(d - g_R) + \sum_{d \in S} (d - g_R)^T.M^T(\frac{\partial M}{\partial m_{ij}}).(d - g_R)$$

$$-$$

$$\sum_{d \in R} (d - g_R)^T.(\frac{\partial M}{\partial m_{ij}})^T M.(d - g_R) + \sum_{d \in R} (d - g_R)^T.M^T(\frac{\partial M}{\partial m_{ij}}).(d - g_R) = 0$$

Where $\dfrac{\partial M}{\partial m_{ij}} = \begin{matrix} \\ \\ i \\ \\ \\ \\ \\ \end{matrix} \begin{pmatrix} 0 \dots 0\ \overset{j}{0}\ 0 \dots 0 \\ \vdots\ \ddots\ \vdots\ \vdots\ \vdots\ \ddots\ \vdots \\ 0 \dots 0\ 1\ 0 \dots 0 \\ 0 \dots 0\ 0\ 0 \dots 0 \\ 0 \dots 0\ 0\ 0 \dots 0 \\ \vdots\ \ddots\ \vdots\ \vdots\ \vdots\ \ddots\ \vdots \\ 0 \dots 0\ 0\ 0 \dots 0 \end{pmatrix}$

And so

$$(\frac{\partial M}{\partial m_{ij}})^T M = \begin{matrix} \\ \\ \\ j \\ \\ \\ \\ \end{matrix} \begin{pmatrix} 0 & \dots & 0 & \dots & 0 \\ \vdots & \ddots & \vdots & \ddots & \vdots \\ 0 & \dots & 0 & \dots & 0 \\ m_{i1} & \dots & m_{ij} & \dots & m_{in} \\ 0 & \dots & 0 & \dots & 0 \\ \vdots & \ddots & \vdots & \ddots & \vdots \\ 0 & \dots & 0 & \dots & 0 \end{pmatrix}$$

And

$$M^T(\frac{\partial M}{\partial m_{ij}}) = \begin{pmatrix} 0 \dots 0\ \overset{j}{m_{i1}}\ 0 \dots 0 \\ \vdots\ \ddots\ \vdots\ \vdots\ \vdots\ \ddots\ \vdots \\ 0 \dots 0\ m_{ij}\ 0 \dots 0 \\ \vdots\ \ddots\ \vdots\ \vdots\ \vdots\ \ddots\ \vdots \\ 0 \dots 0\ m_{in}\ 0 \dots 0 \end{pmatrix}$$

Then the Equation 5 is a linear equation.

To find a solution of the Equation 3: first, we have to solve the following linear system with n^2 in n^2 variables:

$$
\begin{cases}
\dfrac{\partial[\sum\limits_{d \in S} (d - g_R)^T.M^T M.(d - g_R) - \sum\limits_{d \in R} (d - g_R)^T.M^T M.(d - g_R)]}{\partial m_{ij}} = 0 \\[4mm]
\forall\ 1 \leq i \leq j \leq n
\end{cases}
\tag{6}
$$

A solution of this system is called a critical point. Second, we study the concavity of the function f on each solution M of the System 6. If f is convex on the matrix M, then this matrix is a solution of the Equation 3. To study the concavity of f, we have to apply the second derivative test. The *Hessian matrix* is a square matrix of second-order partial derivatives of a function. It describes the local curvature of a function of many variables. If the Hessian matrix is positive definite[4] at a critical point M, then it attains a maximum at M. In this case the function is locally convex.

To apply the second derivative test it suffices to compute the eigenvalues of the Hessian matrix of the function f. The second-order partial derivatives are: for all $1 \leq i \leq j \leq n$ and $1 \leq k \leq l \leq n$

$$
\frac{\partial^2 f}{\partial m_{kl} \partial m_{ij}} =
$$

$$
\sum_{d \in S} (d - g_R)^T.(\frac{\partial M}{\partial m_{ij}})^T \frac{\partial M}{\partial m_{kl}}.(d - g_R) + \sum_{d \in S} (d - g_R)^T.(\frac{\partial M}{\partial m_{kl}})^T (\frac{\partial M}{\partial m_{ij}}).(d - g_R)
$$

$$
-
$$

$$
\sum_{d \in R} (d - g_R)^T.(\frac{\partial M}{\partial m_{ij}})^T \frac{\partial M}{\partial m_{kl}}.(d - g_R) + \sum_{d \in R} (d - g_R)^T.(\frac{\partial M}{\partial m_{kl}})^T (\frac{\partial M}{\partial m_{ij}}).(d - g_R)
$$

If $i \neq k$, then $(\frac{\partial M}{\partial m_{ij}})^T \frac{\partial M}{\partial m_{kl}} = 0$, otherwise

$$
(\frac{\partial M}{\partial m_{ij}})^T \frac{\partial M}{\partial m_{il}} =
\begin{matrix}
 & & l & & j & \\
\end{matrix}
\begin{pmatrix}
0 \dots 0 & 0 \dots 0 & 0 \dots 0 \\
\vdots \ddots \vdots & \vdots \ddots \vdots & \vdots \ddots \vdots \\
0 \dots 0 & 0 \dots 1 & 0 \dots 0 \\
0 \dots 0 & 0 \dots 0 & 0 \dots 0 \\
\vdots \ddots \vdots & \vdots \ddots \vdots & \vdots \ddots \vdots \\
0 \dots 1 & 0 \dots 0 & 0 \dots 0 \\
0 \dots 0 & 0 \dots 0 & 0 \dots 0 \\
\vdots \ddots \vdots & \vdots \dots \vdots & \vdots \ddots \vdots \\
0 \dots 0 & \ddots \ 0 \ 0 & 0 \dots 0
\end{pmatrix}
\begin{matrix} \\ \\ l \\ \\ \\ j \\ \\ \\ \\ \end{matrix}
$$

---

[4] Its eigenvalues are positive.

Then

$$
\begin{cases}
\dfrac{\partial^2 f}{\partial m_{kl} \partial m_{ij}} = 0 & if\, k \neq i, \\[4mm]
\dfrac{\partial^2 f}{\partial m_{il} \partial m_{ij}} = 2(\sum_{d \in S} (d - g_R)^T.(d - g_R) - \sum_{d \in R} (d - g_R)^T.(d - g_R)) & otherwise.
\end{cases}
\tag{7}
$$

4.2 Relevance Feedback

In the VSM, the score of a document vs. a query is often expressed by the inner product:

$$RSV(d, q) = d^T.q$$

If now the document and the query are expressed in an optimal vector space basis whose transition matrix is M, this score becomes:

$$RSV(Md, Mq) = (M.d)^T.M.q = d^T.M^T.M.q$$

This score represents the score of the document d, in the original basis, vs. the query $q' = M^T.M.q$. Hence the vector space basis change has an effect of query reformulation.

5 Experiments

In this section we give the different experiments and results obtained to evaluate our approach. We describe the environnement of evaluation and the experimental conditions.

5.1 Environnement

The two test collections TREC-7 and TREC-8 were used for the experiments. Data was preprocessed through stop-word removal and Porters stemming, and one-word terms were stored; the initial rankings of documents (Baseline Model) were weighted by the $BM25$ formula proposed in [19]. BM25 parameters are $b = 0.5$, $k_1 = 1.2$, $k_2 = 0$ and $k_3 = 8$.

The experiments consist to re-rank the results of the Baseline Model.

For our approach the reformulated query is:

$$Q_{new} = M^T.M.Q_{int} \tag{8}$$

Where M is a solution of the Equation 3, then:

For the Rocchio model, the reformulated query is:

$$Q_{new} = \alpha.Q_{int} + \beta.\frac{1}{|R|} \sum_{d \in R} d + \gamma.\frac{1}{|S|} \sum_{d \in S} d$$

- The initial query Q_{int} is made from the short topic description, and using it the top 1000 documents are retrieved from the collections (weighted $\alpha = 1$).
- R is the set of top ranking 5 documents, assumed to be relevant (weighted $\beta = 1$).
- S is the set of retrieved documents $501 - 1000$, assumed to be irrelevant (weighted $\gamma = 0$).

For our approach and the Rocchio model, the retrieved documents are ranked by the inner product done using:

$$RSV(Q_{new}, d) = Q_{new}^T.d \qquad (9)$$

5.2   Results

To evaluate the performance we executed several runs using the topics provided by TREC. In detail, the TREC-7 and TREC-8 collections has 100 topics. Topics are structured in three fields: title, description and narrative. To generate a query, the title of a topic was used, thus falling into line with the common practice of TREC experiments; description and narrative were not used.

We performed 50 runs by considering all possible combinations of the two parameters involved in our method. In particular, we took into account: n (the cardinality of R), m (the cardinality of S). We selected different ranges for each parameter: n ranges in $(1, 2, 3, 4, 5)$, m ranges in $(1, 5, 10, 15, 20, 25, 30, 35, 40, 50)$.

We evaluate each run in terms of Mean Average Precision (MAP). The experiments and the evaluations are articulated around the comparison between our model and the Vector Space Model (Baseline Model) and the Rocchio Model.

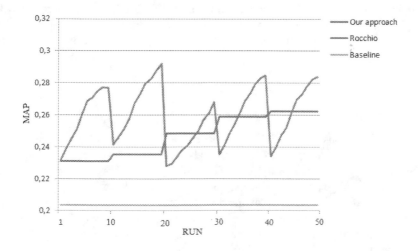

Fig. 1. Plot of MAP values on TREC-7

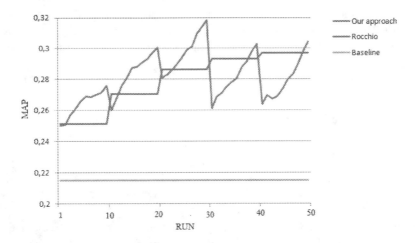

Fig. 2. Plot of MAP values on TREC-8

Figures 1 and 2 plot the MAP values for each run and approach. These graphs highlights as the system performance vary according to parameters changes. The value of MAP of the Baseline model for TREC-7 is 0.2035 and for TREC-8 is 0.215. For the Rocchio model the lowest MAP value for TREC-7 is 0.2312 and for TREC-8 is 0.2513. This value occurs when only one relevant document is involved. For our approach the lowest MAP value for TREC-7 is 0.2321 and for TREC-8 is 0.2501. This value occurs when only one relevant document and one irrelevant document are involved. For the Rocchio model the best MAP value for TREC-7 is 0.2625 and for TREC-8 is 0.2971. This value occurs when 10 relevant documents are involved. For our approach the best MAP value for TREC-7 is 0.2921. This value occurs when 2 relevant documents and 50 irrelevant documents are involved. For TREC-8 the best MAP value, for our approach, is 0.3183. This value occurs when 3 relevant documents and 50 irrelevant documents are involved.

6   Conclusion

This paper proposes a RF algorithm based on vector space basis change. By changing the basis, each vector coordinate changes. In this paper, to guide the RF process we compute a vector space basis which gives a better representation of the documents such that the relevant documents are gathered and the irrelevant ones are kept away from the relevant documents.

What makes our approach different from the previous works is the assumption that each document or query can be re-written onto different vector space bases. This distinguishes from the LSI method which aimed at computing latent descriptors by assuming only one base. In [11], the authors built a vector space basis which minimizes the function quotient of first, the sum of squared distances between each relevant document and g_R and second, the sum of squared distances between each irrelevant document and g_R. In this paper, we study the concavity of the function difference between first, the sum of squared distances between each relevant document and g_R and second, the sum of squared distances between each irrelevant document and g_R.

In [10], the author shows that each document or query can be associated to a distinct basis, which corresponds to one context. Consequently, document ranking can take advantage of the diversity of contexts. The major difference between this work and our approach is that we search, from the infinite set of vector space bases, a basis that provides the best document representation. That is,

This paper reports about using transition matrices (i.e. the algebraic operator responsible for change of basis) to model a RF algorithm. We intend to apply other algebraic operator (like vector product) to build a geometric relevance feedback approach.

References

1. Amati, G., Carpineto, C., Romano, G.: Query difficulty, robustness, and selective application of query expansion. In: McDonald, S., Tait, J.I. (eds.) ECIR 2004. LNCS, vol. 2997, pp. 127–137. Springer, Heidelberg (2004)
2. De Campos, L.M., Fernández-Luna, J.M., Huete, J.F.: Relevance feedback in the bayesian network retrieval model: An approach based on term instantiation. In: Hoffmann, F., Adams, N., Fisher, D., Guimarães, G., Hand, D.J. (eds.) IDA 2001. LNCS, vol. 2189, p. 13. Springer, Heidelberg (2001)
3. Cao, G., Nie, J.-Y., Gao, J., Robertson, S.: Selecting good expansion terms for pseudo-relevance feedback. In: SIGIR 2008: Proceedings of the 31st Annual International ACM SIGIR Conference on Research and Development in Information Retrieval, pp. 243–250. ACM, New York (2008)
4. Croft, W.B., Harper, D.: Using Probabilistic Models of Information without Relevance Information. Journal of Documentation 35(4), 285–295 (1979)
5. Croft, W.B., Lavrendo, S.C.T.V.: Relevance Feedback and Personalization: A Language Modelling Perspective. In: Proceedings of the Joint DELOS-NSF Workshop on Personalization and Recommender Systems in Digital Libraries, CIKM 2006, pp. 49–54 (2001)
6. Deerwester, S., Dumais, S.T., Furnas, G.W., Landauer, T., Harshman, R.: Indexing by Latent Semantic Analysis. Journal of the ASIS 41(6), 391–407 (1990)
7. Ide, E.: New Experiments in Relevance Feedback. In: The SMART Retrieval System-Experiments in Automatic Document Processing, pp. 337–354 (1971)
8. Lv, Y., Zhai, C.: Positional relevance model for pseudo-relevance feedback. In: SIGIR 2010, pp. 579–586. ACM, New York (2010)
9. Melucci, M.: Context Modeling and Discovery using Vector Space Bases. In: Proceedings of the ACM Conference on Information and Knowledge Management (CIKM), Bremen, Germany, pp. 808–815. ACM Press (2005)
10. Melucci, M.: A basis for information retrieval in context. ACM Trans. Inf. Syst. 26(3), 1–41 (2008)
11. Mbarek, R., Tmar, M.: Relevance Feedback Method Based on Vector Space Basis Change. In: Calderón-Benavides, L., González-Caro, C., Chávez, E., Ziviani, N. (eds.) SPIRE 2012. LNCS, vol. 7608, pp. 342–347. Springer, Heidelberg (2012)

12. Manning, C.D., Raghavan, P., Schutze, H.: Introduction to Information Retrieval. Cambridge University Press (2008)
13. van Rijsbergen, C.J.: The Geometry of Information Retrieval. Cambridge University Press, UK (2004)
14. Robertson, S., Sparck Jones, J.K.: Relevance Weighting of Search Terms. Journal of the ASIS 27(3), 129–146 (1976)
15. Rocchio, J.: Relevance Feedback in Information Retrieval. The SMART retrieval system-experiments in automatic document processing, pp. 313–323. Prentice-Hall Inc. (1971)
16. Ruthven, I., Lalmas, M.: A survey on the use of relevance feedback for information access systems. The Knowledge Engineering Review 18(2), 95–145 (2003)
17. Ruthven, I., Lalmas, M., Rijsbergen, K.: Ranking Expansion Terms with Partial and Ostensive Evidence. In: Fourth International Conference on Conceptions of Library and Information Science: Emerging Frameworks and Methods, Seattle WA, USA, pp. 199–219 (2002)
18. Sakai, T., Manabe, T., Koyama, M.: Flexible pseudo-relevance feedback via selective sampling. ACM Transactions on Asian Language Information Processing (TALIP) 4(2), 111–135 (2005)
19. Walker, S., Hancock-Beaulieu, M., Gull, A., Lau, M.: Okapi at TREC. In: TREC, pp. 21–30 (1992)
20. Tao, T., Zhai, C.: Regularized estimation of mixture models for robust pseudo-relevance feedback. In: SIGIR 2006, pp. 162–169. ACM Press, New York (2006)
21. Xu, Y., Jones, G.J., Wang, B.: Query dependent pseudo- relevance feedback based on wikipedia. In: SIGIR 2009, pp. 59–66. ACM, New York (2009)
22. Zhou, D., Lawless, S., Wade, V.: Improving search via personalized query expansion using social media. Information Retrieval 15, 218–242 (2012)
23. Zhou, D., Truran, M., Liu, J., Zhang, S.: Collaborative pseudo-relevance feedback. Expert Systems with Applications 40, 6805–6812 (2013)

How Complementary Are Different Information Retrieval Techniques?
A Study in Biomedicine Domain

Xiangdong An and Nick Cercone

Department of Electrical Engineering & Computer Science
York University, Toronto, ON M3J 1P3, Canada
{xan,ncercone}@yorku.ca

Abstract. In this paper, we make an empirical study on the submitted runs to the TREC Genomics Track, a gathering for information retrieval research in biomedicine. Based on the evaluation criteria provided by the track, we investigate how much relevant information is generally lost from a run, and how well the relevant nominees are actually ranked w.r.t. the level of relevancy and how they are distributed among the irrelevant ones in a run. We examine whether the relevancy or the level of relevancy play a more important role in the performance evaluation. Answering these questions may give us some insight into and help us improve the current IR technologies. The study reveals that the recognition of relevancy is more important than that of level of relevancy. It indicates that on average more than 60% of relevant information is lost from each run w.r.t. to either the amount of relevant information or the amount of aspects (subtopics, novelty or diversity), which suggests the big potential room for performance improvement. The study shows that the submitted runs from different groups are quite complementary, which implies ensemble IRs could significantly improve retrieval performance. The experiments illustrate that a run performs "good" or "bad" mainly due to its performance on its top 10% rankings, and the rest of the run only contributes to the performance marginally.

1 Introduction

Information retrieval (IR) systems automatically find the information that matches the needs of users described by their queries. A lot of efforts have been made to improve the retrieval performance, but much more efforts are anticipated [1]. The Text REtrieval Conference (TREC) supports research within the IR community by providing the infrastructures necessary for large-scale evaluation of text retrieval methodologies. The TREC Genomics track focused on IR evaluations in biomedicine based on large test collections, which ran from 2003 to 2007. For the TREC 2007 Genomics track [2], an IR research group could submit up to 3 runs, each nominating up to 1000 passages for each of the 36 topics from a corpus of 162,259 HTML documents. Here, a passage is a piece of continuous context within a paragraph of a document that could range from a phrase up to the entire paragraph [3]. The passage level retrieval performance was evaluated mainly based on two measures: Passage2 mean average precision (MAP) and Aspect MAP. Passage2 MAP was the average precision of the nominated passages for one topic, averaged across all 36 topics, and Aspect MAP the average precision of the

A. Gelbukh (Ed.): CICLing 2014, Part II, LNCS 8404, pp. 367–380, 2014.

aspects found for one topic, averaged across all 36 topics (see Section 2 for more details). In the rest of the paper, we may simply call them measure Passage2 and measure Aspect. We rerank all 63 runs submitted by 26 groups to make them approximately optimal in terms of Passage2 (which are called *passage2 optimal* runs) and optimal in terms of Aspect (which are called *aspect optimal* runs), respectively. Figure 1 show the performance comparisons between the two optimal runs and the respective submitted runs. It is indicated that both sets of optimal runs perform dramatically better than the submitted runs in general. The average absolute and relative performance improvements of the passage2 optimal runs over the submitted runs are 16.49% and 475% w.r.t. Passage2, respectively. The average absolute and relative performance improvements of the aspect optimal runs over the submitted runs are 38.03% and 359% w.r.t. Aspect, respectively.

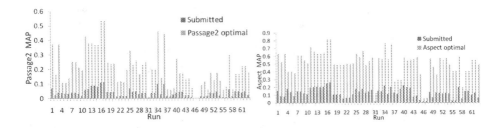

Fig. 1. Left: Passage2 performances of the submitted and the passage2 optimal runs. Right: Aspect performances of the submitted and the aspect optimal runs.

Where do these big performance improvements exactly come from? Is it because we do not rank the relevant passages well in terms of the level of relevancy or because we present too many irrelevant passages in the submitted runs? How the relevant passages are distributed among the submitted runs? The optimal runs are not perfect. The average passage2 performance of the passage2 optimal runs is 20.38% and the average aspect performance of the aspect optimal runs is 51.14%. Their distances with the perfect performances are 79.62% and 48.86% respectively, which is caused by the loss of the relevant passages (from the submitted runs). What are the percentage rates of such losses? Answering these questions may help us understand and improve the current IR technology, particularly in biomedicine. In this paper, we answer these questions from a data mining perspective.

The rest of the paper is organized as follows. In Section 2, the experimental environment is introduced. The empirical study is presented in Section 3. Based on the experimental results, we give some discussions in Section 4. Section 5 concludes.

2   Dataset and Measures

From 2003 to 2005, the TREC Genomics track was on document-level retrieval. In its last two years (2006 & 2007), it focused on passage retrieval, where a passage could range from a phrase to a sentence or a paragraph of a document and must be continuous [3]. The passage retrieval was evaluated based on both the passage-level and the aspect-level retrieval performances.

Algorithm 1, summarized from the TREC 2007 Genomics track scoring program, shows Passage2 evaluation. The list of ordered passages nominated for each topic is processed from the top (ranked highest) to the bottom (ranked lowest). If a nominated passage is not relevant (i.e., does not overlap with any gold passages for the topic), it is penalized by increasing *deno* by the passage length (line 5) since *deno* would be used as the denominator when calculating the sum of precision (*sumPrecision*) of all nominated passages (line 11). Otherwise, the passage would be processed character by character (line 7). If the character is not within the corresponding gold passage range (i.e., an *irrelevant character*), only *deno* is increased by 1 as penalty (lines 8-9); if the character is within the corresponding gold passage range and has not been used for scoring (i.e., a *novel character*), both *deno* and *nume* would be increased by 1 (lines 8-11), which is considered rewarding since *nume* would be used as the numerator in calculating *sumPrecision*. Nothing would be done if the exactly same character has been used for scoring before (i.e., a *redundant character*). Therefore, if all relevant passages are ranked higher than irrelevant ones, and are within the respective gold passages, *sumPrecision* would equal the total length of all relevant passages. Furthermore, if all gold passages for a topic are exactly nominated, *averagePrecision* for the topic would equal 100%.

Algorithm 1. Passage2 evaluation for each topic

Input: {nominatedPassages[Topic]}, {goldPassages[Topic]}.
Output: Passage2 average precision by topic.

```
1  for each Topic do
2      nume=0; deno=0; sumPrecision=0.0;
3      for each nominated Passage do
4          if no any relevant characters then
5              deno += passageLength;
6          else
7              for each character do
8                  if irrelevant or novel then
9                      deno +=1;
10                 if novel then
11                     nume +=1; sumPrecision+= nume/deno;
12                 end
13             end
14          end
15     end
16  end
17  count=numCharactersInGoldPassages[Topic];
18  averagePrecision[Topic]= sumPrecision/count;
19 end
```

From the analysis above, to make a list of passages *Passage2 optimal*, we first need to push all irrelevant passages back behind the relevant ones. We then need to find an order for all relevant passages that maximizes the Passage2 score. This could involve examining a number of $m!$ permutations, where m is the number of gold passages identified for a topic ($1 \leq m \leq 609$ [2]). This is generally intractable. Instead, we may order the relevant passages by always selecting the one that contributes to the performance score most with ties broken arbitrarily. Obviously, this does not guarantee optimal solutions. For example, for the 2 passages $p_1 = c_r c_i$ and $p_2 = c_i c_r c_r$, where c_r denotes a relevant (and novel) character and c_i an irrelevant character, they should be ordered as $p_2 p_1$ based on the method (since $\frac{1}{2} + \frac{2}{3} > 1$), but $p_1 p_2$ gains higher score (since $1 + \frac{2}{4} + \frac{3}{5} > \frac{1}{2} + \frac{2}{3} + \frac{3}{4}$). Intuitively, an optimal order should be the one that puts all relevant characters as forward as possible while the irrelevant characters as backward as possible. In that case, the relevant characters would be penalized least by the irrelevant characters. Based on the intuition, we may order relevant passages simply based on their ratios of the relevant characters:

$$r - ratio = \frac{numRelevantCharacters(p_i)}{len(p_i)}$$

with ties broken arbitrarily, where p_i is a passage. We can actually simply use *r-ratio* to order all nominated passages, which would naturally push all irrelevant passages back after the relevant ones. We will use *r-ratio* to measure the relevancy level of a passage: a passage with higher *r-ratio* is more relevant.

Algorithm 2. Aspect evaluation for each topic

Input: {nominatedPassageSet[Topic]}, {goldPassageSet[Topic]}.
Output: Aspect average precision by topic.
1 **for** *each Topic* **do**
2 | nume=0; deno=0; sumPrecision=0.0;
3 | **for** *each nominated Passage* **do**
4 | | **if** *relevant* **then**
5 | | | **if** *numNewAspects > 0* **then**
6 | | | | nume +=1; deno +=1;
7 | | | | sumPrecision+=$\frac{numNewAspects*nume}{deno}$;
8 | | | **end**
9 | | **else**
10 | | | deno +=1;
11 | | **end**
12 | **end**
13 | count=numUniqueAspects[Topic];
14 | averagePrecision[Topic]=$\frac{sumPrecision}{count}$;
15 **end**

Algorithm 2, summarized from the TREC 2007 Genomics track scoring program, shows the details of the aspect-level evaluation. Any nominated passages containing relevant aspects are considered relevant (line 4); all other passages are considered irrel-evant and are simply penalized by increasing *deno* by 1 (line 10). Any relevant passages that contain new aspects are considered *novel* and would be rewarded by increasing both *nume* and *deno* by 1 (line 6). Nothing would be done for the *redundant* passages that only contain previously seen aspects. The variable *sumPrecision* is updated only upon novel passages (line 7). Here, the *averagePrecision* (line 14) is equivalent to the *weighted recall* [4] of the relevant aspects. If all relevant passages are ordered to be in front of the irrelevant ones, *sumPrecision* would equal the number of aspects all nomi-nated passages contain. Therefore, the maximum value of *averagePrecision* for a topic is 100%. From the algorithm, pushing all relevant passages forward to the top of a run would make the run *Aspect optimal*. The set of 63 aspect optimal runs perform as shown in Figure 1 (Right). Regarding the level of novelty, we say a passage with more unique aspects is more novel. Here, the level of novelty has no impact on the performance of Aspect optimal runs, however.

3 Experiment

In Figure 1, we have shown the significant performance improvements made by the optimal runs on Passage2 and Aspect measures. In this section, we first show how much we lose from the improper ordering of the relevant passages — misrecognition of the level of relevancy, and how much we lose from ranking the irrelevant passages higher than the relevant ones — misrecognition of the relevancy. We then explore what more exactly have happened to the submitted runs to make such loss patterns possible.

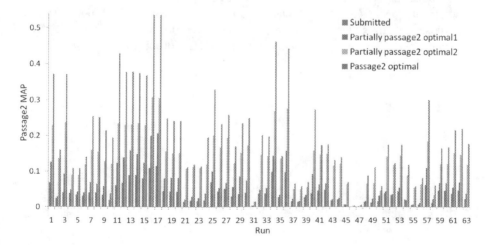

Fig. 2. Performances of the submitted, the partially passage2 optimal1, the partially passage2 optimal2, and the passage2 optimal runs w.r.t. Passage2

3.1   Impacts of Relevancy and Its Levels

To know how much we lose from the misrecognition of the level of relevancy (novelty), we change the positions of the relevant (novel) passages among themselves — the irrelevant passages remain at the original positions — so that only relevant (novel) passages are ordered based on the level of relevancy (novelty) w.r.t. measure Passage2 (Aspect). We call runs obtained such *partially Passage2 (Aspect) optimal1*. Note that the level of novelty would make a difference in Aspect performance when irrelevant passages appear before the novel ones. In such cases, finding an ordering for *relevant* passages that maximizes Aspect performance might be infeasible. In this paper, we take the heuristic that orders relevant passages by always selecting the one with the most new aspects. To understand how much we lose from the misrecognition of the relevancy (novelty), we get another set of 63 runs that are optimized on the relevancy (novelty), called *partially Passage2 (Aspect) optimal2*, by pushing all relevant (novel) passages to the top of the lists while maintaining their original relative ordering. Therefore, the two sets of partially optimal runs are optimal w.r.t. only one of the two concerned factors: the relevancy (novelty) or the level of relevancy (novelty).

Figure 2 shows the performances of the submitted, the partially passage2 optimal1, the partially passage2 optimal2, and the passage2 optimal runs w.r.t. Passage2. It is indicated that, in general, the partially passage2 optimal1 runs perform better while the partially passage2 optimal2 runs much better than the respective submitted runs. Figure 3 (Left) shows that the relative improvements of the two sets of partially passage2 optimal runs over the submitted ones. The average relative improvements over the submitted runs are respectively 55.60% and 319.30%. As discussed above, the performance improvements of the partially passage2 optimal1 runs are contributed by reranking all relevant passages based on their level of relevancy while the performance improvements of the partially passage2 optimal2 runs are caused by the proper recognition of the relevant passages. This indicates that we might have lost much more performance from improper relevancy recognition than from improper level of relevancy recognition.

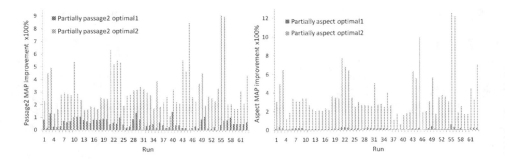

Fig. 3. Left: Relative performance improvements of the partially passage2 optimal1 and optimal2 runs over the submitted runs on Passage2. Right: Relative performance improvements of the partially aspect optimal1 and the partially aspect optimal2 runs over the submitted runs on Aspect.

The performance comparisons on the submitted, the partially Aspect optimal1, the partially Aspect optimal2, and the Aspect optimal runs tell us a similar story. Figure 4 shows their performances w.r.t. Aspect. It is indicated the partially aspect optimal1 runs perform slightly better while the partially aspect optimal2 runs significantly better than the respective submitted runs in general. Note that all partially aspect optimal2 runs perform the same as the respective aspect optimal runs. This is because when all relevant passages are on the top of the list, the list is optimal w.r.t. Aspect. In such a case, the optimization on the level of novelty has no impact on aspect performance. Figure 3 (Right) shows the relative improvements of the partially aspect optimal1 and optimal2 runs over the submitted runs. The average relative improvements are respectively 13.78% and 358.97%. The experimental results on the aspect measure further confirm that the relevancy (novelty) recognition has significantly stronger impacts on performance than the level of relevancy (novelty) recognition.

Next, we see what more exactly have happened to the submitted runs to result in such loss patterns. For example, how are the relevant passages distributed among the 1000 passages allowed to nominate? Are they generally presented at the top of the lists or dispersed anywhere? Are the distributions sensitive to the topic difficulties or the amount of passages allowed to nominate? What is the percentage rate of gold passages (aspects) that are lost from the nominated passages in general?

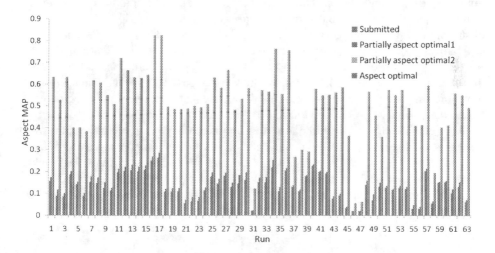

Fig. 4. Performances of the submitted, the partially aspect optimal1, the partially aspect optimal2, and the aspect optimal runs on Aspect MAP

3.2 What Happened?

Overall Distribution. In this section, we look at overall distributions of the relevant passages and the novel passages.

We first look at the overall distribution of the relevant passages, which is obtained as follows. For each topic of each submitted run, we examine all nominated passages against the gold passages for that topic to make a map for the relevant passages in the run, where each relevant passage is represented by its rank number. Note that the rank number stands for the passage position in the run. We then, based on the 36 (topics) x 63 (runs) maps, compute the percentage of relevant passages falling into each range of 100 positions (the 1000 positions are divided into 10 ranges). Figure 5 (Left) shows the distribution of the relevant passages in all ranges. On average, 42.37 relevant passages are correctly nominated for each topic by each run, and are distributed across all ranges. The closer to the top of the list, the more densely the relevant passages are distributed. In the first range, the distribution is significantly dense. In particular, about 36.12% of all relevant passages correctly nominated, about 15.30 relevant passages in other words, are located in the top-most range. This makes a precision of about 15.30% for the relevant passages on the top 100 positions, which is quite low. In the least densely distributed range (the last 100 positions), there still exist about 3.22% of relevant passages.

Next, we look at how novel passages are distributed. The distribution of novel passages is obtained as follows. For each topic of each submitted run, we check all nominated passages against the gold passages for that topic to get a map of novel passages, where each novel passage is represented by its rank number. We then, based on the 36 (topics) x 63 (runs) maps, compute the percentage of novel passages falling into each range of 100 positions. Figure 5 (Right) shows the distribution of the novel passages across all ranges. On average, 14.62 novel passages are correctly nominated for each topic by each run, and are distributed across all ranges. The closer to the beginning of a run, the more densely distributed the novel passages are. On the top 100 positions, the distribution is significantly dense. In particular, about 48.49% of novel passages, about 7.09 novel passages in other words, are located in the most densely distributed range. In the least densely distributed range (the last 100 positions), there exist about 2.20% of novel passages. Note that a higher ratio of novel passages are distributed on the top 100 positions than the relevant passages. This is because that the closer to the beginning of a run, the less likely a relevant passage is redundant.

Although the closer it is to the top of a run, the more densely the relevant (novel) passages are distributed, more than half (63.88% for the relevant passages and 51.51% for the novel passages) of them are located beyond the first 100 positions (i.e., ranked 100+). In particular, these more than half of relevant (novel) passages are relatively evenly distributed among all other ranges. This indicates the current IR techniques are

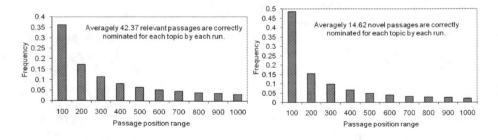

Fig. 5. Left: Overall distribution of the relevant passages. Right: Overall distribution of the novel passages.

still far from satisfactory in distinguishing the relevant passages from the irrelevant ones, and this may also explain why the performance of the submitted runs can be improved so significantly by properly recognizing relevant (novel) passages only. Since both the relevant and the novel passages are most densely distributed among the first 100 positions, we are interested to see how they are specifically distributed in the range.

Distribution among Top 100 Positions. We divide the first 100 positions into 10 ranges of 10 positions. Among the first 100 positions, the 15.30 relevant passages are distributed as shown by Figure 6 (Left). It is still true that the closer it is to the beginning of a list, the more densely the relevant passages are distributed. However, the relevant passages are more evenly distributed in the first 100 positions than they are in the 1000 positions since the relevant passage distribution curve for the top 100 positions looks more flat. In particular, the frequency difference 18.08%-6.01%=12.07% between the most dense and the least dense ranges for the top 100 positions is quite smaller than that for the 1000 positions which is 36.12%-3.22%=32.9%. From another perspective, the standard deviation of the frequency distribution for the top 100 positions is 3.86% while that for the 1000 positions is 10.13%.

Next, we look at how novel passages are distributed in the first 100 positions. Among the first 100 passages, on average 7.09 novel passages are distributed as shown in Figure 6 (Right). It is still true that the closer it is to the beginning of the list, the more densely the novel passages are distributed. Similarly, the novel passages look like more evenly distributed among the first 100 positions than they are among the 1000 positions since the distribution curve for the top 100 positions looks more flat. In particular, the frequency difference 26.19%-4.22%=21.97% between the densest and the least dense ranges for the top 100 positions is quite smaller than that for the 1000 positions which is 48.49%-2.20%=46.29%. The standard deviation of the frequency distribution for the top 100 positions is 6.86% while that for the 1000 positions is 14.13%.

That the relevant or the novel passages are more evenly distributed within the top 100 positions than they are in the 1000 positions indicates that the current IR techniques are less confident when asked to identifying relevant or novel passages more exactly.

Next, we look at how the level of topic difficulty impacts the distribution patterns.

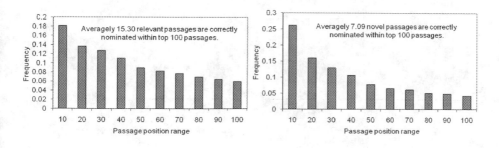

Fig. 6. Left: The average distribution of the relevant passages among the top 100 positions. Right: The average distribution of the novel passages among the top 100 positions.

Distribution on Topics of Different Difficulty. The amounts of gold passages found from the pool of nominated passages for different topics are quite different from as small as 1 to as many as 609 [2]. We assume the amount of gold passages found for each topic reflects the difficulty of the topic. We divide all 36 topics into 3 groups based on the amounts of gold passages found for them: the 8 hard topics having 1 to 20 gold passages, the 14 medium topics having 21 to 100 gold passages, and the 14 easy topics having 101 to 609 gold passages.

On average, 86.40 relevant passages and 30.16 novel passages are correctly nominated for each of the 14 easy topics per submitted run, and are distributed as shown in Figure 7. On average, 19.99 relevant passages and 6.31 novel passages are correctly nominated for each of the 14 medium topics per submitted run, and are distributed as shown in Figure 8. On average, 4.49 relevant passages and 1.96 novel passages are correctly nominated for each of the 8 hard topics per submitted run, and are distributed as shown in Figure 9. All the 6 distribution curves indicate that the closer it is to the beginning of the list, the more densely the relevant (novel) passages are distributed. In the first 100 positions, the distribution is drastically dense. Similar to Figure 5, all these 6 curves show that a higher ratio of novel passages is distributed in the beginning of the list than that of the relevant passages. This is because the closer it is to the end of the list, the more likely a relevant passage becomes redundant.

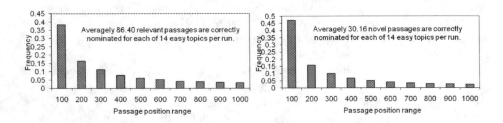

Fig. 7. Left: The average distribution of the relevant passages for the 14 easy topics. Right: The average distribution of the novel passages for the 14 easy topics.

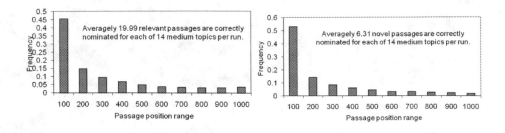

Fig. 8. Left: The average distribution of the relevant passages for the 14 medium topics. Right: The average distribution of the novel passages for the 14 medium topics.

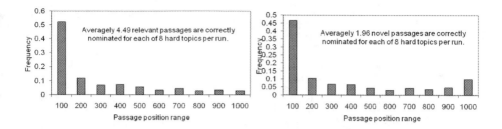

Fig. 9. Left: The average distribution of the relevant passages for the 8 hard topics. Right: The average distribution of the novel passages for the 8 hard topics.

Figures 7-9 also indicate that the harder the topics are, the higher ratio of the relevant (novel) passages are distributed in the beginning of the list. Note the ratio here is actually the *recall* as a function of range k, i.e., the recall at range k (the percent of relevant or novel passages reached within range k). It is different from the *precision* at range k (the fraction of passages reached within range k that are relevant or novel). Higher recall at a range does not necessarily mean higher precision at the same range. For example, the recalls of the relevant passages at the first range for easy, medium, and hard topics are 34.44%, 41.29% and 52.12%, respectively, while their precisions at the same range are respectively 29.76%, 8.25% and 2.34%. The recalls of the novel passages at the first range for easy, medium, and hard topics are 47.22%, 52.86% and 57.98%, respectively, while their precisions at the same range are respectively 14.24%, 3.34% and 1.13%. Therefore, the harder topics are indeed harder. The higher recall at the first range for harder topics is caused by the relatively higher quota (to nominate passages) provided to the harder topics considering their relatively smaller amount of gold passages.

Finally, we look at how many relevant passages and the aspects were lost from the passages nominated for each topic per submitted run on average.

3.3   Amounts of Relevant Passages Lost

For each topic of each submitted run, we calculate the number of relevant passages retrieved by the run. By averaging them across all 63 submitted runs, we can get the average number of relevant passages retrieved for each topic by each submitted run. The average number of relevant passages retrieved for each topic and its difference with the number of gold passages found for the topic are as shown in Figure 10 (Left). Note that the topics are numbered from 200 to 235. Averaging them across all 36 topics, about 42.37 relevant passages were retrieved and about 82.43 relevant passages were lost per topic per run (the average number of gold passages found for each topic is 124.8 [2]). Figure 11 shows the fraction of gold passages that was lost from the nominated passages for each topic on average. Averaging them across all 36 topics, about 63% relevant passages were lost per topic per run. This indicates we lost more than half of the relevant passages from the 1000 passages nominated on average.

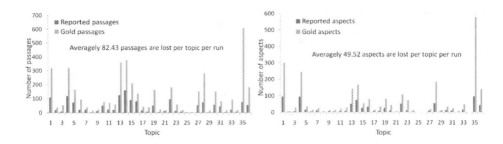

Fig. 10. Left: Comparison between the number of relevant passages retrieved on average and the number of gold passages for each topic. Right: Comparison between the number of aspects retrieved on average and the number of gold aspects for each topic.

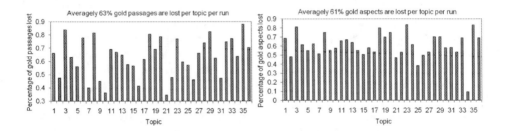

Fig. 11. Left: The fraction of relevant passages lost for each topic on average. Right: The fraction of aspects lost for each topic on average.

3.4   Amounts of Aspects Lost

For each topic of each submitted run, we examine the number of aspects retrieved for the topic by the run. By averaging across all 63 submitted runs, we can get the average number of aspects retrieved for each topic per submitted run. The average number of aspects retrieved for each topic and its distance with the number of aspects assigned to the topic are as shown by Figure 10 (Right). Averaging them across all 36 topics, about 22.75 aspects were retrieved and about 49.52 aspects were lost per topic per run (the average number of assigned aspects for each topic is 72.3 [2]). Figure 11 (Right) shows the fraction of aspects lost from the nominated passages for each topic on average. Averaging them across all 36 topics, about 61% aspects were lost per topic, which is quite similar to the percent of relevant passages lost per topic. This indicates while we lost more than half of the relevant passages, we lost a similar ratio of aspects.

4   Discussion

Obviously, the more passages we are allowed to nominate, the more relevant passages we would be able to reach. When there is no restriction on the number of passages we can nominate, we can reach all relevant passages by nominating an arbitrary amount of passages. As shown by Figure 11, on average 63% relevant passages were lost per topic per run when we are allowed to nominate 1000 passages for each topic (assume the gold passages would not increase or decrease when more or less passages are allowed to nominate). An interesting question is how many more passages we must nominate to include all gold passages based on the current techniques. Figure 5 (Left) gives us some clues. As shown by Figure 5 (Left), 3.22% of all 42.37 relevant passages, 1.36 relevant passages in other words, are distributed in the last 100 positions. In particular, the curve indicates that the distribution is changing very slowly at the last ranges. We therefore assume there are at most 1.36 relevant passages in each 100 passages nominated after the first 1000 ones. Based on the rate, we therefore need to nominate at least 100*(82.43/1.36)=6061 more passages for each topic to recover all gold passages. That we may recover all 124.8 gold passages if we are allowed to nominate 7061 passages may imply the current level of IR techniques. This represents a precision of 1.77% for a recall of 100%. To make current IR perfect, we need to identify and push forward all relevant passages among the 7061 nominated passages to the top of the list and properly rank them based on the level of relevancy. This shows we may still have a long way to go for perfect IR.

The analysis above indicates more than half of the relevant passages are distributed beyond 1000 nominees very sparsely (lower than 1.36%). This from another point of view shows that the current IR technology suffers more problems in relevancy recognition than in the level of relevancy recognition. On the other hand, that 63% relevant passages per topic per run could be obtained from other runs on average indicates existing IR systems and methods are quite complementary (i.e., low biased and highly ambiguous). This from another perspective shows the potential benefits of ensemble IR.

The study also reveals the big potential room we can improve for the retrieval results presented by current IR systems. Compared to the potential room, improvements made by existing reranking techniques are still quite marginal [5–7].

5   Conclusion

In this paper, we first showed the big IR performances loss from the set of retrieval results submitted to TREC 2007 Genomics track — an information retrieval gathering in biomedicine — w.r.t. to different performance measures. We then studied whether the relevancy or the level of relevancy recognition should be responsible more for the losses. The study indicated that the relevancy recognition should take much more responsibility on the performance losses than the level of relevancy recognition. In the future we may put more efforts on relevancy recognition which turns out to be a binary classification problem. We finally empirically investigated what happened to the retrieval results to make such loss pattern possible. The experiments showed that the current IR systems do present more relevant (novel) passages on the top of runs (lists, rankings), and this observation is not affected by the level of topic difficulties in general. The experiments showed that the current IR systems used by different research groups are highly complementary. This implies the potential benefits of ensemble IR. The study revealed that the performance improvements made by current reranking techniques are quite minimal compared to the big potential room existing in the retrieval results. The study also showed we may still have a long way to go for optimal (perfect) IR.

In the future, we would investigate whether we can get similar conclusions from other IR application fields by applying the methodology to those fields. To the best of our knowledge, this is the first IR technology study from a data mining perspective.

Acknowledgment. This research is supported in part by the research grant from the Natural Sciences & Engineering Research Council (NSERC) of Canada, and in part by the Center for Innovation in Information Visualization and Data Driven Design (CIVDDD). We thank anonymous reviewers for their thorough review comments on this paper.

References

1. Manning, C.D., Raghavan, P., Schütze, H.: Introduction to Information Retrieval. Cambridge University Press, Cambridge (2008)
2. Hersh, W., Cohen, A., Roberts, P.: TREC 2007 genomics track overview. In: TREC 2007, pp. 98–115 (2007)
3. Hersh, W., Cohen, A., Roberts, P., Rekapalli, H.K.: TREC 2006 genomics track overview. In: TREC 2006, pp. 68–87 (2006)
4. Zhai, C., Cohen, W.W., Lafferty, J.: Beyond independent relevance: methods and evaluation metrics for subtopic retrieval. In: SIGIR 2003, pp. 10–17 (2003)
5. Yang, L., Ji, D., Tang, L.: Document re-ranking based on automatically acquired key terms in chinese information retrieval. In: COLING 2004, pp. 480–486 (2004)
6. Goldberg, A.B., Andrzejewski, D., Gael, J.V., Settles, B., Zhu, X.: Ranking biomedical passages for relevance and diversity: University of Wisconsin, Madison at TREC genomics 2006. In: TREC 2006, pp. 129–136 (2006)
7. Hu, Q., Huang, X.: A reranking model for genomics aspect search. In: SIGIR 2008, pp. 783–784 (2008)

Performance of Turkish Information Retrieval: Evaluating the Impact of Linguistic Parameters and Compound Nouns

Hatem Haddad[1] and Chedi Bechikh Ali[2]

[1] Mevlana University, Konya, Turkey
Faculty of Engineering
Department of Computer Engineering
[2] ISG, Tunis University, Tunisia
LISI Laboratory, INSAT, Carthage University, Tunisia
{haddad.hatem,chedi.bechikh}@gmail.com,
hhatem@mevlana.edu.tr

Abstract. Turkish is an agglutinative language where linguistic parameters can have significant consequences on the information retrieval performances. In this paper, different Turkish linguistic parameters (truncation, stemming, stop words, etc.) have been studied and their impacts on an information retrieval system performance have been invistiguated. Three word truncations at fixed length (3, 4 and 5 characters) have been studied. The results have been compared using Snowball and Zemberek stemmers. Moreover, the results of using compound nouns, in addition to simple keywords, to index queries and documents have been studied. In the experimental part, *Milliyet* test collectionn have been tested by three information retrieval models. The comparisons of performance analysis have been done by he traditional information retrieval metrics and *bpref* metric since the test collection is build on an incomplete relevance judgments.

Keywords: Turkish information retrieval, natural language processing, truncation, stemming, compound nouns.

1 Introduction

According to the study of "What kind of barriers do Turkish Internet users face when using Turkish on the Internet?" [1], the main information retrieval problem do Turkish Internet users face is the language barrier and difficulty in dealing with the information of other languages. Turkish Information Retrieval (Turkish IR) presents a significant challenge for the information retrieval research community due to the Turkish language characteristics. Indeed, Turkish is an agglutinative language where words are constructed using inflectional and derivation suffixes linked to a root. For example, there are approximately 23,000 stems and 350-400 roots actively used; but if we include inflection of the words, the number raises to millions [2]. For this reason, studies focused on the Turkish natural language processing rather than the information retrieval questions.

A. Gelbukh (Ed.): CICLing 2014, Part II, LNCS 8404, pp. 381–391, 2014.
© Springer-Verlag Berlin Heidelberg 2014

Compared to other languages, Turkish did not have enough attention by the IR community. The reason of that is the absence of a standard test collection in Turkish, except the Milliyet test collection which is an incomplete one [3].

According to [4], the Turkish population is about 80 millions and the number of Internet users is more than 35 millions. Turkish belongs to the Altaic branch of the Ural-Altaic languages family. The Turkish alphabet is based on Latin characters and has 29 letters consisting of 8 vowels and 21 consonants. It has all the letters of the English alphabet, except q, w, and x. In addition, it has extra characters: ç ,ğ, ş, ü, ö, and ı. Authors in [5] and [6] summarize the Turkish language history, alphabet and Turkish language grammar. Because of the complexity of Turkish morphology, Turkish needs a special attention when it comes to text processing and computational linguistics. Turkish IR necessitates the use of special linguistic analysis.

Rest of the paper is organized as follows: Turkish information retrieval Background and related work is discussed in Section 2. Experimental setup and results are presented in Section 3, followed by the discussion of results and perspectives in Section 4.

2   Turkish Information Retrieval

In this section, we survey previous studies on Turkish IR. Due to the lack of standard IR test collections in Turkish, Turkish IR is a field that has not achieved much interest compared to other languages [5]. Most of the studies in Turkish IR have built their own test collections. Thereby, this makes it difficult to compare different studies and results. In [7], past and recent works in Turkish IR have been summarized.

According to [5], the earliest published Turkish IR study is done by [8] using 570 documents on computer science with 12 queries. It evaluated the effectiveness of truncation and concluded, after evaluating different prefix sizes, that the use of the first 5 characters (5-prefix) of words is the best truncation approach.

In [9], authors applied a stemmer in a collection of 533 news articles obtained from a Turkish news agency. The morphological analysis used in that study was implemented using lex and yacc utilities handling a word in three steps; root determination, morphophonemic checks and morphological parsing. Authors concluded that stemming reduces the index dictionary size about 65 percent.

Most of the Turkish IR research studies developed their own stemmer. However, the most used stemmers are Snowball [10] and Zemberek [11]. Snowball is a stemmer developed using Snowball6 string processing language [12] where Turkish words are analyzed with an affix stripping approach without any dictionary lookups. Zemberek is a stemmer designed for Turkish and based on a root dictionary. It provides root forms of given words using a root dictionary-based parser combined with Natural Language Processing algorithms [11]. It handles special cases for suffixes and can be used as a lemmatizer based stemmer.

In [2], evaluating the effectiveness of a stemming algorithm called FINDSTEM was proposed. The stemmer was evaluated using 2,468 law documents and 15 queries in a vector space model. Authors concluded that stemming increases search precision by approximately 25% of the average precision values at 11-point recall levels when compared to no stemming at all.

Authors in [13] studied the effectiveness of queries stemming on Turkish IR. They used a Turkish collection containing the titles and abstracts of 6,289 economic and political news stories extracted from Turkish newspapers in the period from 1991 to 1993. Also 50 queries prepared by 30 Turkish natives providing both natural-language queries and relevance judgments on the stemmed and unstemmed search outputs. In their experiment, only query words were stemmed (document words were used as they are), and authors compared the retrieval effectiveness using stemmed and unstemmed query words. Stemming was achieved using the two-level morphological analyzer; PC-KIMMO. Since different roots can correspond to one query word, the query was extended with all possible roots that can be related to query words. Based on the OKAPI text retrieval system, the stemmed queries provided a number of relevant documents about 32% more than the unstemmed queries at the retrieval cut-off of 10 and 20 documents.

In [5], the results of four different stemming options on Turkish IR effectiveness; no stemming, simple word truncation, the successor variety method adapted to Turkish [15] and a lemmatizer-based stemmer for Turkish [16]; were compared. Authors also studied the impact of using a stop words list. Experiments were conducted on the Milliyet collection. They concluded that, unlike other languages, in the case of the Turkish language a stop words list has no influence on system effectiveness. In addition to that, a simple word truncation approach, a word truncation approach that uses corpus statistics, and an elaborated lemmatizer-based stemmer provide similar performances in terms of effectiveness.

The author [14] compared the retrieval performances of three retrieval models: Lemur TF-IDF, OKAPI and Language Modeling using Milleyet Collection. Two measures were used to evaluate the retrieval results: MAP and bpref. As in [5], the author concluded that stop words do not have great effects on retrieval performance when we look at bpref values of different retrieval methods. The author stressed on the importance of the stemming and concluded that stemming can increase retrieval performances of the three models up to 20%. He also concluded that the retrieval performances of the three retrieval models are quite similar and no conclusion can be done regarding which model is the best for the Turkish language. He suggested that combining the language model with a lemmatizer based stemmer might be a promising approach.

IR systems face the challenge of being effective and ordering the relevant retrieved documents. Most IR models use keywords as "bags of words" and documents are represented as an unordered terms set. To achieve an optimal precision, an IR system must use an expressive documents and queries representation. Earlier works proved that the use of simple terms as keywords is not accurate enough to represent the documents contents due to the words ambiguity. A solution to this problem is to use compound nouns[1] instead of simple

---

[1] By compound nouns, we refer to complex terms and noun phrases.

terms. The assumption is that compound nouns are more likely to identify semantic entities than simple words. So they are a better representation of the semantic documents content [17][18].

While earlier works focused on statistical techniques to extract and index documents with compound nouns [19] [20], recent works propose linguistic techniques for detecting compouns nouns [21]. Light Natural Language Processing showed that it can improve matching results by combining the methods of classical IR with compound nouns recognition [22].

In [23], a method for automating Turkish keyphrase extraction using multi-criteria ranking was proposed. It compares extracted keyphrases to manually assigned keyphrases. Authors did not apply their method to an information retrieval system. Based on a statistical approach to extract words collocations, authors in [24] concluded that Chi-square hypothesis test and mutual information methods have produced better results compared to other statistical methods on Turkish corpus. In [25], feature sets that take the morphological characteristics of Turkish into account to extract bigrams and trigrams are used. Even if authors applied the extracted N-grams to automatic speech recognition, their extraction approach have shown promising results. In a similar way, our approach relies on morphological characteristics of Turkish to extract compound nouns.

3   Experimental Setup and Results

In this section, we describe the dataset and the experimental setup we used. We structure our findings by the type of task: stop words, documents and queries structure, truncation, stemming and compound nouns. In our tables of result, the highest scores for each model/measure pair are shown in bold font.

3.1   Milliyet Collection

For the results in this paper, we have used Bilkent Information Retrieval Group Test Collection called Milliyet [26]. It contains 408,305 news articles and columns of five years, 2001 to 2005, from the Turkish newspaper Milliyet [3]. The collection is about 800 Megabytes and 95.5 million words before stop words elimination. Without stop words elimination, each document contains 234 words on average.

The collection is built on an incomplete relevance judgments. Indeed, to determine the relevant documents of the queries, Bilkent Information Retrieval Group used the pooling concept. The queries are written and evaluated according to the TREC approach by 33 native speaker assessors. The original query owners do the evaluation using binary judgment. During evaluation, the query pool contents are presented to the assessors in random order, and the rest of the collection is assumed to be irrelevant [5].

3.2   Terrier IR System and Evaluation Measures

We have used Terrier (an open source information retrieval framework) [27] [28]. It implements various retrieval models. Our experiments are based on three models: tf-idf, BM25 and the Language Model (LM).

Milliyet collection documents are short documents (documents with maximum 100 words), medium length (documents with 101 to 300 words), and long documents (documents with more than 300 words) [5]. As document length effects the retrieval performances, "ignore.low.idf" option is used in the Terrier system during our experiments in order not to ignore short documents.

Since the collection is built on an incomplete relevance judgments, the bpref is the most adequate measure to evaluate retrieval performances. Indeed, bpref is based on the relative ranks of judged documents only [29]. The mean average precision (MAP) measure represents the overall effectiveness performance of a system. Bpref and MAP are very highly correlated when used with complete judgments. But when judgments are incomplete, rankings of systems by bpref still correlate highly to the original ranking, whereas rankings of systems by MAP do not. For the evaluation, we also use the traditional 11-point interpolated average precision (11pt-avg) and the precision at low recall considering only the top most results returned by the system: precision at 5 documents (P@5), precision at 10 documents (P@10) and precision at 15 documents (P@15).

3.3   Baseline and Stop Words List Results

The retrieval results in Table 1 provide a useful baseline for evaluating the effect of the different linguistic parameters on the IR system. In this run, original queries and documents are used with no pre-processing such as stop words removal or stemming. The baseline run is referred to as BL.

We used the Turkish Stop Word List 1.1 provided by the Natural Language Processing Group, Department of Computer Engineering, Fatih University [30]. The list contains 223 Turkish stop words. Runs using the stop words list are referred to as SW.

Table 1 presents results of the BL and SW runs. Comparing the Baseline results and the Baseline results using a stop words list, we can conclude, as [5][14], that stop words results have a small performance gain. Besides, the bpref results are nearly equal to Baseline results.

3.4   Documents and Queries Structure

Milliyet collection documents are represented using XML. Each document consists of one root node DOC and eight child nodes: DOCNO, SOURCE, URL, DATE, TIME, AUTHOR, HEADLINE, TEXT. HEADLINE and TEXT fields contain textual information. Queries format is composed, similar to TREC query format, from three fields: title, description and narratives.

We tried different runs combining different documents and queries structures. The best results were obtained when using the nodes HEADLINE and TEXT for

Table 1. Baseline results

Measures	BL			SW		
	tf-idf	**BM25**	LM	tf-idf	**BM25**	LM
P@5	0,5806	0,5944	0,5306	0,6056	0,6	0,5306
P@10	0,5278	0,5444	0,5	0,525	0,5389	0,5181
P@15	0,5019	0,5139	0,4676	0,5111	0,5148	0,4796
11pt-avg	0,2598	0,2757	0,2561	0,2673	0,2799	0,2614
MAP	0,2361	0,2522	0,2306	0,2424	0,2567	0,2353
bpref	0,3899	0,4041	0,3769	0,3963	0,4081	0,3799

documents, and the fields title and description for queries. This run is referred to as ST. Then, we used the structure and the stop words list. This run is referred to as ST_SW.

According to Table 2, the use of the structure has more performance gain than stop words. For the bpref measures, stop words and structure results are relatively equal for the 3 models. For the different runs, BM25 performs better than tf-idf and LM.

Table 2. Structure results

Measures	ST			ST_SW		
	tf-idf	**BM25**	LM	tf-idf	**BM25**	LM
P@5	0,5861	0,5833	0,5417	0,6056	0,6111	0,5556
P@10	0,5625	0,5819	0,5139	0,5736	0,5833	0,5222
P@15	0,5565	0,562	0,5	0,5565	0,5583	0,5
11pt-avg	0,3259	0,3357	0,3016	0,3311	0,3379	0,3047
MAP	0,3044	0,3149	0,2779	0,3102	0,3186	0,2813
bpref	0,3981	0,4039	0,3612	0,4028	0,4075	0,3651

3.5 Truncation Results

In this section, we present evaluation results of three word prefix sizes: first 3 characters (3-prefix), first 4 characters (4-prefix) and first 5 characters (5-prefix).

Compared to the Baseline results, the use of the truncation gives better results as shown in Table 3. Best results are obtained when the first 4 characters with the BM25 model is used. The LM model gives the best results according to tf-idf

Table 3. Truncation results

Measures	3-prefix			4-prefix			5-prefix		
	tf-idf	BM25	LM	tf-idf	BM25	LM	tf-idf	BM25	LM
P@5	0,3594	0,4389	0,3244	0,6111	0,6194	0,5583	0,6361	0,6333	0,5722
P@10	0,4639	0,4208	0,4194	0,5667	0,5917	0,5333	0,6083	0,6014	0,5292
P@15	0,4306	0,4019	0,3667	0,5407	0,5537	0,5009	0,5639	0,5704	0,5231
11pt-avg	0,218	**0,2223**	0,2161	0,3225	0,3385	**0,3389**	0,3451	0,3566	**0,3605**
MAP	0,1912	**0,1989**	0,1953	0,3032	0,3213	**0,3242**	0,3278	0,3396	**0,3411**
bpref	**0,3972**	0,3634	0,3426	0,4364	**0,4462**	0,4058	0,4374	**0,4439**	0,4024

and MAP measures using the first 5 characters. The same conclusion mentioned in [8] is seen in our results too.

3.6 Stemming Results

According to the stemming results in Table 4, we can conclude that the use of a stemmer positively affects the Turkish IR system performances. Moreover, Zemberek provides a superior performance to Snowball. When the bpref measure results are close to each other for tf-idf and BM25 models using the Snowball stemmer, the BM25 is performing better than tf-idf and LM model when using the Zemberek stemmer. This is due to the fact that Zemberek stemmer is using a root dictionary-based stemmer opposite to the Snowball stemmer.

Table 4. Stemming results

Measures	Snowball			Zemberek		
	tf-idf	**BM25**	LM	tf-idf	**BM25**	LM
P@5	0,5556	0,5583	0,4944	0,6083	0,6167	0,5361
P@10	0,5319	0,5306	0,4667	0,5903	0,5944	0,5278
P@15	0,5065	0,5083	0,4509	0,5676	0,5676	0,5093
11pt-avg	0,2945	0,3055	0,2726	0,3506	0,3585	0,3513
MAP	0,2755	0,2865	0,252	0,3336	0,344	0,335
bpref	0,4039	0,4104	0,3642	0,4534	0,4603	0,4114

3.7   Compound Nouns Results

To extract Turkish compound nouns, we rely on syntactic information which allow integrating dependencies among words and overcome the paradigm of "bag of words". Since Zemberek stemmer is performing better than the Snowball stemmer, we use it as a lemmatizer based stemmer to extract compound nouns. Our approach is based on two steps:

- we conduct a linguistic analysis on the documents and the queries with Zemberek. This analysis generates a tagged collection where each word is associated to a tag corresponding to the syntactic category of the word: noun, verb, adjective, preposition, etc. For example, if we consider the Headline of the document number 400000: "adalet bakanı pasaport ister kırmızı, ister turuncu olsun"[2]. The tagged sentence, where syntactic categories are in uppercase, is as the following: adalet NOUN bakan NOUN pasaport NOUN iste VERB kırmızı ADJECTIVE, iste VERB turuncu ADJECTIVE ol VERB.
- then, we use the tagged collection to extract out a set of compound nouns. Candidate compound nouns are extracted out by the identification of lexico-syntactic patterns. In our experiments, we use three patterns: <Noun, Noun>, <Noun, Adjective> and <Adjective, Noun>. The extracted compound nouns from the previous example are "adalet bakan" and "bakan pasaport".

The number of compound nouns extracted using the three patterns are: 21353529 compound nouns for the <Noun,Noun> pattern, 1388429 for the <Noun, Adjective> pattern and 1430866 for the <Adjective, Noun> pattern.

In Table 5, we present the evaluation results of using compound nouns in addition to simple keywords to index queries and documents. The CN run refers to the integration of compound nouns to the baseline run. In the run CN_S, the structure is used with compound nouns. The run CN_S_I is the same as CN_S using the "ignore.low.idf" option.

Comparing the compound nouns runs, we can conclude that LM model is performing better that tf-idf and BM25 models. Indeed, LM model results of the CN run are better than those of the tf-idf and BM25 models. But unlike the tf-idf and BM25 models, the use of the "ignore.low.idf" option decreases the bpref value of the LM model. Indeed, the "ignore.low.idf" option can cause some query terms to be omitted in small documents.

4   Discussion of Results and Perspectives

The results of this paper indicate that stop words have no significant influence on the information performances. On the other hand, using the document and query structures, the system is having a performance gain but this is an inherent characteristic of the test collection. Even if the stemming is performing better than truncation, we can not conclude that stemming is more adequate for Turkish IR. Indeed, more experiments of stemming techniques is a suitable matter

---

[2] Word to word translation: justice minister passport want red, want orange be.

Table 5. Compound nouns results

Measures	CN			CN_S			CN_S_I		
	tf-idf	BM25	**LM**	tf-idf	BM25	LM	tf-idf	BM25	LM
P@5	0,5278	0,5583	0,5389	0,5833	0,55	0,5639	0,5694	0,5694	0,5194
P@10	0,4764	0,4972	0,5069	0,5542	0,5444	0,5333	0,5556	0,5486	0,5167
P@15	0,4296	0,4583	0,4907	0,5241	0,5167	0,5231	0,5259	0,5306	0,5065
11pt-avg	0,2206	0,2533	0,2952	0,3043	0,3143	**0,3518**	0,3129	0,3227	**0,3411**
MAP	0,1965	0,231	0,2735	0,2836	0,2971	**0,3351**	0,2934	**0,303**	0,3249
bpref	0,3797	0,4144	0,418	**0,4383**	0,4372	0,4159	0,4407	**0,4472**	0,4086

for further work. Compound nouns results show an improvement compared to Baseline runs when compound nouns are combined with the use of the structure. This result encourages us to work more intensively on compound nouns indexing and to increase the patterns size. In the case of the LM model, "ignore.low.ldf" option decreases the performances. Even if the BM25 model is performing better that the tf-idf and LM models, we can observe that the LM model is the more sensitive to the Turkish linguistic parameters. The LM model performances increase more than the other models when a linguistic analysis is added to the information retrieval process.

Our future work entails a language model for Turkish IR based on compound nouns. When [31] investigated the question of whether natural language processing techniques can improve the effectiveness of information retrieval and studied the morphological, lexico-semantical and syntactic levels of Turkish, we will investigate the question of how to determine which compound noun patters are more informative than the others. Also to determine compound nouns weights. In the other hand, we will evaluate the effect of different smoothing methods and we will use the paired t-test for the statistical significance testing.

References

1. Aytac, S.: Identification of Common Molecular Subsequences. The International Information & Library Review 37(4), 275–284 (2005)
2. Sever, H., Bitirim, Y.: FindStem: Analysis and Evaluation of a Turkish Stemming Algorithm. In: Nascimento, M.A., de Moura, E.S., Oliveira, A.L. (eds.) SPIRE 2003. LNCS, vol. 2857, pp. 238–251. Springer, Heidelberg (2003)
3. Milliyet, http://www.milliyet.com.tr
4. Internet World Stats, http://www.internetworldstats.com
5. Can, F., Kocberber, S., Balcik, E., Kaynak, C., Ocalan, H.C., Vursavas, O.M.: Information Retrieval on Turkish Texts. J. Am. Soc. Inf. Sci. Technol. 59(3), 407–421 (2008)

6. Göknel, Y.: Turkish Grammar (Transformational Generative and Contrastive). Vivatinell Warwick, United Kingdom (2010)
7. Can, F.: Turkish Information Retrieval: Past Changes Future. In: 4th International Conference on Advances in Information Systems, pp. 13–22. Izmir, Turkey (2006)
8. Köksal, A.: Tümüyle Özdevimli Deneysel Bir Belge Dizinleme ve Erisim Dizgesi: TÜDER. In: 3. Ulusal Bilisim Kurultayi, pp. 37–44. Ankara, Turkey (1981)
9. Solak, A., Can, F.: Effects of stemming on Turkish text retrieval. In: 9th Int. Symp. on Computer and Information Sciences, pp. 49–56. Antalya, Turkey (1994)
10. Porter, M. F.: Snowball: A language for stemming algorithms, http://snowball.tartarus.org/texts/introduction.html
11. Akın, A. A., Akın, M. D.: Zemberek , an open source NLP framework for Turkish Languages, http://zemberek.googlecode.com
12. Çilden, E. K.: Snowball: Stemming Turkish Words Using Snowball, http://snowball.tartarus.org/algorithms/turkish/stemmer.html
13. Ekmekioglu, F.C., Willett, P.: Effectiveness of stemming for Turkish text retrieval. Program 34, 195–200 (2000)
14. Yilmazel, O.: A Language Modeling Approach to Turkish text retrieval. Journal of Science and Technology Applied Sciences and Engineering 11(2), 163–172 (2010)
15. Hafer, M.A., Weiss, S.F.: Word segmentation by letter successor varieties. Information Storage and Retrieval 10, 371–385 (1974)
16. Altintas, K., Can, F.: Stemming for Turkish: A comparative evaluation. In: 11th Turkish Symposium on Artificial Intelligence and Neural Networks, Istanbul, Turkey, pp. 181–188 (2002)
17. Mitra, M., Buckley, C., Singhal, A., Cardie, C.: An Analysis of Statistical and Syntactic Phrases. In: 24th International Symposium on Computer and Information Sciences, Montreal, Canada, pp. 200–214 (1997)
18. Salton, G., McGill, M.J.: Introduction to Modern Information Retrieval. McGraw-Hill, Inc., New York (1986)
19. Fagan, J.: Experiments in automatic phrase indexing for document retrieval: A comparison of syntactic and non-syntactic methods. Ph.D. thesis, Dept. of Computer Science, Cornell Univ., Ithaca, N.Y. (1987)
20. Fagan, J.: Automatic Phrase Indexing for Document Retrieval. In: 10th Annual International ACM SIGIR Conference on Research and Development in Information Retrieval, pp. 91–101. ACM Press, New York (1987)
21. Pickens, J., Croft, W.B.: Turkish - An exploratory analysis of phrases in text retrieval. In: Recherche d'Information Assiste par Ordinateur, Paris, France, pp. 1179–1195 (2000)
22. Arampatzis, A., van der Weide, T., Koster, C.H.A., van Bommel, P.: An evaluation of linguistically-motivated indexing schemes. In: 22nd BCS-IRSG Colloquium on IR Research, Cambridge, England, pp. 34–45 (2000)
23. Ozdemir, B., Cicekli, I.: Turkish Keyphrase Extraction Using Multi-Criterion Ranking. In: 24th International Symposium on Computer and Information Sciences, Guzelyurt, Cyprus, pp. 269–273 (2009)
24. Senem Kumova, M., Karaoğlan, B.: Collocation Extraction in Turkish Texts Using Statistical Methods. In: 7th International Conference on Advances in Natural Language Processing, Reykjavik, Iceland, pp. 238–249 (2010)
25. Arisoy, E., Roark, B., Shafran, I., Saraclar, M.: Discriminative N-gram language modeling for Turkish. In: 19th Annual Conference of the International Speech Communication Association, Brisbane, Australia, pp. 825–828 (2008)

26. Can, F., Kocberber, S., Balcik, E., Kaynak, C., Ocalan, H.C., Vursavas, O.M.: First large-scale information retrieval experiments on Turkish texts. In: 29th Annual International ACM SIGIR Conference on Research and Development in Information Retrieval (Poster paper), pp. 627–628. ACM Press, New York (2006)
27. Terrier IR Platform, http://terrier.org
28. Ounis, I., Amati, G., Plachouras, V., He, B., Macdonald, C., Johnson, D.: Terrier Information Retrieval Platform. In: 27th European Conference on IR Research, Spain, pp. 517–519 (2006)
29. Buckley, C., Voorhees, E.M.: Retrieval Evaluation with Incomplete Information. In: 27th Annual International ACM SIGIR Conference on Research and Development in Information Retrieval, pp. 25–32. ACM Press, New York (2004)
30. Turkish Stop Word List 1.1, The Natural Language Processing Group, Department of Computer Engineering, Fatih University, http://nlp.ceng.fatih.edu.tr/blog/tr/?p=31
31. Pembe, F.C., Say, A.C.C.: A linguistically motivated information retrieval system for Turkish. In: Aykanat, C., Dayar, T., Körpeoğlu, İ. (eds.) ISCIS 2004. LNCS, vol. 3280, pp. 741–750. Springer, Heidelberg (2004)

How Document Properties Affect Document Relatedness Measures

Jessica Perrie, Aminul Islam, and Evangelos Milios

Dalhousie University, Faculty of Computer Science
Halifax, NS, Canada B3H 4R2
jssc.perrie@gmail.com, {islam,eem}@cs.dal.ca

Abstract. We address the question of how document properties (word count, term frequency, cohesiveness, genre) affect the quality of unsupervised document relatedness measures (Google trigram model and vector space model). We use three genres of documents: aviation safety reports, medical equipment failure descriptions, and biodiversity heritage library text. Quality of document relatedness is assessed by the accuracy of a classification task using the kNN method. Experiments discover correlations between document property values and document relatedness quality, and we discuss how one approach may perform better depending on property values of the dataset.

Keywords: Document Relatedness Models, kNN-classification, Document attributes, Google trigram model.

1 Introduction

Identifying relatedness between documents has a number of applications: for example, document categorization–assigning a category to test documents depending on their similarity values to other documents in a training set [1] or alternately, in hospital dialog systems, to classify patients' text responses and to generate the correct feedback [2]. Other uses for document relatedness include document clustering [3], text summarization [4], and word sense disambiguation [5]. Our study investigates how two different unsupervised approaches to document relatedness perform in a document classification task depending on the datasets' properties.

Motivation. There are many approaches to document relatedness, which may work better or worse depending on the tested datasets' document properties: document attributes (e.g., word count, term frequency, or cohesion) or dataset genre (e.g., medical data, aviation reports, biodiversity heritage text). For example, the vector space model can only work if common terms are found between documents. Assuming that no single approach will yield the most accurate relatedness values over all types of datasets, it is useful to know which approach is optimal based on the properties of the dataset. This knowledge can be found based on the approaches' strengths and weaknesses, and based on patterns between document attribute values and genre and relatedness accuracies.

A. Gelbukh (Ed.): CICLing 2014, Part II, LNCS 8404, pp. 392–403, 2014.

When considering approaches, *automatic* approaches to calculate document relatedness may be more desired, because manual approaches–having human judges assign relatedness–can be subject to unconscious bias, random human error, inter-rater disagreement, and extra costs, such as time and money. Unsupervised approaches are ideal in certain situations, because they have a number of advantages over supervised approaches: they require no labeled training set–hence, reducing the issues of over-fitting or under-fitting potential datasets; data used in an unsupervised approach need not require the time and effort of an expert or human participants to accurately label; and lacking labeled data means that small datasets or datasets where labels are unknown may also be tested. [6,7]

Contributions. While different unsupervised approaches have been compared in previous research [8,9,10], our analysis represents the beginning of a study to observe how different approaches to text relatedness perform in relation to the average attribute values of tested documents. We observe this relation between classification accuracies using the document relatedness' measurements and the tested documents' attribute values. Hence, we present the results of a classification task that uses a subset of datasets selected based on their defined attributes and genre. We use kNN-classification [11] as an extrinsic way of assessing performance of the relatedness models, (assuming that if relatedness measures are more accurate, then the classification must also be more accurate) because document relatedness values, as measurements, are intrinsically harder to judge. kNN-classification has also produced very good performance compared to other classification techniques [12].

Our study considers two unsupervised approaches–the vector space model (VSM) [13] and the Google trigram model (GTM) [1]–and shows how applying different dataset subsets may favour one approach over the other by comparing their classification accuracy results. Different dataset subsets are tested sets of documents defined by their attribute values; (e.g., a set could be all documents with word counts between 50 and 150). While studies typically present how a proposed approach works better on multiple datasets, we find that different approaches can appear to perform better by applying them on subsets of each dataset–a subset defined by its attribute values. Consequently, we show how such attributes may affect the classification task accuracies in a distinct relationship.

2 Related Work

Document relatedness approaches greatly vary, but some, like the GTM, use an approach to word similarity to measure the relatedness between words that strongly represent each document. These unsupervised approaches generally use a very large unlabeled dataset to generate similarity scores based on co-occurrence of the words being compared [9] or results from a web search engine [14,15]. A detailed survey of document or text relatedness can be found in [16]. We extend treatment of the GTM by using its document relatedness approach in comparison with the VSM to calculate document relatedness scores.

Islam et al.'s evaluations on GTM word similarity in [9] were based on agreement with similarity of human annotators and synonym matching from the Test of English as a Foreign Language. Like our study, they compare document-relatedness, but only by applying the GTM method with different approaches to word-similarity in the back-end. Document-relatedness approach evaluations are carried out by comparing sentence-pairs relatedness to find correlation to human annotators–as opposed to our study, which compares approaches by reviewing their resulting relatedness measurements in a categorization task, and assuming an implicitly higher similarity for document-pairs of the same category. Additionally, unlike Lee et al.'s research to compare text document similarity models to human-annotated document similarities [10], we focus on comparing two document-relatedness approaches over more varying datasets based on attribute values.

3   Methodology

The methodology for the study, as well as general outline for this paper, is presented in Figure 1. Datasets and Relatedness Models are detailed in this Methodology section, while Evaluations and their results are described and discussed in Section 4.

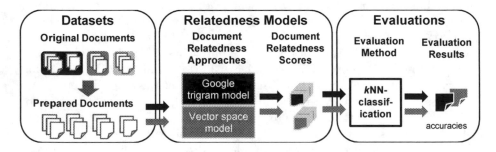

Fig. 1. Each dataset is prepared through cleaning and stop word removal. Then, each document-relatedness model is applied between document-pairs to generate relatedness scores. Selected scores between documents of certain attribute values are then evaluated in the kNN-classification task for each model. Finally, to find which model performs better, the resultant accuracies are compared and analysed.

3.1   Dataset Sources

Datasets selected in this study were taken from collections of documents of three different sources. Because both the VSM and GTM are affected by shared words within document-pairs, and therefore may work poorly when not enough common terms exist, we have restricted the study to datasets of short documents. Shorter documents are less likely to contain the same words than longer documents. However, given the patterns visible between datasets of shorter documents (BHL Titles) and longer documents (BHL Introductions), we believe that our findings are applicable to longer texts to some extent.

Aviation Safety Reporting System (ASRS). ASRS documents came from the SIAM 2007 Text Mining competition training set–preprocessed such that joined words were separated [17]. This study used 399 ASRS reports to assign an unofficial category; documents were about 96 words on average. Each document was written by an aviation personnel to explain a single incident or situation occurring during the flight [18], and according to [19,20], was unofficially categorized into *Incursion (collision hazard)* (165), *Altitude deviation* (59), *Fire or smoke problems* (62), and *Security Concern Threat* (116)–categories chosen because they showed the best results in preliminary tests.

Medical Vigilance Report List (Med). The Med dataset used a subset of descriptions of issues from a vigilance report list generated between 2008 and 2009 that described issues with medical equipment and subsequent categorization from the MEDEVIPAS Greek National project [21]. The dataset used 659 descriptions from the list that were classed as one of two categories: *software* (298) or *hardware* (361). The two other categories in the dataset (*user-interface* and *other*) were not used because they were selected infrequently. Most descriptions were very short; the dataset had documents of 19 words on average. A few examples are listed below:

- *C-arm may tilt*
- *iPulse Console SC1035 displayed a "Low Pressure / Low Flow" alarm and stopped pumping during patient transport in battery operation.*
- *Device may fail to administer therapy during use*

Biodiversity Heritage Library (BHL). BHL is an online resource containing books and journals from natural history and botanical collections [22]. It provides text-based representations of books using optical character recognition on the scanned pages [23]. Within each book subject, terminology of the pages and titles was generally shared. From this source, we selected two different datasets:

- **BHL Titles:** 1152 titles categorized by the book's subject. Categories were generally evenly distributed: *Poultry* (297), *Zoology* (289), *Agriculture* (297), *Botany* (269). Because the datasets were made using the BHL API, some titles were incomplete or included extra information like the authors' names. Titles were, on average, seven words long.
- **BHL Introductions (Intro):** Each document consisted of the first page after the books' title pages (this included a table of contents, introductions, or prefaces) of 338 books to be categorized by the book's subject. Categories were generally evenly distributed: *Sheep* (58), *Biochemistry* (63), *Dairying* (64), *Bacteriology* (94), *Tobacco* (59). The documents were 152 words long on average.

Preprocessing. Before being processed by the different document-relatedness models, each document was preprocessed to clean it and remove stop words. The cleaning process transformed documents to lowercase, removed most punctuation and newline characters, while the stop words came from a list provided by [24].

Single Categorization Justification. Multiple categorizations could potentially upset the evaluation results, which are based on the assumption that
document-pairs within the same selected category should be more similar than
document-pairs where documents are from different categories. Categories in
each dataset (ASRS, Med, BHL Titles and Intro) refer to how documents (reports, descriptions, titles or introductions) were grouped (unofficial categorization, assignment or book subjects, respectively). Hence, because multiple
additional categories could affect the similarity of these supposedly dissimilar
documents, documents classed into multiple categories could not be used in our
evaluations.

3.2   Document Relatedness Models

This section presents the two approaches to document-relatedness that are compared in this study: the vector space model, and the Google trigram model.

Vector Space Model (VSM). VSM is a commonly used and widely accepted
method for calculating document-relatedness within document-pairs. Each document is represented as a vector; relatedness between two documents is calculated
by taking their cosine similarity. Each vector is an ordered list where each list
item represents a word in the dataset by its term-frequency inverse document-
frequency (tf-idf) value within the represented document.

Google Trigram Model (GTM). GTM uses Google trigrams to calculate
word relatedness of representative words from each text to compute text-pair relatedness as described in [1], and it is available on the GTM Web site (http://
ares.research.cs.dal.ca/gtm/). The approach first removes shared instances
between documents being compared (d_1 and d_2, where $|d_1| \leq |d_2|$) before com-
puting the word similarities using Google trigrams–constructing a 'word relatedness matrix' of word-similarities between the remaining words. The elements
of each row i greater or equal to sum of the mean and standard deviation of that
row (A_i) is then averaged ($\mu(A_i)$). Finally, these values for each row are summed
($\sum_{i=1}^{|d_1|-\delta} \mu(A_i)$), added to the number of shared instances (δ), and then, scaled
as shown in Equation 1.

$$\frac{(\delta + \sum_{i=1}^{|d_1|-\delta} \mu(A_i)) \times (|d_1| + |d_2|)}{2|d_1||d_2|} \tag{1}$$

4   Results and Evaluations

4.1 kNN-Classification Set-up

The kNN method was used to measure the ability of each approach to generate
effective relatedness scores by assigning categories to the datasets' entries in
a classification task. Each dataset was randomly divided 30 times for usage
in the kNN-classification task with 10-fold cross-validation. A one-sided t-test–
conducted for the statistical significance at $p < 0.05$–was then carried out on

the 30 outputted accuracies to determine if one approach provided significantly higher accuracy scores than the other.

Because the performance of the kNN-classification is dependent on the value of k (the range of "nearest neighbours" to choose from) different values of k–from 1 to the ceiling of the square root of the number of documents in the dataset– were selected [11]. When comparing the two approaches, the highest accuracy from applying the evaluation to each model's relatedness scores at different k was used.

4.2 Document Attribute Values

To measure the effect of different attribute values on kNN-classification performance, all document relatedness scores were computed using the GTM and the VSM separately. During each evaluation, only scores from document-pairs that matched the attribute restrictions were used in the kNN task. Attribute restrictions were bounds from the lower limit of the attribute to the highest at different incrementing intervals. The average of these attribute values was then observed in relation to the resultant accuracy.

The chosen attributes are different, not strongly correlated to each other, and display some effect on VSM and GTM: word count (longer documents are known to create too much variation for VSM, while shorter documents may not contain enough similar words), term frequency (both GTM and VSM rely on words frequent enough to be shared within document-pairs), and cohesion (GTM uses representative words to find document relatedness; if the document's cohesion is low, the words in the document may be a poor choice). Each attribute assigned a value to each document; if a document's attribute value was between the limits of the subset, then the document was included in the classification test. Attributes for each document are described below:

- **Word Count:** The number of words within a document.
- **Term Frequency:** A normalized average of the frequency of each word within the document compared to its dataset as shown in Equation 2

$$\frac{\sum_{\text{each document word}} \text{frequency of that word in dataset}}{\text{\# document words}} \tag{2}$$

- **Cohesion:** The average word similarity (as found by applying GTM between consecutive words in the document), as shown in Equation 3

$$\frac{\sum_{\text{each document word}} \text{word similarity between word and next}}{\text{\# document words - 1}} \tag{3}$$

4.3 Results and Discussion

From varying the levels of different attributes it was found that for different dataset subsets the document-relatedness approaches yielded significantly different results as shown in Table 1. There were also distinctive relationships between the different resulting accuracies and average attribute values of the tested documents.

Table 1. kNN-classification 10-fold cross-validation result summary for each attribute at limits in the minimum lower bound (Min.), maximum upper bound (Max.), interval (Int.). The percentage of tests in which 1-sided significance is found, is shown under "GTM ? VSM". The correlation coefficients between the average attribute values of each dataset subset and the mean classification accuracy are presented (Attr. Correlation) following different relation patterns: Positive linear (Pl), Negative linear (Nl), Positive parabolic (Pp), and Negative parabolic (Np). Highest correlations of each approach are **bolded**.

Dataset	Limits Min.	Max.	Int.	GTM ? VSM >	<	no diff.	Attr. Correlation GTM		VSM	
Word Count:										
ASRS	6	302	8	36.6	41.7	21.7	Pl	**0.662**	Np	0.366
Med	2	100	2	62.2	26.0	11.8	Pp	0.531	Pp	**0.603**
BHL Titles	0	36	2	67.5	14.2	18.3	Pp	0.004	Pp	**0.031**
BHL Intro	53	539	9	0.0	99.0	0.1	Nl	0.335	Nl	**0.625**
Term Frequency:										
ASRS	0.04	0.36	0.01	17.5	57.3	25.2	Np	**0.713**	Np	0.561
Med	0.01	0.52	0.01	68.0	23.6	8.4	Pl	0.721	Pl	**0.931**
BHL Titles	0.00	1.00	0.05	63.8	30.7	5.5	Np	**0.604**	Np	0.578
BHL Intro	0.03	0.21	0.01	1.0	91.0	8.0	Pp	**0.859**	Pp	0.834
Cohesion:										
ASRS	0.15	0.30	0.01	20.8	65.3	13.9	Np	**0.889**	Np	0.882
Med	0.00	0.37	0.01	74.1	17.3	8.6	Np	0.276	Np	**0.620**
BHL Titles	0.00	0.45	0.01	79.5	9.3	11.2	Np	**0.517**	Np	0.470
BHL Intro	0.05	0.35	0.01	0.0	99.3	0.0	Np	**0.743**	Np	0.719

Word Count. From the kNN-classification, word count was an important factor in determining which approach produced significantly better results, and adjusting its average attribute value affected the classification accuracy in different patterns as shown in Table 1.

In both the ASRS and Med datasets, once the average word count of the tested subset passed a certain threshold, accuracies of the classification task that used the sets' generated document-relatedness measurements were significantly higher in either the GTM or the VSM–demonstrating that as subsets are taken in accordance with an attribute, the resultant classification accuracy preference may be changed. Furthermore, according to the results for ASRS, kNN-classification accuracies generated using GTM relatedness measurements had a moderately strong linear relationship with word count [25]. Evaluation results using VSM had a non-linear relationship: at lower and higher levels of word count, the VSM accuracy dipped–suggesting that documents in these sets did not have enough frequent terms within categories or held too many common words–known disadvantages of the model.

Evaluation results involving the Med dataset did not correlate with average word count linearly, but, as shown in Table 1, tended to follow a positive parabolic relationship; however, accuracies tended to become lower as document

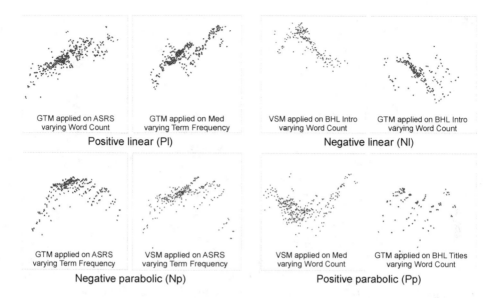

Fig. 2. Select scatterplots of average kNN-classification accuracies and subset average attribute values that represent the general patterns found in results. Linear relations indicate that the classification accuracy changes linearly with an increase of the average attribute values. Parabolic relations indicate a minimum (Positive) or maximum (Negative) accuracy value.

length grew. Results involving the BHL Intro also showed this negative linear relationship. In comparison with the BHL Intro dataset (average 152 words), the Med dataset is smaller (average 19 words)–showing how this relationship may be visible in larger datasets. As both the results involving the GTM and VSM models yield higher accuracy when applied to documents of lower count, shorter documents were more representative of the datasets. As document lengths grew, more similar documents words became less representative of the document to differentiate categories. Also, the Med dataset included many similar documents; hence, as similar documents shared the same length, smaller subsets may have had better accuracy assignment of the majority class was based on neighbouring documents with similar words.

Any relationship between accuracy and word count was less observable in evaluations involving the BHL Titles mainly because book titles were too short to meaningfully divide subsets by length (note their low correlation with the parabolic relation to attributes) although a somewhat negative relationship was observable, as shown in Figure 2. This relationship is likely because longer titles occasionally contained extraneous details like author names or notes not related the book's subject matter (e.g., "With three hundred and eleven illustrations.").

Term Frequency. In the results from the kNN-classification, term frequency can be an important attribute. In evaluations involving Med, BHL Titles and Intro datasets, as term frequency got higher, both the GTM and VSM approaches yield generally better results–expected given how both used frequent terms shared within document-pairs to calculate relatedness. Evaluation results using BHL Titles and Intros are less correlational than Med, potentially because the terms in titles and pages are more varied than medical reports (which may have similar wording)–making less impact on the results.

While most relationships tend to trend upward, evaluations involving the ASRS dataset distinctly yield the worst results at more frequent terms, as shown in Figure 2. Potentially, at too many frequent terms, documents in different classes contain too many similar words–affecting both models' use of common words within document-pairs. This idea is supported because unlike other datasets, ASRS contained words that were found in a majority of its documents: *aircraft* (found in 250 documents), *flight* (245), *airport* (226). On the contrary, at small term frequencies, not enough common terms are found for tf-idf (VSM) or shared instances (GTM) to calculate meaningful document similarity.

In review of the comparison between GTM and VSM, considering term frequency, classification accuracies generally favour one approach for all cases except between two thresholds of term frequency. This is likely because of how each approach computes relatedness when too many or too few frequent terms are found. For example, at lower and higher term frequencies there are too few or too many terms to accurately represent the documents, the GTM yields significantly higher results than the VSM by relying on its word similarity approach.

Cohesiveness. As shown in the t-test significance results of Table 1, for half of the datasets, one approach generally worked better for all levels of cohesion, but in Med and BHL Titles, cohesion played an important factor in determining which approach did better–allowing areas between thresholds of cohesion value to yield better classification accuracy results. This is likely because both Med and Titles on average contained relatively short documents–meaning that with lower cohesion, the selected words were less representative of the overall document.

Interestingly enough, all datasets displayed a negative parabolic relationship between the cohesion accuracy and the attribute value–with a generally negative linear trend. At low levels of cohesion, understandably, the text is less compact–meaning that it should be less cohesive at presenting a united idea or less representative of the document's purpose. This low accuracy grows as the cohesion increases. However, at higher levels of the measured cohesion, accuracy results get lower. This could be a result of how documents with higher cohesion generally had a lower word count, and hence, a lower chance of finding shared instances within the document-pair.

Application to Different Datasets. Through using each dataset elicited relationships, we found that at different properties of the dataset, relationships between the resulting accuracies and average attribute values will work differently depending on the dataset being used. Although similar patterns may be observed (for example, the negative parabolic patterns in tests varying cohesion in Table 1), relations between the classification accuracy and average attribute values generally followed different paths. Consider similar values for cohesion between datasets: if the plots involving each dataset are combined, the correlation becomes weaker, as the classification accuracies between datasets are based on each individual dataset.

The results presented have implications on applying these found relationships to identify which approach will yield better results on a brand new dataset. The new dataset must be compared to known datasets in terms of similarity of genre (e.g., aviation reports), attributes (e.g., word count), and potentially, other document properties not addressed in this study. If the dataset follows the same properties, an appropriate choice of document relatedness approach can be made. Failure in the best choice could be because of identifying wrong relationships leading to an incorrect and biased result. Although our study does not address this ideal task, we have presented a basis for its usage.

4.4 Limitations

Within this study there are a number of limitations that could be overcome in future studies. The datasets selected were very limited in a number of ways: type, lengths, category and size. Only aviation reports, medical data descriptions, and biodiversity literature were used, while different documents yielded different results when using the same document relatedness model. Hence, other datasets could produce findings different than those observed in our studies. Additionally, as experiments showed, document length was an important factor in defining which approach did better, but the documents tested were all relatively short. Also, datasets were only organized into 2 to 5 categories–yielding high accuracies from both relatedness models. And finally, the datasets' numbers of documents were generally low: 338 (BHL Intro) to 1152 (BHL Titles). Future studies could consider using larger datasets with more categories.

There were also limitations based on how the evaluations were set up. Unlike other studies [9], we chose to measure performance based on categorization using the document-relatedness scores. Categorizing can depend on many aspects of the document (e.g., publication date, author, generalness of topic), yet our study assigned categories using relatedness values computed by involving only the document-pairs' word content.

Finally, while the effects of the different attributes were discussed, observations were mainly correlational, and do not imply causation. Additionally, the different attributes were measured separately; future work could focus on observing the important attributes in conjunction and their combined effects in different models.

5   Conclusion

In our study, we applied two different document relatedness models–the vector space model (VSM) and the Google trigram model (GTM)–to various different datasets–ASRS, Medical equipment failure reports data, BHL titles and introductory pages–and evaluated the resultant scores using kNN-classification on subsets of the datasets chosen based on different document attribute values. Based on these ranges of word count, term frequency, and cohesion, we found cases where one approach worked significantly better, and discussed potential reasons for the effect of different attributes on the classification accuracies.

Future Work. Future work in this area should consider overcoming limitations outlined in this study and continue to develop a set of guidelines to help the user to decide which unsupervised document-relatedness approach should be used to produce better results. While our observations between average attribute values and resultant accuracies are correlational and considered using only one attribute type, we could review the results of using document subsets based on multiple attributes (e.g., using only a subset of documents with high word count *and* high term frequency) to select the most accurate document relatedness approach.

Acknowledgements. This research was funded by the Natural Sciences and Engineering Research Council of Canada (NSERC), and the Boeing Company.

References

1. Islam, A., Milios, E., Kešelj, V.: Text similarity using google tri-grams. In: Kosseim, L., Inkpen, D. (eds.) Canadian AI 2012. LNCS, vol. 7310, pp. 312–317. Springer, Heidelberg (2012)
2. Bickmore, T., Giorgino, T.: Health dialog systems for patients and consumers. Journal of Biomedical Informatics 39(5), 556–571 (2006)
3. Liu, T., Liu, S., Chen, Z., Ma, W.Y.: An evaluation on feature selection for text clustering. In: Fawcett, T., Mishra, N. (eds.) Proc. 20th International Conference on Machine Learning (ICML 2003), pp. 488–495. AAAI Press (August 2003)
4. Erkan, G., Radev, D.: Lexrank: Graph-based lexical centrality as salience in text summarization. Journal of Artificial Intelligence Research 22, 457–479 (2004)
5. Schutze, H.: Automatic word sense discrimination. Computational Linguistics 24(1), 97–124 (1998)
6. Wagstaff, K.L.: Chapter 1: Introduction: 1.2 Supervised Learning: Disadvantages of supervised learning (April 2007), `http://www.wkiri.com/research/papers/wagstaff-diss-1.ps` (last accessed August 13, 2013)
7. Liu, Q., Wu, Y.: Supervised Learning (Janurary 2011), `http://www.fxpal.com/publications/FXPAL-PR-11-626.pdf` (last accessed August 13, 2013)
8. Sathy, R., Abraham, A.: Comparison of Supervised and Unsupervised Learning Algorithms for Pattern Classification. International Journal of Advanced Research in Artificial Intelligence 2(2), 34–38 (2013)
9. Islam, A., Milios, E., Kešelj, V.: Comparing Word Relatedness Measures Based on Google n-grams. In: International Conference on Computational Linguistics, pp. 495–506 (December 2012)
10. Lee, M.D., Pincombe, B.M., Welsh, M.B.: An empirical evaluation of models of text document similarity. In: Proceedings of the XXVII Annual Conference of the Cognitive Science Society, Austin, Texas, USA, pp. 1254–1259. Cognitive Science Society (2005)
11. Thirumuruganathan, S.: A Detailed Introduction to K-Nearest Neighbor (KNN) Algorithm (May 2013), `http://saravananthirumuruganathan.wordpress.com/2010/05/17/a-detailed-introduction-to-k-nearest-neighbor-knn-algorithm/` (last accessed July 30, 2013)

12. Colas, F., Brazdil, P.: Comparison of svm and some older classification algorithms in text classification tasks. In: Bramer, M. (ed.) Artificial Intelligence in Theory and Practice. IFIP, vol. 217, pp. 169–178. Springer, Heidelberg (2006)
13. Salton, G., Wong, A., Yang, C.S.: A vector space model for automatic indexing. Commun. ACM 18(11), 613–620 (1975)
14. Bollegala, D., Matsuo, Y., Ishizuka, M.: Measuring semantic similarity between words using web search engines. In: Proceedings of the 16th International Conference on World Wide Web, WWW 2007, pp. 757–766. ACM, New York (2007)
15. Turney, P.D.: Mining the web for synonyms: PMI-IR versus LSA on TOEFL. In: Flach, P.A., De Raedt, L. (eds.) ECML 2001. LNCS (LNAI), vol. 2167, pp. 491–502. Springer, Heidelberg (2001)
16. Gomaa, W.H., Fahmy, A.A.: A survey of text similarity approaches. International Journal of Computer Applications 68(13), 13–18 (2013)
17. Islam, M. A., Inkpen, D.Z., Kiringa, I.: A generalized approach to word segmentation using maximum length descending frequency and entropy rate. In: Gelbukh, A. (ed.) CICLing 2007. LNCS, vol. 4394, pp. 175–185. Springer, Heidelberg (2007)
18. Oza, N.: SIAM 2007 Text Mining Competition dataset (September 2010), https://c3.nasa.gov/dashlink/resources/138/ (last accessed May 30, 2013)
19. Berry, M.W.: Automating the Detection of Anomalies and Trends from Text (2007), http://citeseerx.ist.psu.edu/viewdoc/download?rep=rep1&type=pdf&doi=10.1.1.132.751 (last accessed August 23, 2013)
20. Kiros, R., Soto, A.J., Milios, E., Keselj, V.: Representation learning for sparse, high dimensional multi-label classification. In: Yao, J., Yang, Y., Słowiński, R., Greco, S., Li, H., Mitra, S., Polkowski, L. (eds.) RSCTC 2012. LNCS, vol. 7413, pp. 463–470. Springer, Heidelberg (2012)
21. Bliznakov, Z., Stavrianou, K., Pallikarakis, N.: Medical devices recalls analysis focusing on software failures during the last decade. In: Roa Romero, L.M. (ed.) XIII Mediterranean Conference on Medical and Biological Engineering and Computing 2013. IFMBE Proceedings, vol. 41, pp. 1174–1177. Springer International Publishing (2014)
22. BHL consortium: About (June 2013), http://biodivlib.wikispaces.com/About (last accessed August 7, 2013)
23. BHL consortium: Download All File Types and Descriptions (August 2010), http://biodivlib.wikispaces.com/Download+All+File+Types+and+Descriptions (last accessed August 13, 2013)
24. Lewis, D.D.: RCV1-v2/LYRL2004: The LYRL2004 Distribution of the RCV1-v2 Text Categorization Test Collection (April 12 2004 Version) (April 2004), http://www.ai.mit.edu/projects/jmlr/papers/volume5/lewis04a/lyrl2004_rcv1v2_README.htm (last accessed August 12, 2013)
25. Taylor, R.: Interpretation of the correlation coefficient: A basic review. Journal of Diagnostic Medical Sonography 1(6), 35–39 (1990)

Multi-attribute Classification of Text Documents as a Tool for Ranking and Categorization of Educational Innovation Projects

Alexey An[1], Bakytkan Dauletbakov[1], and Eugene Levner[2]

[1] The Al-Farabi Kazakh National University, Almaty, Republic of Kazakhstan
analexey1@gmail.com; dauletbakovb@mail.ru
[2] Ashkelon Academic Colege, Ashkelon, and Holon Institute of Technology, Holon, Israel
levner@hit.ac.il

Abstract. We suggest a semi-automatic text processing method for ranking and categorization of educational innovation projects (EIP). The EIP is a nation-wide program for strategic development of an university or a group of academic institutions. Our approach to the EIP evaluation is based on the multi-dimensional system ranking that uses quantitative indicators for three main missions of higher education institutions, namely, education, research, and knowledge transfer. The main part of this paper is devoted to the design of a semi-automatic method for ranking the EIPs exploiting multi-attribute document classification. The ranking methodology is based on the generalized Borda voting method.

Keywords: Text classification, numerical classifiers, Borda ranking method, educational innovation projects.

1   Introduction

The text classification problem consists in assigning a document to one or several predefined *classes,* or *categories,* corresponding to different topics of interest. This can be done in several modes: either *manually* by an expert (for instance, by a librarian doing book classification), or *automatically*, where the classification is done entirely by a computer without any human interference, or *semi-automatically*, where some part of the document process (for example, document weighting) is done by an external expert whereas a computational evaluation part is done by a computer algorithm. In this paper we shall use the latter classification mode exploiting, on the one hand, the know-how and intelligence of human experts for document weighting and, on the other hand, powerful computational resources of modern computers.

A variety of different data mining and machine learning techniques have been applied for text categorization, including nearest neighbour classifiers, decision trees, Bayesian classifiers, and support vector machines. Excellent surveys of modern text classification methods and algorithms are presented in [1-3]. However, many current methods and algorithms are too complicated for their practical implementation. New

A. Gelbukh (Ed.): CICLing 2014, Part II, LNCS 8404, pp. 404–416, 2014.
© Springer-Verlag Berlin Heidelberg 2014

computerized systems that quickly filter, pre-sort and rank the documents are highly desirable and they are of great scientific and practical interest.

Today text classification is of great practical importance due to the rapid growth of information and the explosion of electronic texts from the Internet. The text classification problem finds many direct applications to automatically classify articles and web pages [4, 5], to classify scientific abstracts [6, 7] and real-world data sets [8], to automatically learn the interests of users [9], and to sort electronic mail [10]. A variety of other applications are presented in [1, 2].

In recent years, researchers have become increasingly concerned over using bibliometric and webometric data for ranking and categorization of universities and higher education institutions [11-15]. Continuing and extending this research venue, in this paper we demonstrate that the text classification techniques can be used for ranking and categorizing educational innovation projects.

We define an *educational innovation project* (EIP) as a nation-wide program for strategic development of an university or a group of academic institutions, which includes the following features: (1) preparing students for their future careers by developing an educational program corresponding to the standards of the best world universities; (2) infusing innovative information technology (IT) solutions throughout teaching and training, such as e-learning methods, problem-based learning, computer simulation, web-based modules, etc.; (3) promoting pedagogical tools and IT means for continuous lifelong learning and education; (4) transferring new knowledge into practice; and (5) enhancing permanent active links of academia with industry, business and public organizations.

The information that decision makers and researchers use in their work for ranking the EIPs can come from many text sources, such as scientific reports, published literature, web pages, technical prospects, electronic databases, etc. As the amount and volume of text sources exponentially grows in time and a growing portion of them becomes available more easily from the internet websites, the decision makers are faced with the difficult task of ranking and evaluating innovation projects relevant to their interests.

Most users of practical systems for ranking innovation projects, however, would not have the patience to manually process thousands of text documents. They would obviously prefer algorithms that provide accurate classifications after hand-processing of only a few dozen test documents. The impetus for developing the methodology developed in this paper was a desire to suggest a computationally efficient, user-friendly and transparent method of the EIP classification understandable to the user.

In this paper, we consider a *numerical type* of a classifier, which provides a numerical rating of the EIPs. The novelty of the suggested methodology is that we suggest a Borda type ranking method for the computer-based project classification computation generalizing the original Borda ranking menthod. The Borda method is known to be simple and provides the minimum of voting paradoxes [16,17]. (The basics and properties of the Borda method can be found in [16-22]).

The following basic assumptions (BA) are used in our approach:

BA-I. Our methodology follows 16 Berlin Principles on Ranking of Higher Education Institutions (HEI) [23]. The principles emphasize the purposes and goals of rankings HEIs, the design and weighting of indicators, the collection and processing of data, and the presentation of ranking results. They articulate important standards of practical ranking of the EIPs.

BA-II. We restrict ourserves to considering three main missions of the EIPs (namely, education, research, and knowledge transfer), and evaluating their numerical indicators.

BA-III. The quantitative estimation of the EIP mission indicators can be done (automatically or semi-automatically) with the help of the analysis of document attributes.

BA-IV. The estimates of the individual misssion indicators can be integrated into the global EIP ranking using the generalized Borda method.

The rest of the paper is organised as follows. Section 2 gives an overview of the suggested classification method and the individual IEP indicators. Section 3 provides an illustrative example of the EIP classification method and reports the results of an experiment. Section 4 draws the conclusion.

2   Description of a New Classification Method

The project classification method developed in this paper is an extension of the text classification method proposed by Levner et al. [7]. The essence of the extension is the following.

First, the goal of the categorization problem in paper [7] has been to classify individual documents into a fixed number of predefined categories. In contrast, the ultimate goal of the novel methodology is to rank and classify the EIPs rather than the documents. Second, the mentioned ultimate goal implies that the suggested classification method has a more complicated (multi-stage) computational structure than the method suggested in [7].

The multi-stage ranking process starts with the text document classification and terminates with the ultimate ranking/classification of the EIPs. It can be presented schematically as follows:

$$\{Document\ attributes \rightarrow Document\ fitness\ estimation \rightarrow Indicator\ values \rightarrow Project\ ranks\}$$

For the sake of completeness and clarity, at this point we briefly remind the above-mentioned notions used in the new classification method.

Document attributes are the main elements or bibliometric characteristics of the document. In this paper, we consider the following document attributes: *title, keywords, abstract, conclusion, bibliography, authors' bio; number of citations; h-index; impact factor of the journal.*

EIP's indicators are integrated numerical characteristics representing main missions of the EIP's. In this paper, we consider three main indicators corresponding to three main missions of EIPs, namely, education, research, and knowledge transfer).

Document fitness is a function $(i,j,k) \rightarrow f_{ij}^k$, which for each triple (i,j,k) measures the correspondence f_{ij}^k of document i to the EIP's mission indicator j with respect to attribute k.

The new method works in several sequential stages: at the first stage, the quantitative attributes of the considered documents are processed to obtain the numerical values for three groups of EIP's indicators corresponding to three main missions of EIPs (i.e., education, research, and knowledge transfer); at the final stage, the obtained mission indicators' values are exploited for ranking and classification of the EIPs. Such an approach permits to efficiently use experts´ knowledge and the information contained in the text documents.

A main practical contribution, in comparison with the earlier known approaches of ranking educational institutions using the bibliometric data, is that we use a Borda type voting method as a tool for computing the ranks of the EIPs and grouping the EIPS with similar or close Borda counts into separate clusters (categories). The corresponding definition of the Borda count and main features of the Borda method are given in Section 2.3. The method is illustrated with a numerical example with several EIPs in Section 3.

2.1   Overview of the Mission Indicators

Consider the individual indicators of EIPs corresponding to the main missions of the EIPs, namely, education, research, and knowledge transfer. Evidently, the complete set of the indicators is lengthy and strongly depends upon the interests and goals of the users and stakeholders of the ranking/classification procedure. To illustrate the suggested methodology, we select below typical subsets, each containing ten universal and standard indicators.

Education Indicators. Education is the core mission (function) of a typical EIP. Educational (teaching and learning) processes aim to transmit knowledge, skills and educational values to students.

For the sake of definity, we select the following indicators:

1.1. Expenditure on teaching activities as a percentage of total expenditure.
1.2. The rate of unemployment of bachelor graduates and master graduates as a percentage of the national rate of unemployment of graduates.
1.3. The number of students per FTE academic staff.
1.4. The qualification of the academic staff (the rate of PhD degree holders and professors).
1.5. The access to IT, internet and computer support.
1.6. Overall satisfaction of students with their higher education institution (HEI).
1.7. Satisfaction of students with teaching and lecturers.
1.8. Satisfaction of students with HEI's facilities and service.
1.9. Satisfaction of students with courses and programs.
1.10. Satisfaction of students with the information and web page of HEI.

Research Indicators. The term *research* covers three dimensions: basic research, applied research and experimental works. In this study, we select the following indicators.

2.1. Expenditure on research in the reference year.
2.2. Number of articles published in the refereed journals.
2.3. Number of publications in top 10% most highly ranked (according to ICI) journals
2.4. Number of chapters in books.
2.5. Number of papers in refereed conference proceedings.
2.6. Number of papers in non-refereed proceedings and books of abstracts.
2.7. Number of finished Ph.D theses.
2.8. Number of post-doc positions /FTE academic staff
2.9. Area-normalized citation impact score, according to the Thomson Reuters journal categories.
2.10. Number of prizes, medals, awards and scholarships won for the research work.

Knowledge-Transfer Indicators. *Knowledge transfer* is the process by which the new knowledge, expertise and intellectual achievements of EIP are applied beyond the projects for the benefit of the economy, business and society.

We select the following knowledge-transfer indicators.

3.1. Annual income from research projects and programs.
3.2. Number of university-industry and university-business joint publications.
3.3. Number of international joint publications.
3.4. Number of patent applications and awarded patents by EIP's staff.
3.5. Number of international cooperation projects and programs.
3.6. Budget and number of joint research projects with private sector.
3.7. Budget and number of joint research projects with industries.
3.9. Budget and number of joint research projects with banks and business.
3.10. Number of start-ups created by EIP staff over the last three years.

2.2   Data Collection from Text Documentation

Text documentation is one of the most important sources of knowledge related to the EIP ranking and categorization. Estimation of the EIP mission indicators listed above is most frequently taken through two basic text sources: bibliometric data and questionnaires. The bibliometric data include data about publications and citations. The (weighted) number of publications and citations and other text information can serve for estimating the EIP índices, especially those related to the research mission, whereas the text information extracted from experts', students' and other type questionnaires can be especially helpful for estimating the quality of the education and knowledge transfer missions of the EIPs.

The text document databases relelvant to the EIP classification are diverse and multi-purpose. They may include journal articles, books, conference proceedings, Ph.D theses, patents, research reports, technical prospects, webpages, citations, and other document types. Accoring to our methodology, the classification algorithm processes large sets of documents related to the considered set of educational innovation projects, estimates each document from the point of view of its belonging/non-belonging to an EIP's mission category and extracts a qualitative measure of belonging/non-belonging by estimating documents' attributes (number of relevant words, qualitative characteristics, etc.) The individual estimates are weighted and "summed-up", establishing, as a result, first, the indicators' evaluation for each EIP's mission. Then the EIPs are ranked and their integrated quality index (IQI) is found basing on the extended Borda's method.

2.3   Text Classification and a New EIP Classification Method

The new EIP classification method uses, first, the document classification method suggested in [4] and, then, basing on this classification, integrates the obtained knowledge for ranking and classifying the EIPs. The objective is to develop a computer-aided procedure for categorizing the EIPs automatically.

The proposed EIP classification method consists of five steps.

The first step is the identification of attributes and indicators. We identify six sets:
(a) *the set of attributes* characterizing the documents;
(b) *the set of predefined document categories* (each category here corresponds to a mission indicator);
(c) *the set of classifying words and expressions* within each category;
(d,e, and f) *the set of weights* w_k, u_i, and v_j, provided by experts and establishing the relational importance of the document attributes, the documents, and the mission's indicators, respectively.

The weights can be either numerical, or interval-valued.

The second step is data collection and computing of the attribute-dependent fitness, for each attribute of each document. At this step, for each available EIP s to be ranked, we collect the attribute values and compute the *fitness* measure (correspondence) f_{ij}^k of document i to the EIP's mission indicator j with respect to attribute k. In order to avoid unnecessary complications in the notation, at this point we omit the EIP index s in the latter symbol, though, strictly speaking, it should be of the form $f_{ij}^{k,s}$.

The fitness f_{ij}^k is a function of several (in our case, three) arguments:
(1) the number of words in the attribute k of document i that coincide with predefined classifying words and expressions for mission indicator j;
(2) the size (the number of words) of attribute k of document i, and
(3) the number of the classifying words in mission indicator j. For simplicity of notation, at steps 1-3, we omit the indices of EIPs.

The third step is computing of the whole document-indicator (attribute-independent) fitness f_{ij}. At this step, for each considered EIP for each document, we define the complete (integrated with respect to the attributes) the fitness f_{ij}^s of each document i related to EIP s in respect to the mission indicator j, by using an additive approach as in [7]:

$$f_{ij}^s = \sum_k w_k f_{ij}^k = \sum_k w_k f_{ij}^{k,s},$$

where the attribute weights w_k are defined by experts at the first step.

The fourth step is the computation of the individual indicator (document-independent) values V_{js}. At this step, for each EIP, indexed s, we first define the complete (integrated with respect to all the available documents) value (importance) f_j^s of each individual mission indicator j (for each EIP s), by using an additive approach:

$$f_j^s = f_j^s(\text{EIP}_{(s)}) = \sum_i u_i f_{ij}^s = \sum_i u_i^{(s)} f_{ij}^s,$$

where $u_i = u_i^{(s)}$ is document's i weight (relational importance) defined by the experts at the first step. Notice that, generally speaking, the weights u_i can be dependent on s.

Next, the weighted indicator's value V_{js} with respect to EIP s is defined as follows:

$$V_{js} = v_j f_j^s, j=1,\dots, J; s = 1,\dots, N,$$

where v_j is indicator's j weight (relational importance) also defined by the experts at the first step. These weights play an important role in the suggested Borda-type ranking method. Indeed, they permit us not only to simply rank the EIPs but also to assign a measure of integrated quality to any one of them. This role will be demonstrated in the example below wherein the EIPs not only ranked but also can be grouped into clusters with a similar quality.

The fifth step. Classifying and ranking the EIPs. At this step, we use the obtained numerical indicator values V_{js}. For each EIP s ($s =1,\dots, N$) we have a J-dimensional vector of J different criteria (indicators) ($V_{1s}, V_{2s}, \dots, V_{Js}$) which we may use for finding a consistent coordinated ranking of all the EIPs. The presence of indicator *weights* (relative importance values) v_j lying in different intervals within a given scale (which can be, e.g. [0; 10], or [0; 100]) permits us to consider the indicators of different importance.

An approach pursued at this step of the algorithm is the generalized Borda ranking method that works as follows. Suppose that a finite number of *indicators* (also called criteria or measures of effectiveness) are used to evaluate a given number of EIPs (the latters are also called 'alternatives' [17, 18]). Our goal is to aggregate information for several indicators and to obtain an overall consensus ranking of the EIPs. Our ranking method permits to express the *degree* of preference of one alternative (EIP) over another, for the case of multiple criteria and generalizes the Borda method for heterogeneous criteria. The main idea is that we calibrate the levels of alternatives. As far as we allow to add numeric or linguistic weights to the criteria, this permits us to integrate information obtained from the indicators of different types.

Suppose that there are N EIPs and J *criteria* (indicators, MOEs), and that the jth indicator has its associated weight v_j, possibly different for different indicators. A preference order (sometimes called Borda's individual order) is supplied by indicator values $\{V_{js}\}$ that rank the EIPs (with respect to each individual indicator j) from the most preferred to the least preferred, possibly with ties. A preference order in which an EIP_1 is ranked first, EIP_2 is ranked second, and so forth, is denoted by $(1,2,\dots,N)$, and this order may be different for different j.

Let us briefly describe the Borda's method for aggregating information provided by different indicators and obtaining an overall preference order integrating the information from all the indicators. The method works as follows. Given N documents, the points ("Borda's grades") $N - 1, N - 2, \dots, 0$ are assigned to the first-ranked, second-ranked, $\dots$, and last-ranked EIP in each indicator's preference order;

then, the Borda grades are summed up, and, finally, the winning document is the one with the greatest total number of points. Formally, if r_{ik} is the rank of EIP s by indicator j, the *Borda count* for EIP s is $b_i = \sum_k (N - r_{ik})$.

The EIPs are then ordered according to these counts. We refer to [7, 16-22] for further details of the Borda method's properties and applications.

3 Classification of the IEPs: An Illustrative Example

Consider a numerical example containing four EIPs and using a bibliometric database consisting of 100 documents. The EIP ranking is to be obtained, based solely on the data extracted from this database.

Our EIP ranking/classification method works as follows.

The first step. Identification of document attributes and EIP indicators. At this step, we identify five sets:

(a) the set of document attributes. This is a set of one hundred of 6-dimensional vectors:

$$DAS_i = \{title, \ key\text{-}words, \ abstract, \ conclusion, \ bibliography, \ and \ authors' \ bio\}.$$
$$i=1,\ldots, 100.$$

(b) the set of predefined EIP's missions and corresponding indicators (that are categories). For simplicity, we consider only four indicators in each of the three missions, as follows:

Education_Mission Indicators =
{Expenditure rate; Students/teachers rate; Staff qualification; Students' satisfaction};

Research_Mission Indicators =
{Expenditure; Number of journal articles; Number of citations; Number of awards};

Knowledge_Transfer_Mission Indicators =
{Income, Number of industrial projects; Number of patents; Number of start-ups}

(c) the set of classifying words and expressions, for all the document attributes and all the indicators.

This set, which is the same for all indicators (categories) for all the four EIPs, contains 1,000 words and expressions selected by experts. This list is omitted here because of lack of space.

(d) The weights for the doc attributes and the EIP's indicators. For simplicity, all they are taken to be 1.

The second step. Data collection and finding the attribute-dependent fitness of each document. At this step, separately for each EIP, we collect the attribute values and compute the *fitness* measure (correspondence) f_{ij}^k of document i to the EIP-mission indicator j with respect to attribute k.

Totally, for all the four EIPs, the algorithm performs $|i| \times J \times |k| = 100 \times 12 \times 6 = 720$ elementary computations of the *fitness* value f_{ij}^k of document i with respect to category j with respect to attribute k (where $i=1,\ldots, 100$; $j=1,\ldots,12$; $k=1,\ldots, 6$) employing the following relation suggested in [7]:

$$f_{ij}^k = \min \{1, g_{ij}^k / \alpha_j^k\}, \tag{1}$$

where g_{ij}^{k} is the number of words in attribute k of document i that coincide with the pre-specified classifying words for category j; α_{j}^{k} is the minimal number of classifying words in category j with respect to attribute k whose presence in k is sufficient for classifying a document as belonging to the category j (the α_{j}^{k} value is defined by experts). The maximum possible fitness is 1.

For example, assume that attribute k is *abstract* of some document i^{*}, that $\alpha_{j}^{k} = 10$, and four words in the considered *abstract* are found among the classifying words of category j; then the fitness $f_{ij}^{k} = 0.4$. Because of lack of space, we omit here the intermediate calculations of the f_{ij}^{k}.

The third step. Computing the integrated fitness f_{ij} of each document i in respect to the mission indicator j, by using an additive relation $f_{ij} = \sum_{k} w_{k} f_{ij}^{k}$, where all the $w_{k} = 1$.

The fourth step. Computing the individual indicator (document-independent) values V_{js}, for each indicator for each EIP, $s = 1,2,3,4$; $j = 1,2,3$ by using the following formulas: $V_{js} = v_{j} f_{j}^{s} = v_{j} \sum_{i} u_{i}^{s} f_{ij}^{s}$, $j = 1, \ldots, 3$; $s = 1, \ldots, 4$, where all v_{j} are taken to be 1. The coefficients $u_{i}^{(s)}$ reflecting the relative importance and experts' confidence are taken to be either 1, or 2, or 3, the same for all EIPs and are the following: $u_{1}^{(s)} = u_{1} = 3$; $u_{2}^{(s)} = u_{2} = 2$; $u_{3}^{(s)} = u_{3} = 3$; $u_{4}^{(s)} = u_{4} = 2$; $u_{5}^{(s)} = u_{5} = 3$; $u_{6}^{(s)} = u_{6} = 1$; $u_{7}^{(s)} = u_{7} = 1$; $u_{8}^{(s)} = u_{8} = 1$; $u_{9}^{(s)} = u_{9} = 3$; $u_{10}^{(s)} = u_{10} = 2$; $u_{11}^{(s)} = u_{11} = 1$; $u_{12}^{(s)} = u_{12} = 2$.

Omitting the intermediate calculations of f_{ij}^{k} and f_{ij}, the resulting V_{js} values obtained at Step 4 are presented in Table 1 and in the radar charts in Figure 1. The values V_{js} are normalized to be in the interval [0; 1] and set as follows: $V_{js} := V_{js}/\max_{s} V_{js}$.

The fifth step. Classifying and ranking the EIPs. First, we find the preference ordering defined by each indicator that rank the EIPs (with respect to each individual indicator j) from 1 to 4. The corresponding ranks are presented in Table 2.

Table 1. The individual indicator values for four selected EIPs

Missions	Indicators V_{js}	EIP1	EIP2	EIP3	EIP4
Education	Expenditure on education V_{1s}	0.80	0.26	0.40	1.0
	Teachers/students rate V_{2s}	0.82	0.50	0.80	0.90
	Staff qualification V_{3s}	1.00	0.55	0.46	0.90
	Students' satisfaction V_{4s}	0.60	0.85	0.84	0.90
Research	Expenditure on research V_{5s}	0.44	0.70	0.15	0.72
	Number of journal articles V_{6s}	0.70	0.40	0.96	1.00
	Number of citations V_{7s}	0.70	0.80	0.30	0.75
	Number of awards V_{8s}	0.40	0.20	0.30	0.90
Knowledge Transfer	Income from research projects	0.70	0.80	0.60	0.40
	Number of joint industrial projects $V_{10,s}$	0.80	0.10	0.50	0.70
	Number of patents $V_{11,s}$	0.94	1.00	0.96	0.95
	Number of start-ups $V_{12,s}$	0.96	0.60	0.15	1.00

The numerical data of Table 1 are presented as radar charts in Fig.1.

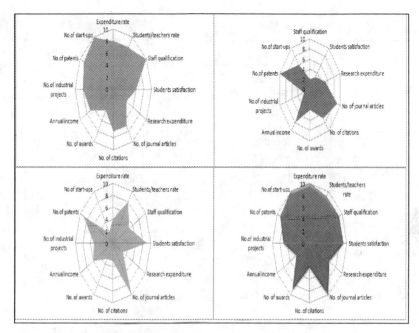

Fig. 1. The radar charts for four EIPs

Table 2. The ranks of the EIPs with respect to the individual indicators

Missions	Indicators V_{is}	EIP1	EIP2	EIP3	EIP4
Education	Expenditure on education V_{1s}	2	4	3	1
	Teachers/students rate V_{2s}	2	4	3	1
	Staff qualification V_{3s}	1	3	4	2
	Students' satisfaction V_{4s}	4	2	3	1
Research	Expenditure on research V_{5s}	3	2	4	1
	Number of journal articles V_{6s}	3	4	2	1
	Number of citations V_{7s}	2	3	4	1
	Number of awards V_{8s}	2	4	3	1
Knowledge Transfer	Income from research projects	2	1	3	4
	Number of industrial projects $V_{10,s}$	1	4	3	2
	Number of patents $V_{11,s}$	4	1	2	3
	Number of start-ups $V_{12,s}$	2	3	4	1

Next, knowing the intermediate ranks r_{ik} of the EIP for each individual indicator enlisted in Table 2, we compute the so-called *Borda counts* for each EIP *s* as follows: $b_i = \sum_k (N - r_{ik})$. The corresponding values are presented in Tables 3 and 4.

Table 3. The intermediate partial Borda counts for the selected EIPs

Missions	Indicators V_{js}	EIP1	EIP2	EIP3	EIP4
Education	Expenditure on education V_{1s}	2	0	1	3
	Teachers/students rate V_{2s}	2	0	1	3
	Staff qualification V_{3s}	3	1	0	2
	Students' satisfaction V_{4s}	0	2	1	3
Research	Expenditure on research V_{5s}	1	2	0	3
	Number of journal articles V_{6s}	1	0	2	3
	Number of citations V_{7s}	2	1	0	3
	Number of awards V_{8s}	2	0	1	3
Knowledge Transfer	Annual income from research projects	2	3	1	0
	Number of joint industrial projects $V_{10,s}$	3	0	1	2
	Number of patents $V_{11,s}$	0	3	2	1
	Number of start-ups $V_{12,s}$	2	1	0	3

Table 4. The overall Borda counts for selected EIPs ($b_i = \sum_k (N - r_{ik})$, N=4)

EIPs	EIP1	EIP2	EIP3	EIP4
$b_i = \sum_k (N - r_{ik})$	18	13	10	29

The final operation of the fifth step is the overall ranking of the EIPs according to the found Borda counts, presuming that the best EIP is that one whose overall Borda count is maximum. In this example, we can observe that the EIP4 has an essentially better overall Borda count than any other project. The EIP1 is the second-best one, and two other projects, EIP2 and EIP3, with close Borda count values, can be considered as constituting the third category that is of a worse integral quality than the first two categories.

It is worth noticing that the radar charts in Figure 1 yield a partial qualitative visual information indicating, in particular, that the EIP2 and EIP3 are worse than the EIP1 and EIP4. However, these charts do not reveal that the EIP4 has, in fact, an essentially better overall quality than the EIP1. This is a specific property and an advantage of the suggested quantitative project classification method.

We can summarize the obtained numerical results as follows. The suggested quantitative Borda-type method provides both the ranking and classification of the considered EIPs. It discloses that all the considered EIPs can be grouped into several different categories, where an EIP (or a group of several EIPs) with the highest Borda count is the first (the best) category, the EIP (or several EIPs) having the next highest Borda count constitutes the second-best category, and so on, so that the EIP or several EIPs with the lowest Borda count value constitute the least-quality category. In our future research we intend to verify the applicability of the new method for large-size educational projects in a real-world situation.

4   Conclusion

We proposed a new method for the ranking and classification of educational innovation projects (EIP) using an extended version of the classic Borda ranking method. A multi-stage presentation of the ranking process, that is, {*Document attributes – Document fitness – Indicator values – Project ranks*} permitted to use experts´ knowledge and the information contained in text documents. This method has been illustrated with a numerical example containing several EIPs.

The new method employs the advantages of human-computer interaction. No human can read and process such vast volumes of text information as a computer can. The method strongly relies on expert-defined weights which are used at several steps of the evaluation process. This involves quite a lot of work on the part of experts but this is a better way of exploiting their know-how rather than if we simply ask them to evaluate the EIPs directly, which, in fact, inevitably leads to heavily subjective and intuitive evaluations and takes much more time.

The Borda type method developed in this paper has several advantages, such as simplicity, user-friendliness, computational efficiency and minimum of voting paradoxes. At the same time, we aware of its limitations, for example, it does not permit to handle linguistic and interval-valued data. An extension of the suggested ranking method for the case of linguistic and interval parameters is a perspective direction for future research. Other prospective directions could be to apply the suggested approach for ranking and classifying nation-wide real-world EIPs and to compare this method with other semi-automatic and empiric project ranking/classification methods.

Acknowledgement. The authors wish to thank two anonymous referees for their critical remarks and helpful suggestions that helped improve the paper.

References

1. Maimon, O., Rokach, L.: Data Mining and Knowledge Discovery Handbook. Springer, Berlin (2005)
2. Feldman, R., Sanger, J.: The Text Mining Handbook: Advanced Approaches in Analyzing Unstructured Data. Cambridge University Press, Cambridge (2006)
3. Pawar, P.Y., Gawande, S.H.: A comparative study on different types of approaches to text categorization. International Journal of Machine Learning and Computing 2, 423–426 (2012)
4. Chung, E.K., Miksa, S., Hastings, S.K.: A framework of automatic subject term assignment for text categorization: An indexing conception-based approach. Journal of the American Society for Information Science and Technology 61, 688–699 (2010)
5. Du, R., Safavi-Naini, R., Susilon, W.: Web filtering using text classification. In: The 11th IEEE International Conference on Networks, September 28-October 1, pp. 325–330 (2003)
6. Alexandrov, M., Gelbukh, A., Rosso, P.: An Approach to Clustering Abstracts. In: Montoyo, A., Muñoz, R., Métais, E. (eds.) NLDB 2005. LNCS, vol. 3513, pp. 275–285. Springer, Heidelberg (2005)

7. Levner, E., Alcaide, D., Sicilia, J.: Multi-attribute Text Classification Using the Fuzzy Borda Method and Semantic Grades. In: Masulli, F., Mitra, S., Pasi, G. (eds.) WILF 2007. LNCS (LNAI), vol. 4578, pp. 422–429. Springer, Heidelberg (2007)

8. Makagonov, P., Alexandrov, M., Gelbukh, A.: Selection of typical documents in a document flow. In: Mastorakis, N., Kluev, V. (eds.) Advances in Communications and Software Technologies, pp. 197–202. WSEAS Press (2002)

9. Pazzani, M.J., Muramatsu, J., Billsus, D.: Syskill & Webert: Identifying interesting Web sites. In: Proceedings of the 13 National Conference on Artificial Intelligence, pp. 54–56 (1996)

10. Yu, H., Han, J., Chang, K.C.-C.: PEBL: Positive example based learning for Web page classification using SVM. In: The International Conference on Knowledge Discovery and Data mining, KDD (2002)

11. Dvir, D.: S Lipovetsky, A Shenhar, A Tishler: In search of project classification: a non-universal approach to project success factors. Research Policy 27, 915–935 (1998)

12. Artto, K., Martinsuo, M., Gemunden, H.G.: Foundations of program management: A bibliometric view. International Journal of Project Management 27, 1–18 (2009)

13. Aguillo, I.F., Ortega, J.L., Fernandez, M.: Webometric ranking of world universities: Introduction, methodology, and future developments. Higher Education in Europe 33, 234–244 (2008)

14. Aguillo, I.F., Bar-Ilan, J., Levene, M., Ortega, J.L.: Comparing university rankings. Scientometrics 85(10), 243–256 (2010)

15. Waltman, L., Calero-Medina, C., Kosten, J., Noyons, E.C.M., Tijssen, R.J.W., van Eck, N.J., van Leeuwen, T.N., van Raan, A.F.J., Visser, M.S., Wouters, P.: The Leiden ranking 2011/2012: Data collection, indicators, and interpretation. Journal of the American Society for Information Science and Technology 63, 2419–2432 (2012)

16. Lansdowne, Z.F.: Ordinal ranking methods for multi-citerion decision making. Naval Research Logistics 43, 613–627 (1996)

17. Saari, D.G.: Mathematical structure of voting paradoxes: II. Positional voting. Journal of Economic Theory 15, 55–102 (2000)

18. M.: v. Erp and L. Schomaker: Variants of the Borda count method for combining ranked classifier hypotheses. In: The Seventh International Workshop on Frontiers in Handwriting Recognition, Amsterdam, pp. 443–452 (2000)

19. Chen, L., Ding, D., Wang, D., Lin, F., Zhang, B.: AP-based borda voting method for feature extraction in TRECVID-2004. In: Losada, D.E., Fernández-Luna, J.M. (eds.) ECIR 2005. LNCS, vol. 3408, pp. 568–570. Springer, Heidelberg (2005)

20. Levner, E., Alcaide Lòpez de Pablo, D., Benayahu, J.: Environmental risk ranking: Theory and applications for emergency planning. In: Skanata, D., Byrd, D.M. (eds.) Computational Models of Risks to Infrastructure, pp. 307–317. Springer (2007)

21. Sevillano, X., Socoro, J.C., Alıas, F.: Fuzzy clusterers combination by positional voting for robust document clustering. Procesamiento del Lenguaje Natural 43, 245–253 (2009)

22. Pacuit, E.: Voting methods, in The Stanford Encyclopedia of Philosophy. In: Zalta, E. N. (ed.)   (2012),
    http://plato.stanford.edu/archives/win2012/entries/voting-methods/

23. UNESCO International Ranking Expert Group: Berlin Principles on Ranking of Higher Education Institutions (2006), http://www.che.de/downloads/ Berlin_Principles_IREG_534.pdf (retrieved February 06, 2014)

Named Entities as New Features
for Czech Document Classification

Pavel Král

Dept. of Computer Science & Engineering, and
NTIS - New Technologies for the Information Society
Faculty of Applied Sciences
University of West Bohemia
Plzeň, Czech Republic
pkral@kiv.zcu.cz

Abstract. This paper is focused on automatic document classification. The results will be used to develop a real application for the Czech News Agency. The main goal of this work is to propose new features based on the Named Entities (NEs) for this task. Five different approaches to employ NEs are suggested and evaluated on a Czech newspaper corpus. We show that these features do not improve significantly the score over the baseline word-based features. The classification error rate improvement is only about 0.42% when the best approach is used.

1 Introduction

Nowadays, the amount of electronic text documents and the size of the World Wide Web are extremely rapidly growing. Therefore, automatic document classification is particularly important for information organization, storage and retrieval.

This work is focused on a real application of the document classification for the Czech News Agency (CTK).[1] CTK produces daily about one thousand text documents, which belong to different classes such as sport, culture, business, etc. In the current application, documents are manually annotated. Unfortunately, the manual annotation represents a very time consuming and expensive task. Moreover, this annotation is often not sufficiently accurate. It is thus beneficial to propose and implement an automatic document classification system.

Named Entity (NE) Recognition was identified as a main research topic for automatic information retrieval around 1996 [1]. The objective is identification of expressions with special meaning such as person names, organizations, times, monetary values, etc. The named entities can be successfully used in many fields and applications, e.g. question answering, information filtering, etc.

In this paper, we propose new features for document classification of the Czech newspaper documents based on the named entities. We believe that NEs bring some additional information, which can improve the performance of our document classification system. Our assumptions are supported by the following observations.

---

[1] http://www.ctk.eu

A. Gelbukh (Ed.): CICLing 2014, Part II, LNCS 8404, pp. 417–427, 2014.
© Springer-Verlag Berlin Heidelberg 2014

- NEs can be used to differentiate some similar words according to the context. For example, "Bush" can be American president or popular British band. Using information about the NE, the documents can be classified correctly to one of the two different categories: politics or culture.
- It is possible to use named entities to discover synonyms, e.g. "USA" and "United States" are two different words. However, the word-sense is similar and they represent the same NE, the *country*. This additional information should help to classify two different documents containing "USA" and "United States" words, respectively, into the same category.
- Named entities shall be also used to identify and connect individual words in the multiple-words entities to one token. For example, the words in the expression "Mladá fronta dnes" (*name of the Czech newspaper*) do not have any sense. They can, used separately, produce a mismatch in document classification because the word "dnes" (*today*) is mostly used in the class weather. Using one token can avoid this issue.

Five different approaches to employ this information are proposed and evaluated next.

1. add directly the named entities to the feature vector (which is composed of words (or lemmas)) as new tokens
2. concatenate words related to multiple-word entities to one individual token
3. combine (1) and (2)
4. concatenate words and named entities to one individual token
5. replace words related to the named entities by their NEs

Note that, to the best of our knowledge, named entities were never used previously as features for the document classification task of the Czech documents. Moreover, we have not found another work which uses NEs similarly for document classification.

Section 2 presents a short review about the document classification approaches with the particular focus on the use of the NE recognition in this field. Section 3 describes our approaches of the integration of named entities to the feature vector. Section 4 deals with the realized experiments on the CTK corpus. We also discuss the obtained results. In the last section, we conclude the research results and propose some future research directions.

2   Related Work

Document clustering is an unsupervised approach that aims at automatically grouping raw documents into clusters based on their words similarity, while document classification relies on supervised methods that exploit a manually annotated training corpus to train a classifier, which in turn identifies the class of new unlabeled documents. Mixed approaches have also been proposed, such as semi-supervised approaches, which augment labeled training corpus with unlabeled data [2], or methods that exploit partial labels to discover latent topics [3]. This work focuses on document classification based on the Vector Space Model (VSM), which basically represents each document with a vector of all occurring words weighted by their Term Frequency-Inverse Document Frequency (TF-IDF).

Several classification algorithms have been successfully applied [4, 5], e.g. Bayesian classifiers, decision trees, k-Nearest Neighbour (kNN), rule learning algorithms, neural networks, fuzzy logic based algorithms, Maximum Entropy (ME) and Support Vector Machines (SVMs). However, the main issue of this task is that the feature space in VSM is highly dimensional which negatively affects the performance of the classifiers.

Numerous feature selection/reduction approaches have been proposed [6] in order to solve this problem. The successfully used feature selection approaches include Document Frequency (DF), Mutual Information (MI), Information Gain (IG), Chi-square test or Gallavotti, Sebastiani & Simi metric [7, 8]. Furthermore, a better document representation may lead to decreasing the feature vector dimension, e.g. using lemmatization or stemming [9]. More recently, advanced techniques based on Labeled Latent Dirichlet Allocation (LDA) [10] or Principal Component Analysis (PCA) [11] incorporating semantic concepts [12] have been introduced.

Multi-label document classification [13, 14][2] becomes a popular research field, because it corresponds usually better to the needs of the real applications than one class document classification. Several methods have been proposed as presented for instance in surveys [15, 16].

The most of the proposed approaches is focused on English and is usually evaluated on the Reuters,[3] TREC[4] or OHSUMED[5] databases.

Only little work is focused on the document classification in other languages. Yaoyong et al. investigate in [17] learning algorithms for cross-language document classification and evaluate them on the Japanese-English NTCIR-3 patent retrieval test collection.[6] Olsson presents in [18] a Czech-English cross-language classification on the MALACH[7] data set. Wu et al. deals in [19] with a bilingual topic aspect classification of English and Chinese news articles from the Topic Detection and Tracking (TDT)[8] collection.

Unfortunatelly, only few work about the classification of the Czech documents exits. Hrala et al. proposes in [20] a precise representation of Czech documents (lemmatization and Part-Of-Speech (POS) tagging included) and shown that mutual information is the most accurate feature selection method which gives with the maximum entropy or support vector machines classifiers the best results in the single-label Czech document

---

[2] One document is usually labeled with more than one label from a predefined set of labels.

[3] http://www.daviddlewis.com/resources/
testcollections/reuters21578

[4] http://trec.nist.gov/data.html

[5] http://davis.wpi.edu/xmdv/datasets/ohsumed.html

[6] http://research.nii.ac.jp/ntcir/permission/perm-en.html

[7] http://www.clsp.jhu.edu/research/malach/

[8] http://www.itl.nist.gov/iad/mig//tests/tdt/

classification task[9]. It was further shown [21] that the approach proposed by Zhu et al. in [22] is the most effective one for multi-label classification of the Czech documents.

To the best of our knowledge, only little work on the use of the NEs for document classification has been done. Therefore, we will focus on the use of the named entities in the closely related tasks. Joint learning of named entities and document topics has mainly been addressed so far in different tasks than document clustering. For instance, the authors of [23] exploit both topics and named entity models for language model adaptation in speech recognition, or [24] for new event detection. Topic models are also used to improve named entity recognition systems in a number of works, including [25–27], which is the inverse task to our proposed work. Joint entity-topic models have also been proposed in the context of unsupervised learning, such as in [28] and [29].

The lack of related works that exploit named entity recognition to help document classification is mainly explained in [30], which has precisely studied the impact of several NLP-derived features, including named entity recognition, for text classification, and concluded negatively. Despite this very important study, we somehow temper this conclusion and show that our intuition that suggests us that named entity features cannot be irrelevant in the context of document classification, might not be completely wrong. Indeed, nowadays NLP tools have improved and may provide richer linguistic features, and the authors of [30] only use a restricted definition of named entities, which are limited to proper nouns, while we are exploiting more complex types of named entities.

3   Document Classification with Named Entities

3.1   Preprocessing, Feature Selection and Classification

The authors of [20] have shown that morphological analysis including lemmatization and POS tagging with combination of the MI feature selection method significantly improve the document classification accuracy. Therefore, we have used the same preprocessing in our work.

Lemmatization is used in order to decrease the feature number by replacing a particular word form by its *lemma* (base form) without any negative impact to the classification score. The words that should not contribute to classification are further filter out from the feature vector according to their POS tags. The words with approximately uniform distribution among all document classes are removed from the feature vector. Therefore, only the words having the POS tags noun, adjective or adverb remain in the feature vector.

Note that the above described steps are very important, because irrelevant and redundant features can degrade the classification accuracy and the algorithm speed.

In this work, we would like to evaluate the importance of new features. Absolute value of the recognition accuracy thus does not play a crucial role. Therefore, we have chosen the simple Naive Bayes classifier which has usually an inferior classification score. However, it will be sufficient for our experiments to show whether new features bring any supplementary information.

---

[9] One document is assigned exactly to one label from a predefined set of labels.

3.2   Named Entity Integration

For better understanding, the features obtained by the proposed approaches will be demonstrated on the Czech simple sentence "Český prezident Miloš Zeman dnes navštívil Spojené státy." (*The Czech president Miloš Zeman visited today the United States*) (see Table 1). The baseline features after lemmatization and POS-tag filtration are shown in the first line of this table. The second line corresponds to the English translation and the third line illustrates the recognized named entities.

Table 1. Examples of the NE-based features obtained by the five proposed approaches

Český	Prezident	Miloš		Zeman	dnes		Spojené	státy.		
Czech	president	Miloš		Zeman	today		United	States		
O	O	Figure-B		Figure-I	Datetime-B		Country-B	Country-I		
1.	Český	Prezident	Miloš	Zeman	Figure	dnes	Datetime	Spojené	státy	Country
2.	Český	Prezident	Miloš-Zeman	dnes	Spojené-státy					
3.	Český	Prezident	Miloš-Zeman	Figure	dnes	Datetime	Spojené-státy	Country		
4.	Český	Prezident	Miloš-Zeman-Figure	dnes-Datetime	Spojené-státy-Country					
5.	Český	Prezident	Figure	Datetime	Country					

Named Entities as New Tokens in the Feature Vector - (1).   The baseline feature vector is composed of words (lemmas in our case) and their values are calculated by the TF-IDF approach. In this approach, we insert directly the named entity labels to the feature vector as new tokens. The values of the NE features are calculated similarly as the values of the word features using the TF-IDF method. One example of the resulting features of this approach is shown in the first line of the second section of Table 1.

Note that the feature values in all following approaches will be also computed by the TF-IDF weighting.

Concatenation of Words (lemmas) Related to Multiple-word Entities to One Individual Token - (2).   As mentioned previously, the individual words of the multiple-word entities have usually the different meaning than connected to one single token. In this approach, all words which create a multiple-word NE are connected together and the NE labels are further discarded. The second line of the second section of Table 1 shows the features created by this approach.

Combination of the Approach (1) and (2) - (3).   We assume that the NE labels can bring other information than the connected words of the multiple-word NEs. Therefore, in this approach we combine both previously proposed methods as illustrated in the third line of the second section of Table 1.

Concatenation of Words (lemmas) and Named Entities into One Individual Token - (4).   The concatenated words of the multiple-word NEs and their NE labels are used in the previous approach as two separated tokens. In this approach, they are linked together to create one token. This approach should play an important role for word sense disambiguation (e.g. "Bush-Figure" vs. "Bush-Organization"). One example of the features obtained by this approach is shown in the fourth line of the second section of Table 1.

Named Entities Instead of the Corresponding Words - (5). The previously proposed approaches increase the size of the feature vector. In this last approach, the size of the vector is reduced replacing the words corresponding to the named entities by their NE labels. The last line of Table 1 shows the features created by this approach.

Weighting of the Named Entities. We assume that named entities represent the most important words in the documents. Therefore, we further slightly modify the TF-IDF weighting in order to increase the importance of the named entities. The original weight is multiplied by K when named entity identified.

Note that we often use the term "words" in the text while in the experiment we use rather their "lemmas" instead.

4 Experiments

4.1 Tools and Corpora

We used the mate-tools[10] for lemmatization and POS tagging. The lemmatizer and POS tagger were trained on 5853 sentences (94.141 words) randomly taken from the PDT 2.0[11] [31] corpus. The performance of the lemmatizer and POS tagger are evaluated on a different set of 5181 sentences (94.845 words) extracted from the same corpus. The accuracy of the lemmatizer is 81.09%, while the accuracy of our POS tagger is 99.99%. Our tag set contains 10 POS tags as shown in Table 2.

We use the top scoring Czech NER system [32]. It is based on Conditional Random Fields. The overall F-measure on the CoNLL format version of Czech Named Entity Corpus 1.0 (CNEC) is 74.08%, which is the best result so far. We have used the model trained on the private CTK Named Entity Corpus (CTKNEC). The F-measure obtained on this corpus is about 65%.

For implementation of the classifier we used an adapted version of the MinorThird[12] tool. It has been chosen mainly because of our experience with this system.

As already stated, the results of this work shall be used by the CTK. Therefore, for the following experiments we used the Czech text documents provided by the CTK. Table 2 shows the statistical information about the corpus. This corpus is available only for research purposes for free at http://home.zcu.cz/~pkral/sw/ or upon request to the authors.

In all experiments, we used the five-folds cross validation procedure, where 20% of the corpus is reserved for the test. All experiments are repeated 10 times with randomly reshuffled documents in the corpus. The final result of the experiment is then a mean of

---

[10] http://code.google.com/p/mate-tools/
[11] http://ufal.mff.cuni.cz/pdt2.0/
[12] http://sourceforge.net/apps/trac/minorthird

Table 2. Corpus statistical information

Unit name	Unit number	Unit name	Unit number
Document	11,955	Numeral	216,986
Category	60	Verb	366,246
Word	5,145,788	Adverb	140,726
Unique word	193,399	Preposition	346,690
Unique lemma	152,462	Conjunction	144,648
Noun	1,243,111	Particle	10,983
Adjective	349,932	Interjection	8
Pronoun	154,232		

Table 3. NE tag-set and distribution in the CTK document corpus

NE	No.	NE	No.	NE	No.	NE	No.
City	55,370	E-subject	5,447	Number	160,633	Religion	24
Country	56,081	Figure	133,317	Organization	119,021	Sport	12,524
Currency	25,429	Geography	7,418	Problematic	17	Sport-club	38,745
Datetime	108,594	Nationality	5,836	Region	14,988	Uknown	1,750

all obtained values. For evaluation of the classification accuracy, we used, as frequently in some other studies, a standard *Error Rate (ER)* metric. The resulting error rate has a confidence interval of $< 0.5\%$.

Our NE tag-set is composed of 16 named entities (see Table 3). This table further shows the numbers of the NE occurrences in the corpus. The total number of the NE occurrences is about 700,000 which represents a significant part of the corpus (approximately 13%).

Note that, this named entities have been identified fully-automatically. Some labeling errors are thus available. This fact can influence the following experiments negatively.

4.2 Analysis of the Named Entity Distribution According to the Document Classes and Classification with Only NEs

This experiment should support our assumption that named entities bring useful information for document classification. Therefore, we realize a statistical study of the distribution of the named entities according to the document classes in the corpus (see Figure 1). This figure shows that some NEs (e.g. E-subject, Region, Sport, etc.) are clearly discriminant across the document classes. The analysis supports our assumption that the NEs can have a positive impact to the document classification.

We further realize another experiment in order to show whether only named entities (without the word features) are useful for the document classification. The results of this experiment (see the first line of Table 4) shows that NEs bring some information for document classification. However, their impact is small.

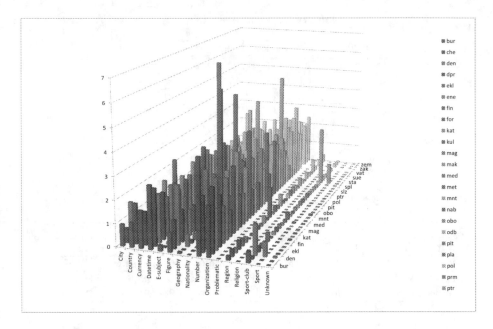

Fig. 1. Distribution of the named entities according to the document classes

4.3   Classification Results of the Proposed Approaches

The Table 4 further shows the recognition error rates of the proposed approaches. We evaluate the NE weights $K \in \{1, 2, 3\}$. The greater weight values are not used because the classification scores is decreasing according to this value in all experiments.

This table shows that the named entities help for document classification only slightly and this improvement is unfortunately statistically not significant. The best score is obtained by the second approach when the words are concatenated across the NEs and the information about the NE labels is completely removed from the feature vector.

Table 4. Document classification error rates [in %] of the different implementations of the named entity features (NE weights $K \in \{1, 2, 3\}$)

	Approach	NE weights		
		1	2	3
	NEs only	84.25		
	Lemmas (baseline)	17.83		
1.	Lemmas + NEs	17.60	17.75	17.79
2.	Concatenated lemmas	**17.41**	18.01	18.71
3.	1 + 2 together	17.48	18.02	18.61
4.	Concatenated Lemmas + NEs as one token	17.54	18.20	18.59
5.	NEs instead of the corresponding words	17.81	18.27	19.03

4.4   Analysis of the Confusion Matrices

In this experiment, we analyze the confusion matrices in order to compare the errors when the baseline word-bases features and the proposed features used (see Table 5). This table illustrates the number of errors, number of different errors in absolute value and in %, respectively. It is depicted that about 27% of errors (except the first proposed method) is different. Therefore, this experiment confirms that named entities bring some additional information. Unfortunately, this information is not sufficient to improve significantly the document classification accuracy on the CTK document corpus.

Note that the error number of the baseline approach is 2,131.

Table 5. Analysis of the confusion matrices errors between the baseline and five proposed approaches

Baseline vs. Proposed Approach	Error no.	Diff. err. no.	Diff. err. no [in %]
1. Lemmas + NEs	2,104	357	16.97
2. Concatenated lemmas	2,081	577	27.73
3. 1 + 2 together	2,089	581	27.81
4. Concatenated Lemmas + NEs as one token	2,097	569	27.13
5. NEs instead of the corresponding words	2,129	594	27.9

5   Conclusions and Future Work

In this paper, we have proposed new features for the document classification based on the named entities. We have introduced five different approaches to employ NEs in order to improve the document classification accuracy. We have evaluated these methods on the Czech CTK corpus of the newspaper text documents. The experimental results have shown that these features do not improve significantly the score over the baseline word-based features. The improvement of the classification error rate was only about 0.42% when the best approach is used. We have further analyzed and compared the confusion matrices of the baseline approach with our proposed methods. This analysis has shown that named entities bring some additional information for document classification. Unfortunately, this information is not sufficient to improve significantly the document classification accuracy.

However, we assume that this information could play more important role on smaller corpora with more unknown words in the testing part of the corpus. The first perspective thus consists in evaluation of the proposed features on the other (smaller) corpora including more European languages. Then, we would like to propose other sophisticated features which introduce the semantic similarity of word-based features. These features should be useful for example for word-sense disambiguation and can be created for instance by the semantic spaces.

Acknowledgements. This work has been partly supported by the European Regional Development Fund (ERDF), project "NTIS - New Technologies for Information Society", European Centre of Excellence, CZ.1.05/1.1.00/02.0090. We also would like to thank Czech New Agency (CTK) for support and for providing the data.

References

1. Grishman, R., Sundheim, B.: Message understanding conference-6: a brief history. In: Proceedings of the 16th Conference on Computational Linguistics, COLING 1996, Copenhagen, Denmark, vol. 1, pp. 466–471. Association for Computational Linguistics (1996)
2. Nigam, K., McCallum, A.K., Thrun, S., Mitchell, T.: Text Classification from Labeled and Unlabeled Documents Using EM. Mach. Learn. 39, 103–134 (2000)
3. Ramage, D., Manning, C.D., Dumais, S.: Partially labeled topic models for interpretable text mining. In: Proceedings of the 17th ACM SIGKDD International Conference on Knowledge Discovery and Data Mining, KDD 2011, pp. 457–465. ACM, New York (2011)
4. Bratko, A., Filipič, B.: Exploiting structural information for semi-structured document categorization. In: Information Processing and Management, pp. 679–694 (2004)
5. Della Pietra, S., Della Pietra, V., Lafferty, J.: Inducing features of random fields. IEEE Transactions on Pattern Analysis and Machine Intelligence 19, 380–393 (1997)
6. Forman, G.: An extensive empirical study of feature selection metrics for text classification. The Journal of Machine Learning Research 3, 1289–1305 (2003)
7. Yang, Y., Pedersen, J.O.: A comparative study on feature selection in text categorization. In: Proceedings of the Fourteenth International Conference on Machine Learning, ICML 1997, pp. 412–420. Morgan Kaufmann Publishers Inc., San Francisco (1997)
8. Galavotti, L., Sebastiani, F., Simi, M.: Experiments on the use of feature selection and negative evidence in automated text categorization. In: Proceedings of the 4th European Conference on Research and Advanced Technology for Digital Libraries, ECDL 2000, pp. 59–68. Springer, London (2000)
9. Lim, C.S., Lee, K.J., Kim, G.C.: Multiple sets of features for automatic genre classification of web documents. Information Processing and Management 41, 1263–1276 (2005)
10. Ramage, D., Hall, D., Nallapati, R., Manning, C.D.: Labeled lda: A supervised topic model for credit attribution in multi-labeled corpora. In: Proceedings of the 2009 Conference on Empirical Methods in Natural Language Processing, EMNLP 2009, pp. 248–256. Association for Computational Linguistics, Stroudsburg (2009)
11. Gomez, J.C., Moens, M.F.: Pca document reconstruction for email classification. Computer Statistics and Data Analysis 56, 741–751 (2012)
12. Yun, J., Jing, L., Yu, J., Huang, H.: A multi-layer text classification framework based on two-level representation model. Expert Systems with Applications 39, 2035–2046 (2012)
13. Novovičová, J., Somol, P., Haindl, M., Pudil, P.: Conditional mutual information based feature selection for classification task. In: Rueda, L., Mery, D., Kittler, J. (eds.) CIARP 2007. LNCS, vol. 4756, pp. 417–426. Springer, Heidelberg (2007)
14. Forman, G., Guyon, I., Elisseeff, A.: An extensive empirical study of feature selection metrics for text classification. Journal of Machine Learning Research 3, 1289–1305 (2003)
15. Sebastiani, F.: Machine learning in automated text categorization. ACM Computing Surveys (CSUR) 34, 1–47 (2002)
16. Tsoumakas, G., Katakis, I.: Multi-label classification: An overview. International Journal of Data Warehousing and Mining (IJDWM) 3, 1–13 (2007)
17. Yaoyong, L., Shawe-Taylor, J.: Advanced learning algorithms for cross-language patent retrieval and classification. Information Processing & Management 43, 1183–1199 (2007)
18. Olsson, J.S.: Cross language text classification for malach (2004)
19. Wu, Y., Oard, D.W.: Bilingual topic aspect classification with a few training examples. In: Proceedings of the 31st Annual International ACM SIGIR Conference on Research and Development in Information Retrieval, pp. 203–210. ACM (2008)
20. Hrala, M., Král, P.: Evaluation of the document classification approaches. In: Burduk, R., Jackowski, K., Kurzynski, M., Wozniak, M., Zolnierek, A. (eds.) CORES 2013. AISC, vol. 226, pp. 875–884. Springer, Heidelberg (2013)

21. Hrala, M., Král, P.: Multi-label document classification in czech. In: Habernal, I., Matoušek, V. (eds.) TSD 2013. LNCS, vol. 8082, pp. 343–351. Springer, Heidelberg (2013)
22. Zhu, S., Ji, X., Xu, W., Gong, Y.: Multi-labelled classification using maximum entropy method. In: Proceedings of the 28th Annual International ACM SIGIR Conference on Research and Development in Information Retrieval, pp. 274–281. ACM (2005)
23. Liu, Y., Liu, F.: Unsupervised language model adaptation via topic modeling based on named entity hypotheses. In: Proceedings ASSP (2008)
24. Kumaran, G., Allan, J.: Text classification and named entities for new event detection. In: Proceedings of the 27th Annual International ACM SIGIR Conference on Research and Development in Information Retrieval, pp. 297–304. ACM (2004)
25. Guo, H., Zhu, H., Guo, Z., Zhang, X., Wu, X., Su, Z.: Domain adaptation with latent semantic association for named entity recognition. In: Proceedings of Human Language Technologies: The 2009 Annual Conference of the North American Chapter of the Association for Computational Linguistics, NAACL 2009, pp. 281–289. Association for Computational Linguistics, Stroudsburg (2009)
26. Knopp, J., Frank, A., Riezler, S.: Classification of named entities in a large multilingual resource using the Wikipedia category system. PhD thesis, Masters thesis, University of Heidelberg (2010)
27. Zhang, Z., Cohn, T., Ciravegna, F.: Topic-oriented words as features for named entity recognition. In: Gelbukh, A. (ed.) CICLing 2013, Part I. LNCS, vol. 7816, pp. 304–316. Springer, Heidelberg (2013)
28. Newman, D., Chemudugunta, C., Smyth, P.: Statistical entity-topic models. In: Proceedings of the 12th ACM SIGKDD International Conference on Knowledge Discovery and Data Mining, KDD 2006, pp. 680–686. ACM, New York (2006)
29. Vosecky, J., Jiang, D., Leung, K.W.T., Ng, W.: Dynamic multi-faceted topic discovery in twitter. In: Proceedings of the 22nd ACM International Conference on Conference on Information & Knowledge Management, CIKM 2013, pp. 879–884. ACM, New York (2013)
30. Moschitti, A., Basili, R.: Complex linguistic features for text classification: A comprehensive study. In: McDonald, S., Tait, J.I. (eds.) ECIR 2004. LNCS, vol. 2997, pp. 181–196. Springer, Heidelberg (2004)
31. Hajič, J., Böhmová, A., Hajičová, E., Vidová-Hladká, B.: The Prague Dependency Treebank: A Three-Level Annotation Scenario. In: Abeillé, A. (ed.) Treebanks: Building and Using Parsed Corpora, pp. 103–127. Kluwer, Amsterdam (2000)
32. Konkol, M., Konopík, M.: CRF-based czech named entity recognizer and consolidation of czech NER research. In: Habernal, I., Matoušek, V. (eds.) TSD 2013. LNCS, vol. 8082, pp. 153–160. Springer, Heidelberg (2013)

A Knowledge-Poor Approach to Turkish Text Categorization

Savaş Yıldırım

Department of Computer Engineering, Faculty of Engineering
Istanbul Bilgi University
Santral Istanbul Campus, Eyüp, Istanbul / Turkey

Abstract. Document categorization is a way of determining a category for a given document. Supervised methods mostly rely on a training data and rich linguistic resources that are either language-specific or generic. This study proposes a knowledge-poor approach to text categorization without using any sets of rules or language specific resources such as part-of-speech tagger or shallow parser. Knowledge-poor here refers to lack of a reasonable amount of background knowledge. The proposed system architecture takes data as-is and simply separates tokens by space. Documents represented in vector space models are used as training data for many machine learning algorithm. We empirically examined and compared a several factors from similarity metrics to learning algorithms in a variety of experimental setups. Although researchers believe that some particular classifiers or metrics are better than others for text categorization, the recent studies disclose that the ranking of the models purely depends on the class, experimental setup and domain as well. The study features extensive evaluation, comparison within a variety of experiments. We evaluate models and similarity metrics for Turkish language as one of the agglutinative language especially within poor-knowledge framework. It is seen that output of the study would be very beneficial for other studies.

Keywords: Text Categorization, Vector Space Model, Machine Learning.

1 Introduction

Document categorization (DC) is a major field of information retrieval (IR). Documents are mostly identified by lists of terms to represent their contents, which is called Vector Space Model, [2]. A given document can be assigned a category, where categories are generally human defined for the needs of a system. In vector space, some statistical models can be simply applied. An essential difficulty is the enormous number of terms that prohibits many approaches from applying efficiently and properly. As the number of features highly increases, it is getting hard to apply some machine learning algorithms due to computation cost and machine memory limitation. Thus, feature elimination is a crucial pre-process for document classification. A simple assumption is that rare terms have

A. Gelbukh (Ed.): CICLing 2014, Part II, LNCS 8404, pp. 428–440, 2014.

non-informative properties or are not globally effective in system performance. This could be contrary to many others natural language processing such as word sense disambiguation (WSD). Less frequent terms could be informative and discriminative in WSD. Senses (category) in the document classification are so common and coarse such as politics, economics, sports etc. Thus elimination of rare terms is not considered an odd way, but even does ease the system. Taking the most frequent terms into account is usually called document frequency (DF) thresholding.

Moreover, it has recently shown that the selections relying on some other metrics could be more effective. Information gain (IG) and chi-square (CHI) are the most important selection criteria. Latent semantic indexing (LSI) is also applied and evaluated in some studies, [25]. It is very effective particularly for reducing the high-dimensional space. It can be extremely useful in IR.

Once the system reduces the number of features, then one of the supervised machine learning algorithms could easily assign the labels to the documents. A classifier maps documents to predetermined classes. To obtains word space and data to which a classifier can apply, general preprocessing phase consists of following steps: it morphologically analyzes the discourses, determines lemma and part-of-speech, suffixes, removes the punctuations, digits, specify noun phrases, proper nouns, and finally remove stop-words. Some studies especially select only nouns or/and verbs. Word Space could rely on unigram or any n-gram. However building n-gram ($n > 1$) space could be costly. Some studies, [1,20], use WordNet ontology in order to improve accuracy. They propose term aggregation methods to consolidate original terms with higher frequencies for improvements by exploiting hypernymy or snyset relation. However, [20] find the effect of term aggregation is not significant as expected.

A system utilizing these resources such as WordNet, stemmer etc. may not be suitable for some domain, language or be very costly in computational time. For instance, Turkish as a less resourced languages has no any consistent word-net like dictionary. Therefore, we discuss the problem in less studied language environment. Our objective is how well the metrics and approaches works for multi-class datasets within a knowledge-poor environment. No prior knowledge such as a morphologically parsed text, a list of lexicon, any WordNet-like dictionaries, stopword list are needed. The only resource required is a relatively large training documents in more than one category. Thus, this work conducts many experiment to find a solution to overcome the challenges and exploit the opportunities.

2   Related Works

The first example of feature elimination using CHI and IG in DC is done by [4]. [6] also uses the these utility functions to select the features. [7] review feature selection methods and their impact on classification accuracy. They refer to expected mutual information as IG. [10] applies different methods for reducing word vocabulary for multiple classifier.

[7] has shown that there are strong correlations between DF, IG and CHI. The excellent performance of them indicates that common terms are indeed informative for text categorization problem. Even though DF-based feature selection is found a proper way, it has recently shown that the selections relying on CHI or IG could be more effective. They also provided an evaluation of feature selection methods. They found IG and CHI most effective in aggressive term removal without losing categorization accuracy in the experiments with kNN and LLSF. Their maximum accuracy is 89.2%. [11] evaluates six popular feature selection metrics on topic-based and sentiment classification problem. They check the methodologies against both low and high feature numbers separately. Supervised techniques can be used not only to assign a category to a document but also to weight the terms[12].

Working on imbalanced data is indisputably harder than balanced one especially in high-dimensionality. Thus, feature selection methods are crucial to achieve better performance [13,10]. Some works measure the effect of three types of metrics on imbalanced data [14]. They compare metrics and discuss their way of combining positive and negative features.

[25] presents a work related to tf-idf, LSI and multi-word for text representation. They found that LSI has better performance than other two methods in both language domain. The purpose was to study the effectiveness of different representation methods. The worst one was multi-word representation method at all. However, [3] show that is not an optimal representation for text representation. It always damages the text classification performance when it is being applied to the entire training data that is called global LSI, Thus, some local LSI approaches have been successfully applied.

[15] gives useful introduction to statistical natural language processing with elaborate examples and methodologies. There are a variety of approaches mainly relying on corpus statistics. [16] comprehensively provide a review of information retrieval methodologies with a lot of meaningful results. The survey discusses the main approaches to text categorization that fall in the machine learning approaches.

Most studies in the literature use machine learning to assign the labels to a given documents. They compare the performances of the algorithms in DC [16,17]. Some studies especially emphasize that SVM is better than other algorithms, [18,8,19,5]. However,[26] states that the performances of the system mainly depends on the domain.

3   Feature Space

3.1   Feature Selection Criteria

The aim of feature seleciton is to more efficiently apply a classifier in terms of accuracy and speed by aggressively decreasing the size of vocabulary. In other words, it focuses on removing the terms increasing classification error. Term goodness score U(t,c) between each term and each category is computed based on a term goodness criterion. Then an averaging score, $U_{avg}(t)$ across categories is computed by combining category scores of each term as follows

$$U_{avg}(t) = \sum_{i=1}^{m} P_r(C_i)U(t, C_i) \tag{1}$$

Or alternatively maximum based score $U_{max}(t)$ is to be

$$U_{max}(t) = max_{i=1}^{m}U(t, C_i) \tag{2}$$

Where U could be any selection criterion such as χ^2 or IG. Some studies select maximum score rather than average. However, we observe that there is no significant differences between them. Five different utility functions or *term goodness criteria* are tested in the study. These criteria are as defined below:

– **Information Gain (IG)**
  IG is used to measure how much number of bits of information the presence or absence of a term in a document contributes to making the correct classification decision on a category. It is also called expected mutual information, [26]. The formula is defined to be.

$$IG(t) = -\sum_{i=1}^{m} P_r(C_i)logP_r(C_i) + P_r(t)\sum_{i=1}^{m} P_r(C_i|t)logP_r(C_i|t)$$

$$+ P_r(\tilde{t})\sum_{i=1}^{m} P_r(C_i|\tilde{t})logP_r(C_i|\tilde{t}) \tag{3}$$

– χ^2 **statistic (CHI):** It measures the lack of independence between a term and a category using two way contingency table. The events A and B are defined to be independent if $P(A, B) = P(A)P(B)$ or, equivalently, $P(A|B) = P(A)$ and $P(B|A) = P(B)$. The formula is defined to be

$$\chi^2 = \frac{N(N_{11}N_{00} - N_{01}N_{10})^2}{(N_{11} + N_{01})(N_{11} + N_{10})(N_{10} + N_{00})(N_{01} + N_{00})} \tag{4}$$

  where N_{11} is the number of times term t and category c co-occur, N_{10} is number of times t appears without c, N_{01} is number of times c appears without t, N_{00} is number of times neither t nor c occurs, N is the total number of documents.
– **PMI:** There is terminological confusing on point-wise mutual information. Some use term Mutual Information to refer to pointwise mutual information, however some studies use same term to refer to Information Gain. In this study we share the same terminology with (Yang,1997).
  It is defined to be.

$$PMI(t, c) = \log \frac{N_{11}N}{(N_{11} + N_{01})(N_{11} + N_{10})} \tag{5}$$

- **DICE:** The dice, jaccard and cosine coefficients have the attractions of simplicity and have mostly been used in IR. It is very similar to pmi criterion. While pmi is theoretical measure, dice is empirical one. The dice coefficient of two sets is a measure of their intersection scaled by their size. The formula is defined to be:

$$DICE(t,c) = \log \frac{2N_{11}}{(2N_{11} + N_{01} + N_{10})} \tag{6}$$

- **Frequency based selection (DF):** The idea is to rank terms by means of document frequency or collection frequency. Then first N most frequent terms are attained as feature space.

3.2   Term Weighting Schema

Weighting is a way of numerical statistic which reflects how important a word to a document. There are various term weighting schema derived from the different assumptions and the probabilistic models. Term frequency, document frequency, total number of documents, term-document joint probabilities are the factors that play a significant roles. We select widely used term weighting metrics. **Binary bi(t,d)**: It refers to absence or presence of a given term in related document. Thus, the possible values is either 0 or 1. **Term-Frequency tf(t,d)**:it is how many times of a term appears in a given document d. **Log-tf (t,d)** It is simply logarithm of term frequency. **Tf-idf :** the score is tf * idf, where idf refers to inverse document frequency and it is formulated as log (N / df). N is number of document and df is document frequency of a term.

4   Experimental Setup

4.1   Data

For English language, almost all the machine learning algorithms are tested against a well prepared and reliable labeled corpus such as Reuters collection. In this work, Turkish corpus consists of a set of newswire documents classified under many coarse grain categories such as economy, politics etc. The data is publicly available under the link [1]. We select seven different categories and more than 700 documents for each. The categories are world, economy, culture-art, health, politics, sports and technology. We split the data into tree parts; one is used to select features, one is to build classifier and last one is to test that classifier. Using the same data to select the features and train the model leads to over-fitting and trivial models that can reach F1 of 99%. The tokens separated by white space character are taken and all digits, punctuations are removed. In knowledge-poor framework, we only use surface form of the terms. Eventually, a word can appear in many alternations. Total number of unique terms is over 100K.

---

[1] http://www.kemik.yildiz.edu.tr

4.2 Classification

The performance of feature selection methodologies are evaluated applying four widely used machine learning algorithms; multi nominal *Naive Bayes Classifier* (NB) based on multi-nominal distribution is an apparently effective classification method as shown in many studies. *SVM* is another popular machine learning algorithms. The drawback is that training SVM with large dataset is very expensive. It is clear that the best kernel for SVM is linear in IR. Many researchers also prove that performance of linear kernel better than that of radial and Gaussian. We also observe and measure same results. *K-nearest neighbor algorithm* (kNN), as a lazy-Learner measures the distance between the test case and other events and select the major class of the closest K neighbors. Some studies apply it successfully. *Decision Trees* (DT) relies on information theory. The main advantage is its human readable output and divide-and-conquer paradigm. However, it sometimes suffers from over-fitting. It eliminates many features when building and pruning tree. Therefore it has difficulties in applying tree to low-frequent term space. Thanks to [24], WEKA: open-source data mining project and [27], libsvm: a library for SVM, we easily adopt our data into the their format and properly apply all these algorithm with different settings and do cross-classifier comparison.

We exponentially select size of feature set as 500, 1000, 2000, and 4000. At the end, we apply four different machine learning algorithms (Bayes, SVM, IBK, J48) with five different feature selection methods (IG, CHI, DICE, PMI, FREQ) across four different term weighting schema (binary, tf, log-tf, tf-idf) within 4 different size of feature space. It is equal to 4 x 4 x 5 x 4 = 320 different setups. However, relying on the observations we only provide and demonstrate meaningful figures and results.

5 Results and Evaluation

Micro-averaging computes a simple average over classes. It pools per-documents accuracy across classes and compute simple averaging the scores. In this works, all results are evaluated in terms of F1-measure and micro-averaging. Although some researchers believe that SVM is better than kNN, kNN betters NB for IR, [26] emphasizes that the ranking of the models purely depends on the class and domain as well. Looking at the results obtained, we can also agree with the idea of domain-class dependency.

5.1 Performance of Feature Selection Methodology

Table 1 anf Figure-1-A shows us many immediate results; IG feature selection method has the best F1 performance of 74% , and 72 % among five different feature selection criteria at all for feature size of 2000, 4000 respectively. CHI behaves similarly, but it is slightly worse. With small number of feature space, the third successful criteria is DICE. Contrary to some studies in DC, DF does not

Table 1. All results

	# of terms 500						# of terms 1000				
	nb	svm	ibk	tree	AVG		nb	svm	ibk	tree	AVG
ig	89	78	65	71	76		88	79	61	72	75
chi	85	77	63	72	74		87	78	61	71	74
dice	81	73	50	68	68		85	75	50	71	70
freq	53	50	46	26	44		65	60	53	20	49
pmi	53	50	46	25	44		64	59	52	19	49
AVG	72	66	54	52	61		78	70	56	51	63

	# of terms 2000						# of terms 4000				
	nb	svm	ibk	tree	AVG		nb	svm	ibk	tree	AVG
ig	89	80	56	72	74		90	81	45	72	72
chi	89	80	54	72	74		90	82	44	71	72
dice	87	78	47	70	70		88	81	42	69	70
freq	80	74	62	41	64		89	82	54	73	75
pmi	79	74	61	39	64		89	82	54	73	75
AVG	85	77	56	59	69		89	82	48	72	73

show significant performance under smaller space. PMI behaves very similarly. As Figure-1 (A) suggests, they perform well only in bigger feature space at least 4000 terms. We observe two reasons for the failure; One is the feature space built with df and pmi are hardly capable of representing the cases in test data. Thus many cases are left undecided since their vectors have all-zero cells. Secondly, DT learning algorithm does not perform especially with these two selection criteria even though DT globally has better performance. We will discuss it in later section.

5.2   Cross-Classifier Comparison

Looking at the performance of learning algorithms across feature selection criteria, the most successful is obviously Multinomial NB Classifier. With IG feature selection, NB has 89% and 90% F1 scores at feature selection size of 2000 and 4000 respectively, where SVM has 80% and 81%, kNN (lazy-learner) has 56% and 45%, and the DT has 72%. The results indicate that a setup with IG feature selection and NB classifier does always show better performance. Looking at Figure 1-B, It is clearly said, as the feature size increases, while NB, SVM and DT give better results, kNN does not perform well at all. No matter what feature criterion, term weighting and feature dimension is selected, kNN is the poorest classifier in our domain.

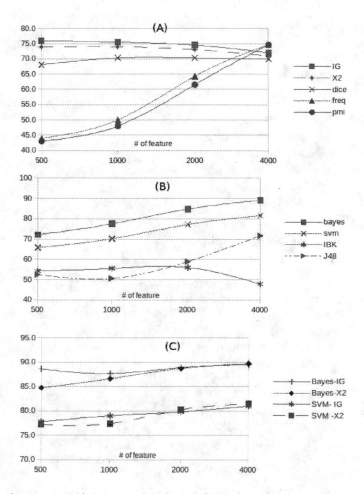

Fig. 1. Performance of A) Selection Methods, B) ML Algorihtms and C) Some Settings in terms of F1

The most interesting observation is that DT has poorest performance with DF and PMI under smaller feature space. This comes from subset selection paradigm of DT. Looking at the following tree and its performance with DF term selection, the tree is built upon lack of some features. It means that any query does not contain any of six terms as seen below, it is classified as C1. This leads to a bias favoring absence. As seen, while 1618 cases are correctly classified, 1354 ones are incorrectly labeled due to lack of these six terms.

```
F-maçında = 0
|   F-parti = 0
|   |   F-lider = 0
|   |   |   F-hastalarda = 0
|   |   |   |   F-ağrısı = 0
|   |   |   |   |   F-diyabet = 0: C1 (1618.0/1354.0)
|   |   |   |   |   F-diyabet > 0: C6 (26.0/3.0)
|   |   |   |   F-ağrısı > 0: C6 (29.0/1.0)
|   |   |   F-hastalarda > 0: C6 (27.0)
|   |   F-lider > 0: C5 (28.0)
|   F-parti > 0: C5 (47.0/1.0)
F-maçında > 0: C4 (106.0)
```

5.3   Evaluation of Term Weighting

Term weighting is another techniques in order to improve the performance. However, we do not observe a significant weighting criterion that efficiently works both with any particular term selection and any machine learning algorithm. Table-2 demonstrates that under IG and CHI feature selection, term weightings are compared across NB and SVM. Bayes with tf-idf and SVM with log-tf have very slightly better performance. We conclude here term weighting does not effectively contribute system performance.

Table 2. Term Weighting Assessment

Term Weighting				
	IG		CHI	
	NB	**SVM**	NB	**SVM**
TF	89.5	81.0	89.7	81.5
Log-tf	89.7	**82.9**	89.8	**83.3**
Tf-idf	**90.0**	82.5	**90.0**	82.3
binary	88.6	82.7	89.0	83.0

5.4   Evaluation of Feature Selection Size

Changing the size of vocabulary deeply affects the performance of all the setups as shown in Figure-1, where the improvements are observable. All peak at about feature size of 2000 and later. As figure-1(B) shows, only negatively affected and decreased classifier is kNN by increasing the term size.

Figure-1 suggests that IG is slightly better than CHI at all. Dice requests more feature space. Df and PMI is not recommended within especially small number of terms. Figure1-C shows another important observation that is two classifier NB and SVM and two feature selection criteria CHI and IG have a robust characteristics. Under even a smaller dimension they give a good performance. Changing the size of dimension has slight impact on them. Additionally the most successful setting could be done with NB and IG feature selection with any size of feature.

5.5   Impact of Document Frequency on Accuracy

To measure effect of document frequency of terms over success, all terms are ranked by df as done in DF selection. After skipping first N most frequent term, we always keep the feature size as 500. Suppose, when N is 100, while first 100 most frequent terms are removed, the terms between 100 and 600th level are selected. Looking at different level of N, we clearly see the importance of document frequency as shown in Figure-2. It plots two different curves. One is F1 of NB classifier under selected space. As the number of skipping frequent terms increases, the system performance are getting worse as expected. The observation shows that until 400 terms, the system can survive in terms of precision but not recall, then sharply decreases at 1000 terms. Another measurement is number of document represented by terms. A feature space including low frequent terms neither guarantees to represent all train and test document, nor overlaps any term in a given document. The second curve shows feature representation capacity percentage over all documents remains until 500 terms. At about 20000, it represents only 8% of the document space. When selecting last 2000 low frequent terms as feature space, the capacity is about 2%. So, it is not possible to run a model.

It is also worth analyzing random feature selection algorithms. When we randomly select 1000 features evenly distributed over document frequency, many experiments shows that random setup with NB gives at most F1 measure of 50

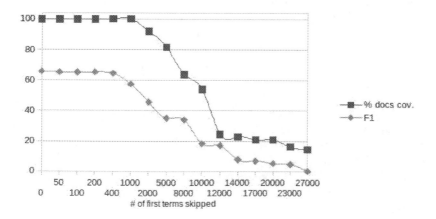

Fig. 2. Impact of term frequency

%. This gives a baseline reference and any proposed model can be acceptable if and only if it outperforms that baseline performance.

For a comparison for Turkish domain, two studies, [28,29] can be taken into account. They both apply their methodologies in knowledge-rich framework. While former obtains success rate of at best 93% for three class genre detection problem with 200 examples for each category, latter achieves at best 95.8 %for six class category detection with 100 documents for each class. For a meaningful comparison, equalizing the size of both categories and train set in the experiment we get 96.1% and 90.8% success rate for three class and six class categorization respectively.

Using coarse grain categories and taking better results can be deceptive. This is very similar to word sense disambiguation problem. Evaluation on coarse grain and fine grain senses must be totally differentiated. Even though such a comparison must be exactly done within same documents, categories and environment, we want to only emphasize that a knowledge-poor framework proposed here shows more or less similar performance with a knowledge rich one. In other words, it proves pre-processing is not a indispensable phase in DC. Therefore the approach could be considered valid and consisted. As a future work, we want to apply the approach in language-independent sense.

6   Conclusion

In this study, we proposed a Document Categorization model for Turkish language as one of the agglutinative language especially within knowledge-poor framework. We evaluate five feature selection formulas among which IG feature selection method has the best F1 performance. CHI behaves very similarly, but it is slightly worse than it. Other metric does not show significant benefit. Contrary to other studies, DF does not show significant performance under smaller space.

Among machine learning algorithms, the best performance is achieved by Multinomial Naive Bayes classifier. Support Vector Machine is the second robust algorithm in terms of F1 measure. Although Decision Tree gives relatively better results, It suffers from its divide-conquer mechanism that leads to a bias favoring absence. The poorest algorithm is the k-nearest neighbor algorithm. We also evaluate feature weighting schema across learning algorithms and feature selections. However, we do not observe a significant weighting criterion that efficiently works either with any particular term selection or any machine learning algorithm.

Another observation is that changing the size of vocabulary deeply and positively affects the performance of all the setups as shown in Figure-1, where the improvements are observable. The peak point is at about feature size of 2000.To measure effect of document frequency of terms over success, all words are ranked by their document frequencies. As eliminating the most frequent words, the system performance is getting worse within all setup. It means that frequent terms are very informative in document classification contrary to many other IR problem such as Word Sense Disambiguation.

To validate the approach, we randomly select 1000 features evenly distributed over document frequency. This gives a baseline reference and any proposed model can be acceptable if and only if it outperforms that baseline performance. Baseline gets about F1 of 51% as its best. We observed that NB with X2 or IG feature selection could get 89-90 % F1 score with sufficient size of feature word.

For a meaningful comparison with other studies especially for Turkish domain, we equalized the size of both categories and train set in the experiment with respect to other studies' settings. We get 96.1% and 90.8% success rate for three class and six class categorization respectively. This means that knowledge-poor design proposed here is more or less similar performance with other studies. It proves that pre-processing is not a indispensable phase in DC. To validate the idea we want to extend and apply the approach in language-independent sense as a future work.

References

1. Scott, S., Matwin, S.: Text Classification Using WordNet Hypernyms. The Workshop on usage of WordNet in NLP Systems. In: COLING-ACL (1998)
2. Salton, G., Wong, A., Yang, C.-S.: A Vector Space Model for Automatic Indexing. Communications of the ACM 18, 613–620 (1975)
3. Liu, T., Chen, Z., Zhang, B., Ma, W.-Y., Wu, G.: Improving Text Classification using Local Latent Semantic Indexing. In: International Conference on Data Mining (ICDM 2004), pp. 162–169. IEEE Computer Society, Washington, DC (2004)
4. Lewis, D., Ringuette, M.: A comparison of two learning algorithms for text categorization. In: Proceedings of SDAIR 1994, pp. 81–93 (1994)
5. Lewis, D.D., Yang, Y., Rose, T.G., Li, F.: RCV1: A New Benchmark Collection for Text Categorization Research. J. Mach. Learn. Res., 361–397 (2004)
6. Schtze, H., Hull, D.A., Pedersen, J.O.: A comparison of classifiers and document representations for the routing problem. In: ACM SIGIR 1995, New York, NY, USA, pp. 229–237 (1995)
7. Yang, Y., Pedersen, J.: A comparative study on feature selection in text categorization. In: The Fourteenth International Conference on Machine Learning, pp. 412–420 (1997)
8. Yang, Y., Liu, X.: A re-examination of text categorization methods. In: Proceedings of the 22nd Annual International ACM SIGIR Conference on Research and Development in Information Retrieval, Berkeley, pp. 42–49 (1999)
9. Forman, G.: An extensive empirical study of feature selection metrics for text classification. Journal of Machine Learning Research, 1289–1305 (2003)
10. Forman, G.: A pitfall and solution in multi-class feature selection for text classification. In: International Conference on Machine Learning, pp. 297–304 (2004)
11. Li, S., Xia, R., Zong, C., Huang, C.-R.: A framework of feature selection methods for text categorization. In: ACL, pp. 692–700 (2009)
12. Debole, F., Sebastiani, F.: Supervised term weighting for automated text categorization. In: Proceedings of the 18th ACM Symposium on Applied Computing, pp. 784–788 (2003)
13. Chen, X., Wasikowski, M.: FAST: A roc-based feature selection metric for small samples and imbalanced data classification problems. In: Proceedings of the 14th ACM SIGKDD, Las Vegas, pp. 124–132 (2008)

14. Ogura, H., Amano, H., Kondo, M.: Comparison of metrics for feature selection in imbalanced text classification. Expert Systems with Applications 38, 4978–4989 (2011)
15. Manning, C.D., Schtze, H.: Foundations of Statistical Natural Language Processing. MIT Press, Cambridge (1999)
16. Sebastiani, F.: Machine learning in automated text categorization. ACM Comput. Surv. 34, 1–47 (2002)
17. Sriurai, W.: Improving text categorization by using a topic model. Advanced Computing: An International Journal 2(6), 21–27 (2011)
18. Joachims, T.: Text categorization with support vector machines: Learning with many relevant features. In: Proceedings of the European Conference on Machine Learning, pp. 137–142 (1998)
19. Lan, M., Tan, C.L., Su, J., Lu, Y.: Supervised and traditional term weighting methods for automatic text categorization. IEEE Transactions on Pattern Analysis and Machine Intelligence 31(4), 721–735 (2009)
20. Chen, Y.-T., Chen, M.C.: Using chi-square statistics to measure similarities for text categorization. Expert Syst. Appl. 38(4), 3085–3090 (2011)
21. Singhal, A.: Modern Information Retrieval: A Brief Overview. IEEE Data Eng. Bull. 24(4), 35–43 (2001)
22. Hersh, W., Buckley, C., Leone, T.J., Hickam, D.: OHSUMED: An Interactive Retrieval Evaluation and New Large Test Collection for Research. In: SIGIR 1994, pp. 192–201 (1994)
23. Lang, K.: NewsWeeder: learning to filter netnews. Paper Presented at the Meeting of the Proceedings of the 12th International Conference on Machine Learning (1995)
24. Witten, I.H., Frank, E.: Data mining: Practical machine learning tools with java implementations. Morgan Kaufmann, San Francisco (2000)
25. Zhang, W., Yoshida, T., Tang, X.: A comparative study of TF*IDF, LSI and multi-words for text classification. Expert Syst. Appl., 2758–2765 (2011)
26. Manning, C.D., Raghavan, P., Schtze, H.: Introduction to Information Retrieval. Cambridge University Press, New York (2008)
27. Chang, C.C., Lin, C.J.: LIBSVM: A library for support vector machines. ACM Trans. Intell. Syst. Technol. 2(3) (2011)
28. Amasyalı, M.F., Diri, B.: Automatic turkish text categorization in terms of author, genre and gender. In: Kop, C., Fliedl, G., Mayr, H.C., Métais, E. (eds.) NLDB 2006. LNCS, vol. 3999, pp. 221–226. Springer, Heidelberg (2006)
29. Guran, A., Akyokus, S., Bayazit, N.G., Gurbuz, M.Z.: Turkish Text Categorization using N-Gram words. In: International Symposium on Innovations in Intelligent Systems and Applications, pp. 369–373 (2009)

Credible or Incredible?
Dissecting Urban Legends

Marco Guerini[1] and Carlo Strapparava[2]

[1] Trento-RISE, Via Sommarive 18, Trento - I-38123 Italy
marco.guerini@trentorise.eu
[2] FBK-Irst, Via Sommarive 18, Trento - I-38123 Italy
strappa@fbk.eu

Abstract. Urban legends are a genre of modern folklore, consisting of stories about rare and exceptional events, just plausible enough to be believed. In our view, while urban legends represent a form of "sticky" deceptive text, they are marked by a tension between the credible and incredible. They should be credible like a news article and incredible like a fairy tale. In particular we will focus on the idea that urban legends should mimic the details of news (*who, where, when*) to be credible, while they should be emotional and readable like a fairy tale to be catchy and memorable. Using NLP tools we will provide a quantitative analysis of these prototypical characteristics. We also lay out some machine learning experiments showing that it is possible to recognize an urban legend using just these simple features.

1 Introduction

Urban legends are a genre of modern folklore consisting of stories told as true – and plausible enough to be believed – about some rare and exceptional events that supposedly happened to a real person or in a real place. Like memes [1], urban legends tend to propagate across communities and exhibit variation over time; for example the sewer alligator, that originally "appeared" in New York City [2], also appeared in different cities to suit regional variations. With the advent of the Internet, urban legends gained new lifeblood, as they began to be circulated by e-mail.

In [3], the authors discuss the idea of "stickiness" popularized by the book "The Tipping Point" [4], seeking to explain what makes an idea or concept memorable or interesting. They also focus on urban legends and claim that, by following the acronym "SUCCES" (each letter referring to a characteristic that makes an idea "sticky"), it is possible to describe their *prototypical* structure:

– Simple – find the core of any idea
– Unexpected – grab people's attention by surprising them
– Concrete – make sure an idea can be grasped and remembered later
– Credible – give an idea believability
– Emotional – help people see the importance of an idea
– Stories – empower people to use an idea through narrative

A. Gelbukh (Ed.): CICLing 2014, Part II, LNCS 8404, pp. 441–453, 2014.

Such features are allegedly placed at the core of persuasive and viral language; urban legends constitute an ideal framework with which to computationally verify these assertions. Table 1 displays a few examples of urban legends claims.

In particular we will investigate some of the prototypical characteristics that can be found in urban legends as compared to similar literary genres. In our view, urban legends are viral since they are stressed by a tension between credible and incredible: credible like a *news* and incredible like a *fairy tale*. We will focus on the idea that UL should mimic the details of news (*who*, *where*, *when*) to be credible, and they should be *emotional* and *readable* like the story of a fairy tale to be catchy and memorable. We will verify these psychological hypotheses – appeared in the literature – using NLP tools, to drive a *quantitative* analysis of these *qualitative* theories. For example, the idea that urban legends derive much of their credibility from details concerning the location where the situation took place, is presented in [5]. Anecdotally, the television series "1000 Ways to Die" – that recreates unusual supposed deaths and debunked urban legends in a way similar to the Darwin Awards[1] – introducing each story with the location and date of each supposed incident, to render it more credible.

Table 1. Examples of Urban Legend Claims

A tooth left in a glass of Coca-Cola will dissolve overnight.
A stranger who stopped to change a tire on a disabled limo was rewarded for his efforts when the vehicle's passenger, Donald Trump, paid off his mortgage.
Walt Disney arranged to have himself frozen in a cryonic chamber full of liquid nitrogen upon his death, and he now awaits the day when medical technology makes his re-animation possible.
Drugged travelers awaken in ice-filled bathtubs only to discover one of their kidneys has been harvested by organ thieves.
Facebook users can receive a $5,000 cash reward from Bill Gates for clicking a share link.

In the tension between credible and incredible, details should be neither too specific, like in the news, nor too few, as in fairy tales: effective urban legends should be credible but not verifiable. Similarly, emotions should be enough to make it sticky/catchy but not too much to render it not-credible. Finally urban legends should be easy to read, similar to fairy tales, to render them more memorable.

---

[1] The Darwin Awards are an ironical honor, granted to individuals who have contributed to human evolution by "self-selecting themselves out of the gene pool" via incredibly foolish actions; Darwin Awards explicitly try to disallow urban legends from the awards. See `darwinawards.com`

In the following sections we first review relevant work that addresses the problem of deceptive language and behavior both in online and offline scenarios, followed by an overview of work that addresses the virality of online content. Then we describe the data collected for our experiments and the features extracted to model the aforementioned prototypical characteristics of urban legends. We use these features in both descriptive statistics and generalization tasks and we report the best performing features. Finally we discuss future research on further prototypical characteristics of urban legends.

2   Related Work

The topic of deceptive and/or false messages is a burning topic within the NLP community. A seminal work on the linguistic recognition of lies can be found in [6]. Still, defense from subtle persuasive language in broadcast messages, including social networks, is needed in many applied scenarios. Viral messages have become a very important factor for persuasion and are currently almost entirely out of control. So, protection from fraudulent communication is needed, especially in competitive commercial situations. Two main approaches are currently under investigation in the literature:

1) Recognizing the linguistic characteristics of deceptive content in the social web: for example preventing deceptive consumer reviews [7] on sites like Trip Advisor is fundamental both for consumers seeking genuine reviews, and for the reputation of the site itself. Deceptive consumer reviews are fictitious opinions that have been deliberately written to sound authentic. Another example concerns online advertising [8]: detecting fraudulent ads is in the interest of users, of service providers (e.g. Google AdWords system), and other advertisers.

2) Recognizing on-line behavioral patterns of deceptive users: For example recognizing groups of propagandists or fake accounts that are used to push the virality of content [9]. Four main patterns are recognized: (i) sending high volumes of tweets over short periods of time, (ii) retweeting while publishing little original content, (iii) quickly retweeting, and (iv) colluding with other, seemingly unrelated, users to send duplicate or near-duplicate messages on the same topic simultaneously. Another example is [10] where the authors hypothesize that there is a set of representative distributions of review rating scores. Deceptive business entities that hire people to write fake reviews can then be recognized since they will necessarily distort distribution of review scores, leaving "distributional footprints" behind.

We want to consider a third point, which is linked to the previous two but different at the same time: deceptive content that spreads quickly but without an explicit strategy of making them spread, which is the case with urban legends.

Finally, the spreading dynamics of an urban legend on one hand closely resembles those of memes that undergo many variations while spreading [11]; on the other hand their characteristics resemble those of viral content. Several researchers have studied information flow, community building and similar processes using Social Networking sites as a reference [12–14]. However, the great majority concentrate on network-related features without taking into account the actual content spreading within the network [15]. A hybrid approach focusing on both product characteristics and network related features is presented in [16]: in particular, the authors study the effect of passive-broadcast and active-personalized notifications embedded in an application to foster word of mouth.

Recently, the correlation between content characteristics and virality has begun to be investigated, especially with regard to textual content; in [17], for example, features derived from sentiment analysis of comments are used to predict stories' popularity. The work in [18] uses *New York Times* articles to examine the relationship between emotions evoked by the content and virality, using semi-automated sentiment analysis to quantify the affectivity and emotionality of each article. Results suggest a strong relationship between affect and virality, where virality corresponds to the number of people who emailed the article.

The relevant work in [19] measures a different form of content spreading by analyzing which features of a movie quote make it "memorable" online. Another approach to content virality, somehow complementary to the previous one, is presented in [11], and takes the perspective of understanding which modification dynamics make a meme spread from one person to another (while movie quotes spread remaining exactly the same). More recently, some works tried to investigate how different textual contents give rise to different reactions in the audience: the work presented in [20] correlates several viral phenomena with the wording of a post, while [21] shows that specific content features variations (like the readability level of an abstract) differentiate among virality level of downloads, bookmarking, and citations.

3 Datasets

To explore the characteristics of urban legends and understand the effectiveness of our ideas we collected a specific dataset. It is composed of roughly 8000 textual examples: 2518 Urban Legends (UL), 1860 Fairy Tales (FT) and 3575 Google News articles (GN). The description of how the datasets have been created follows.

- **Urban Legends** have been harvested from the website `snopes.com`. While almost 5 thousand urban legends were collected and discussed on the website, we considered only those that were reported along with a textual example (usually e-mail circulated on the Internet), and extracted the textual example only when it was at least thirty tokens long.
- **News Articles** have been selected from a corpus of about 400.000 Google News articles, from the years 2009-2012. We collected those with the highest similarity among the titles of the Urban Legends, to grant that textual content is comparable. The similarity scores were computed in a Latent Semantic space, built from the British National Corpus using 400 dimensions. The typical categories of GN articles are science, health, entertainment, economy and sports.

- **Fairy Tales.** We exploit a corpus of fairy tales collected and preprocessed by [22] that were downloaded from Project Gutenberg [23]. Since the corpus ranges from very short tales (the shortest is 75 words) to quite long ones (the longest is 15,000 words) we split the longest tales to get a total of 1860 documents. The mean length of the resulting documents is about 400 words.

4   Feature Extraction

After collecting the datasets we extracted four different groups of features, relevant to the prototypical characteristics we want to analyze.

Named Entities, (*NE*). To annotate named entities we used the TextPro toolkit [24], and in particular its Named Entities recognition module. The output of the tool is in the IOB2 format and includes the tags Person (*PER*), Organization (*ORG*), Location (*LOC*) and Miscellaneous (*MISC*).

Temporal Expressions, (*TIMEX*). To annotate temporal expressions we used the toolkit TTK [25]. The output of the tool is in TimeML annotation language format [26]. In particular time expressions are flagged with TIMEX3 tags (tern.mitre.org). The tags considered are *DATE*, *DURATION* and *TIME*.

Sentiment (*SENT*). Since the three corpora have different characteristics, rather than computing word polarity using specialized bag-of-words approaches, we resort to words' *prior polarity* - i.e. if a word out of context evokes something positive or something negative. This technique, even if less precise, guarantee that the same score is given to the same word in different contexts, and that none of the corpora is either overestimated or underestimated. To this end, we follow the methodology proposed in [27], using SentiWordNet 1.0 [28], that assigns prior polarities to words starting from their posterior polarities. In particular we choose the best performing approach. This formula uses a weighted mean, i.e. each sense weight is chosen according to a harmonic series. The rationale behind this choice is based on the assumption that more frequent senses should bear more "affective weight" than very rare senses when computing the prior polarity of a word. In particular, for each word we returned its positive (*POS*) and negative (*NEG*) prior polarity score:

$$POS = \frac{\sum_{i=1}^{n}(\frac{1}{i} \times posScore_i)}{\sum_{i=1}^{n}(\frac{1}{i})} \tag{1}$$

where $posScore_i$ represents the modulus of the positive polarity of the ith sense of that word. The *NEG* score is computed following the same procedure.

To compute the importance of these features, and to explore the characteristics of urban legend texts, we used the method proposed in [6]. We calculate a score associated with a given set of entities (features), as a measure of saliency for the given word class inside the text, called *coverage*.

More formally, given a set of feature instances present in a text, C={W_1, W_2, ..., W_N}, we define the feature coverage in that text (or corpus) A as the percentage of words from A belonging to the feature set C:

$$Coverage_A(C) = \frac{\sum_{W_i \in C} Frequency_A(W_i)}{Words_A} \qquad (2)$$

where $Frequency_A(W_i)$ represents the total number of feature occurrences W_i inside the text A, and $Words_A$ represents the total size (in words) of the text. Note that we computed features' coverage regardless of their actual length: "New York City" or "Paris" both count as one LOC even if the former is composed of three tokens while the latter only of one. Note also that this approach normalizes according to text length, avoiding biases due to different corpus characteristics.

Readability ($READ$). We further analyzed the texts in the three datasets according to readability indices, to understand whether there is a difference in the language difficulty among them. Basically, the task of readability assessment consists of quantifying how difficult a text is for a reader. This kind of assessment has been widely used for several purposes, such as evaluating the reading level of children and impaired persons and improving Web content accessibility, see for example what reported in [29].

We use three indices to compute the difficulty of a text: the Gunning Fog [30], Flesch [31] and Kincaid [32] indices. These metrics combine factors such as word and sentence length that are easy to compute and approximate the linguistic elements that have an impact on readability. In the following formulae, $Sent_A$ represents the number of sentences in text A, Cpx_A the number of complex words (those with three or more syllables), and $Syll_A$ the total number of syllables.

The *Fog index* is a rough measure of how many years of schooling it would take someone to understand the content; higher scores indicate material that is harder to read. Texts requiring near-universal understanding have an index less than 8. Academic papers usually have a score between 15 and 20. The score, for a given text A, is calculated according to the formula:

$$Fog_A = 0.4\left(\frac{Words_A}{Sent_A} + 100\frac{Cpx_A}{Words_A}\right) \qquad (3)$$

The *Flesch Index* rates texts on a 100-point scale. Higher scores indicate material that is easier to read while lower numbers mark passages that are more difficult to read. Scores can be interpreted as: 90-100 for content easily understood by an average 11-year-old student, while 0-30 for content best understood by university graduates. The score is calculated with the following formula:

$$Flesch_A = 206.835 - 1.015\frac{Words_A}{Sent_A} - 84.6\frac{Syll_A}{Words_A} \qquad (4)$$

The *Kincaid Index* or "Flesch–Kincaid Grade Level Formula" translates the 0-100 score of the Flesch Index to a U.S. grade level. It can be interpreted as the number of years of education required to understand this text, similar to the Gunning Fog index. The grade level is calculated with the following formula:

$$Kincaid_A = 0.39 \frac{Words_A}{Sent_A} + 11.8 \frac{Syll_A}{Words_A} - 15.59 \qquad (5)$$

5   Descriptive Statistics

As can be seen from Tables 2 (Named Entities) and 3 (Temporal Expressions), urban legends place half-way between fairy tales and news, as we expected. While fairy tales represent out-of-time, out-of-place and always-true stories, news represent circumstantial description of events. This is reflected by the overall use of named entities (respectively almost three and four times more in UL and GN) and of temporal expressions (respectively almost two and three times more). Interestingly the use of person names is the only case where FT reduce the lead of UL and GN, and can be explained by the fact that characters in FT are usually addressed with proper names (e.g. *"Hansel* and *Gretel"*).

Table 2. Coverage of Named Entities

	PER		LOC		ORG		MISC		ALL	
	μ	σ	μ	σ	μ	σ	μ	σ	μ	σ
FT	0.86%	1.03%	0.31%	0.44%	0.27%	0.45%	0.15%	0.30%	1.58%	1.42%
UL	1.29%	1.49%	1.12%	1.40%	1.17%	1.56%	0.57%	0.90%	4.15%	3.18%
GN	1.65%	1.19%	2.02%	1.59%	1.63%	1.28%	0.93%	0.95%	6.22%	2.91%
Average	1.26%	1.24%	1.15%	1.14%	1.02%	1.10%	0.55%	0.72%	3.98%	2.50%

Table 3. Coverage of Temporal Expressions

	DATE		DURATION		TIME		ALL	
	μ	σ	μ	σ	μ	σ	μ	σ
FT	0.42%	0.39%	0.08%	0.16%	*0.01%	0.05%	0.51%	0.43%
UL	0.70%	0.93%	*0.20%	0.43%	*0.02%	0.12%	0.92%	1.08%
GN	1.20%	0.85%	*0.20%	0.32%	0.06%	0.20%	1.46%	0.96%
Average	0.77%	0.72%	0.16%	0.30%	0.03%	0.12%	0 96%	0.82%

In Table 4, statistics for sentiment coverage are reported. As can be seen, in this group of features the differences are less marked and, quite surprisingly, ULs have the lowest scores. As we would expect, FTs have the highest score. We will discuss possible interpretation of *SENT* results in the following sections.

In Table 5, statistics for readability are reported. As can be seen, ULs are readable in a way similar to fairy tales. Still, depending on the readability indices, that grasp different aspects of text difficulty, ULs are either slightly easier than FTs or half-way between FTs and ULs similar to the cases of Tables 2 and 3.

Table 4. Coverage of Prior Polarities Sentiment

	POS		NEG		ALL	
	μ	σ	μ	σ	μ	σ
FT	2.40%	0.69%	-2.62%	0.69%	5.03%	1.12%
UL	2.13%	0.74%	-2.20%	0.90%	4.33%	*1.35%
GN	2.27%	0.87%	-2.29%	0.73%	4.56%	*1.33%
σ	2.27%	0.77%	-2.37%	0.77%	4.64%	1.27%

Table 5. Readability difficulty Scores

	Fog		Flesch		Kincaid		$\frac{Cpx_A}{Words_A}$		$\frac{Syll_A}{Words_A}$		$\frac{Words_A}{Sent_A}$	
	μ	σ	μ	σ	μ	σ	μ	σ	μ	σ	μ	σ
FT	12.71	3.68	67.04	10.92	10.37	3.48	5.77	2.41	1.34	0.07	26.01	8.43
UL	12.13	*5.00	58.56	17.31	9.65	*4.70	11.67	5.13	1.51	0.19	18.74	10.96
GN	15.69	*4.95	41.10	16.39	12.90	*4.64	17.26	4.68	1.67	0.21	22.12	10.77
μ	13.51	4.54	55.57	14.87	10.98	4.28	11.57	4.07	1.51	0.15	22.29	10.05

This behavior can be explained by the fact that ULs have a simpler syntax than FTs but a more complex lexicon. In fact, inspecting the individual elements of the formulae, as reported in the second part of Table 5, we see that while the percentage of complex words (either $\frac{Cpx_A}{Words_A}$ or $\frac{Syll_A}{Words_A}$) puts UL halfway between FT and GN, the average length of sentences ($\frac{Words_A}{Sent_A}$) is surprisingly higher for FT than GN and in turn UL. So, depending on the weight given either to complex words or to sentence length, the results in Table 5 can be interpreted.

All differences in the means reported in the tables are statistically significant (Student's t-test, $p < 0.001$) apart from *TIME*, between UL and FT, and *DURATION*, between UL and GN, (signalled with * in Table 3).

Turning to the analysis of variance, we see that FT is – on average – a more cohesive genre, with lower standard deviations, while GN and UL have higher and closer standard deviations. In fact, all differences in the standard deviations reported in the tables are statistically significant (f-test, $p < 0.001$) apart between UL and GN in *Fog*, *Kincaid* and in *ALL* sentiment (signalled with * in the respective Tables).

6   Classification Experiments

The goal of our experiments is to understand to what extent it is possible to assign a text to one of the aforementioned classes using just the prototypical characteristics (features) discussed above, and whether there is a subset of features that stands out among the others in this classification task. For every feature combination we conducted a binary classification experiment with ten-fold cross validation on the dataset. We always randomly downsampled the majority class in order to make the dataset balanced, i.e. 50% of positive examples and 50% of negative examples; this accounts for a random baseline of 0.5. We also normalized all features according to z-score. Experiments were carried out using SVM [33], in particular libSVM [34] under its default settings. Results are reported in Table 6; all significance tests discussed below are computed using an approximate randomization test [35].

Urban Legends vs News. In the UL vs. GN classification task, while all the features together performed well (F1 = 0.833), improving over all other subgroups of features ($p < 0.001$), no single group of features performed so well, apart from READ (F1 = 0.763, $p < 0.001$). Particularly, the sentiment features (*SENT*) – that performed slightly better than the random baseline ($p < 0.001$) – performed worse than *TIMEX* and *NE* ($p < 0.001$) that, in turn, had smaller differences and lower statistical significance between them ($p < 0.05$). Still, all features improved over the baseline ($p < 0.001$).

Table 6. Classification Results

Features	UL vs. GN			UL vs. FT			GN vs. FT		
	Prec	Rec	F1	Prec	Rec	F1	Prec	Rec	F1
NE	0.694	0.694	0.694	0.787	0.768	0.777	0.897	0.896	0.896
TIMEX	0.677	0.676	0.676	0.666	0.666	0.666	0.775	0.767	0.766
SENT	0.573	0.572	0.572	0.661	0.656	0.658	0.606	0.601	0.603
READ	0.765	0.762	0.763	0.869	0.868	0.868	0.973	0.973	0.973
ALL	0.834	0.833	0.833	0.897	0.897	0.897	0.978	0.978	0.978

Urban Legends vs Fairy Tales. In the UL vs. FT classification task, all the features together performed better than the previous experiment (F1 = 0.897), again improving over all the other subgroups of features alone ($p < 0.001$). Interestingly, the best discriminative subgroup of features (still *READ*, F1 = 0.868) in this case reduces the lead with respect to all the features together (*ALL*) and improves over all the others subgroups ($p < 0.001$). The *SENT* group of features – that has no significant difference with *TIMEX* – in this case performs better than in the previous experiment, but still poorly. This is because the *TIMEX* group had similar performances as the previous experiment, while *NE* and *SENT* improved their performance. Finally, all groups of features had a statistically significant improvement over the baseline ($p < 0.001$).

News vs Fairy Tales. Finally, we wanted to check whether UL being "halfway" between GN and FT can be observed in our classification experiments as well. If this hypothesis is correct, by classifying GN vs. FT we would expect to find higher performance than previous experiments. Results show that this is in fact the case. All features together performed better than all previous experiment and incredibly well (F1= 0.978), again improving over all the other subgroups of features alone ($p < 0.001$) apart from READ that performs equally well (F1=0.973, no statistically significant difference). Notably, the *SENT* group of features (F1=0.603) improves over the UL vs. GN task, but not over the UL vs. FT task showing again to be not so good in characterizing UL as compared to the other two genres. Finally, all groups of features had a statistically significant improvement over the random baseline ($p < 0.001$).

Three Class Classification. Finally we also tested feature predictivity on a three class classification task (UL vs GN vs FT). Since in this case we did not performed downsampling, we use the ZeroR classifier as a baseline. For the sake of interpretability of results, along with precision, recall and F1 we also provide the Matthews Correlation Coefficient (MCC) which is useful for unbalanced datasets, as presented in [36] for the multiclass case. MCC returns a value between -1 and +1, where +1 represents a perfect prediction, 0 no better than random and -1 indicates total disagreement. Results are consistent with previous experiments. In Table 7, all feature configurations show an improvement over the baseline ($p < 0.001$) but the sentiment features (*SENT*) have far lower discriminative power as compared to others groups of features (MCC=0.069).

7 Discussion

While between UL and GN the discrimination is given by a skillful mixture of all the prototypical features together, where none has a clear predominance over the others, between UL and FT, readability (*READ*) plays a major role. From the summary in Table 8 we see that while ALL features together have the highest averaged F1, *READ* is the best performing subset of features in all experiments, followed by *NER* and *TIMEX* that perform reasonably well. *SENT*, instead, has reasonable performance only on UL vs FT, while its predictive power is low in the other cases (this is reflected also by the lowest $F1\mu$), especially for the three class classification task (see Table 7).

Table 7. Results for UL vs FT vs GN

Features	Prec	Rec	F1	MCC
NE	0.630	0.650	0.640	0.449
TIMEX	0.570	0.577	0.573	0.339
SENT	0.446	0.461	0.453	0.069
READ	0.746	0.754	0.750	0.611
ALL	0.820	0.822	0.821	0.721
ZeroR	0.202	0.450	0.279	0

Table 8. Overall Feature performances

Features	$F1\mu$	$F1\sigma$
ALL	0.868	0.070
READ	0.819	0.100
NE	0.740	0.100
TIMEX	0.675	0.069
SENT	0.589	0.085

The fact that sentiment does not work so well overall in discriminating UL seems a contrast with previous works – see for example what is reported in [37] on UL and evoked emotions. On one hand this can be explained by the fact that a UL tend to be as objective as possible in order to be credible. Furthermore, this result can be potentially described in light of the distinction between *emotional impact* and *emotional language*. In fact, emotional impact can either derive from the wording of the text itself (usage of strong affect words), or from the depicted situation (i.e. emotions are evoked by describing a vivid situation with plain language). In our experiment we tested the wording rather than the evoked emotions (a concept that is harder to be modeled from an NLP point of view). Let us consider as an example the following excerpt, taken from the "Kidney Theft" UL, as reported by `snopes.com`:

Dear Friends:
I wish to warn you about a new crime ring that is targeting business travelers. This ring is well organized [. . .] and is currently in most major cities and recently very active in **New Orleans**. The crime begins when a business traveler goes to a lounge for a drink [. . .] A person in the bar walks up as they sit alone and offers to buy them a drink. The last thing the traveler remembers until they wake up in a hotel room bath tub, their body submerged to their neck in ice, is sipping that drink. There is a note taped to the wall instructing them not to move and to call 911. [. . .] The business traveler is instructed by the 911 operator to very slowly and carefully reach behind them and feel if there is a tube protruding from their lower back. The business traveler finds the tube and answers, "Yes." The 911 operator tells them to remain still, having already sent paramedics to help. The operator knows that both of the business traveler's kidneys have been harvested. This is not a scam, it is real. It is documented and confirmable. If you travel, please be careful.
Regard
Jerry Mayfield

There is no very strong emotional wording here, it is the situation itself that is scary; on the contrary the email contains locations, the signature of a presumed Jerry Mayfield, and – noticeably – credibility is also explicitly addressed in the text with the adjectives "real", "documented" and "confirmable".

8   Conclusions

In this paper we have presented a study on urban legends, a genre of modern folklore consisting of stories about some rare and exceptional events plausible enough to be believed. We argued that urban legends represent a form of "sticky" deceptive text, marked by a tension between the credible and incredible. To be credible they should resemble a news article while being incredible like a fairy tale. In particular we focused on the idea that ULs should mimic the details of news (*who*, *where*, *when*) to be credible, while being *emotional* and *readable* like a fairy tale to be catchy and memorable. Using NLP tools we presented a quantitative analysis of these simple yet effective features and provided some machine learning experiments showing that it is possible to recognize an urban legend using just these prototypical characteristics. In the future we want to explore other prototypical aspects of urban legends like, for example, linguistic style [38, 39]. With regard to sentiment, besides the simple word polarities we used, we will explore the emotions expressed in UL, FT and GN, using an approach similar to the one described in [40]. Exploiting knowledge-based and corpus-based methods, that approach deals with automatic recognition of affect, annotating texts with six basic emotions. We believe that fine-grained emotion annotation of urban legends could shed more light in the understanding the mechanisms behind persuasive language.

References

1. Dawkins, R.: The Selfish Gene, vol. 199. Oxford University Press, USA (2006)
2. Coleman, L.: Alligators-in-the-sewers: a journalistic origin. Journal of American Folklore, 335–338 (1979)
3. Heath, C., Heath, D.: Made to stick: Why some ideas survive and others die. Random House (2007)
4. Gladwell, M.: The tipping point: How little things can make a big difference. Little, Brown (2000)
5. Brunvand, J.: The vanishing hitchhiker: American urban legends and their meanings. WW Norton & Company (1981)
6. Mihalcea, R., Strapparava, C.: The lie detector: Explorations in the automatic recognition of deceptive language. In: Proceedings of ACL 2009, Singapore, pp. 309–312 (2009)
7. Ott, M., Choi, Y., Cardie, C., Hancock, J.T.: Finding deceptive opinion spam by any stretch of the imagination. In: Proceedings of the 49th Annual Meeting of the Association for Computational Linguistics: Human Language Technologies, HLT 2011, vol. 1, pp. 309–319. Association for Computational Linguistics, Stroudsburg (2011)
8. Sculley, D., Otey, M.E., Pohl, M., Spitznagel, B., Hainsworth, J., Zhou, Y.: Detecting adversarial advertisements in the wild. In: Proceedings of the 17th ACM SIGKDD International Conference on Knowledge Discovery and Data Mining, pp. 274–282. ACM (2011)
9. Lumezanu, C., Feamster, N., Klein, H.: # bias: Measuring the tweeting behavior of propagandists. In: Sixth International AAAI Conference on Weblogs and Social Media (2012)
10. Feng, S., Xing, L., Gogar, A., Choi, Y.: Distributional footprints of deceptive product reviews. In: Proceedings of the 2012 International AAAI Conference on Web-Blogs and Social Media (June 2012)
11. Simmons, M., Adamic, L.A., Adar, E.: Memes online: Extracted, subtracted, injected, and recollected. In: ICWSM (2011)
12. Lerman, K., Ghosh, R.: Information contagion: an empirical study of the spread of news on digg and twitter social networks. In: Proceedings of 4th International Conference on Weblogs and Social Media, ICWSM 2010 (2010)
13. Khabiri, E., Hsu, C.F., Caverlee, J.: Analyzing and predicting community preference of socially generated metadata: A case study on comments in the digg community. In: ICWSM (2009)
14. Aaditeshwar Seth, J.Z., Cohen, R.: A multi-disciplinary approach for recommending weblog messages. In: The AAAI 2008 Workshop on Enhanced Messaging (2008)
15. Lerman, K., Galstyan, A.: Analysis of social voting patterns on digg. In: Proceedings of the First Workshop on Online Social Networks, WOSP 2008, pp. 7–12. ACM, New York (2008)
16. Aral, S., Walker, D.: Creating social contagion through viral product design: A randomized trial of peer influence in networks. Management Science 57, 1623–1639 (2011)
17. Jamali, S.: Comment mining, popularity prediction, and social network analysis. Master's thesis, George Mason University, Fairfax, VA (2009)
18. Berger, J.A., Milkman, K.L.: Social Transmission, Emotion, and the Virality of Online Content. Social Science Research Network Working Paper Series (2009)

19. Danescu-Niculescu-Mizil, C., Cheng, J., Kleinberg, J., Lee, L.: You had me at hello: How phrasing affects memorability. In: Proceedings of the ACL (2012)
20. Guerini, M., Strapparava, C., Özbal, G.: Exploring text virality in social networks. In: Proceedings of ICWSM 2011, Barcelona, Spain (2011)
21. Guerini, M., Pepe, A., Lepri, B.: Do linguistic style and readability of scientific abstracts affect their virality. In: Proceedings of ICWSM 2012 (2012)
22. Lobo, P.V., de Matos, D.M.: Fairy tale corpus organization using latent semantic mapping and an item-to-item top-n recommendation algorithm. In: Language Resources and Evaluation Conference, LREC (2010)
23. Hart, M.: Project gutenberg. Project Gutenberg (2000)
24. Pianta, E., Girardi, C., Zanoli, R.: The textpro tool suite. In: Proceedings of LREC (2008)
25. Verhagen, M., Pustejovsky, J.: Temporal processing with the tarsqi toolkit. In: 22nd International Conference on on Computational Linguistics: Demonstration Papers, pp. 189–192. Association for Computational Linguistics (2008)
26. Pustejovsky, J., Castano, J., Ingria, R., Sauri, R., Gaizauskas, R., Setzer, A., Katz, G., Radev, D.: Timeml: Robust specification of event and temporal expressions in text. New Directions in Question Answering 3, 28–34 (2003)
27. Gatti, L., Guerini, M.: Assessing sentiment strength in words prior polarities. In: Proceedings of the 24th International Conference on Computational Linguistics, COLING 2012 (2012)
28. Esuli, A., Sebastiani, F.: SentiWordNet: A publicly available lexical resource for opinion mining. In: Proceedings of LREC 2006, Genova, IT, pp. 417–422 (2006)
29. Tonelli, S., Manh, K.T., Pianta, E.: Making readability indices readable. In: Proceedings of the First Workshop on Predicting and Improving Text Readability for target reader populations, pp. 40–48. Association for Computational Linguistics (2012)
30. Gunning, R.: The technique of clear writing. McGraw-Hill (1952)
31. Flesch, R.: The Art of plain talk. Harper (1946)
32. Kincaid, J., Fishburne, R., Rogers, R., Chissom, B.: Derivation of new readability formulas (automated readability index, fog count, and flesch reading ease formula) for navy enlisted personnel. Research branch report 8-75, Chief of Naval Technical Training: Naval Air Station Memphis (1975)
33. Vapnik, V.: The Nature of Statistical Learning Theory. Springer, New York (1995)
34. Chang, C., Lin, C.: Libsvm: a library for support vector machines. ACM Transactions on Intelligent Systems and Technology (TIST) 2, 27 (2011)
35. Yeh, A.: More accurate tests for the statistical significance of result differences. In: Proceedings of the 18th Conference on Computational Linguistics, vol. 2, pp. 947–953. Association for Computational Linguistics (2000)
36. Gorodkin, J.: Comparing two K-category assignments by K-category correlation coefficient. Computational Biology and Chemistry 28, 367–374 (2004)
37. Heath, C., Bell, C., Sternberg, E.: Emotional selection in memes: The case of urban legends. Journal of Personality and Social Psychology 81, 1028–1041 (2001)
38. Pennebaker, J., Francis, M.: Linguistic inquiry and word count: LIWC. Erlbaum Publishers (2001)
39. Louis, A., Nenkova, A.: What makes writing great? first experiments on article quality prediction in the science journalism domain. Transactions of ACL (2013)
40. Strapparava, C., Mihalcea, R.: Learning to identify emotions in text. In: Proceedings of the 23rd Annual ACM Symposium on Applied Computing (APPLIED COMPUTING 2008), Fortaleza, Brazil, pp. 1556–1560 (2008)

Intelligent Clustering Scheme
for Log Data Streams

Basanta Joshi, Umanga Bista, and Manoj Ghimire

Immune APS Nepal, Lalitpur, Nepal
{baj,umb,mg}@dev.immunesecurity.com

Abstract. Mining patterns from the log messages is valuable for real-time analysis and detecting faults, anomaly and security threats. A data-streaming algorithm with an efficient pattern finding approach is more practical way to classify these ubiquitous logs. Thus, in this paper the authors propose a novel online approach for finding patterns in log data sets where a locally sensitive signature is generated for similar log messages. The similarity of these log messages is identified by parsing log messages and then, logically analyzing the signature bit stream associated with them. In addition to that the approach is intelligent enough to reflect the changes when a totally new log appears in the system. The validation of the proposed method is done by comparing F-measure of clustering results for labeled datasets and the word order matched percentage of the log messages in a cluster for unlabeled case with that of SLCT.

Keywords: event log mining, similarity search, log clustering, local sensitive hashing, sketching.

1 Introduction

Log messages are generated to record events from one service or different services within the network and are an important indication of the current status of the system(s) they monitor. Modern computer systems and networks are increasing in size and complexity at an exponential rate and the massive amount of log messages are generated from these networks. These messages can be used for network management and systems administration to monitoring and trouble shooting the system(s) behavior and and even for security analysis. Therefore, efficient methods for processing the ever growing volumes of log data has taken an increased importance and mining patterns from these log messages are indispensable. Numerous log file analysis tools and techniques are available to carry out a variety of analyses.

There have been several researches providing insights of varying degrees to log file analysis and tools developed applied in the areas of network management, network monitoring, etc. [1–6]. One of the simplest tool for log message clustering, Simple Log file Clustering Tool (SLCT) was proposed by Risto Vaarandi[1]. The basic operation is inspired by apriori algorithms for mining frequent item

A. Gelbukh (Ed.): CICLing 2014, Part II, LNCS 8404, pp. 454–465, 2014.
© Springer-Verlag Berlin Heidelberg 2014

sets. In this iterative approach, the clustering algorithm is based on finding frequent item sets from log file data and requires human interaction; an automatic approach is not mentioned [7]. The clustering task is accomplished in three passes over the log file. The first is to build a data summary, i.e. the frequency count of each word in the log file according to its position in each line. In the second step, the cluster candidates by choosing log line with words that occur more than the threshold, specified by the user. In the third step, the clusters are chosen from these candidates that occur at a frequency higher than the user specified threshold. The words in each candidate that have a frequency lower than the threshold are considered as the variable part. This algorithm was designed to detect frequently occurring patterns and this frequency is a user specified threshold. All log lines that don't satisfy this condition are considered as outliers [8]. The clusters of log files produced by SLCT can be viewed by a visualization tool called LogView [2]. This method utilizes tree maps to visualize the hierarchical structure of the clusters produced by SLCT. It basically speeds up the analysis of vast amount of data contained in the log files by summarization of different application log files and can be used to detect any security issues on a given application.

There are other tools proposed for mining of temporal patterns from logs with various association rule algorithms [9–11]. In most of these algorithms, there is an assumption that the event log has been normalized, i.e., all events in the event log have a common format. But, in reality log coming from different sources might have different formats and the temporal correlation of data's can only be established by normalizing the log received from different sources. To achieve this, many machine-learning techniques have been applied to aid the process of extracting actionable information from logs. Splunk with Prealert [5] is one of popular log management tool using intelligent algorithms. Another important aspect of log analysis system is real-time analysis of millions of log data per second. This can only be possible if the system knows the pattern of log message it receives or can create the pattern for future references if it receives new data.

To address these issues, the authors have proposed locally sensitive streaming data algorithm where a similarity search is done to create clusters of log messages from big volume of log data. The clustered logs have a single signature pattern and the list of cluster pattern is maintained as a universal signature. Inspired by the cluster evaluation strategy used in Makunju *et al.* [12], the performance of the proposed algorithm is compared with widely used Simple log clustering tool.

2   Related Literature

Efficient processing over log messages has taken an increased importance due to the growing availability of large volumes of log data from a variety of resources connected to a system. In particular, monitoring huge and rapidly changing streams of data that arrive online has emerged as the data-streaming model and has recently received a lot of attention. This model differs from computation over traditional stored data sets since algorithms must process their input by making one or a small number of passes over it, using only a limited amount of working memory. The streaming model applies to settings where the size of the input far exceeds the size of the main memory available and the only feasible access to the data is by making one or more passes over it [13].

A fundamental computational primitive for dealing with massive dataset is the Nearest Neighbor (NN) problem. The problem is defined as follows: given a collection of n objects, build a data structure, given arbitrary query object, reports the dataset object that is most similar to the query [14]. Nearest neighbor search is inherently expensive due to the curse of dimensionality. For high dimensional spaces, there are often no known algorithms for nearest neighbor search that are more efficient than simple linear search. As linear search is too costly for many applications, this has generated an interest in algorithms that perform approximate nearest neighbor search, in which non-optimal neighbors are sometimes returned. Such approximate algorithms can be orders of magnitude faster than exact search, while still providing near optimal accuracy. A detail study of exact search schemes and approximate search schemes has been presented by Panigrahy[15].

One of the method for dramatic performance gains are obtained using approximate search schemes, such as the popular Locality Sensitive Hashing (LSH) [16]. Several extensions have been proposed to address the limitations of this algorithm, in particular, by choosing more appropriate hash functions to better partition the vector space[17, 18]. One of the methods to generate the hash function is by using, sketches, a space-efficient approach. A sketch is a binary string representation of an object. The similarity of the objects can be estimated by using the hamming distance of their respective sketches [3]. A fast hash function and sketches of an object can be created using a Bloom filter which is a simple space-efficient randomized data structure for representing a set in order to support membership queries [19, 20].

Inspired by the methods, the authors are proposing a hybrid method for clustering log messages in real time, in which LSH method is combined with a method of generating the hash map by using sketching. The details of this method is discussed in Section. 3.

3   Log Clustering Scheme

Common sources of log information are facilities, such as Syslog under Unix like operating systems and Windows Event Logs under Microsoft Windows. Several log facilities collect events with a text field, which is used in many ways and not further standardized. A special scheme is necessary for analyzing this unstructured text field. So, locally sensitive streaming data algorithm is a viable option, where a similarity search is done to create clusters of log messages from big volume of log data. A signature for each log message is generated using sketching scheme and similarity is identified by analyzing the difference of bit patterns

Table 1. Signature pattern for log messages

		Tokenized word			
Log1:	User	Manoj	Ghimire	logged	out.
H1	5	11	4	37	23
Hashed function H2	14	20	20	34	1
H3	10	38	19	25	38
Signature	0100110000110010000110010100000000100110				
Log2:	User	Basanta	Joshi	logged	out.
H1	5	1	22	37	23
Hashed function H2	14	19	10	34	1
H3	10	33	17	25	38
Signature	0100010000100010010100110100000001100110				

between current log message and that of the bit pattern stored in the system database. The steps are described in the section below.

3.1 Signature Generation

In this algorithm, at first, a log data is tokenized such that a log message is represented as a list of words and each word is assigned a numerical value by using a non-cryptographic hash function specifically Murmur hash function [21]. The seeding value of this Murmur hash function is changed to generate 3-4 random hashed values for each tokenized word. Then, a log message is represented as a set of vectors; each vector represents a particular word. Inspired by algorithms proposed by Müller *et al.* [3], an n-bit signature pattern is generated for each log message. Only those bits in the pattern are set which are indicated by 3-4 random hashed values of the tokenized word in a log message [19]. The range of randomized hashed function is defined by the number of the bits in the signature pattern.

For example: Lets say we have two log messages and three hashed functions and a log message is represented by 40-bits signature as shown in Table. 1. Here, for each log message, there are five tokenized words. The hashed value for word user with three randomized hashed function is 5, 14 and 10 respectively which sets 6th, 15th and 11th bit of the signature. Similarly, all those bits in the signature are set, which are indicated by random hashed values of the other tokenized words.

3.2 Similarity Search

Lets assume that we are creating lists of universal signatures for all the log files. In this case, a log signature is matched with the universal signature list. If the log signature is similar to a universal signature in the list then it is assigned to that group, otherwise a new universal signature is created as shown in Fig. 1.

The criteria for choosing whether the log message belong to a particular pattern is given by the ratio of the number of common set bits between log signature

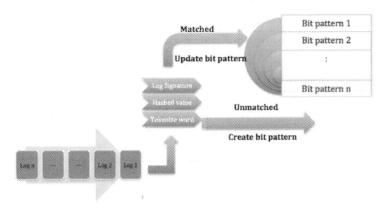

Fig. 1. Log clustering algorithm

and universal pattern to the number of the set bits in the log signature and universal pattern. The ANDing of the bits of log signature and universal pattern represents the common set bits while ORing of the bits of log signature and universal pattern represents the set bits in both as shown in Table. 2.

Table 2. Similarity estimation terms

Term	Symbol	Operation
No of common set bits	N_c	ANDing of universal signature and log signature
Total no of set bits	N_s	ORing of universal signature and log signature

As seen in Eq. 1, the ratio must be greater than some threshold percentage, and then the log message can be considered similar to a universal signature. In this case, the universal signature is updated with ANDing of the bits, which is considered to be learning in the algorithm. Ideally, and after a rigorous learning, the set bit in the universal bit stream reflects the exact pattern for a log cluster.

$$\frac{N_c}{N_s} \geq T_p \tag{1}$$

4   Results and Discussion

The performance assessment of the proposed algorithm for clustering of log messages was done by using data collected from different servers and comparing the results with a simple log clustering tool. The F-measure was compared for the clusters generated with proposed algorithm against the clusters by SLCT, for the datasets in which events (or classes) were known. While the results with unlabeled log datasets, where the manual identification of events (or classes) is time consuming or almost impossible, is evaluated using our validation algorithm. In this validation algorithm, the percentage of the matched word order between the messages in a particular cluster is used as a evaluation measure. The validation algorithm is only used for the proposed clustering algorithm. All the process was carried out in Mac OSX Lion with i7 processor 2.9 GHz speed and 8 GB RAM.

4.1   Test Data Set

The test data comprises of variety of log data collected from different servers. It includes Syslog, Windows server logs, Windows firewall logs, Authentication logs, SSHD login logs, Linux server logs, Webserver logs, Mail server logs etc. For the present test, nine samples of data are used and listed in the order of message volume as shown in Table. 3. Identifier used in this table identifies the different types of log files with varying number of log messages and log types. The first four sets are classified as labeled with known events (or classes), whereas the other five are classified as unlabeled with unknown events (or classes).

Table 3. Test data statistics

SN.	Identifier	Description	No. of Logs	Data size (MB)	Average Word Count	No. of Classes
Datasets with known Labels for Log messages						
1	auth	Authentication logs	132204	16.4	12	14
2	win	Windows firewall	59531	52.8	134	42
3	sshd	SSHD logs	50661	5.3	11	7
4	misc	Mixed of Windows server, and Linux logs	91375	16.3	18	20
Datasets with unlabeled Log messages						
5	unlabeled1		465	0.091	20	Unknown
6	unlabeled2		260386	53	23	Unknown
7	unlabeled3		235381	254	118	Unknown
8	unlabeled4		684035	430	70	Unknown
9	unlabeled5		798459	568	10	Unknown

4.2   Clustering and Validation of Labeled Datasets

In the first test, the performance of SLCT and proposed algorithm was evaluated by F-measure. SLCT has various parameters, but the test was run by tuning support threshold only, and other parameters were left to default. Similarly, the computation of F-measure in case of clustering is not straightforward because the classes always do not conform with the clusters, and there may be split, i.e. same class may split to multiple clusters. In these type of clusters, the appropriate formula for calculating F-measure[22] is summarized in Fig. 2.

Let N be the number of data points, C be the set of Classes, K be the number of Clusters, and n_{ij} be the number of data points of class $c_i \in C$, that are elements of cluster $k_j \in K$.

$$F(C, K) = \sum_{c_i \in C} \frac{|c_i|}{N} \max_{k_j \in K} \{F(c_i, k_j)\}$$

$$F(c_i, k_j) = \frac{2 * P(c_i, k_j) * R(c_i, k_j)}{P(c_i, k_j) + R(c_i, k_j)}$$

$$R(c_i, k_j) = \frac{n_{ij}}{|c_i|}$$

$$P(c_i, k_j) = \frac{n_{ij}}{|k_j|}$$

Fig. 2. Computation of F-measure

A summary of comparison of the proposed method with the SLCT method is shown in Fig. 3. This F-measure evaluation was done on labeled datasets. From the results, it is revealed that the algorithm performs better than SLCT. Also, it is non trivial to tune the support threshold in case of SLCT for optimum F-measure due to dependency of support threshold on number of log messages in the dataset. However, the proposed method has no such weakness, and performs well with F-measure around 0.7 irrespective of number of logs in datasets. Further, SLCT is a batch processing scheme and require entire dataset to be available before analysis, whereas proposed algorithm is an online scheme and hence can be deployed for real time log analytics.

Another parameter that has to be analyzed for the proposed method is the number of bits used to represent a log message. For the labeled datasets, the performance, and number of clusters were evaluated with varying bit signature as shown in Fig. 4. It can be observed that the proposed algorithm performs well for 256 bit signature, in spite of the dataset and is a good trade-off between performance and space requirement.

4.3   Clustering and Validation of Unlabeled Datasets

For the unlabeled datasets, the result of clustering is summarized in Table. 4. The number of clusters due to SLCT is always more than the proposed method, and the choice for appropriate support threshold is non trivial. However, the

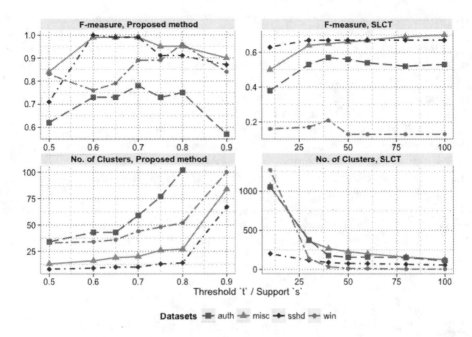

Fig. 3. Clustering comparison of SLCT and proposed method

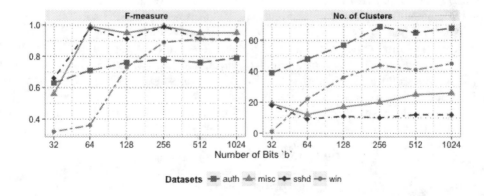

Fig. 4. Algorithm performance for varying signature bit length

threshold choice was fixed to be 0.65 in case of proposed method. Table. 4 demonstrates the number of clusters created due to both methods for some choice of support/threshold. In the case of the unlabeled datasets, we do not have apriori information about the number of events (or classes) as in the case of labeled dataset. So, the clusters created by the proposed method can easily be validated with validation scheme described below.

At first, a log message in a particular cluster is tokenized such that a log message is represented as list of words and this message is matched against all the other messages included in that cluster to identify the common words. For all the common words, we find the individual index of the words in the strings and then the difference of the index between the string is calculated. If the index difference is continuous, then it means that the words are occurring in order else the word is either not present or not in the order it appears in previous log message. As shown in Table. 5, the common words in two different log message falling in the same cluster is identified and the difference of index of the common word is calculated. Here, first four common words are in order such that the continuous index difference 0 occurs four times. So for all the continuous index difference, we calculate the number of occurrence $N_1, N_2....N_n$. The measure of the order of words to common words between two log messages is expressed as a ratio of the sum of square of all these values to the square of number of the common words N_c. Special care is taken to generate a normalized value. The dynamic range of numerator and denominator is changed further by taking the natural log of them and the ratio is manipulated to generate the word order match percent as shown in Eq. 2.

$$M_p = \frac{\log(N_1^2 + N_2^2.........+ N_n^2)}{\log N_c} * 100 \tag{2}$$

The validation of all clusters generated for the unlabeled datasets as shown in Table. 3 is done by this algorithm. The process similar to above is used to identify the number of log messages within a threshold for word order matched percentage. The percentage of the valid messages in a cluster i.e similarity score for a cluster is identified by dividing it with total no of messages falling to the cluster. The average percentage for the valid messages for a sample i.e. similarity score for a sample is calculated by dividing the similarity score for a cluster by number of clusters in a sample. Similarly, the similarity score is calculated for all the samples and are summarized in Table. 6. In this table, the similarity score for unlabeled dataset 1,2 are calculated with word order matched percentage of 80% while the results for remaining two unlabeled dataset are calculated with word order match percent of 75%.

Table 4. Clustering of log messages

Dataset	Approach			
	SLCT		Proposed method	
	Support Threshold	No of Clusters	Threshold	Clusters
Unknown1	10	14	0.65	50
Unknown2	30	394	0.65	307
Unknown3	60	965	0.65	35
Unknown4	60	1785	0.65	619
Unknown5	30	21	0.65	18

Table 5. Method for cluster validation

				Tokenization						
Log1	Conn attempt from machine XYZ				192.168.1.1 to 192.168.1.2 on port 80					
Index 1	0	1	2	3	4	5	6	7	8	9 10
Log2	Conn attempt from machine	ram kr.	192.168.2.1 to 192.168.2.2 on port 23							
Index 2	0	1	2	3	4	5	6	7	8	9	10 11
Common	✓	✓	✓	✓		✓		✓	✓	
Index diff	0	0	0	0		1		1	1	

Table 6. Validation of generated clusters

	Log files				
Unlabeled dataset No.	1	2	3	4	5
Percentage word order match	96	99	85	89	95

It can be observed that the average percentage for the valid message is within the appreciable range even if the samples files are with varying number of log messages and message length. This proves the robustness of the proposed clustering scheme.

5   Conclusion

An intelligent algorithm for signature generation of a log message has been proposed and the clustering of log messages is done based upon their percentage similarity with the signature of the patterns stored in the system database. If the percentage of similarity of log message signature is within the limit, the signature in the database is modified to reflect the effect of that log in the cluster otherwise a new signature is stored in the database for the pattern. From the evaluation of F-measure, it was observed that proposed method performs better than SLCT. For the unlabeled datasets, it was observed that the percentage of word order match was greater than 80% for most of the tests. It was observed that the results with SLCT tool is highly dependent on threshold and regards log messages as anomaly if fails to comply the criterion. Further, SLCT being batch processing method is a limitation for real-time applications. But, the method proposed by present authors not only generates clusters better than SLCT but also applicable for real-time log message pattern identification. The identified pattern can give insights to anomalies and security threats. However, some level of modification is desired in the proposed method to use it in fully distributed environment.

Acknowledgment. The authors would like to appreciate the valuable inputs from the members of LogPoint family at different times during this research work.

References

1. Vaarandi, R.: A data clustering algorithm for mining patterns from event logs. In: Proceedings of the IEEE IPOM 2003, pp. 119–126 (2003)
2. Makanju, A., Brooks, S., Zincir-Heywood, A.N., Milios, E.E.: Logview: Visualizing event log clusters. In: Sixth Annual Conference on Privacy, Security and Trust, PST 2008, pp. 99–108 (2008)
3. Muller-Molina, A.J., Shinohara, T.: Efficient similarity search by reducing i/o with compressed sketches. In: Proceedings of the Second International Workshop on Similarity Search and Applications, SISAP 2009, pp. 30–38. IEEE Computer Society, Washington, DC (2009)
4. Hansen, S.E., Atkins, E.T., Todd, E.: Automated system monitoring and notification with swatch. In: Proceedings of the 7th Systems Administration Conference, NMonterey, CA, pp. 145–155 (1993)
5. Stearley, J., Corwell, S., Lord, K.: Bridging the gaps: Joining information sources with splunk. In: Proceedings of the Workshop on Managing Systems via Log Analysis and Machine Learning Techniques (2010)
6. Yamanishi, K., Maruyama, Y.: Dynamic syslog mining for network failure monitoring. In: Proceedings of the Eleventh ACM SIGKDD International Conference on Knowledge Discovery in Data Mining, KDD 2005, pp. 499–508. ACM, New York (2005)
7. Seipel, D., Neubeck, P., Köhler, S., Atzmueller, M.: Mining complex event patterns in computer networks. In: Appice, A., Ceci, M., Loglisci, C., Manco, G., Masciari, E., Ras, Z.W. (eds.) NFMCP 2012. LNCS, vol. 7765, pp. 33–48. Springer, Heidelberg (2013)
8. Nagappan, M., Vouk, M.A.: Abstracting log lines to log event types for mining software system logs. In: 2010 7th IEEE Working Conference on Mining Software Repositories (MSR), pp. 114–117 (2010)
9. Mannila, H., Toivonen, H., Inkeri Verkamo, A.: Discovery of frequent episodes in event sequences. Data Mining and Knowledge Discovery 1, 259–289 (1997)
10. Zheng, Q., Xu, K., Lv, W., Ma, S.: Intelligent search of correlated alarms from database containing noise data. In: 2002 IEEE/IFIP Network Operations and Management Symposium, NOMS 2002, pp. 405–419 (2002)
11. Wen, L., Wang, J., Aalst, W., Huang, B., Sun, J.: A novel approach for process mining based on event types. Journal of Intelligent Information Systems 32, 163–190 (2009)
12. Makanju, A.A., Zincir-Heywood, A.N., Milios, E.E.: Clustering event logs using iterative partitioning. In: Proceedings of the 15th ACM SIGKDD International Conference on Knowledge Discovery and Data Mining, KDD 2009, pp. 1255–1264. ACM, New York (2009)
13. Demetrescu, C., Finocchi, I.: Algorithms for data streams. Handbook of Applied Algorithms: Solving Scientific, Engineering, and Practical Problems, 241 (2007)
14. Andoni, A.: Nearest Neighbor Search: the Old, the New, and the Impossible. PhD thesis, Massachusetts Institute of Technology (2009)
15. Panigrahy, R.: Hashing, Searching, Sketching. PhD thesis, Stanford University (2006)
16. Paulev, L., Jgou, H., Amsaleg, L.: Locality sensitive hashing: A comparison of hash function types and querying mechanisms. Pattern Recognition Letters 31, 1348–1358 (2010)

17. Slaney, M., Lifshits, Y., He, J.: Optimal parameters for locality-sensitive hashing. Proceedings of the IEEE 100, 2604–2623 (2012)
18. Andoni, A., Indyk, P.: Near-optimal hashing algorithms for approximate nearest neighbor in high dimensions. In: 47th Annual IEEE Symposium on Foundations of Computer Science, FOCS 2006, pp. 459–468 (2006)
19. Broder, A., Mitzenmacher, M.: Network applications of bloom filters: A survey. Internet Mathematics 1, 485–509 (2004)
20. Song, H., Dharmapurikar, S., Turner, J., Lockwood, J.: Fast hash table lookup using extended bloom filter: an aid to network processing. SIGCOMM Comput. Commun. Rev. 35(4), 181–192 (2005)
21. Appleby., A.: Murmurhash 2.0 (2010),
    http://sites.google.com/site/murmurhash/
22. Fung, B.C., Wang, K., Ester, M.: Hierarchical document clustering using frequent itemsets. In: Proceedings of the Third Siam International Conference on Data Mining (2003)

Graph Ranking on Maximal Frequent Sequences for Single Extractive Text Summarization

Yulia Ledeneva[1], René Arnulfo García-Hernández[1], and Alexander Gelbukh[2]

[1] Universidad Autónoma del Estado de México
Unidad Académica Profesional Tianguistenco
Instituto Literario #100, Col. Centro, Toluca, 50000, Estado de México
yledeneva@yahoo.com, renearnulfo@hotmail.com
[2] Centro de Investigación en Computación, Instituto Politécnico Nacional, 07738, DF, Mexico
www.Gelbukh.com

Abstract. We suggest a new method for the task of extractive text summarization using graph-based ranking algorithms. The main idea of this paper is to rank Maximal Frequent Sequences (MFS) in order to identify the most important information in a text. MFS are considered as nodes of a graph in term selection step, and then are ranked in term weighting step using a graph-based algorithm. We show that the proposed method produces results superior to the-state-of-the-art methods; in addition, the best sentences were found with this method. We prove that MFS are better than other terms. Moreover, we show that the longer is MFS, the better are the results. If the stop-words are excluded, we lose the sense of MFS, and the results are worse. Other important aspect of this method is that it does not require deep linguistic knowledge, nor domain or language specific annotated corpora, which makes it highly portable to other domains, genres, and languages.

1 Introduction

A summary of a document is a short text that communicates briefly the most important information from this document. The text summarization tasks can be classified into single-document and multi-document summarization. In single-document summarization, the summary of only one document is to be built, while in multi-document summarization the summary of a whole collection of documents is built. In this work, we have experimented only with single-document summaries, as a future work we apply this idea to multi-document summarization.

The text summarization methods can be classified into abstractive and extractive methods. An abstractive summary is an arbitrary text that describes the contexts of the source document. Abstractive summarization process consists of "understanding" the original text and "re-telling" it in fewer words. Namely, an abstractive summarization method uses linguistic methods to examine and interpret the text and then to find new concepts and expressions to best describe it by generating a new shorter text that conveys the most important information from the original document. While this may seem the best way to construct a summary (and this is how human beings do it), in

A. Gelbukh (Ed.): CICLing 2014, Part II, LNCS 8404, pp. 466–480, 2014.

real-life setting immaturity of the corresponding linguistic technology for text analysis and generation currently renders such methods practically infeasible.

An extractive summary, in contrast, is a selection of text parts (phrases, sentences, paragraphs, etc.) from the original text, usually presented to the user in the same order—i.e., a copy of the source text with most text parts omitted. An extractive summarization method only decides, for each sentence, whether it should be included in the summary. The resulting summary reads rather awkward; however, simplicity of the underlying statistical techniques makes extractive summarization an attractive, robust, language-independent alternative to more "intelligent" abstractive methods. In this paper, we consider extractive summarization.

A typical extractive summarization method consists in several steps, at each of them different options can be chosen. We will assume that the text parts of selection are sentences. Therefore, final goal of the extractive summarization process is *sentence selection*. One of the ways to select the appropriate sentences is to assign some numerical measure of usefulness of a sentence for the summary and then select the best ones; the process of assigning these usefulness weights is called *sentence weighting*. One of the ways to estimate the usefulness of a sentence is to sum up usefulness weights of individual terms of which the sentence consists; the process of estimating the individual terms is called *term weighting*. For this, one should decide what the terms are: for example, they can be words; deciding what objects will count as terms is the task of *term selection*. Different extractive summarization methods can be characterized by how they perform these tasks [1].

There are a number of scenarios where automatic construction of such summaries is useful. For example, an information retrieval system could present an automatically built summary in its list of retrieval results, for the user to decide quickly which documents are interesting and worth opening for a closer look—this is what Google models to some degree with the snippets shown in its search results. Other examples include automatic construction of summaries of news articles or email messages to be sent to mobile devices as SMS; summarization of information for government officials, executives, researches, etc., and summarization of web pages to be shown on the screen of a mobile device, among many others.

The main proposal consists in detecting Maximal Frequent Sequences, and ranks them using a graph-based algorithm. The main contribution of this paper is the proposal of using MFS as nodes of a graph in term selection step, and the second contribution is using a graph-based algorithm in sentence weighting step.

The paper is organized as follows. Section 2 summarizes the state-of-the-art text summarization and graph-based methods. In Section 3, a graph-based algorithm is presented. Section 4 describes Maximal Frequent Sequences. In Section 5, a new method is presented. The experimental setting is described, and some conclusions are discussed in Section 6. Section 7 concludes the paper.

2   Graph-Based Algorithm

Recently, graph-based algorithms are applied successfully to different Natural Language Processing tasks. For example, a linear time graph-based soft clustering algorithm was introduced for Word Sense Induction [2]. Given a graph, vertex pairs are assigned to the same cluster if either vertex has maximal affinity to the other. Clusters of varying size, shape, and density are found automatically making the algorithm suited to tasks such, where the number of classes is unknown and where class distributions may be skewed.

Other example of such applications consists of quantifying the limits and success of extractive summarization systems across domains [3]. The topic identification stage of single-document automatic text summarization across four different domains: newswire, literary, scientific and legal documents. The summary space of each domain is explored using an exhaustive search strategy, and finds the probability density function (pdf) of the ROUGE score distributions for each domain. Then this pdf is used to calculate the percentile rank of extractive summarization systems. The results introduce a new way to judge the success of automatic summarization systems and bring quantified explanations to questions such as why it was so hard for the systems to date to have a statistically significant improvement over the lead baseline in the news domain.

In [4], a hybrid graph-based method was presented annotating relationship maps with cross-document Structure Theory [5], and using network metrics [6]. It helped for Portuguese multi-document summarization.

Graph is a data structure that permits to model the meaning and structure of a cohesive text of many text-processing applications in a natural way. Particularly relevant in this paper is the application of random walks to text processing, as done in TextRank system [7]. TextRank has been successfully applied to three natural language processing tasks [8]: document summarization [3; 7], word sense disambiguation [9], and keyword extraction, and text classification [10] with results competitive with those of state-of-the-art methods. The strength of the model lies in the global representation of the context and its ability to model how the co-occurrence between features might propagate across the context and affect other distant features. The description of TextRank is given below.

2.1   Text Representation Using Graphs

Graph Representation. A text represented with a graph, interconnects words or other parts of a text with meaningful relations.

Depending on the application, nodes in the graph can be parts of a text of various sizes and characteristics. For example, words, ngrams, collocations, entire sentences, complete documents, etc. Note that the vertices can belong to different categories in the same graph.

To draw an edge between two vertices of a graph is done in a way of connection, which represent, for example, lexical or semantic relations, measures of text cohesiveness, contextual overlap, membership of a word in a sentence, etc.

Algorithm. After determining the type and characteristics of the elements added to the graph, the main algorithm of the ranking algorithms consists of the following steps [11]:

- Identify text units that best define the task, and add them as vertices in the graph.
- Identify relations that connect such text units, and use these relations to draw edges between vertices in the graph. Edges can be directed or undirected, weighted or unweighted.
- Apply a graph-based ranking algorithm to find a ranking over the nodes in the graph. Iterate the graph-based ranking algorithm until convergence. Sort vertices based on their final score. Use the values attached to each vertex for ranking/selection decisions.

2.2   Graph-Ranking Algorithms

The basic idea implemented by a random-walk algorithm is that of "voting" or "recommendation." When one vertex links to another one, it votes for that other vertex. The higher the number of votes that are cast for a vertex, the higher the importance of the vertex. Moreover, the importance of the vertex casting a vote determines how important the vote itself is; this information is also taken into account by the ranking algorithm.

A random-walk algorithm called PageRank [Bri98] has been recently found successful in several text-processing applications such as text summarization and word sense disambiguation.

Given a directed graph $G = (V, E)$ with the set of vertices V and the set of edges E, where E is a subset of V x V. For a given vertex V_a, let In(V_a) be the set of vertices that point to it (predecessors), and let Out(V_a) be the set of vertices that vertex V_a points to (successors). The PageRank score associated with the vertex V_a is defined using a recursive function that integrates the scores of its predecessors:

We describe below two graph-based ranking algorithms:

$$S(V_a) = (1-d) + d \times \sum_{V_b \in In(V_a)} \frac{S(V_b)}{|Out(V_b)|}, \tag{1}$$

where d is a parameter set between 0 and 1.

The score of each vertex is recalculated upon each iteration based on the new weights that the neighboring vertices have accumulated. The algorithm terminates when the convergence point is reached for all the vertices, meaning that the error rate for each vertex falls below a pre-defined threshold.

This vertex-scoring scheme is based on a random-walk model, where a walker takes random steps on the graph, with the walk being modelled as a Markov process. Under certain conditions (namely, that the graph should be aperiodic and irreducible), the model is guaranteed to converge to a stationary distribution of probabilities associated with the vertices in the graph. Intuitively, the stationary probability associated with a vertex represents the probability of finding the walker at that vertex during the random walk, and thus it represents the importance of the vertex within the graph.

PageRank [Bri98] is perhaps one of the most popular ranking algorithms, which was designed as a method for Web link analysis. Unlike other ranking algorithms, PageRank integrates the impact of both incoming and outgoing links into one single model, and therefore it produces only one set of scores:

$$PR(V_i) = (1-d) + d \times \sum_{V_j \in In(V_i)} \frac{PR(V_j)}{|Out(V_j)|}. \tag{2}$$

In matrix notation, the PageRank vector of stationary probabilities is the principal eigenvector for the matrix A_{row}, which is obtained from the adjacency matrix A representing the graph, with all rows normalized to sum to 1: $P = A^T_{row}P$.

A ranking process starts by assigning arbitrary values to each node in the graph, followed by several iterations until convergence below a given threshold is achieved. Convergence is achieved when the error rate for any vertex in the graph falls below a given threshold, where the error rate of a vertex V_i is approximated with the difference between the scores computed at two successive iterations: $S^{k+1}(V_i) - S^k(V_i)$ (usually after 25-35 iteration steps). After running the algorithm, a score is associated with each vertex, which represents the "importance" (*rank*) of the vertex within the graph. Note that for such iterative algorithms, the final value obtained for each vertex is not affected by the choice of the initial value; only the number of iterations to convergence may be different.

Undirected Graphs: Although traditionally applied on directed graphs, algorithms for node activation or ranking can be also applied to undirected graphs. In such graphs, convergence is usually achieved after a larger number of iterations, and the final ranking can differ significantly compared to the ranking obtained on directed graphs.

Weighted Graphs: When the graphs are built from natural language texts, they may include multiple or partial links between the units (vertices) that are extracted from text. It may be therefore useful to indicate and incorporate into the model the "strength" of the connection between two vertices V_i and V_j as a weight w_{ij} added to the corresponding edge that connects the two vertices. Consequently, we introduce new formulae for graph-based ranking that take into account edge weights when computing the score associated with a vertex in the graph, e.g.

$$PR^W(V_i) = (1-d) + d \times \sum_{V_j \in In(V_i)} w_{ij} \frac{PR^W(V_j)}{\sum_{V_k \in Out(V_j)} w_{jk}} \tag{3}$$

3   Maximal Frequent Sequences

An ngram is a sequence of n words. We say that an ngram occurs in a text if these words appear in the text in the same order immediately one after another. For example, a 4-gram (ngram of length 4) *words appear in the text* occurs once in the previous sentence, while *appear immediately after another* does not (these words do not appear on adjusting positions), neither does *the text appear in* (order is different).

The definition of ngram depends on what one considers words. For example, one can consider capitalized (*Mr. Smith*) and non-capitalized (*a smith*) words as the same word or as different words; one can consider words with the same morphological stem (*ask, asked, asking*), the same root (*derive, derivation*), or the same meaning (*occur, appear*) as the same word; one can omit the stop-words (*the, in*) when counting word positions, etc. Say, one can consider that in our example sentence above there occur the ngrams *we say* (capitalization ignored), *word appear* (plural ignored), *appear text* (*in the* ignored). This can affect counting the ngrams: if one considers *occur* and *appear* as equivalent and ignores the stop-words, then in our example sentence the bigram *appear text* occurs twice.

We call an ngram frequent (more accurately, β-frequent) if it occurs more than β times in the text, where β is a predefined threshold. Frequent ngrams—we will also call them frequent sequences (FSs)—often bear important semantic meaning: they can be multiword expressions (named entities: *The United States of America*, idioms: *kick the basket*) or otherwise refer to some idea important for the text (*the President's speech, to protest against the war*).

Our hypothesis is that FSs can express ideas both important and specific for the document. This can be argued in terms of *tf-idf* (term frequency—inverse document frequency, a notion well known in information retrieval [12]). On the one hand, the idea expressed by an FS is important for the document if it repeatedly returns to it (high term frequency). On the other hand, the corresponding idea should be specific for this document, otherwise there would exist in the language a single word or at least an abbreviation to express it (high inverse document frequency). It is important to note that this argument does not apply to 1-grams, i.e., single words. Therefore, we do not consider 1-grams as ngrams in the rest of this paper.

An ngram can be a part of another, longer ngram. All ngrams contained in an FS are also FSs. However, with the arguments given above one can derive that such smaller ngrams may not bear any important meaning by their own: e.g., *The United States of America* is a compound named entity, while *The United* or *States of America* are not. Exceptions like *The United States* should not affect much our reasoning since they tend to be synonymous to the longer expression, and the author of the document would choose one or another way to refer to the entity, so they should not appear frequently both in the same document.

FSs that are not parts of any other FS are called Maximal Frequent Sequences (MFSs) [13, 14]. For example, in the following text

> ... *Mona Lisa is the most beautiful picture of Leonardo da Vinci* ...
> ... *Eiffel tower is the most beautiful tower* ...
> ... *St. Petersburg is the most beautiful city of Russia* ...
> ... *The most beautiful church is not located in Europe* ...

the only MFS with $\beta = 3$ is *is the most beautiful*, while the only MFS $\beta = 4$ is *the most beautiful* (it is not an MFS with $\beta = 3$ since it is not maximal with this β). As this example shows, the sets of MFSs with different thresholds do not have to, say, contain one another.

One of our hypotheses was that only MFSs should be considered as bearing important meaning, while non-maximal FSs (those that are parts of another FS) should not be considered. Our additional motivation was cost vs. benefit considerations: there are too many non-maximal FSs while their probability to bear important meaning is lower. In any case, MFSs represent all FSs in a compact way: all FSs can be obtained from all MFSs by bursting each MFS into a set of all its subsequences. García [13] proposed an efficient algorithm to find all MFSs in a text, which we also used to efficiently obtain and store all FSs of the document.

The notions of FSs and MFSs are closely related to that of repeating bigrams; see Section 5. This set is conceptually simpler, but for computational implementation, MFSs could be more compact.

4   Proposed Method

In this section, the proposed method is presented.

TextRank. We use a graph-based ranking algorithm *TextRank* to find a ranking over the nodes in the graph. Iterate the graph-based ranking algorithm until convergence. Sort vertices based on their final score. Use the values attached to each vertex for ranking/selection decisions.

- *Vertices.* We propose to use MFSs as vertices of a graph (see Section 3).
- *Nodes.* Relations that connect MFSs are term weighting relations such as (1) frequency of MFSs in a text: f, (2) length of MFS: l, and its presence as 1 or its absence as 0 (see Section 3).
- *Configuration of algorithm (TextRank)*: for this task, the goal is to rank MFSs, and therefore a *vertex* is added to the graph for each MFS in the text. To draw *nodes* between vertices, we are defining a term weighting relation, where "term weighting" can be defined in various ways. In the experiments realized in this paper, we use a term weighting described in below (in term weighting step). Such a relation between two sentences can be seen as a process of recommendation: a sentence that addresses certain concepts in a text, gives the reader a recommendation to refer to other sentences in the text that address the same or similar concepts. The resulting graph is highly connected, with a weight associated with each edge, and thus we use again the weighted version of the graph algorithms.

Term Selection. We experiment with MFSs and other term selection options derived from them. Namely, we considered the following variants of term selection:

- M: the set of all MFSs, i.e., an ngram $m \in M$ if it is an MFS with some threshold β (recall that MFSs are of 2 words or longer and $\beta \geq 2$).[1] In the example from Section 3, M = {*is the most beautiful, the most beautiful*}. Also, we denote by M_2 the set of all MFSs with $\beta = 2$.
- W: single words (unigrams) from elements of M. Namely, a word $w \in W$ iff there exists an MFS $m \in M$ such that $w \in m$.
  In our example, $W = \{is, the, most, beautiful\}$.

---

[1] In practice, we only considered the MFSs with the thresholds $\beta = 2$, 3, and 4, since MFSs with higher thresholds were very rare in our collection, except for those generated by stop-words.

The set W are naturally derived from the notion of MFS and at the same time can be efficiently calculated.

Optionally, stop-words were eliminated at the pre-processing stage; in this case our MFSs could span more words in the original text, as explained in Section 4.

Term Weighting. Different formulae were considered containing the following values:

- f: frequency of the term in MFSs, i.e., the number of times the term occurs in the text within some MFS. In our example, $f(is) = 3$ since it occurs 3 times in the text within the MFS *is the most beautiful*. If the term itself is an MFS, then this is just the frequency of this term in the text (e.g., for M, f is the same as term weight in Section 5; for W and N it is not). Under certain realistic conditions (MFSs do not intersect in the text, words do not repeat within one MFS) f is the number of times the term occurs in the text as part of a repeating bigram. In our example, $f(is) = 3$ since it occurs 3 times in a repeating bigram *is the* (and one time in a non-repeating context *church is not*).
- l: the maximum length of a MFS containing the term. In our example, $l(is) = 4$ since it is contained in a 4-word MFS *is the most beautiful*.
- 1: the same weight for all terms.

Sentence Weighting. The sum of the weights of the terms contained in the sentence was used. For sentence selection, the following options were considered:

- best: sentences with greater weight were selected until the desired size of the summary (100 words) is reached. This is the most standard method.
- $k$best+first: k best sentences were selected, and then the first sentences of the text were selected until the desired size of the summary was reached. This was motivated by a hard-to-beat baseline mentioned in Section 2: only very best sentences according to our weighting scheme could prove to be above this baseline.

5   Experimental Setting and Results

Main Algorithm. We have conducted several experiments to verify our hypotheses formulated in the previous section. In each experiment, we followed the standard sequence of steps:

- *TextRank algorithm*: we use undirected version of PageRank;
- *Term selection*: decide which features are to be used to describe the sentences;
- *Term weighting*: decide how the importance of each feature is to be calculated;
- *Sentence weighting*: decide how the importance of the features is to be combined into the importance measure of the sentence;
- *Sentence selection*: decide which sentences are selected for the summary.

Test Data Set. We used the DUC collection provided [15]. In particular, we used the data set of 567 news articles of different length and with different topics. Each document in the DUC collection is supplied with a set of human-generated summaries provided by two different experts. While each expert was asked to generate summaries of different length, we used only the 100-word variants.

Evaluation Procedure. We used the ROUGE evaluation toolkit [16], which was found to correlate highly with human judgments [17]. It compares the summaries generated by the program with the human-generated (gold standard) summaries. For comparison, it uses n-gram statistics. Our evaluation was done using n-gram $(1, 1)$ setting of ROUGE, which was found to have the highest correlation with human judgments, namely, at a confidence level of 95%.

Experiment 1. We conducted this experiment in two phases: first, we tried sentence weighting using option *best* and then option *k*best+first. In each experiment, we test two term selection options M and W. We excluded stop-words. The results are shown in Table 1.

Table 1. Results for different term selection, term weighting and sentence selection options using the proposed method (options: M, W, *excluded, best,* and *kbest+first*)

Term Selection		Term weighting	Sentence Weighting	Results		
Terms	Stop-words			Recall	Precision	F-measure
W	excluded	f	1best+first	47.603	47.518	47.543
			2best+first	47.718	47.621	47.652
M	excluded	l	1best+first	47.783	47.699	47.724
			2best+first	48.212	48.088	48.132
		f	1best+first	47.797	47.712	47.737
			2best+first	**48.211**	**48.093**	**48.134**
M	excluded	1	best	46.668	48.337	47.474
		f		48.009	47.757	47.865
		f^2		48.056	47.801	47.910
		l		48.025	47.773	47.881
		l^2		48.058	47.812	47.917
		$f \times l$		48.060	47.810	47.916
		f^l		48.079	47.831	47.937
W	excluded	1	best	47.682	47.604	47.626
		f		**48.659**	**48.324**	**48.473**
		f^2		48.705	48.235	48.451

We started our experiment from modifying term weighting parameters for the term selection scheme W with the term weighting option f, which showed good performance in the first experiment; and then we tried the term selection options M with the term weighting option 1 and the option f. We use *kbest+first* option for the sentence-weighting, see the first part of Table 1. The last line of the first part of the table represents the best result from the first part of Table 1. The best results are highlighted in boldface.

In the second part of Table 2, we change the sentence selection option using *best* option. In addition, we tried more term weighting option related to *f*. First, we tried the term selection scheme *M*, because of the better results obtained from the first part of the table, and then we tried the term selection *W*.

Term selection *W* gave a better result than *M*. Finally, with the best combinations obtained from the first two experiments, are highlighted in boldface and underlined.

Experiment 2. In the second part of experiments (see Table 2), we included stop-words and tried sentence weighting using *best* and *k*best+first options.

Table 2. Results for different term selection, term weighting and sentence selection options using the proposed method (options: *M*, *W*, *included*, *best*, and *kbest+first*)

Term Selection	Term weighting	Sentence Weighting	Results		
			Recall	Precision	F-measure
W	*f*	1*best+first*	47.694	47.612	47.635
		2*best+first*	47.870	47.761	47.798
M	*l*	1*best+first*	47.711	47.623	47.650
		2*best+first*	48.064	47.923	47.976
	f	1*best+first*	47.738	47.649	47.676
		2*best+first*	**48.148**	**48.016**	**48.065**
M	1	*best*	47.484	49.180	48.283
	f		48.803	48.533	48.626
	f²		48.746	48.482	48.572
	l		<u>**48.823**</u>	<u>**48.577**</u>	<u>**48.658**</u>
	l²		48.741	48.518	48.587
	f × l		48.796	48.529	48.620
	fˡ		48.716	48.497	48.564
W	1	*best*	47.529	47.483	47.489
	f		**48.821**	**48.424**	**48.604**
	f²		48.784	48.322	47.489

Taking into account the best combination obtained from the first experiment, we tried different sentence selection variants including stop-words; see Table 2. From Table 1, we knew that term selection scheme *M* with stop-words removed, gave the best results with other parameters fixed (term weighting, sentence weighting, and sentence selection); see the first part of Table 1. Therefore, we started from modifying these parameters for the same term selection scheme; see the first part of Table 2. The last line of the first part of the table represents the best result from the upper first part of Table 1. The best results are highlighted in boldface.

Then we tried the term selection option *W* with the term weighting option 1 and the options related to *f*, which showed good performance in the first experiment. The results are shown in the first below part of Table 2. Term selection *M* gave a better result than *W*. The option for term weighting *l* represents the best result from Table 2. The best results are highlighted in boldface and underlined.

Conclusion 1. In this conclusion, we discuss the best term selection option. We show in Table 3 the comparison of the best results for the term selection options of M and W from Table 2. The best result was obtained with MFSs (option M): it means that the proposed method ranks better on MFSs (option M) than words derived from MFSs (option W).

In [1] was shown that W are better than M for single text extractive summarization using options $W, f, best$ and $W, f, 1best+first$. Thus, we conclude that the proposed method benefits MFSs (option M).

Table 3. Comparison for different term selection, term weighting and sentence selection options using the proposed method (options: M, W, *included, best,* and *kbest+first*). Comparison of the best results using the term selection option of M and W from the Table 2.

Term Selection	Term Weighting	Sentence Weighting	Results		
			Recall	Precision	F-measure
M	f	$2best+first$	**48.148**	**48.016**	**48.065**
W	f	$2best+first$	47.870	47.761	47.798
M	l	$best$	**48.823**	**48.577**	**48.658**
W	f	$best$	48.821	48.424	48.604

Conclusion 2. In this conclusion, we discuss the best term weighting options from Table 2. We take the best results from Table 2, and compose Table 4. The best result was obtained with the term weighting option l: length of the corresponding MFSs (option M); it means that the longer MFSs the better for single text summarization. In addition, it means that the proposed method benefits the length of MFS.

Table 4. Comparison for different term selection, term weighting and sentence selection options using the proposed method (options: M, W, *included, best,* and *kbest+first*). Comparison of the term weighting options l and f.

Term Selection	Term Weighting	Sentence Weighting	Results		
			Recall	Precision	F-measure
M	L	$best$	**48.823**	**48.577**	**48.658**
W	F	$best$	48.821	48.424	48.604
M	F	$2best+first$	48.148	48.016	48.065
W	F	$2best+first$	47.870	47.761	47.798

Conclusion 3. In this conclusion, we discuss the state-of-the-art methods that use different pre-processing options, see Table 5. The results are getting better, if for options with the term selection option M, stop-words are included. Contrary to the option W, the results are getting better, if stop-words are excluded. For the reason that the stop-words do not try any sense, when we eliminate stop-words, the results are getting better because the words with meaning are kept.

The option M considers MFSs (multiword expressions [20]), so when we exclude stop-words from multiword expressions, the sense is lost and the results are worsened.

Table 5. Comparison of the state-of-the-art methods that use different pre-processing options (sentence selection options: *best* and *kfirst+best*)

Method	Term Selection		Term weight-ing	Results		
	Terms	Stop-words		Recall	Precision	F-measure
Related work [1]	W	Included	f	46.523	48.219	47.344
Related work [1]	W	Excluded	f	46.576	48.278	47.399
Related work [1]	W	Excluded	f	46.536	48.230	47.355
Related work [1]	W	Excluded	f	46.622	48.407	47.486
Related work [1]	W	Excluded	f	46.788	48.537	47.634
Pre-processing [21]	M	Excluded	l	46.266	47.979	47.094
Pre-processing [21]	M	excluded stemming	l	46.456	48.169	47.285
Pre-processing [21]	M	included stemming	l	46.508	48.233	47.343
Proposed	W	Excluded	f	**48.659**	**48.324**	**48.473**
Proposed	M	Included	l	**48.823**	**48.577**	**48.658**

Conclusion 4. One can observe from [1] that any *k*best+first sentence selection option not outperformed any combination that used the standard sentence selection scheme, with bigger *k* always giving better results—that is, only the slightest correction to the baseline deteriorate it. See the comparison of MFS1 and MFS2 [1]. For the proposed method, the result with the option *kbest+first* is better than with the option *best*.

It is very important result because no one state-of-the-art-method could beat baseline configuration described in Section 5: only the very best sentences according to our weighting scheme might prove to be above this baseline. Therefore, the proposed method could beat this baseline configuration. The proposed method finds better sentences than baseline (baseline for the configuration of news article was not possible to improve until now because the structure of news article where the first sentences are the most important). Observe in Table 6 that the result with options **Proposed:** *M, l, best* is better than **Proposed:** *M, l, 1best+first* and **Proposed:** *M, l, 2best+first*, contrary to *W, f, best* not outperformed *W, f, 1best+first* and *W, f, 2best+first*.

See in Table 1 and 2 details for more comparison. For example, *W, f, best* also are better than *W, f, kbest+first*.

Table 6. Comparison of methods to show the difference between *best* and *kbest+first* options

Comparison	Method	Recall	Precision	F-measure
Related work [1]	*W, f, best*	44.609	45.953	45.259
	W, f, 1best+first	46.576	48.278	47.399
	W, f, 2best+first	46.158	47.682	46.895
Proposed	*M, l, best*	48.823	48.577	48.658
	M, f, 1best+first	47.711	47.623	47.650
	M, f, 2best+first	48.064	47.923	47.976

Conclusion 5. The comparison to the state-of-the-art methods is given in Table 7. We group methods depending on additional information were used for that methods: none, order of sentences, pre-processing, clustering. Even though the proposed method does not use any additional information, outperforms the other methods.

The best overall result is for the proposed method that does not use the additional information for generating text extractive summaries. It means that this method is completely independent (domain independent and text position independent). In addition, this method is language independent because does not use any lexical information, and can be used for different languages.

Table 7. Comparison of the results with other state-of-the-art methods

Additional info used	Method	Recall	Precision	F- measure
None	Baseline: *random*	37.892	39.816	38.817
	TextRank: [7]	45.220	43.487	44.320
	MFS: [19]	44.609	45.953	45.259
	Proposed: *M, l, best*	**48.823**	**48.577**	**48.658**
Order of sentences	Baseline: *first*	46.407	48.240	47.294
	MFS: [19]	46.576	48.278	47.399
	Proposed: *M, f, 2best+first* (without pre-processing)	**48.148**	**48.016**	**48.065**
Pre-processing	TextRank: [A]	46.582	48.382	47.450
	TextRank: [3]	47.207	48.990	48.068
	Proposed: *W, f, best*	**48.659**	**48.324**	**48.473**
Clustering	MFS (*k*-best) [19]	47.820	47.340	47.570
	MFS (EM-5) [18]	47.545	48.075	47.742

Table 8 shows the results according to their relevance. Considering that Topline is the best obtained result and Baseline: *random* as the worse obtained result.

Table 8. Comparison of the results with other state-of-the-art methods (F-measure and significance)

Method	F- measure	Significance
Baseline: *random*	38.817	0%
TextRank: [7]	44.320	26.47%
MFS: [19]	45.259	31.28%
Baseline: *first*	47.294	40.78%
MFS: [19]	47.399	41.29%
TextRank: [A]	47.450	41.53%
MFS (*k*-best) [19]	47.570	42.11%
MFS (EM-5) [18]	47.742	42.94%
Proposed: $M, f, 2best+first$ (without pre-processing)	**48.065**	**44.47%**
TextRank: [3]	48.068	46.20%
Proposed: $M, l, best$	**48.658**	**47.35%**
Topline [20]	59.600	100%

6   Future Work

As a future work, this method can be applied on different Natural Language Processing Tasks such as Word Sense Disambiguation, Text Classification, Collocation Extraction, and others. We believe that in the future the proposed method can improve various state-of-the-art methods and contribute with even better results.

In particular, we plan to extend the notion of MFS to that of syntactic n-gram [22, 23], and extend our method to multi-document summarization, especially in the context of social networks [24, 25].

References

1. Ledeneva, Y.N., Gelbukh, A., García-Hernández, R.A.: Terms Derived from Frequent Sequences for Extractive Text Summarization. In: Gelbukh, A. (ed.) CICLing 2008. LNCS, vol. 4919, pp. 593–604. Springer, Heidelberg (2008)
2. Hope, D., Keller, B.: MaxMax: A Graph-Based Soft Clustering Algorithm Applied to Word Sense Induction. In: Gelbukh, A. (ed.) CICLing 2013, Part I. LNCS, vol. 7816, pp. 368–381. Springer, Heidelberg (2013)
3. Ceylan, H., Mihalcea, R., Ozertem, U., Lloret, E., Palomar, M.: Quantifying the Limits and Success of Extractive Summarization Systems Across Domains. In: Proc. of the North American Chapter of the ACL (NACLO 2010), Los Angeles (2010)
4. Ribaldo, R., Akabane, A.T., Rino, L.H.M., Pardo, T.A.S.: Graph-based Methods for Multi-document Summarization: Exploring Relationship Maps, Complex Networks and Discourse Information. In: Caseli, H., Villavicencio, A., Teixeira, A., Perdigão, F. (eds.) PROPOR 2012. LNCS, vol. 7243, pp. 260–271. Springer, Heidelberg (2012)
5. Maziero, E.G. and Pardo, T.A.S. Automatic Identification of Multi-document Relations. In the (on-line) Proceedings of the PROPOR 2012 PhD and MSc/MA Dissertation Contest, Coimbra, Portugal, April 17-20, pp. 1–8 (2012)

6. Antiqueira, L., Oliveira Jr., O.N., Costa, L.F., Nunes, M.G.V.: A Complex Network Approach to Text Summarization. Information Sciences 179(5), 584–599 (2009)
7. Mihalcea, R.: Random Walks on Text Structures. In: Gelbukh, A. (ed.) CICLing 2006. LNCS, vol. 3878, pp. 249–262. Springer, Heidelberg (2006)
8. Mihalcea, R., Radev, D.: Graph-based Natural Language Processing and Information Retrieval. Cambridge University Press (2011)
9. Sinha, R., Mihalcea, R.: Unsupervised Graph-based Word Sense Disambiguation. In: Nicolov, N., Mitkov, R. (eds.) Current Issues in Linguistic Theory: Recent Advances in Natural Language Processing. John Benjamins Publishers (2009)
10. Hassan, S., Mihalcea, R., Banea, C.: Random-Walk Term Weighting for Improved Text Classification. In: IEEE International Conference on Semantic Computing (ICSC 2007), Irvine, CA (2007)
11. Mihalcea, R., Tarau, P.: TextRank: Bringing Order into Texts. In: Proceedings of the Conference on Empirical Methods in Natural Language Processing (EMNLP 2004), Barcelona, Spain (2004)
12. Salton, G., Buckley, C.: Term-weighting approaches in automatic text retrieval. Information Processing & Management 24, 513–523 (1988)
13. García-Hernández, R.A., Martínez-Trinidad, J.F., Carrasco-Ochoa, J.A.: A Fast Algorithm to Find All the Maximal Frequent Sequences in a Text. In: Sanfeliu, A., Martínez Trinidad, J.F., Carrasco Ochoa, J.A. (eds.) CIARP 2004. LNCS, vol. 3287, pp. 478–486. Springer, Heidelberg (2004)
14. García-Hernández, R.A., Martínez-Trinidad, J.F., Carrasco-Ochoa, J.A.: A New Algorithm for Fast Discovery of Maximal Sequential Patterns in a Document Collection. In: Gelbukh, A. (ed.) CICLing 2006. LNCS, vol. 3878, pp. 514–523. Springer, Heidelberg (2006)
15. DUC. Document understanding conference (2002), http://www-nlpir.nist.gov/projects/duc
16. Lin, C.Y.: ROUGE: A Package for Automatic Evaluation of Summaries. In: Proceedings of Workshop on Text Summarization of ACL, Spain, (2004)
17. Lin, C.Y., Hovy, E.: Automatic Evaluation of Summaries Using N-gram Co-Occurrence Statistics. In: Proceedings of HLT-NAACL, Canada, (2003)
18. Ledeneva, Y., Hernández, R.G., Soto, R.M., Reyes, R.C., Gelbukh, A.: EM Clustering Algorithm for Automatic Text Summarization. In: Batyrshin, I., Sidorov, G. (eds.) MICAI 2011, Part I. LNCS, vol. 7094, pp. 305–315. Springer, Heidelberg (2011)
19. Soto, R.M., Hernández, R.G., Ledeneva, Y., Reyes, R.C.: Comparación de Tres Modelos de Representación de Texto en la Generación Automática de Resúmenes. Procesamiento del Lenguaje Natural 43, 303–311 (2009)
20. Ledeneva, Y.: PhD. Thesis: Automatic Language-Independent Detection of Multiword Descriptions for Text Summarization, Mexico: National Polytechnic Institute (2008)
21. Ledeneva, Y.N.: Effect of preprocessing on extractive summarization with maximal frequent sequences. In: Gelbukh, A., Morales, E.F. (eds.) MICAI 2008. LNCS (LNAI), vol. 5317, pp. 123–132. Springer, Heidelberg (2008)
22. Sidorov, G.: Syntactic Dependency Based N-grams in Rule Based Automatic English as Second Language Grammar Correction. International Journal of Computational Linguistics and Applications 4(2), 169–188 (2013)
23. Sidorov, G.: Non-continuous Syntactic N-grams. Polibits 48, 67–75 (2013)
24. Bora, N.N.: Summarizing Public Opinions in Tweets. International Journal of Computational Linguistics and Applications 3(1), 41–55 (2012)
25. Balahur, A., Kabadjov, M., Steinberger, J.: Exploiting Higher-level Semantic Information for the Opinion-oriented Summarization of Blogs. International Journal of Computational Linguistics and Applications 1(1-2), 45–59 (2010)

A Graph Based Automatic Plagiarism Detection Technique to Handle Artificial Word Reordering and Paraphrasing

Niraj Kumar

TCS Innovation Lab, Tata Consultancy Services,
New Delhi, India
niraj.kumar11@tcs.com

Abstract. Most of the plagiarism detection techniques are based on either string based matching or semantic matching of adjacent strings. However, due to the use of artificial word re-ordering and paraphrasing, the detection of plagiarism has become a challenging task of significant interest. To solve this issue, we concentrate on identification of overlapping adjacent plagiarized word patterns and overlapping non-adjacent/reordered plagiarized word patterns from target document(s). Here the main aim is to capture the simple cases and the complex cases (i.e., artificial word reordering and/or paraphrasing) of plagiarism in the target document. For this first of all we identify the relation between all overlapping word pairs with the help of controlled closeness centrality and semantic similarity. Next, to extract the plagiarized word patterns, we introduce the use of minimum weighted bipartite clique covers. We use the plagiarized word patterns in the identification of plagiarized texts from the target document. Our experimental results on publicly available and annotated dataset like: 'PAN 2012 plagiarism detection dataset' and 'Student answer related plagiarism dataset' shows that it performs better than state-of-arts systems in this area.

Keywords: Plagiarism detection, normalized pointwise mutual information, controlled closeness centrality, minimum weighted bipartite graph clique cover, semantic graph.

1   Introduction

Plagiarism is the re-use of someone else's prior ideas, processes, results, or words without explicitly acknowledging the original author and source [2]. Due to the new arrival of the huge volume of texts in digital form with different levels of impacts on business, research and other fields, plagiarism detection has become a highly demanding research area. [10], showed an astonishing growth of commercial plagiarism detection systems available online.

Although paraphrasing is the linguistic mechanism underlying many plagiarism cases, little attention has been paid to its analysis in the framework of automatic plagiarism detection. Therefore state-of-the-art plagiarism detectors find it difficult to detect cases of paraphrase plagiarism [1]. According to [3], paraphrases are sentences or phrases that convey the same meaning using different wording.

A. Gelbukh (Ed.): CICLing 2014, Part II, LNCS 8404, pp. 481–494, 2014.

In this paper, we basically focus on the development of a common plagiarism detection technique for both, i.e., (1) traditional copy paste and (2) paraphrasing and/or artificial word re-ordering in text.

Our Contributions: Our contributions toward the development of the entire system can be summarized as:

➢ For plagiarism detection task, we introduce the use of bigram based model for semantic word graph construction. Next, we populate it with semantic and statistical information of bigrams.

➢ We introduce the combined use of controlled closeness centrality calculated on a semantic word graph of text and strict semantic relatedness score to identify the statistical and semantic closeness of overlapping word pairs in the source and target text.

➢ We introduce the use of minimum weighted bipartite graph clique covering based pattern identification technique to detect the plagiarism in the case of paraphrasing and artificial text filling.

Paper Organization: In section 2, we present the major issues and motivation in this area. In section 3, we present a brief literature survey of this area. In section 4, we calculate normalized pointwise mutual information score of word pairs. In section 5, we present the way to construct the word graph of text. In section 6, we calculate controlled closeness centrality score of co-occurring words. In section 7, we extract the plagiarized word patterns by using "minimum weighted bipartite clique cover". In section 8, we identify the plagiarized passages from the target document. Section 9, represents the pseudocode for the entire system. In section 10, we present the experimental evaluation of the system. Finally we concluded the work in section 11.

2   Problem Statements and Motivation

From the above discussion, it is clear that, paraphrasing and artificial word reordering is a major problem in this area. Addition to this problem, we faced some research issues towards the development of an efficient plagiarism detector by using the graph based technique. The following contains problem statements and motivations behind the techniques proposed to solve them.

I. Generating the word graph for such techniques is a challenging task. As, several times, adjacent words may not show significant semantic cohesion, so, adding such word pairs does not seem useful in the case of the bigram based word graph of text. To solve this issue, we introduce the use of semantic relatedness in preparation of word graph. Finally, we populate the word graph by using the product of statistical and semantic relatedness score of words (similar to that used in [11]).

II. The next important issue is, how to identify the relation between words after word reordering and/or paraphrasing cases. As, after such changes, words may not occur adjacent to each other. To solve this issue we require a system which can stalibsh the relation between words. We use closeness centrality score of words on the word graph of text to identify the initial closeness of all overlapping word pairs in source and target text. The closeness centrality score of a node shows its closeness w.r.t., all other nodes in the graph. Thus, it captures the role of words in the source and target texts. Based on the change in closeness centrality score, it is possible to identify the changes in the role of words in text. However, such statistical measures may misguide us. To solve this issue, we calculate the normalized pointwise mutual information score [4] of such word pairs, w.r.t., DBpedia extended abstracts. The normalized pointwise mutual information score is introduced and used by [4] in co-location extraction. This two level filtration gives us information that, whether given overlapping word pairs in the target text (may be adjacent or non-adjacent) show similar relation as given in source text.

III. Now the final task is how to identify the plagiarized patterns in the given target text. Here the words of plagiarized patterns may not be adjacent/very near to each other as given in the source text. To solve this issue, we introduce the use of minimum weighted bipartite graph clique cover. Here, we use inverse of semantic and statistical closeness of vertexes to calculate the weight of an edge. Thus, minimum weighted bipartite clique covers identify all highly similar/plagiarized patterns. We use these patterns in plagiarism detection. Based on the extracted patterns and their importance we extract some features and apply SVM based classification to identify the plagiarized passages.

3   Literature Survey

Based on the way to handle the problem of text based plagiarism detection, we can divide the research efforts into the following categories.

Overlapping word n-grams: Texts in the dataset are extracted as sequences of n-grams, and the similarity between texts is calculated by applying a similarity metric. For example, [12]; [13]; [14]. Next, [8], tries to recognize the clusters of N-grams matches as matching passages in the pair of documents.

Frequency-based method: This method is based on the hypothesis that similar documents should contain words with a similar number of occurrences [15]. Tf-idf and its variants are used to calculate the weight of word to identify the similarity.

Fingerprinting: The fingerprint presents the document and comparison is based on the fingerprint instead of the actual document, thereby it reduces the need to perform exhaustive comparison [16].

Structural method: It is based on structural methods, which identifies patterns between the query and collection based on indexing and retrieval metrics [6].

NLP based methods: Such methods use several linguistic features. For example, [17], [18] uses the thesaurus to generalize synonyms. [19] applied synonym, hypernym and hyponym substitutions using WordNet and incorporated these into ROUGE [20], a metric which measures the similarity by n-gram frequency, skip-bigram and longest common subsequence. [10], also uses multiple NLP features. Additionally, [23] describes semantically oriented method based on semantic networks which are derived by a syntactico-semantic parser for plagiarism detection.

4   Normalized PMI Score

We use the scheme given in [4], to calculate the normalized PMI score of bigrams. The value of normalized pointwise mutual information lies in the range of [-1, 1]. The greater than zero value shows the positive co-location strength or strong semantic relatedness between two words. The calculation scheme is given below:

$$nPMI(T_i, A_j) = \begin{cases} -1 & if \ p(T_i, A_j) = 0 \\ \dfrac{\log p(T_i) + \log p(A_j)}{\log p(T_i, A_j)} - 1 & otherwise \end{cases} \tag{1}$$

Where, $p(T_i, A_j)$ = is the joint probability, and can be calculated by counting the number of observations of words T_i and A_j in a window of size 100 words in DBpedia articles. We use this small window size to get the strict semantic similarity. $p(T_i)$ = probability of occurrence of T_i in DBpedia extended abstracts and so on.

Using the Normalized PMI Score: We use normalized PMI score of all bigrams in the calculation of link weight. Next, we again normalize the scores lying in the range of [0, 1] to a new range [0, 10]. The main aim behind this normalization is to overcome the difference between co-occurrence frequency and semantic score of bigrams.

5   Word Graph Construction

We prepare separate word graphs for the given source and target text passages. We use these word graphs in identifying the relation between each overlapping word pair in the target text passage, w.r.t., source text passage.

5.1   Input Cleaning and Pre-processing

Our input cleaning step includes: (a) removal of noisy symbols and stopwords and (b) stemming of text (by using the Porter stemming algorithm[1]). In the preprocessing step, we filter the passages.

5.2   Preparing Word Graph of Text

We treat each distinct word of given text passage as node of graph and prepare an undirected word graph of text. Let $G = (V, E)$ represents an undirected graph, where, $V = \{V_1, V_2, V_3, ..., V_n\}$ represents the vertexes or nodes of the graph. The edge of the graph 'G' can be defined as $E = (V_i, V_j)$, if there exist link between nodes V_i and V_j,

---

[1] http://tartarus.org/martin/PorterStemmer/

where $\{i \neq j; 1 \leq i \leq n; 1 \leq j \leq n\}$. We move a sliding window of size two through the entire text, to add links between any two nodes.

We prepare a word graph of text for both, i.e. source and target texts. Now the next task is to calculate the link weight and the path length for each link of the graph.

Calculating Link Weight: To calculate the link weight of a link between two connected nodes, we multiply the co-occurrence frequency of adjacent word pairs and their semantic strength (obtained through the normalized PMI score).

$$Link_{wt}(V_i, V_j) = Freq(V_i, V_j) \times nPMI(V_i, V_j) \tag{2}$$

Where, $Link_{wt}(V_i, V_j)$ = Link weight of link between V_i and V_j. $Freq(V_i, V_j)$ = Co-occurrence frequency of adjacent words (nodes) V_i and V_j in the given text text passage. $nPMI(V_i, V_j)$ = Normalized Pointwise mutual information score of the word (node) pair V_i and V_j. If $nPMI(V_i, V_j)$ shows score zero, then we take $nPMI(V_i, V_j)$ = Minimum of positive normalized pointwise mutual information score for any adjacent word pairs.

Calculating *Path Length*: Now to calculate the path length between any two adjacent connected vertexes, we take the inverse of link weight between them. The main idea is: "word pair having higher link weight will show more similarity between them and it will reduce the path length"[11].

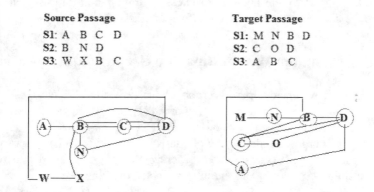

Fig. 1. Preparing word graph for source and target texts, circle represents the overlapping words or co-occurring words in source and target passages

Our Concept of Semantic Graph: Here we use the term, semantic graph in the sense that, (1) we discard the link between adjacent word pairs, if they do not show good semantic relatedness, i.e., normalized semantic relatedness score lies in the range [-1, 0) and (2) for the rest of the links, link weight is calculated by using the product of frequency of adjacent word pairs and their normalized semantic relatedness score (see eq-2). We use the semantic graph to calculate the controlled closeness centrality score of nodes in the graph.

6   Calculating Controlled Closeness Centrality Score

We use the controlled closeness centrality score of words to identify the statistical role of overlapping words. To overcome the writing gap/modification related issues at word level, we use synonym and hyponyme lists given in WordNet (similar to that used in [21]). Before going into details, we first discuss about closeness centrality.

Calculating Closeness Centrality Score: The closeness centrality of any node V_i is defined as the mean geodesic distance (i.e., the shortest path) between a node V_i and all of the nodes reachable from V_i as follows, where n≥2 is the size of the connected component reachable from V_i . [22]

$$C_C(V_i) = \frac{(n-1)}{\sum_{t \in V/V_i} d_G(V_i, t)}$$

(3)

Where, $C_C(V_i)$= Closeness centrality of node / vertex V_i , $d_G(V_i, t)$= Sum of geodesic distance from V_i to 't', we use the *path length* obtained from above step in the calculation of all geodesic distances.

Normalizing the closeness centrality score: The difference in length of text may create huge differences in closeness centrality scores. So, we normalize the closeness centrality score in a range of [0, 1]. For this we divide the closeness centrality score of all words of each text by the highest closeness centrality score in that text. We use normalized the closeness centrality score of words in all calculations.

7   Using Minimum Weight Bipartite Graph Clique Covering

We use minimum weighted bipartite graph clique covering to identify all sets of words which shows a certain level of plagiarism. Here sets of words may contain adjacent plagiarized word patterns and/or non-adjacent plagiarized word patterns. Thus it serves the purpose of identifying plagiarisms in copy-paste based cases and in the cases of paraphrasing/artificial word reordering. Before going into details, we first go through a simple description of minimum weighted bipartite graph clique cover.

Minimum Weighted Bipartite Graph Clique Cover: Let, $B = (L, R, E)$ be a bipartite graph, where, 'L' represents the left set of nodes and 'R' represents the right set of nodes and 'E' represents the edges of bipartite graphs. Now for, $\forall u \in (L \cup R)$, let C_u be the cost associated to u . The problem is to find a set of disjoint bi-cliques covering B of minimum weight. i.e., $\{K_i\}$ s.t. $\sum_i \sum_{u \in K_i} C_u$ is minimal and K_i is a biclique of B .

The Mean of Covering: All the vertices and edges should belong to one K_i and conversely all edges and vertex in each K_i should be present in the original bipartite graph B. In other words, K_i can be considered as an encoding of B in terms of non-overlapping bi-cliques.

Bi-Clique: A bi-clique is a complete bipartite sub-graph. That is, given a bipartite graph $G = (X, Y, E)$ and a subset $A \subseteq (X \cup Y)$, the sub-graph of G formed by 'A' is a bi-clique, if for all $x \in (A \cap X)$ and $y \in (A \cap Y)$, $\{x, y\} \in E$.

Creating bipartite graph: We add bipartite edges between nodes of two different sets, by using two different strategies. First of all, we add an edge between two nodes related to two different sets, if they represent the same word/node. (e.g., adding edge between same words on both sides, see Figure 2). Next, we add an edge between non matching but overlapping words, which shows a certain level of semantic and statistical similarity. See the additional links in Figure 3 (w.r.t., Figure 2). The procedure to add a bipartite edge between such nodes is given below.

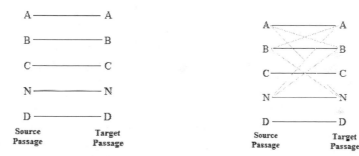

Fig. 2. Adding link between matching words in both text passages

Fig. 3. Adding non-matching words (in addition to matching words), which shows a certain level of semantic and statistical similarity

Adding an Edge between Semantically and Statistically Similar Terms: For this we take each pair of non-matching but overlapping words from source and target passages and check the percentage difference in weighted closeness centrality score. If the percentage difference in closeness centrality is less than 50% then we check the semantic closeness of the same pair of words. To check the semantic closeness score of word pairs, we use normalized pointwise mutual information score calculated by considering the co-occurrences of both words in a window of size 100 words. If both scores satisfy following thresholds, then we consider both words i.e., W_i and W_j of the given pair as statistically and semantically similar words.

$$\left. \begin{array}{c} if \left(\%Diff_{CC} \left(W_i, W_j \right) < 50 \right) \\ and \left(nPMI \left(W_i, W_j \right) > 0 \right) \end{array} \right\} \Rightarrow \left\{ W_i \text{ and } W_j \text{ are similar, given} \left(W_i \neq W_j \right) \right\} \quad (4)$$

Where, $\%Diff_{CC}(W_i, W_j)$ = Modulo of the percentage difference between the controlled closeness centrality score of words W_i and W_j. Here W_i represents i^{th} words from source passage and W_j represents the jth word from target passage.

Calculating the Weight of the Edges in the Bipartite Graph: To calculate the weight of the edges of bipartite graph, we use a semantic and statistical closeness of words/vertexs. To calculate the link weight we use following relation.

$$Link_{wt}(V_i, V_j) = avg(CC(V_i), CC(V_j)) \times nPMI(V_i, V_j) \tag{5}$$

Where, $Link_{wt}(V_i, V_j)$ = Link weight of bipartite edge between vertexes V_i and V_j.
$avg(CC(V_i), CC(V_j))$ =Average of closeness centrality score of vertexes V_i and V_j. We normalize this score in the range [0-10] (See eq-3, to calculate closeness centrality score of nodes). $nPMI(V_i, V_j)$ = Normalized pointwise mutual information score of vertexes V_i and V_j. We normalize the score of $nPMI(V_i, V_j)$ from [0,1] to [0, 10] scale. For all *mathing words* (i.e. as used in Figure 2), we use $nPMI(V_i, V_j)$ =10.00. (i.e., highest score achieved after normalizing in the range [0,10]).

Finally, we use the inverse of *link weight, i.e.,* $1/Link_{wt}(V_i, V_j)$ in the extraction of minimum weighted bipartite graph clique covers (see, the definition above: "Minimum weighted bipartite graph clique cover"). Here inverse of link weight represents the path length and minimum path length will show maximum similarity [11].

8   Extracting Plagiarized Texts

We used the minimum weighted bipartite graph clique covers to identify all plagiarized word patterns for target passage, w.r.t., given source passage. Now, the main aim here is to identify the coverage and importance of plagiarized patterns in the candidate passage from the target document. As bigrams[2] are more common and may be frequent, but higher length plagiarized word patterns may be more important but may not be so frequent. So, to properly exploit such features, we apply the separate treatment strategy for the plagiarized word patterns of different length. We also capture their importance. For this, we use the extracted patterns and their weight as features and apply SVM based classification. The final procedure is given below:

Input: (1) plagiarized word patterns for the given target candidate passage w.r.t., source passage, (2) candidate passages from target document, (3) Tf-idf score of all words of the target document.

---

[2] In this section 8, we use term 'bigram' to represent plagiarized word patterns of length two. It may or may not be adjacent to each other.

Output: plagiarized passage from target document.

Algorithm:

Step1. By using plagiarized word patterns for the given target candidate passage w.r.t, given source passage and the candidate passage from the target document, we collect the following features.

    a. *Feature-1*: Count of overlapping bigrams in the given target candidate passage.

    b. *Feature-2*: Tf-Idf score of each word of all overlapping bigrams in the given target candidate passage. We normalize this score in the range of [0-1].

    c. *Feature-3*: Count of overlapping higher length plagiarized word patterns (number of words more than two) in the given target candidate passage.

    d. *Feature-4*: Tf-Idf score of each word of all overlapping higher length plagiarized word patterns in the given target candidate passage. We normalize this score in the range of [0-1].

    e. *Feature-5*: We calculate the sum of Tf-Idf scores of all words in the given target candidate passage. Let us denote it by 'A'. Next, calculate the sum of Tf-Idf scores of the words of overlapping plagiarized word patterns in the given target candidate passage. Let us denote it by 'B'. Next calculate the percentage (%) of 'B' w.r.t. 'A'. We normalize the score in the range [0-1].

Step2. Use the given class level(s) for the given target candidate passage. For example, with 'PAN 2012 plagiarism detection dataset', we use two class levels, i.e., plagiarized or not-plagiarized and with 'student answer related plagiarism dataset', we use four class levels, i.e., Near copy, Light revision, Heavy revision and Non-plagiarism.

Step3. We repeat this process, i.e., Step1 and Step2 with all source passages and candidate passages and collect all the above discussed features.

Step4. Finally we apply SVM based classification. We use the same feature set (i.e., Feature-1 to 5 and class-Label in entire SVM based classification.

9   Pseudo Code

The pseudo-code for entire scheme is given below:

Input: (1) ASCII text documents (Including source and target documents), (2) DBpedia corpus

Output: Plagiarized passage(s) from target document.

Algorithm:

St1. Apply input cleaning and preprocessing for given documents. (subsection 4.1).

St2. Prepare semantic word graph of text for passages of both source and target document (Section-5). Calculate the link weight of links in both graphs by using semantic score (i.e., normalized pointwise mutual information score, section 5) between word pairs and their co-occurrence frequency in the corresponding text passage.

St3. Calculate controlled closeness centrality score of words related to passages of source and target documents, which either matches or synonyms or hyponyms of each other. Finally, normalize the closeness centrality score of words in the range of [0, 1]. (section 6)

St4. Prepare bipartite graph by using source and target passages and identify all minimum weighted bipartite graph clique covers. (section 7)

St5. Use identified minimum weighted bipartite graph clique covers and identify the plagiarized passage(s) in target document w.r.t., passage in given source document and finally identify all plagiarized passages. (section 8).

10   Evaluation

We use two different datasets, (1) PAN 2012 plagiarism detection dataset [9] and (2) student answer related plagiarism dataset [5]. The details of datasets are given below:

PAN 2012 Plagiarism Detection Dataset: The corpus comprises "susp": 3000 suspicious documents as plain text in test set and 1804 documents in training set. Next it contains "src": 3500 source documents as plain text (incl. the translations of 500 non-English source documents) in test set and 4210 documents in training set. The suspicious documents contain passages taken from the source documents, obfuscated with one of five different obfuscation techniques. The details of plagiarism types used in evaluation are given below. Each type contains 500 document pairs in test set and 1000 document pairs in training set.

✓ *01_no_plagiarism*: XML files for document pairs without any plagiarism.

✓ *02_no_obfuscation*: XML files for document pairs where the suspicious document contains exact copies of passages in the source document.

✓ *03_artificial_low*: XML files for document pairs where the plagiarized passages are obfuscated by the means of moderate word shuffling.

✓ *04_artificial_high*: XML files for document pairs where the plagiarized passages are obfuscated by the means of not so moderate word shuffling.

✓ *05_translation*: XML files for document pairs where the plagiarized passages are obfuscated by translation from a European language.

✓ *06_simulated_paraphrase*: XML files for document pairs where the plagiarized passages are obfuscated by humans via Amazon Mechanical Turk.

Student Answer Related Plagiarism Dataset: We used the corpus developed by [5], which provides samples of plagiarized short passages associated with different levels of plagiarism. It consists of a total of 100 documents containing 5 Wikipedia articles as the original texts and 95 suspicious plagiarized short passages. The latter were written by students to answer 5 questions, each related to one original document. The answers were based on the original texts (except for non-plagiarized cases), with various degrees of text overlapping, according to the instructions given by the corpus creators. We use 10-fold cross validation (same as used by [10]) to evaluate the performance of the system. The 4 classes of suspicious documents are classified as follows:

(1) Near copy: copy-and-paste from the original text;

(2) Light revision: minor alteration of the original text by substituting words with synonyms and performing some grammatical changes;

(3) Heavy revision: rewriting of the original by paraphrasing and restructuring;

(4) Non-plagiarism: based on participants 'own knowledge as the original texts were not given. The short passages were between 200-300 words and 57 samples were marked as heavy revision, light revision or near copy levels whereas the remaining 38 cases were non-plagiarized.

10.1 Systems Used in Evaluation and Evaluation Metrics

Baseline Systems: We consider top 3 systems from PAN 2012 contest as baseline systems with "PAN 2012 plagiarism detection dataset". Next, we use a latest published work, i.e., "Chong [10]" as baseline system for "Student answer related plagiarism dataset".

➤ **"kong"** [6]. To extract all plagiarism passages from the suspicious document and their corresponding source passages from the source document, [6] combined the semantic similarity and structure similarity based features. It uses information retrieval to get candidate pairs of sentences from suspicious document and potential source document. Finally, [6], uses "Bilateral Alternating Sorting" to merge pairs of sentences.

➤ **"Suchomel"** [7]'s detailed comparison system detects common features of the input document pair, computing valid intervals from them, and then merging some detections in the post processing phase.

➤ **"Grozea"** [8]. In the first phase [8] computes a matrix of kernel values, which gives a similarity value based on n-grams between each source and each suspicious document. In the second phase, each promising pair is further investigated, in order to extract the precise positions and lengths of the subtexts that have been copied and may be obfuscated using encoplot, "a novel linear time pairwise sequence matching technique".

➤ **"Chong"**, [10], tried to improve the accuracy of plagiarism detection by incorporating Natural Language Processing (NLP) techniques into existing approaches. [10], tested the performance of the system on "student answer related plagiarism dataset" [5].

Evaluation Metric: To evaluate the results with the PAN-2012 dataset, we use (1) overall plagiarism detection score "PlagDet" [5], (2) Precision score [5], (3) Recall score [5], and (4) Granularity score [5]. Next to evaluate the performance on "student answer related plagiarism dataset", we use (1) Precision, (2) Recall and (3) F-measure score, as used in [10].

10.2 Analysis of Result

All results are given in Table-1, 2, 3,4, and 5. Table 1,2,3 and 4 contains the published results of corresponding systems as given in PAN-2012 overview paper [9]. Next, Table 5 contains the best published results of [10]. Highest results are represented by bold font.

Table 1. Comparative "PlagDet scores" of baseline systems and our devised system

System	Overall	Real Cases	Simulated	Translation	Artificial High	Artificial Low	No Obfuscation
Kong	0.738	0.756	0.758	**0.771**	0.396	0.822	0.898
Suchomel	0.682	0.624	0.611	0.748	0.153	0.801	**0.938**
Grozea	0.678	0.636	0.668	0.713	0.380	0.753	0.848
Our System	**0.745**	**0.761**	**0.774**	0.763	**0.425**	**0.847**	0.901

Table 2. Comparative "Precision scores" of baseline systems and our devised system

System	Overall	Real Cases	Simulated	Translation	Artificial High	Artificial Low	No Obfuscation
Kong	0.842	0.964	0.900	0.820	0.750	0.928	0.834
Suchomel	0.893	0.869	0.915	**0.956**	0.793	0.948	0.885
Grozea	0.774	0.667	0.888	0.709	0.818	0.879	0.753
Our System	**0.898**	**0.971**	**0.921**	0.843	**0.841**	**0.949**	**0.891**

Table 3. Comparative "Recall scores" of baseline systems and our devised system

System	Overall	Real Cases	Simulated	Translation	Artificial High	Artificial Low	No Obfuscation
Kong	0.687	0.621	0.657	0.727	0.277	0.777	0.973
Suchomel	0.552	0.487	0.458	0.614	0.085	0.694	**0.999**
Grozea	0.635	0.607	0.536	0.716	0.272	0.778	0.969
Our System	**0.696**	**0.640**	**0.661**	**0.730**	**0.309**	**0.839**	**0.999**

Table 4. Comparative "Granularity-scores" of baseline systems and our devised system

System	Overall	Real Cases	Simulated	Translation	Artificial High	Artificial Low	No Obfuscation
Kong	1.01	1.00	1.00	1.00	1.02	1.04	1.00
Suchomel	1.00	1.00	1.00	1.00	1.00	1.00	1.00
Grozea	1.03	1.00	1.00	1.00	1.10	1.13	1.00
Our System	1.00	1.00	1.00	1.00	1.00	1.00	1.00

Table 5. Comparative evaluation on "student answer related plagiarism dataset"

Class	System [10] (with best features)			Our System		
	Precision	Recall	F-Measure	Precision	Recall	F-Measure
Near copy	0.667	0.526	0.588	**0.701**	**0.583**	**0.637**
Light revision	0.5	0.474	0.486	**0.570**	**0.510**	**0.538**
Heavy revision	0.55	0.579	0.564	**0.580**	**0.611**	**0.595**
Non-plagiarism	0.881	0.974	0.925	**0.941**	**0.983**	**0.962**
Average	0.649	0.638	0.641	**0.698**	**0.672**	**0.674**

Analysis: From the results given in table 1,2,3 and 4, it is clear that our devised system performs better than top systems of PAN-2012 plagiarism detection task in three important cases, i.e., (1) 03_artificial_low, (2) 04_artificial_high and (3) 06_simulated_paraphrase. These three cases contain plagiarism with different levels of paraphrasing. The main goal of the devised system was to imrove the plagiarism detection in the case of paraphrasing and artificial word reordering. However, system also performs well other cases like: 01_no_plagiarism, 02_no_obfuscation and 05_translation.

From the results given in Table-6, it is clear that system, shows better precision recall and F-measure score from baseline system of this area.

11   Conclusion and Future Scope

In this paper, we targeted the paraphrasing and artificial word reordering based cases for palagiarism detection. However, the system is comparable/better than other systems in other simple cases of plagiarism detection. From the experimental results, it is clear that minimum weighted bipartite graph clique covers, effectively captured the plagiarized text patterns of (1) adjacent words and (2) non adjacent words of any given sequence. This strategy is effective in capturing the notion of word reordering and paraphrasing. Next, the way to construct the graph and a combination of global statistical and semantic measures also used in this work, which are the integral part of key technique i.e., "minimum weighted bipartite graph clique covering".

We can extend this scheme towards the enhancements of pattern based evaluation strategies, like: (1) automatic answer evaluation, (2) automatic summarization evaluation and (3) evaluation of quality of summarization etc. Due to high flexiblity, we can easily extend this system to a wide range of multilingual environments and different domains.

References

1. Barron-Cedeno, A., Vila, M., Martí, M.A., Rosso, P.: Plagiarism Meets Paraphrasing: Insights for the Next Generation in Automatic Plagiarism Detection. Computational Linguistics. MIT Press (2013), doi: 10.1162/COLI_a_00153.
2. IEEE. 2008. A Plagiarism FAQ, http://www.ieee.org/publicationsstandards/publications/rights/plagiarismFAQ.html (last accessed November 25, 2012)
3. Bhagat, R., Hovy, E.: What Is a Paraphrase? Computational Linguistics. MIT Press (2013), doi:10.1162/COLI_a_00166.
4. Bouma, G.: Normalized (Pointwise) Mutual Information in Collocation Extraction. In: Proceedings of the International Conference of the German Society for Computational Linguistics and Language Technology, pp. 31–40 (2009)
5. Clough, P., Stevenson, M.: Developing a corpus of plagiarised short answers. In: Language Resources and Evaluation, LREC 2010, vol. 2010 (2009)
6. Kong, L., Qi, H., Wang, S., Du, C., Wang, S., Han, Y.: Approaches for Candidate Document Retrieval and Detailed Comparison of Plagiarism Detection—Notebook for PAN at CLEF 2012, http://www.clef-initiative.eu/publication/working-notes, ISBN 978-88-904810-3-1
7. Suchomel, Š., Kasprzak, J., Brandejs, M.: Three Way Search Engine Queries with Multi-feature Document Comparison for Plagiarism Detection—Notebook for PAN at CLEF (2012). In: Forner et al [6], http://www.clef-initiative.eu/publication/working-notes, ISBN 978-88-904810-3-1

8. Grozea, C., Popescu, M.: Encoplot - Tuned for High Recall (also proposing a new plagiarism detection score). In Forner et al. [6], http://www.clef-initiative.eu/publication/working-notes, ISBN 978-88-904810-3-1

9. Potthast, M., Gollub, T., Hagen, M., Kiesel, J., Michel, M., Oberländer, A., Tippmann, M., Barrón-Cedeño, A., Gupta, P., Rosso, P., Stein, B.: Overview of the 4th International Competition on Plagiarism Detection. CLEF (Online Working Notes/Labs/Workshop) (2012)

10. Chong, M., Specia, L., Mitkov, R.: Using Natural Language Processing for Automatic Detection of Plagiarism. In: Proceedings of the 4th International Plagiarism Conference (IPC 2010), Newcastle-upon-Tyne, UK (2010)

11. Kumar, N., Srinathan, K., Varma, V.: A Knowledge Induced Graph-Theoretical Model for Extract and Abstract Single Document Summarization. In: Gelbukh, A. (ed.) CICLing 2013, Part II. LNCS, vol. 7817, pp. 408–423. Springer, Heidelberg (2013)

12. Nahnsen, T., Uzuner, O., Katz, B.: Lexical chains and sliding locality windows in content-based text similarity detection. In: Proceedings of the 2nd International Joint Conference on Natural Language Processing (IJCNLP 2005), Jeju Island, Korea, pp. 150-154 (2005)

13. Zini, M., Fabbri, M., Moneglia, M., Panunzi, A.: Plagiarism Detection through Multilevel Text Comparison. In: 2006 Second International Conference on Automated Production of Cross Media Content for Multi-Channel Distribution (AXMEDIS 2006), December 2006, pp. 181–185 (2006), doi:10.1109/AXMEDIS.2006.40.

14. Lancaster, T., Culwin, F.: A Visual Argument for Plagiarism Detection using Word Pairs. In: Proceedings of the 1st International Plagiarism Conference, Newcastle, UK, vol. 4, pp. 1–14 (2004a)

15. Hoad, T., Zobel, J.: Methods for identifying versioned and plagiarized documents. Journal of the American Society for Information Science and Technology 54(3), 203–215 (2003)

16. Shivakumar, N., Garcia-Molina, H.: SCAM: A copy detection mechanism for digital documents. In: Proceedings of the Second Annual Conference on the Theory and Practice of Digital Libraries, Texas, USA, pp. 1–13 (1995)

17. Ceska, Z.: Automatic Plagiarism Detection Based on Latent Semantic Analysis. Doctoral thesis, University of West Bohemia (2009)

18. Alzahrani, S., Salim, N.: Fuzzy Semantic-Based String Similarity for Lab Report for PAN at CLEF 2010. In: Proceedings of the International Conference of the Cross-Language Evaluation Forum (CLEF 2010), Uncovering Plagiarism, Authorship, and Social Software Misuse Worksop (PAN 2010), Padua,Italy (2010)

19. Chen, C.-Y., Yeh, J.-Y., Ke, H.-R.: Plagiarism Detection using ROUGE and WordNet. Journal of Computing 2(3), 34–44 (2010)

20. Kohler, K., Weber-Wul, D.: Plagiarism Detection Test 2010. Technical report, HTW Berlin (2010)

21. Scott, S., Matwin, S.: Text classification using WordNet hypernyms.In: Use of WordNet in Natural Language Processing Systems: Proceedings of the Conference (1998)

22. Tang, Liu: Community Detection and Mining in Social Media. Morgan & Claypool Publishers (2010)

23. Hartrumpf, S., vor der Brück, T., Eichhorn, C.: Semantic duplicate identification with parsing and machine learning. In: Sojka, P., Horák, A., Kopeček, I., Pala, K. (eds.) TSD 2010. LNCS, vol. 6231, pp. 84–92. Springer, Heidelberg (2010)

Identification of Plagiarism Using Syntactic and Semantic Filters

R. Vijay Sundar Ram[1], Efstathios Stamatatos[2], and Sobha Lalitha Devi[1]

[1] AU-KBC Research Centre, MIT Campus of Anna University, Chennai, India
[2] Dept. of Information and Communication Systems Eng. University of the Aegean,
83200–Karlovassi, Greece
{sundar,sobha}@au-kbc.org, stamatatos@aegean.gr

Abstract. We present a work on detection of manual paraphrasing in documents in comparison with a set of source documents. Manual paraphrasing is a realistic type of plagiarism, where the obfuscation is introduced manually in documents. We have used PAN-PC-10 data set to develop and evaluate our algorithm. The proposed approach consists of two steps, namely, identification of probable plagiarized passages using dice similarity measure and filtering the obtained passages using syntactic rules and lexical semantic features extracted from obfuscation patterns. The algorithm works at sentence level. The results are encouraging in difficult cases of plagiarism that most of the existing approaches fail to detect.

Keywords: Manual paraphrasing, Syntactic rules and Lexical Semantics, Plagiarism detection.

1   Introduction

Manual paraphrasing concerns the transformation of an original text, so that the resulted text has the same meaning as the original, but with significant differences in wording and phrasing. Detecting manual paraphrasing in plagiarism cases is challenging, since the similarity of the original text with the re-written text is purposefully hidden. The exponential increase of unstructured data on the web provides multiple sources for plagiarism. The need for accurate plagiarism detection is vital to ensure originality of text with applications in areas such as publishing, journalism, patent verification, academics etc. There are many commercial plagiarism detection tools such as Turnitin and prototype systems such as COPS (Copy Protection System), SCAM (Stanford Copy Analysis Mechanism), and MOSS (Measure of Software similarity) [6]. Yet identification of manual paraphrasing is a challenge to available plagiarism detection tools since most of them are only able to detect easy, copy-and-paste cases. The detection of plagiarism when paraphrasing is used requires understanding of re-ordering of phrases governed by syntactic rules and use of synonym words governed by lexical semantics.

Several researchers have studied plagiarism detection for the past few decades. Plagiarism detection gained more focus and geared up with the yearly automatic

A. Gelbukh (Ed.): CICLing 2014, Part II, LNCS 8404, pp. 495–506, 2014.

plagiarism detection competition started in 2009 in the framework of PAN evaluation campaigns. In these competitions, various text re-use issues such as external and intrinsic plagiarism detection as well as cross-lingual plagiarism detection [10]. The competition started in 2009 (PAN 2009) with a small corpus. In 2010 (PAN 2010), it evolved with huge set of suspicious and source data (PAN-PC-10) and continued similarly in the year 2011 (PAN 2011). The overview report of PAN 2010 shows, 18 participants participated in the competition. They have followed similar steps in the detection of plagiarized passages, namely, candidate (or source) retrieval, detailed analysis (or text alignment) and post processing.

In the candidate retrieval step, most of the participants reduced the search space by removing the source document which does not have significant similarities with the suspicious documents. In this step, the used techniques are based on such as finger prints and position of data, finger prints with threshold are used. Information retrieval, document fingerprinting, string kernel matrices, word and character n-grams,. In the detailed analysis step some of the techniques used were winnowing fingerprinting, FastDoCode technique, cosine similarity, and jaccard similarity. Post processing techniques were focused on filtering the detected passages using chunk ratio with a predefined threshold and n-gram match, using different similarity measures [10].

In PAN-2011, the participants used approaches similar to those used in PAN-2010. They have improved the efficiency and the processing time of their approaches in comparison with systems in PAN-2010 [10]. In PAN-2012 and PAN-2013, the competition was remodeled with smaller datasets to focus on individual steps. Evaluation was done at different steps [8, 9]. We will look in detail the various plagiarism detection approaches using semantic, syntactic and structural information, focused on identifying manual paraphrasing.

Palkovskii et al. [7] have used a semantic similarity measure to detect the plagiarized passages. Semantic similarity measure used hyponym taxonomy in WordNet for calculating the path length similarity measure. Uzner et al. [15] used a low-level syntactic structure to show linguistic similarities along with the similarity measure based on tf-idf weighted keywords. Here they have used Levin's verb classes. A fuzzy semantic string similarity algorithm for filtering the plagiarized passages was used by Alzahrani et al [1]. The authors had mentioned that their approach didn't work for all levels especially for higher obfuscation, which includes manual paraphrasing. Chong and Specia [3] used a lexical generalization approach using WordNet to generalize the words and performed n-gram based similarity measure, namely overlap co-efficient, to identify the plagiarized passages. Stamatatos [14] used structural information for identifying the plagiarized passages. He used the stop-words to identify the structure of sentence. This approach varied from most of the other approaches where the content words (nouns, verb, adjective, and adverb) were considered important and the stop-words were removed as these words occur frequently.

In our work, we try to identify the manual paraphrasing using lexical semantics and syntactic rules. The paper is organized as follows: In the next section, we describe our approach, where we elaborate on identification of probable plagiarized passages and filtering the irrelevant passages with lexical semantics and syntactic rules. In the third section, we demonstrate our experiment and describe our results. The paper ends with the concluding section.

2   Our Approach

We present an approach for identifying the manual paraphrasing. Our approach works at sentence level. We detect the plagiarized passages in a two step approach. In the first step, we try to find all probable plagiarized passages using dice similarity measure. In the second step, we identify the correct passages from the probable plagiarized passages obtained in the first level using lexical semantics and syntactic rule based filters. The approach is described in detail in the following sections.

2.1   Retrieval of Probable Plagiarized Passages

In this step, we try to identify all probable plagiarized passages from a given suspicious document in comparison with the source documents. Here we identify the plagiarized sentences in the suspicious documents using dice similarity measure and group the sentences into plagiarized passages using heuristic rules. We start by pre-processing the suspicious and source documents with a tokeniser and sentence splitter. We remove the function words, connectives, pronouns from the text, making the sentences incoherent.

Identification of Similar Sentences

We have collected the sentences with common bigram words from the suspicious and the source documents and performed similarity comparison using dice similarity measure. Dice similarity (Q_s) is defined as follows,

$$Q_s = 2C/(A+B)$$

where A and B are the number of words in sentence A and B, respectively, and C is the number of words shared by the two sentence; Q_S is the quotient of similarity and ranges from 0 to 1 [4].

We have used dice similarity measure instead of cosine similarity measure with tf-idf as weights. The cosine similarity measure provides a smaller score, if a sentence has many common words and a rarely occurred word.
Consider the following sentences,

1. *This was written by E.F. in 1627 and printed exactly as the original.*
2. *Written by E.F. in the year 1627, and printed verbatim from the original.*

The dice similarity score between sentence 1 and 2 is 0.769, whereas the cosine similarity score is 0.462. The cosine similarity score is smaller as the term 'verbatim' has very high inverse-document frequency value and the denominator value in cosine similarity becomes big.

The steps followed in this task are described in the algorithm below.

1. Sentences with common bigram words in the suspicious and source documents are collected.
2. Between the pair of sentences from the suspicious and source documents having bigram words, dice similarity comparison is performed. Comparison is done between the following set of suspicious and source pair sentences.
   (a) One suspicious and source sentence having a common word bigram.
   (b) Two consecutive suspicious sentences and one source sentence having a common word bigram.
   (c) One suspicious sentence and two consecutive source sentences having a common word bigram.
3. Those suspicious-source pairs with similarity greater than a predefined threshold t are collected to form plagiarized passages.

Formation of Probable Plagiarized Passages

In this step, we try to group the suspicious-source sentence pairs which are greater than the predefined threshold t into plagiarized passages. We group the suspicious-source sentence pairs with another pair which have neighboring sentence in both suspicious document and source document into a plagiarized passage. Here we also consider the next neighboring sentence to form a passage. We describe the steps in detail in the algorithm given below.

Passage Forming Algorithm

1. For each of the suspicious–source document pair which have suspicious-source sentence pairs having dice score greater than the predefined threshold t, do steps 2-7.
2. For each pair from the probable pairs of suspicious-source sentences collected using dice similarity, do step 3-6.
3. Compare this pair with the rest of the probable pairs, consider the pair has suspicious sentence x and source sentence y.
4. If the current suspicious-source sentence pair has another suspicious-source sentence pair which is consecutive to the current pair i.e., $(x+1, y+1)$, group these suspicious sentences and source sentences to form a new passage, $(x, x+1; y, y+1)$.
5. If a suspicious-source pair has a consecutive suspicious sentence $(x+1)$ and source sentence which is the sentence following the consecutive sentence $y+2$, to the current pair, group the suspicious sentences and source sentences to form a new passage $(x, x+1; y, y+1, y+2)$.
6. If a suspicious-source pair has a suspicious sentence which is the sentence following the consecutive sentence $x+2$ and consecutive source sentence $(y+1)$, to the current pair, group the suspicious sentences and source sentences to form a new passage $(x,x+1,x+2; y, y+1)$.
7. Repeat steps 3 to 7 for the set of new passages till there are no pairs to group.

By performing the steps in the algorithm above, we form passages having from one sentence to n sentences. Here n is determined based on the clustering of neighboring sentences having similarity score greater than the threshold t.

2.2 Filtering of Plagiarized Passages based on Syntactic Rules and Lexical Semantics

The methods mentioned in previous work for filtering out the irrelevant suspicious-source passages were using similarity measures such as jaccard, cosine similarity or dice similarity with a given threshold. In manual paraphrasing, these similarity measures fail to filter out the irrelevant passages without drastically affecting the true positive passages, as these plagiarized passages are well rewritten. To filter out the irrelevant passages, we manually analysed the manual paraphrased passages in PAN-PC-10 training data and identified a set of patterns of changes performed by people, while plagiarizing a text. From these patterns we came up with a set of syntactic rules and lexical semantic features. These lexical semantics and syntactic rules are helpful in identifying the manual paraphrased passages. These rules obtained by analysing the training corpus are then used in identifying the correct passages in the test corpus.

The set of identified patterns and rules to handle them are presented below in this section. As the rules are based on syntactic information, all probable passages are preprocessed with a POS tagger [2] and a text chunker [13]. Pleonastic 'it' is also identified in the passages using a CRFs engine [5].

Synonym Substitution

While plagiarizing a sentence, people tend to substitute nouns, verbs, adjectives and adverbs by its equivalent synonym words. Verbs are also substituted by combinations of verbs and prepositional phrases. For example 'kill' can be substituted by 'put to death'. Phrases are also replaced by its semantically-equivalent phrases. By observing the data, this type of synonym substitution covers 75.71% of changes.

Consider the following sentences:

3.a *This question is linked closely to the often-debated issue of the Pointed Style's beginnings.*

3.b *This Query is, of course, intimately connected with the much-disputed question of the origin of the Pointed Style itself.*

Here, 3.a is the suspicious sentence and 3.b is the source sentence. By comparing sentences 3.a and 3.b, we see that the noun 'query' is substituted with 'question', 'origin' is substituted with 'beginning'. The noun phrase 'much-disputed question' is substituted with 'often-debated issue' and the verb phrase 'intimately connected' is substituted with 'linked closely'.

We handle this synonym substitution using Rule 1.

Rule1:

      a. Check for common words in the preprocessed suspicious and source sentence.

      b. if common words exists then do the following steps:

1. The words in the suspicious sentences which are not common with the source sentences are examined and their synonyms are obtained from the WordNet, taking into account the POS category of the word. The synset words obtained are matched with words in the source text and the matched synonyms replace initial words in the suspicious sentence. While performing comparison with the source sentence, synset words are matched with words in the source sentence in the position corresponding to the suspicious sentence.

2. Similarly, we obtain equivalent phrases from the phrase dictionary, which is built using the WordNet synonym dictionary, for the phrases in the suspicious sentence and match with the equivalent phrases in the source sentence. If the equivalent phrase exists in the source sentence, then in the suspicious sentence the phrase is replaced by its equivalent phrase.

Re-ordering of Phrases

When plagiarizing a sentence, people tend to re-order prepositional phrases in sentences by moving it to the end of the sentence or from the end of the sentence to the start of the sentence. Moreover, the position of adverbs are re-ordered within verb phrases, the position of adjectives and nouns are re-ordered within noun phrases and possessive nouns are changed into prepositional phrase following its head noun and vive-versa. This re-ordering of phrases cover 10% of the introduced changes. The re-ordering of phrases is explained with the example sentences below.

4.a *After the treaty of Tilsit, Emporer Alexander took control of Finland.*

4.b *The Emperor Alexander possessed himself of Finland after the treaty of Tilsit.* Sentence 4.a is the suspicious sentence and 4.b is the source sentence. In the sentence 4.a and 4.b, the prepositional phrase, 'after the treaty of Tilsit' which occurs in the end of the source sentence is moved to the start of the suspicious sentence.

5.a *I saw a little of the Palmyra's inner life.*

5.b *I saw something of the inner life of Palmyra.*

Here sentences 5.a and 5.b are suspicious and source sentence respectively. In the source sentence, a noun phrase followed by a prepositional phrase "the inner life of Palmyra" is replaced in the suspicious sentence by a possessive noun following a head noun "Palmyra's inner life".

This re-ordering of phrases is handled by Rule 2.

Rule 2

1. If the suspicious and source sentences have common prepositional phrases, check its position.
   (a) If the position of the prepositional phrase varies then re-order the phrase exactly as it has occurred in the source sentence.
2. If either the suspicious or the source sentence has a possessive noun phrase and the other sentence has the head noun of a possessive noun phrase with prepositional phrase, then normalize possessive noun phrase as it has occurred in the source sentence.

3. If a noun phrase exists in suspicious and the source sentences with common words but different ordering, then re-order the adjectives and nouns as it has occurred in the source sentence.

4. If a verb phrase exists in suspicious and the source sentences with common words but different ordering, then re-order the adverb and verb as it has occurred in the source sentence.

Introduction of Pleonastic 'it'

If the source sentence starts with a phrase such as 'by pure luck', in the plagiarized sentence it may be re-written with pleonastic 'it' as subject 'It was pure luck'. This type of introduction of pleonastic 'it' is 2.14% of the total changes.

This is explained with sentence 6.a and 6.b, where 6.a is the suspicious sentence and 6.b is source sentence.

6.a *It was pure luck that I made that shot.*

6.b *By good luck I hit.*

In the above sentence 6.a and 6.b., the source sentence starts with a prepositional phrase and does not have a subject and in the suspicious sentence a null subject 'it' is introduced. This is handled by Rule 3.

Rule 3:

a. Check if occurrences of 'it' in the suspicious or source sentence is marked as pleonastic 'it'.

b. If the pleonastic 'it' occurs in the suspicious sentence and not in the source sentence, then remove 'it'.

Antecedent Replacement

The pronouns in source sentences may be replaced by its antecedent while plagiarizing the sentence. This covers 2.85% of the total changes.

Consider the following sentences:

7.a *Madame Omar was a lovely German woman who's husband kept her all but locked away in the harim.*

7.b *She was a charming German lady ; but her husband kept her secluded in the harim like a Moslem woman.*

Here the sentence 7.a is the suspicious sentence and 7.b is the source sentence. In the suspicious sentence, the pronoun 'she' is replaced by 'Madame Omar'. The suspicious sentence also has adjective (charming -> lovely), noun (lady -> woman) and verb (secluded -> locked) synonym substitutions and an adjunct drop (like a Moslem woman). The antecedent replacement is handled by Rule 4.

Rule 4:

If in a given pair of suspicious and source sentences a noun phrase exists in the place of a pronoun and if the context of the noun phrase and the pronoun are same, then replace the pronoun with the noun phrase.

Other Observed Patterns

In this section we list the set of extracted patterns which we have not handled in this study. These cover 9.26% of the changes.

Rewriting the sentence completely

When plagiarizing the sentence, people assimilate the sense conveyed in the sentences and they re-write in a completely different manner.

Consider the example 8.a and 8.b. Sentence 8.a is plagiarized from sentence 8.b. Here the sentence 8.a and 8.b does not have any structural similarity.

8.a *Even if the man is skilled enough to hook the fish, he could catch the fish from the rough sea merely because of his luck.*

8.b *No mortal skill could have killed that fish.*

Addition / Reduction of the descriptions in Sentences

While plagiarizing a sentence, people tend to add more or reduce the descriptions in the plagiarized sentence, as in example sentences 9.a and 9.b.

9.a *The proof that a wire may be stretched to long that the current will no longer have enough strength to bring forward at the station to which the despatch is made known.*

9.b *It is evident, therefore, that the wire may be continued to such a length that the current will no longer have sufficient intensity to produce at the station to which the despatch is transmitted those effects by which the language of the despatch is signified.*

Here sentence 9.a is the suspicious sentence and 9.b is the source sentence. In the suspicious sentence, there is a reduction in the description.

The Proposed Algorithm

Using the lexical semantics and syntactic rules mentioned in the previous section, we try to filter out the irrelevant passages from the retrieved probable passages. The steps involved in this process are described in detail in the algorithm given below.

1. For each sentence in the suspicious passage do step 2.
2. Compare the suspicious sentence with all sentences in the source passage.
   (a) Compare suspicious sentence and source sentence, check if the suspicious sentence has prepositional phrase re-ordering, possessive noun re-ordering, adverb re-ordering or adjective – noun re-ordering, then apply Rule 2.
   (b) To correct the synonym substitutions in the suspicious sentence, apply Rule 1.
   (c) If the suspicious sentence has pleonastic 'it', then apply Rule 3.
   (d) Compare the suspicious and the source sentence, if there is a pronoun in a sentence and noun phrase in another sentence in the same position, then apply Rule 4.
   (e) Similarity between suspicious and source sentences after applying the above rules is measured based on similarity of syntactic features and their positions such as noun phrase, verb phrase, proportional phrase, and their order of occurrence.

3. Similar suspicious and source sentences are identified.

4. If there exists a set of similar sentences in the suspicious and source passage, then this plagiarized passage is considered as probable plagiarized passage.

3   Experiments and Results

We have used PAN-PC-10 test data set for evaluating our algorithm for detecting manual paraphrasing [12]. This dataset has 15,925 suspicious documents and 11,147 source documents. In the suspicious dataset 50% of the documents do not have plagiarized passages. In the suspicious documents with plagiarism, the plagiarized passages are classified into three major types: passages with artificial obfuscation, passages where the obfuscation was introduced by translated passages, and passages with manually paraphrased obfuscation. In this evaluation, we have considered suspicious document with only manual paraphrasing, which counts to 397 suspicious documents. We have tested our algorithm with 397 suspicious documents having manual paraphrasing and 11,147 source documents. The evaluation is done using the performance metrics used in PAN-2010 evaluation campaign [11]. First, macro-average precision and recall are calculated at the passage level. In addition, granularity measures the ability of the the plagiarism detection algorithm to detect a plagiarized passage as a whole or as several pieces. Then, precision, recall and granularity are combined in the overall performance score called Plagdet.

In table 1, we have presented the scores obtained by the top 9 out of 18 participants in PAN2010 competition for these 397 suspicious documents with manual paraphrasing along with the scores obtained by our approach. The scores in table 1 are generated using the runs of each participants and the evaluation program provided at the PAN-2010 website.

Table 1. Scores of PAN 10 participants on suspicious documents with manual paraphrasing

Participant	Plagdet Score	Recall	Precision	Granularity
Our Approach	*0.4804*	*0.3914*	*0.8234*	*1.15*
Muhr	0.387	0.244	0.938	1.0
Grozea	0.322	0.198	0.944	1.0
Zou	0.290	0.172	0.948	1.0
Oberreuter	0.283	0.178	0.691	1.0
Kasprzak	0.231	0.131	0.954	1.0
Torrejon	0.230	0.133	0.836	1.0
Palkovskii	0.115	0.062	0.938	1.02
Sobha	0.061	0.031	0.889	1.0
Gottron	0.016	0.009	0.809	1.25

The low performance scores in table 1 show the need of an effective algorithm for detecting manual paraphrasing, which is the more realistic scenario in plagiarism detection. Our algorithm substantially improves recall while precision remains relatively high. Identification of probable plagiarized passages with a low similarity score helps in getting better recall.

Table 2. Scores of our approach after step 1 (without filtering)

	PlagdetScore	Recall	Precision	Granularity
Without lexical semantics and syntactic filtering	0.1954	0.4314	0.1553	1.25

The syntactic rule and lexical semantics filtering of the probable plagiarized passages helps in getting a high precision without largely disturbing the recall. This is shown from table 2. Before filtering the probable plagiarized passages using lexical semantics and syntactic rules, the precision is low. The filtering helps in removing the irrelevant passages without drastically disturbing the recall.

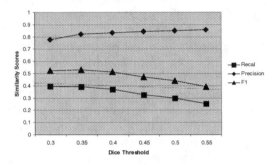

Fig. 1. Performance with varying similarity thresholds

Figure 1 shows the performance of our approach, in terms of recall, precision and F1 for varying threshold values used in retrieving probable sentences. Choosing the dice similarity threshold as 0.35 enhances the recall by extracted most of the probable plagiarized passages. We empirically derived and used this threshold in the reported experiments. In this study, we have not handled completely rewritten sentences and sentences with additional description. This affects the recall slightly, as the filtering algorithm filters out the complete rewritten sentences, though the passages may have many common words, which are identified in the first step. Improving the phrase to word and phrase to phrase dictionaries will help in improving the precision as well as the recall.

Figure 2 shows precision and recall while using different features in filtering the probable passages. The results show an increase in recall when both lexical semantics and syntactic rules are used in filtering the probable passages. When one of the features either lexical semantics or syntactic rules is used in filtering there is a drastic reduction in recall as compared to the recall achieved when both features are used together in filtering. We also observe that the use of lexical semantic features in filtering gives greater recall than filtering using syntactic rules.

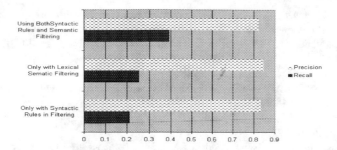

Fig. 2. Comparison of performance with different filtering features

4   Conclusion and Future Work

We have presented a two step approach to detect simulated plagiarized passages. In the first step, we extract the probable plagiarized passages using dice similarity measure from the sentences in the suspicious and source documents having common word bigrams. We maintain the similarity score between the suspicious and source sentence as low as 0.35 to find most of the probable plagiarized passages, which helps in getting better recall. In the second step, we have used lexical semantics and syntactic rules for filtering the relevant plagiarized passages, which helps in achieving high precision without largely harming the recall. Normalising the words using synonym dictionaries at the initial stage will help in boosting the recall drastically. We have not handled sentences which are completely rewritten. We are planning to take up the above mentioned tasks as our future work.

References

1. Alzahrani, S., Salim, N.: Fuzzy Semantic-Based String Similarity for Extrinsic Plagiarism Detection: Lab Report for PAN at CLEF 2010. In: Notebook Papers of Labs and Workshops CLEF 2010, Padua, Italy (2010)
2. Brill, E.: Some Advances in transformation Based Part of Speech Tagging. In: Proceedings of the Twelfth International Conference on Artificial Intelligence (AAAI 1994), Seattle, WA (1994)
3. Chong, M. and Specia. L.: Lexical Generalisation for Word-level Matching in Plagiarism Detection. In: Recent Advances in Natural Language Processing, pp 704–709, Hissar, Bulgaria, (2011)
4. Dice, L.R.: Measures of the Amount of Ecologic Association Between Species. Ecology 26(3), 297–302 (1945)
5. Lalitha Devi, S., Ram, V.S., Rao, P.R.K.: Resolution of Pronominal Anaphors using Linear and Tree CRFs. In: 8th DAARC, Faro, Portugal (2011)
6. Aimmanee, P.: Automatic Plaiarism Detection Using Word-Sentence Based S-gram. Chiang Mai Journal of Science 38 (special issue), 1–7 (2011)
7. Palkovskii, Y., Belov, A., Muzyka, I.: Using WordNet-based Semantic Similarity Measurement in External Plagiarism Detection - Notebook for PAN at CLEF (2011)

8. Potthast, M., Hagen, M., Gollub, T., Tippmann, M., Kiesel, J., Rosso, P., Stamatatos, E., Stein, B.: Overview of the 5th International Competition on Plagiarism Detection. In: Forner, P., Navigli, R., Tufis, D. (eds.), Notebook Papers of CLEF 2013 LABs and Workshops, CLEF-2013, Valencia, Spain, September 23-26 (2013)
9. Potthast, M., Gollub, T., Hagen, M., Graßegger, J., Kiesel, J., Michel, M., Oberländer, A., Tippmann, M., Barrón-Cedeño, A., Gupta, P., Rosso, P., Stein, B.: Overview of the 4th International Competition on Plagiarism Detection. In: Forner, P., Karlgren, J., Womser-Hacker, C. (eds.), CLEF 2012 Evaluation Labs and Workshop – Working Notes Papers (September 2012)
10. Potthast, M., Eiselt, A., Barrón-Cedeño, A., Stein, B., Rosso, P.: Overview of the 3rd International Competition on Plagiarism Detection. In: Petras, V., Forner, P., Clough, P.D. (eds.) Notebook Papers of CLEF 11 Labs and Workshops (2011)
11. Potthast, M., Barrón-Cedeño, A., Stein, B., Rosso, P.: An Evaluation Framework for Plagiarism Detection. In: Proc. of the 23rd Int. Conf. on Computational Linguistics, COLING 2010, Beijing, China, August 23-27, pp. 997–1005 (2010)
12. Potthast, M., Barrón-Cedeño, A., Eiselt, A., Stein, B., Rosso, P.: Overview of the 2nd International Competition on Plagiarism Detection. In: Braschler, M., Harman, D., Pianta, E. (eds.), Notebook Papers of CLEF 10 Labs and Workshops (September 2010)
13. Ngai, G., Florian, R.: Transformation-Based Learning in the Fast Lane. In: NAACL 2001, Pittsburgh, PA, pp. 40–47 (2001)
14. Stamatatos, E.: Plagiarism Detection Using Stopword n-grams. Journal of the American Society for Information Science and Technology 62(12), 2512–2527 (2011)
15. Uzuner, O., Katz, B., Nahnsen, T.: Using Syntactic Information to Identify Plagiarism. In: 2nd Workshop on Building Educational Applications using NLP (2005)

Readability Classification of Bangla Texts

Zahurul Islam, Md. Rashedur Rahman, and Alexander Mehler

WG Text-Technology
Computer Science
Goethe-University Frankfurt
{zahurul,mehler}@em.uni-frankfurt.de, kamol.sustcse@gmail.com

Abstract. Readability classification is an important application of *Natural Language Processing*. It aims at judging the quality of documents and to assist writers to identify possible problems. This paper presents a readability classifier for Bangla textbooks using information-theoretic and lexical features. All together 18 features are explored to achieve an *F*-score of 86.46%. The paper is an extension of our previous work [1].

Keywords: Bangla, text readability, information-theoretic features.

1 Introduction

Readability classification aims at measuring how well and easy a text can be read and understood [2]. It deals with mapping texts onto degrees of readability. Thus, readability classification can be reconstructed as a sort of automatic text categorization [3]. Various factors influence the readability of a text including simple features such as type face, font size and text vocabulary as well as more complex features relating to the syntax, semantics, or rhetorical structure of a text [1].

Professionals, such as teachers, journalists, or editors, produce texts for specific audiences. They need to check the readability of their output. Readability classifiers are also used as a means of pre-processing in the framework of *natural language processing* (NLP) [1].

A lot of research on readability classification exists for English [4–9], German [10], French [11], Japanese [12] and Chinese [13]. All these languages are considered as high-resourced languages. They are contrasted with low-resourced languages which are spoken by members of a small community or for which only few resources (corpora, tools etc.) exist [14]. Bangla is a low-resourced language in the latter sense. As an Indo-Aryan language it is spoken in Southeast Asia, specifically in present day Bangladesh and the Indian states of West Bengal, Assam, Tripura and Andaman and on the Nicobar Islands. With nearly 250 million speakers [15], Bangla is spoken by a large speech community. Nevertheless, it is low-resourced because of the lack of appropriate corpora and tools. Thus, though many texts are produced in Bangla everyday, authors can hardly measure their readability due to the lack of appropriate readability classifiers.

Recently, some approaches addressed the readability of Bangla text. Das and Roychudhury [16, 17] experimented with two classical readability measures for English nd applied them to Bangla texts. Sinha et al. [18] proposed two alternative readability measures for Bangla. Islam et al. [1] built a readability classifier using a corpus of Bangla

A. Gelbukh (Ed.): CICLing 2014, Part II, LNCS 8404, pp. 507–518, 2014.

textbooks. Although the classifier achieves an *F-score* of 72.10%, classifiers that produce better *F*-scores are still required. In this paper, we provide such a better performing readability classifier for Bangla. This is done by example of an extended version of the corpus used in [1]. The corpus is extracted from textbooks used in consecutive grades of the school system of Bangladesh.

Syntactic, semantic and discourse related features are now broadly explored for building readability classifiers for high-resourced languages. Obviously, it is a challenge to do the same for low-resourced languages that lack preprocessing tools. Thus, in this paper, we explore lexical and information-theoretic features which do not require (much) linguistic preprocessing.

The paper is organized as follows: Section 2 discusses related work followed by a description of the underlying corpus (Section 3). The operative readability features are described in Section 4. An experiment based on these features is the topic of Section 5. Its results are discussed in Section 6. Finally, a conclusion is given in Section 7.

2 Related Work

Since the early twentieth century, researchers proposed different readability measures for English [4–9]. All of them explore simple surface-structural features such as *average sentence length* (ASL), *average word length* (AWL) and *average number of syllables in a word*. Many commercial readability tools use these classical measures. Fitzsimmons et al. [19] stated that the SMOG [9] readability measure should be preferred to assess the readability of texts on health care.

Petersen & Ostendorf [20] and Feng et al. [21] show that the classical models have significant drawbacks. Due to recent achievements in linguistic data processing, models of linguistic features are now in the focus of readability studies. [1] summarizes related work regarding language model-based features [22–26], PoS-related features [21, 24, 27, 28], syntactic features [27, 29–32], and semantic features [21, 32].

Recently, Hancke et al. [10] measured the readability of German texts using lexical and syntactic features in conjunction with language models. According to their findings, morphological features influence the readability of German texts. Vajjala and Meurers [33] used lexical features from the field of *Second Language Acquisition* (SLA). In our study, we use *type token ration* (TTR) related readability measures as studied by Vajjala and Meurers [33].

Only few approaches consider the readability of Bangla texts. Das and Roychudhury [16, 17] show that readability measures proposed by Kincaid et al. [7] and Gunning [6] work well for Bangla. However, the measures were tested only for seven documents, mostly novels.

In our previous study [1], we proposed a readability classifier for Bangla using *entropy* and *relative entropy*-based features. We achieved an *F-Score* of 72.10% by combining these features with lexical ones. Recently, Sinha et al. [18] proposed two readability measures that are similar to classical readability measures for English. They conducted a user experiment to identify important structural parameters of Bangla texts.

Table 1. Examples of conversion problems

Misspelled word	Correct spelling
়ুদই	দুই
়লিস্ভ	স্লিভ
ত্রস়ীর	স্ত্রীর
প্রস়াব	প্রস্তাব

The measures are based on the *average word length*, the *number of poly-syllabic words* and the *number of consonant–conjuncts*.

To make our feature model comparable, we build a baseline that explores five classical readability measures for English. These measures are used by many readability classification tools. We additionally compare our findings with the ones of our previous study and the results reported by Sinha et al. [18].

3   Corpus

We analyze an update of the corpus used in [1]. It contains text samples of 27 additional books. Table 2 compares the update with its predecessor.

As noted in Islam et al. [1], we needed to convert non-standard texts into Unicode texts. The conversion was not straightforward; it produced several kinds of noise in the converted texts. The original textbooks not only contain descriptive texts but also poems, religious hymns, texts from other languages (e.g., Arabic, Pali) and transcriptions of Arabic texts (e.g., Surah). These non-descriptive texts were manually removed. The conversion process left many sentences with unexpected spaces between letters, which we also removed manually. Further, we detected many spelling errors possibly produced by the conversion. The converter mostly affected dependent vowels or consonant conjuncts. In Bangla, a dependent vowel can only occur with a consonant. A consonant conjunct becomes invalid due to the misplacement of the Bangla *virama* sign *hoshonto*. Table 1 shows some spelling errors generated by the converter. All these mistakes have been manually corrected. There were many English terms in the original Bangla texts that were converted to meaningless strings. For example: the English word *hepatitis* was converted to ঈৎফবনরৎ়য়ৎধ়ব. The converter cannot identify the English alphabet because the code points of the Bangla alphabet in *Bijoy* are similar to the code points of the English alphabet. We manually corrected all erroneous strings.

Our classification distinguishes four readability classes: *very easy*, *easy*, *medium* and *difficult*. Documents of (school) grade *two*, *three* and *four* are included into the class *very easy*. Class *easy* covers texts of grade *five* and *six*. Texts of grade *seven* and *eight* were subsumed under the class *medium*. Finally, all texts of grade *nine* and *ten* were mapped onto the class *difficult*. Table 2 informs about the classes and their statistics. Note that the original grades could not be used as target classes due to problems of data sparseness.

Table 2. Statistics of the new Bangla Readability Corpus

Classes	#documents	Document length	Sentence length	Word length
very easy	210	90.75	7.38	5.31
easy	60	176.82	8.22	5.44
medium	199	193.77	10.37	5.47
difficult	113	251.30	12.19	5.66

4 Features

Our aim is to compare a classifier based on lexical and information-theoretic features with a classifier based on five classical measures of readability for English texts. We start with describing the lexical features.

4.1 Lexical Features

Different lexical features have been used from the beginning of text readability research. These features are still popular because they are language independent and do not require much linguistic preprocessing.

We use the lexical features ASL, AWL and the *average number of complex words* as described in [1]. The *type token ratio* (TTR), which indicates the lexical density of a text, has been considered as a readability feature, too. See [33] for such an approach. Low lexical densities involve a great deal of repetition with the same words being highly repeated. Conversely, a high lexical density reflects the diversification of the vocabulary of a text. A lexically diversified text is supposed to be difficult for readers. In a diversified text, synonyms may be used to represent similar concepts. Temnikova [34] emphasizes two problems caused by rich vocabularies that affect readability. Non-native speakers or non-specialists may face problems to detect the relationships between the synonyms [34].

There exist many versions of the TTR in the literature. Carrol [35] proposes a variant of the TTR to reduce effects of sample size. Herdan [36] studies a version of TTR called bi-logarithmic TTR. Köhler and Galle [37] define a time series-related variant of the TTR (see Equation 1) that considers text positions. In Equation 1, x refers to a position in the text, t_x is the number of types up to position x, T is the overall number of types in the text and N denotes the number of tokes in the text. We also propose a variant of the TTR. By focusing on the document-level and the sentence-level TTR, it computes the difference of the TTR on both levels. The lexical features used in our study are listed in Table 4.

– Köhler-Gale method

$$TTR_x = \frac{t_x + T - \frac{x^T}{N}}{N} \tag{1}$$

– Root TTR

$$\frac{T}{\sqrt{N}} \tag{2}$$

– Corrected TTR

$$\frac{T}{\sqrt{2N}} \tag{3}$$

– Bi-logarithmic TTR

$$\frac{\log T}{\log N} \tag{4}$$

– Text-sentence-deviation TTR

$$\sum_{i=0}^{n} \left(\frac{T}{N} - \frac{t_i}{n_i} \right) \tag{5}$$

4.2   Information-Theoretic Features

Information theory studies statistical laws of how information can be optimally coded [38]. The use of information theory for deriving statistical measures of readability is in line with this approach. It allows us to explore conditional probabilities of random text variables as described in the following.

Entropy Based Features.   We already introduced this feature set in our previous work [1, 32]. The entropy of a random variable is related to the difficulty of correctly guessing the value of the corresponding random variable. A text is more readable if follow-up words are easier to guess. The uncertainty of guessing different text properties can be measured by entropy. The entropy of a random variable X is defined as

$$H(X) = - \sum_{i=1}^{n} p(x_i) \log p(x_i) \tag{6}$$

The more the outcome of X converges towards a uniform distribution, the higher $H(X)$. Our hypothesis is that the higher the entropy, the less readable the text along the feature represented by X. In our experiment, we consider the following random variables: *word probability*, *character probability*, *word length probability* and *word frequency probability*. Note that there is a correlation between the probability distribution of words and the corresponding frequency spectrum of word frequencies. As we use *Support Vector Machines* (SVM) for classification, these correlations are reflected.

Information Transmission-Based Features.   There is a relation between readability, sentence length and word length. Generally speaking, longer sentences tend to describe more entities and thus affect the readability negatively. The same can be said about longer words. Features of this sort can be captured by *joint* and *conditional* probabilities of corresponding random variables.

The joint probability of two random variables measures the likelihood of two events occurring together. The conditional probability measures the probability of an event given that another event occurred. In terms of Shannon's joint entropy we get the following features:

$$H(X, Y) = - \sum_{<x,y> \in X \times Y} p(x_i, y_i) \log p(x_i, y_i) \tag{7}$$

The two conditional entropies can be defined as:

$$H(X|Y) = -\sum_{y \in Y} P(y_i) \sum_{x \in X} p(x_i|y_i) \log p(x_i|y_i) \qquad (8)$$

$$H(Y|X) = -\sum_{x \in X} P(x_i) \sum_{y \in Y} p(y_i|x_i) \log p(y_i|x_i) \qquad (9)$$

From the equation 6, 7, 8 and 9, it can be shown that:

$$T_s(X,Y) = H(X) + H(Y) - H(X,Y) \qquad (10)$$

Function 10 is called information transmission; it measures the strength of the relationship between two random variables X and Y [38, 39]. Borst and Theunissen [40] use this feature to measure the amount of information about a stimulus carried in a neural response. They showed how to use this feature to validate simple stimulus-response models of neural coding of dynamic stimuli. We utilize two information transmission-related features as listed in Table 6: the *transmission of sentence length and word length* and the *transmission of sentence length and the number of long words in sentence*. Long words are words of more than 10 letters – they are supposed to be difficult because of their length. Our hypothesis is that a text with higher values for these features is less readable.

5   Experiments

In order to experiment with different features, we divided the experimental data (see Section 2) into training and test data. Note that twenty data sets were randomly generated such that 80% of the corpus was used for training and 20% for testing. The weighted average of *Accuracy* and *F-score* is computed by considering all data sets. We used the SMO [41, 42] classifier model implemented in WEKA [43] together with the Pearson VII function-based universal kernel PUK [44].

5.1   A Baseline Classifier

In order to make our feature set comparable, we implemented a baseline using five classical readability measures: the *Gunning fog readability index* [6], the *Dale–Chall readability formula* [4, 5], the *Automated readability index* [8], the Flesch-Kincaid readability index [7] and the SMOG readability measure [9]. The latter two measures explore syllable information. Note that no *syllable identification system* is freely available for Bangla texts. Thus, we approximate the number of syllables in a Bangla word by counting its number of vowels. The *Gunning fog readability index* and the *Dale-Chall readability measure* both analyze complex or difficult words, while differing in the definition of these words. We apply a similar strategy as [1] in that we define any word of at least 10 letters to be difficult. Table 3 shows the evaluation of the baseline. The evaluation shows that the baseline features do not perform well for Bangla texts. Among the measures, the *Gunning fog readability index* performs best.

Table 3. Evaluation of a baseline system based on 5 classical readability measures

Features	Accuracy	F-Score
Gunning fog readability index	55.24%	55.19%
Dale–Chall readability formula	55.21%	50.36%
Automated readability index	53.59%	53.29%
Flesch-Kincaid readability index	31.56%	25.40%
SMOG readability	40.60%	33.96%
All together	60.60%	56.49%

Table 4. Evaluation of lexical features

Features	Accuracy	F-Score
Average sentence length	65.55%	61.21%
Type-Token ratio per sentence	48.67%	40.33%
Type-Token ratio per document	53.53%	47.76%
Average difficult words per sentence	58.63%	54.98%
Number of difficult words per document	62.66%	57.12%
Avg. word length	41.93%	35.52%
Corrected TTR	57.84%	50.90%
Koh TTR	53.15%	47.90%
Log TTR	46.85%	41.30%
Root TTR	57.60%	50.55%
Text sentence deviation TTR	51.62%	43.83%
11 Lexical features	76.31%	75.87%

5.2   A Classifier Based on Lexical Features

The lexical features used in our study perform better than the classical readability measures (which also include but are not limited to lexical features). Table 2 shows that the *average sentence length* and the difficulty levels correlate: sentence length increases for higher readability classes. This characteristic is reflected in our experiment. Although Table 2 shows the same for the *average word length*, our experimental results show that this is not a good indicator of readability. Table 4 shows the evaluation of the system using only lexical features. Now, *difficult words* are a good indicator of readability. Although the individual accuracy of some of the lexical features is similar to the classical measures, the combination of all lexical features performs far better than the baseline.

5.3   A Classifier Using Entropy-Related Features

Genzel and Charniak [45, 46] provide evidence for the hypothesis that for certain random variables, the entropy rate is constant in a text. Textbooks are also a medium of communication between book authors and students. Information flow in a book of Grade 2 will be different than information flow in a book of grade nine. Thus, we expect an impact of entropy on measuring readability. Analyzed in isolation, entropy-related features perform on an equal footing as lexical features or classical models of

Table 5. Evaluation of entropy based features

Features	Accuracy	F-Score
Word probability	57.69%	48.44%
Character probability	48.46%	40.37%
Word length probability	48.34%	45.11%
Word frequency probability	50.64%	42.80%
Character frequency probability	58.50%	50.23%
5 Entropy–based features	70.43%	66.60%

readability. However, as a collective, entropy-related features outperform their classical counterpart. Among all entropy-related Features, the *character frequency probability* performs best. Table 5 shows the results of the respective experiment.

5.4   A Classifier Based on Information Transmission-Related Features

Table 6 gives results regarding the evaluation of two information transmission-related features. The transmission of *sentence length and word difficulty* is the best performing feature among all features considered here. Also as group, transmission-related features perform better than the baseline and the entropy-based classifier. However, it is outperformed by the classifier based on lexical features. The *average sentence length* and the *number of difficult words* both perform well as individual features. Their combined joint and conditional probabilities give a good indicator of readability. We get an *F*-score of 86.46% when combining 11 lexical and 7 information-theoretic features. Our experiment indicates that these features are very useful for measuring text readability.

Table 6. Evaluation of information transmission-based features

Features	Accuracy	F-Score
Sentence length and word length probability	61.33%	53.56%
Sentence length and difficult word probability	66.30%	61.73%
2 Information transmission-based features	71.52%	67.02%
7 Information–theoretic features	72.95%	68.80%
18 Lexical + Information–theoretic features	86.60%	86.46%

6   Discussion

Classical readability models as proposed for English texts are not useful for Bangla texts. Sinha et al. [18] and Islam et al. [1] found similar results. However, Das and Roychudhury [16, 17] found that classical readability measures are useful for Bangla readability classification. The reason behind the poor performance of classical measures could be that Bangla scripture contains glyphs which represent clusters and ligatures.

Table 7. Evaluation of features proposed by Sinha et al. [18]

Features	Accuracy	F-Score
Sinha et al. [18] model 3	64.99%	60.61%
Sinha et al. [18] model 4	67.58%	66.18%
Together	74.57%	73.73%

Lexical features constitute the best performing subset of features. Further, the *average sentence length* is a good indicator of readability. As an individual feature, entropy-based features perform on an equal footing as other features. However, the combination of information-theoretic features does not perform better than the one of lexical features. It should be noted that we only used seven information-theoretic features. Adding more information-theoretic features may give a better performance. The classification performance increases up to 86% when information-theoretic features are combined with lexical ones.

Sinha et al. [18] proposed two computational models for Bangla text readability. They proposed models by performing user experiments in which users identified structural characteristics of Bangla texts. We also compared their models with our readability measures. Table 7 summarizes the results. It shows that polysyllabic words and consonant-conjuncts should be considered as features when building readability classifiers for Bangla. These features together perform better than information-theoretic features. However, information-theoretic and lexical features perform far better when being combined. It remains an open question what happens with performance if we additionally consider the feature list of Sinha et al. [18].

7 Conclusion

We described and evaluated features for text readability classification by example of Bangla texts. We distinguished 18 quantitative features that can be extracted from texts without (much) linguistic preprocessing. Seven of them are information-theoretically motivated. Recent advances in NLP tools suggest that linguistic features are useful for readability classification. Regarding this finding, our experimental results show that lexical and information-theoretic features are both very effective and easy to compute. There are still many low-resourced languages around the world. Thus, our feature set may be used for readability classification by example of these languages – not only if preprocessing rarely exist for them.

Acknowledgments. We thank Andy Lücking and the anonymous reviewers for their helpful comments. This work is funded by the LOEWE Digital-Humanities project at Goethe-University Frankfurt.

References

1. Islam, Z., Mehler, A., Rahman, R.: Text readability classification of textbooks of a low-resource language. In: Proceedings of the 26th Pacific Asia Conference on Language, Information, and Computation (2012)
2. Mikk, J.: Text comprehensibility. In: Quantitative Linguistics: An International Handbook, pp. 909–921. Walter de Gruyter (2005)
3. Sebastiani, F.: Machine learning in automated text categorization. ACM Computing Surveys 34(1), 1–47 (2002)
4. Dale, E., Chall, J.S.: A formula for predicting readability. Educational Research Bulletin 27(1), 11–20+28 (1948)
5. Dale, E., Chall, J.S.: Readability Revisited: The New Dale-Chall Readability formula. Brookline Books (1995)
6. Gunning, R.: The Technique of clear writing, Fourh Printing Edition. McGraw-Hill (1952)
7. Kincaid, J., Fishburne, R., Rodegers, R., Chissom, B.: Derivation of new readability formulas for Navy enlisted personnel. Technical report, US Navy, Branch Report 8-75, Cheif of Naval Traning, Millington (1975)
8. Senter, R., Smith, E.A.: Automated readability index. Technical report, Wright-Patterson Air Force Base (1967)
9. McLaughlin, G.H.: SMOG grading – a new readability formula. Journal of Reading 12(8), 639–646 (1969)
10. Hancke, J., Vajjala, S., Meurers, D.: Readability classification for German using lexical, syntactic, and morphological features. In: 24th International Conference on Computational Linguistics (COLING), Mumbai, India (2012)
11. François, T., Fairon, C.: An AI readability formula for french as a foreign language. In: Proceedings of the 2012 Joint Conference on Empirical Methods in Natural Language Processing and Computational Natural Language Learning, pp. 466–477. Association for Computational Linguistics (2012)
12. Sato, S., Matsuyoshi, S., Kondoh, Y.: Automatic assessment of japanese text readability based on a textbook corpus. In: LREC (2008)
13. Chen, Y.T., Chen, Y.H., Cheng, Y.C.: Assessing chinese readability using term frequency and lexical chain. Computational Linguistics and Chinese Language Processing 18(2), 1–17 (2013)
14. Islam, M.Z., Tiedemann, J., Eisele, A.: English to bangla phrase-based machine translation. In: The 14th Annual Conference of The European Association for Machine Translation, Saint-Raphaël, France, May 27-28 (2010)
15. Karim, M., Kaykobad, M., Murshed, M.: Technical Challenges and Design Issues in Bangla Language Processing. IGI Global (2013)
16. Das, S., Roychoudhury, R.: Testing level of readability in Bangla novels of Bankim Chandra Chattopodhay w.r.t the density of polysyllabic words. Indian Journal of Linguistics 22, 41–51 (2004)
17. Das, S., Roychoudhury, R.: Readabilit modeling and comparison of one and two parametric fit: a case study in Bangla. Journal of Quantative Linguistics 13(1) (2006)
18. Sinha, M., Sakshi, S., Dasgupta, T., Basu, A.: New readability measures for Bangla and Hindi texts. In: Proceedings of COLING, pp. 1141–1150 (2012)
19. Fitzsimmons, P., Michael, B., Hulley, J., Scott, G.: A readability assessment of online Parkinson disease information. The Journal of the Royal College of Physicians of Edinburgh 40, 292–296 (2010)
20. Petersen, S.E., Ostendorf, M.: A machine learning approach to reading level assesment. Computer Speech and Language 23(1), 89–106 (2009)

21. Feng, L., Elhadad, N., Huenerfauth, M.: Cognitively motivated features for readability assessment. In: Proceedings of the 12th Conference of the European Chapter of the ACL (2009)
22. Collins-Thompson, K., Callan, J.P.: A language modeling approach to predicting reading difficulty. In: HLT-NAACL (2004)
23. Schwarm, S.E., Ostendorf, M.: Reading level assessment using support vector machines and statistical language models. In: The Proceedings of the 43rd Annual Meeting on Association for Computational Linguistics (ACL 2005) (2005)
24. Aluisio, R., Specia, L., Gasperin, C., Scarton, C.: Readability assessment for text simplification. In: NAACL-HLT 2010: The 5th Workshop on Innovative Use of NLP for Building Educational Applications (2010)
25. Kate, R.J., Luo, X., Patwardhan, S., Franz, M., Florian, R., Mooney, R.J., Roukos, S., Welty, C.: Learning to predict readability using diverse linguistic features. In: 23rd International Conference on Computational Linguistics, COLING 2010 (2010)
26. Eickhoff, C., Serdyukov, P., de Vries, A.P.: A combined topical/non-topical approach to identifying web sites for children. In: Proceedings of the fourth ACM International Conference on Web Search and Data Mining (2011)
27. Pitler, E., Nenkova, A.: Revisiting readability: A unified framework for predicting text quality. In: Proceedings of the Conference on Empirical Methods in Natural Language Processing, EMNLP (2008)
28. Feng, L., Janche, M., Huenerfauth, M., Elhadad, N.: A comparison of features for automatic readability assessment. In: The 23rd International Conference on Computational Linguistics, COLING (2010)
29. Barzilay, R., Lapata, M.: Modeling local coherence: An entity-based approach. Computational Linguistics 21(3), 285–301 (2008)
30. Heilman, M., Collins-Thompson, K., Eskenazi, M.: Combining lexical and grammatical features to improve readavility measures for first and second language text. In: Proceedings of the Human Language Technology Conference (2007)
31. Heilman, M., Collins-Thompson, K., Eskenazi, M.: An analysis of statistical models and features for reading difficulty prediction. In: Proceedings of the Third Workshop on Innovative Use of NLP for Building Educational Applications, EANL (2008)
32. Islam, Z., Mehler, A.: Automatic readability classification of crowd-sourced data based on linguistic and information-theoretic features. Computación y Sistemas 17(2), 113–123 (2013)
33. Vajjala, S., Meurers, D.: On improving the accuracy of readability classification using insights from second language acquisition. In: Proceedings of the Seventh Workshop on Building Educational Applications Using NLP, pp. 163–173. Association for Computational Linguistics (2012)
34. Temnikova, I.: Text Complexity and Text Simplification in the Crisis Management Domain. PhD thesis, University of Wolverhampton (2012)
35. Carroll, J.B.: Language and thought. Prentice-Hall, Englewood Cliffs (1964)
36. Herdan, G.: Quantitative linguistics. Butterworths (1964)
37. Köhler, R., Galle, M.: Dynamic aspects of text characteristics. Quantitative Text Analysis, 46–53 (1993)
38. Cover, T.M., Thomas, J.A.: Elements of Information Theory. Wiley Interscience, Hoboken (2006)
39. Klir, G.J.: Uncertainty and Information. Wiley Interscience (2005)
40. Borst, A., Theunissen, F.E.: Information theory and neural coding. Nature Neuroscience 2, 947–957 (1999)
41. Platt, J.C.: Fast training of support vector machines using sequential minimal optimization. MIT Press (1998)

42. Keerthi, S., Shevade, S.K., Bhattacharyya, C., Murthy, K.R.K.: Improvements to Platt's SMO algorithm for SVM classifier design. Neural Computation 13(3), 637–649 (2001)

43. Hall, M., Frank, E., Holmes, G., Pfahringer, B., Reutemann, P., Witten, I.H.: The WEKA data mining software: an update. ACM SIGKDD Explorations 11(1), 10–18 (2009)

44. Üstün, B., Melssen, W., Buydens, L.: Facilitating the application of support vector regression by using a universal Pearson VII function based kernel. Chemometrics and Intelligent Laboratory Systems 81(1), 29–40 (2006)

45. Genzel, D., Charniak, E.: Entropy rate constancy in text. In: Proceedings of the 40st Meeting of the Association for Computational Linguistics, ACL 2002 (2002)

46. Genzel, D., Charniak, E.: Variation of entropy and parse trees of sentences as a function of the sentence number. In: Proceedings of the Conference on Empirical Methods in Natural Language Processing, EMNLP (2003)

State-of-the-Art in Weighted Finite-State Spell-Checking

Tommi A. Pirinen and Krister Lindén

University of Helsinki
Department of Modern Languages
{tommi.pirinen,krister.linden}@helsinki.fi

Abstract. The following claims can be made about finite-state methods for spell-checking: 1) Finite-state language models provide support for morphologically complex languages that word lists, affix stripping and similar approaches do not provide; 2) Weighted finite-state models have expressive power equal to other, state-of-the-art string algorithms used by contemporary spell-checkers; and 3) Finite-state models are at least as fast as other string algorithms for lookup and error correction. In this article, we use some contemporary non-finite-state spell-checking methods as a baseline and perform tests in light of the claims, to evaluate state-of-the-art finite-state spell-checking methods. We verify that finite-state spell-checking systems outperform the traditional approaches for English. We also show that the models for morphologically complex languages can be made to perform on par with English systems.

Keywords: spell-checking, weighted finite-state technology, error models.

1   Introduction

Spell-checking and correction is a traditional and well-researched part of computational linguistics. Finite-state methods for language models are widely recognized as a good way to handle languages which are morphologically more complex [3]. In this article, we evaluate weighted, fully finite-state spell-checking systems for morphologically complex languages. We use existing finite-state models and algorithms and describe some necessary additions to bridge the gaps and surpass state-of-the-art in non-finite-state spell-checking. For the set of languages, we have chosen to study North Sámi and Finnish from the complex, agglutinative group of languages, Greenlandic from the complex poly-agglutinative group, and English to confirm that our finite-state formulations of traditional spelling correction applications are working as described in the literature.

As contemporary spell-checkers are increasingly using statistical approaches for the task, weighted finite-state models provide the equivalent expressive power, even for the morphologically more complex languages, by encoding the probabilities as weights in the automata. As the programmatic noisy channel models [6] can encode the error probabilities when making the corrections, so can the weighted finite-state automata encode these probabilities.

A. Gelbukh (Ed.): CICLing 2014, Part II, LNCS 8404, pp. 519–532, 2014.

The task of spell-checking is split into two parts, error detection and error correction. Error detection by language model lookup is referred to as non-word or isolated error detection. The task of detecting isolated errors is often considered trivial or solved in many research papers dealing with spelling correction, e.g. [19]. More complex error detection systems may be used to detect words that are correctly spelled, but are unsuitable in the syntactic or semantic context. This is referred to as real-word error detection in context [14].

The task of error-correction is to generate the most likely correct word-forms given a misspelled word-form. This can also be split in two different tasks: generating suggestions and ranking them. Generating corrections is often referred to as error modeling. The main point of error modeling is to correct spelling errors accurately by observing the causes of errors and making predictive models of them [9]. This effectively splits the error models into numerous sub-categories, each applicable to correcting specific types of spelling errors. The most used model accounts for typos, i.e. the slip of a finger on a keyboard. This model is nearly language agnostic, although it can be tuned to each local keyboard layout. The other set of errors is more language and user-specific—it stems from the lack of knowledge or language competence, e.g., in non-phonemic orthographies, such as English, learners and unskilled writers commonly make mistakes such as writing *their* instead of *there*, as they are pronounced alike; similarly competence errors will give rise to common confusable words in other languages, such as missing an accent, writing a digraph instead of its unigraph variant, or confusing one morph with another.

A common source of the probabilities for ranking suggestions related to competence errors are the neighboring words and word-forms captured in a language model. For morphologically complex languages, part-of-speech information is needed [19, 23], which can be compared with the studies on isolating languages [14, 29]. Context-based models like these are, however, considered to be out of scope for spell-checking, rather being part of grammar-checking.

Advanced language model training schemes, such as the use of morphological analyses as error detection evidence [14], require large manually verified morphologically analyzed and disambiguated corpora, which do not exist as open, freely usable resources, if at all. In addition, for polysynthetic languages like Greenlandic, even a gigaword corpus is usually not nearly as complete as an English corpus with a million word-forms.

As we compare existing finite-state technologies with contemporary non-finite-state string algorithm solutions, we use Hunspell[1] as setting the current de facto standard in open-source spell-checking and the baseline for the quality to achieve. Taken together this paper demonstrates for the first time that *using weighted finite-state technology, spell-checking for morphologically complex languages* can be made to *perform on par with English systems* and *surpass the current de facto standard*.

This article is structured as follows: In Subsection 1.1, we briefly describe the history of spell-checking up to the finite-state formulation of the problem. In Subsection 1.2, we revisit the notations behind the statistics we apply to our language and error

---

[1] http://hunspell.sf.net

models. In Section 2, we present existing methods for creating finite-state language and error models for spell-checkers. In Section 3, we present the actual data, the language models, the error models and the corpora we have used, and in Section 4, we show how different languages and error models affect the accuracy, precision, and speed of finite-state spell-checking. In Section 5, we discuss the results, and finally, in Section 6, we conclude our findings.

1.1   A Brief History of Automatic Spell-Checking and Correction

Automatic spelling correction by computer is in itself, an old invention, with the initial work done as early as in the 1960's. Beginning with the invention of the generic error model for typing mistakes, the Levenshtein-Damerau distance [8, 12] and the first applications of the noisy channel model [28] to spell-checking [25], the early solutions treated the dictionaries as simple word lists, or later, word-lists with up to a few affixes with simple stem mutations and finally some basic compounding processes. The most recent and widely spread implementation with a word-list, stem mutations, affixes and some compounding is Hunspell, which is in common use in the open-source world of spell-checking and correction and must be regarded as the reference implementation. The word-list approach, even with some affix stripping and stem mutations, has sometimes been found insufficient for morphologically complex languages. E.g. a recent attempt to utilize Hunspell for Finnish was unsuccessful [24]. In part, the popularity of the finite-state methods in computational linguistics seen in the 1980's was driven by a need for the morphologically more complex languages to get language models and morphological analyzers with recurring derivation and compounding processes [2]. They also provide an opportunity to use arbitrary finite-state automata as language models without modifying the runtime code, e.g. [21].

Given the finite-state representation of the dictionaries and the expressive power of the finite-state systems, the concept of a finite-state based implementation for spelling correction was an obvious development. The earliest approaches presented an algorithmic way to implement the finite-state network traversal with error-tolerance [18] in a fast and effective manner [10, 26]. Schulz and Mihov [27] presented the Levenshtein-Damerau distance in a finite-state form such that the finite-state spelling correction could be performed using standard finite-state algebraic operations with any existing finite-state library. Furthermore, e.g., Pirinen and Lindén [22] have shown that the weighted finite-state methods can be used to gain the same expressive power as the existing statistical spellchecking software algorithms.

1.2   Notations and Some Statistics for Language and Error Models

In this article, where the formulas of finite-state algebra are concerned, we assume the standard notations from Aho et al. [1]: a finite-state automaton $\mathcal{M}$ is a system Q, Σ, δ, Q_s, Q_f, W, where Q is the set of states, Σ the alphabet, δ the transition mapping of form $Q \times \Sigma \rightarrow Q$, and Q_s and Q_f the initial and final states of the automaton, respectively. For weighted automata, we extend the definition in the same way as Mohri [15] such that δ is extended to the transition mapping $Q \times \Sigma \times W \rightarrow Q$, where W is the weight, and the system additionally includes a final weight mapping $\rho : Q_f \rightarrow W$.

The structure we use for weights is systematically the tropical semiring ($R_+U+\infty$, min, +, $+\infty$, 0), i.e. weights are positive real numbers that are collected by addition. The tropical semiring models penalty weighting.

For the finite-state spell-checking, we use the following common notations: $\mathcal{M}_D$ is a single tape weighted finite-state automaton used for detecting the spelling errors, $\mathcal{M}_S$ is a single tape weighted finite-state automaton used as a language model when suggesting correct words, where the weight is used for ranking the suggestions. On many occasions, we consider the possibility that $\mathcal{M}_D = \mathcal{M}_S$. The error models are weighted two-tape automata commonly marked as $\mathcal{M}_E$. A word automaton is generally marked as $\mathcal{M}_{word}$. A misspelling is detected by composing the word automaton with the detection automaton:

$$\mathcal{M}_{word} \circ \mathcal{M}_D \tag{1}$$

which results in an empty automaton on a misspelling and a non-empty automaton on a correct spelling. The weight of the result may represent the likelihood or the correctness of the word-form. Corrections for misspelled words can be obtained by composing a misspelled word, an error model and a model of correct words:

$$\mathcal{M}_{word} \circ \mathcal{M}_E \circ \mathcal{M}_S \tag{2}$$

which results in a two-tape automaton consisting of the misspelled word-form mapped to the spelling corrections described by the error model $\mathcal{M}_E$ and approved by the suggestion language model $\mathcal{M}_S$. Both models may be weighted and the weight is collected by standard operations as defined by the effective semiring.

Where probabilities are used, the basic formula to estimate probabilities from discrete frequencies of events (word-forms, mistyping events, etc.) is as follows: $P(x) = c(x)/corpussize$, where x is the event, c is the count or frequency of the event, and $corpussize$ is the sum of all event counts in the training corpus. The encoding of probability as tropical weights in a finite-state automaton is done by setting $Q_{\pi_x} = -\log P(x)$, where Q_{π_x} is the end weight of path π_x, though in practice the weight may be distributed along the path depending on the specific implementation. As events not appearing in corpora should have a larger probability than zero, we use additive smoothing $P(\hat{x}) = (c(x) + \alpha)/(corpussize \times (1 + \alpha))$, so for an unknown event $\hat{x}$, the probability will be counted as if it had α appearances. Another approach would be to set $P(\hat{x}) < 1/corpussize$, which makes the probability distribution leak but may work under some conditions [5].

1.3   Morphologically Complex Resource-Poor Languages

One of the main reasons for going fully finite-state instead of relying on word-form lists and affix stripping is the claim that morphologically complex languages simply cannot be handled with sufficient coverage and quality using traditional methods. While Hunspell has virtually 100 % domination of the open-source spell-checking field, authors of language models for morphologically complex languages such as

Turkish (cf. Zemberek[2]) and Finnish (cf. Voikko[3]) have still opted to write separate software, even though it makes the usage of their spell-checkers troublesome and the coverage of supported applications much smaller.

Another aspect of the problems with morphologically complex languages is that the amount of training data in terms of running word-forms is greater, as the amount of unique word-forms in an average text is much higher compared with morphologically less complex languages. In addition, the majority of morphologically complex languages tend to have fewer resources to train the models. For training spelling checkers, the data needed is merely correctly written unannotated text, but even that is scarce when it comes to languages like Greenlandic or North Sámi. Even a very simple probabilistic weighting using a small corpus of unverified texts will improve the quality of suggestions [22], so having a weighted language model is more effective.

2   Weighting Finite-State Language and Error Models

The task of spell-checking is divided into locating spelling errors, and suggesting the corrections for the spelling errors. In finite-state spell-checking, the former task requires a language model that can tell whether or not a given string is correct. The error correction requires two components: a language model and an error model.

The error model is a two-tape finite-state automaton that can encode the relation between misspellings and the correctly typed words. This relation can also be weighted with the probabilities of making a specific typo or error, or arbitrary hand-made penalties as with many of the traditional non-finite-state approaches, e.g. [16].

The rest of this section is organized as follows. In Subsection 2.1, we describe how finite-state language models are made. In Subsection 2.2, we describe how finite-state error models are made. In Subsection 2.3, we describe some methods for combining the weights in language models with the weights in error models.

2.1   Compiling Finite-State Language Models

The baseline for any language model as realized by numerous spell-checking systems and the literature is a word-list (or a word-form list). One of the most popular examples of this approach is given by Norvig [17], describing a toy spelling corrector being made during an intercontinental flight. The finite-state formulation of this idea is equally simple; given a list of word-forms, we compile each string as a path in an automaton [20]. In fact, even the classical optimized data structures used for efficiently encoding word lists, like tries and acyclic deterministic finite-state automata, are usable as finite-state automata for our purposes, without modifications. We have: $\mathcal{M}_D = \mathcal{M}_S = \cup_{wf \in corpus} wf$, where wf is a word-form and *corpus* is a set of word-forms in a corpus. These are already valid language models for Formula 1 and 2, but in practice any finite-state lexicon [3] will suffice.

---

[2] http://code.google.com/p/zemberek
[3] http://voikko.sf.net

2.2   Compiling Finite-State Versions of Error Models

The baseline error model for spell-checking is the Damerau-Levenshtein distance measure. As the finite-state formulations of error models are the most recent development in finite-state spell-checking, the earliest reference to a finite-state error model in an actual spell-checking system is by Schulz and Mihov [27]. It also contains a very thorough description of building finite-state models for different edit distances. As error models, they can be applied in Formula 2. The edit distance type error models used in this article are all simple edit distance models.

One of the most popular modifications to speed up the edit distance algorithm is to disallow modifications of the first character of the word [4]. This modification provides a measurable speed-up at a low cost to recall. The finite-state implementation of it is simple; we concatenate one unmodifiable character in front of the error model.

Hunspell's implementation of the correction algorithm uses configurable alphabets for the error types in the edit distance model. The errors that do not come from regular typing mistakes are nearly always covered by specific string transformations, i.e. confusion sets. Encoding a simple string transformation as a finite-state automaton can be done as follows: for any given transformation $S : U$, we have a path $\pi_{S:U} = S_1:U_1 S_2:U_2 \ldots S_n:U_n$, where $n = \max(|S|, |U|)$ and the missing characters of the shorter word substituted with epsilons. The path can be extended with arbitrary contexts $L, R \in \Sigma^*$, by concatenating those contexts on the left and right, respectively. To apply these confusion sets on a word using a language model, we use the following formula: $\mathcal{M}_E = \cup_{S:U \in CP} S:U$, where CP is a set of confused string pairs. The error model can be applied in a standard manner in Formula 2. For a more detailed description of a finite-state implementation of Hunspell error models, see Pirinen and Linden [22].

2.3   Combining Weights from Different Sources and Different Models

As both our language and error models are weighted automata, the weights need to be combined when applying the error and the language models to a misspelled string. Since the application performs what is basically a finite-state composition as defined in Formula 2, the default outcome is a weight semiring multiplication of the values; i.e., a real number addition in the tropical semiring. This is a reasonable way to combine the models, which can be used as a good baseline. In many cases, however, it is preferable to treat the probabilities or the weights drawn from different sources as unequal in strength. For example, in many of the existing spelling-checker systems, it is preferable to first suggest all the corrections that assume only one spelling error before the ones with two errors, regardless of the likelihood of the word forms in the language model. To accomplish this, we scale the weights in the error model to ensure that any weight in the error model is greater than or equal to any weight in the language model: $\hat{w}_e = w_e + \text{maxw}(\mathcal{M}_S)$, where $\hat{w}_e$ is the scaled weight of error model weights, w_e the original error model weight and $\text{maxw}(\mathcal{M}_S)$ the maximum weight found in the language model used for error corrections.

3   The Language and Error Model Data Used for Evaluation

To evaluate the weighting schemes and the language and the error models, we have selected two of the morphologically more complex languages with little to virtually no corpus resources available: North Sámi and Greenlandic. Furthermore, as a morphologically complex language with moderate resources, we have used Finnish. As a comparative baseline for a morphologically simple language with huge corpus resources, we use English. English is also used here to reproduce the results of the existing models to verify functionality of our selected approach.

This section briefly introduces the data and methods to compile the models; for the exact implementation, for any reproduction of results or for attempts to implement the same approaches for another language, the reader is advised to utilize the scripts, the programs and the makefiles available at our source code repository.[4]

Table 1. The extent of Wikipedia data per language

Data: Language	Train tokens	Train types	Test tokens	Test types
English	276,730,786	3,216,142	111,882,292	1,945,878
Finnish	9,779,826	1,065,631	4,116,896	538,407
North Sámi	183,643	38,893	33,722	8,239
Greenlandic	136,241	28,268	7,233	1,973

In Table 1, we show the statistics of the data we have drawn from Wikipedia for training and testing purposes. In case of English and Finnish, the data is selected from a subset of Wikipedia test tokens. With North Sámi and Greenlandic, we had no other choice but to use all Wikipedia test tokens.

For the English language model, we use the data from Norvig [17] and Pirinen and Hardwick [20], which is a basic language model based on a frequency weighted wordlist extracted from freely available Internet corpora such as Wikipedia and project Gutenberg. The language models for North Sámi, Finnish and Greenlandic are drawn from the free/libre open-source repository of finite-state language models managed by the University of Tromsø.[5] The language models are all based on the morphological analyzers built in the finite-state morphology [3] fashion. The repository also includes the basic versions of finite-state spell-checking under the same framework that we use in this article for testing. To compile our dictionaries, we have used the makefiles available in the repository. The exact methods for this are also detailed in the source code of the repository.

The error models for English are combined from a basic edit distance with English alphabet a-z and the confusion set from Hunspell's English dictionary containing 96 confusion pairs[6]. The error models for North Sámi, Finnish and Greenlandic are the

---

[4] https://github.com/flammie/purplemonkeydishwasher/
tree/master/fst-spell-journal
[5] http://giellatekno.uit.no/
[6] The file en-US.aff is found in the Ubuntu Linux LTS 12.04 distribution.

edit distances of English with addition of åäöšžčŋđ and ŧ for North Sámi and åäöšž for Finnish. For North Sámi we also use the actual Hunspell parts from the divvun speller[7]; for Greenlandic, we have no confusion sets or character likelihoods for Hunspell-style data, so only the ordering of the Hunspell correction mechanisms is retained. For English, the Hunspell phonemic folding scheme was not used. This makes the English results easier to compare with those of other languages, which do not even have any phonemic error sources.

The keyboard adjacency weighting and optimization for the English error models is based on a basic qwerty keyboard. The keyboard adjacency values are taken from the CLDR Version 22[8], modified to the standard 101—104 key PC keyboard layout.

The training corpora for each of the languages are based on Wikipedia. To estimate the weights in the models, we have used the correct word-forms of the first 90 % of Wikipedia for the language model and the non-words for the error model. We used the remaining 10 % for extracting non-words for testing. The error corpus was extracted with a script very similar to the one described by Max and Wisniewski [13]. The script that performs fetching and cleaning can be found in our repository.[9] We have selected the spelling corrections found in Wikipedia by only taking those, where the incorrect version does not belong to the language model (i.e. is a non-word error), and the corrected word-form does.

3.1   The Models Used for Evaluation

The finite-state language and error models described in this article have a number of adjustable settings. For weighting our language models, we have picked a subset of corpus strings for estimating word form probabilities. As both North Sámi and Greenlandic Wikipedia were quite limited in size, we used all strings except those that appear only once (hapax legomena) whereas for Finnish, we set the frequency threshold to 5, and for English, we set it to 20. For English, we also used all word-forms in the material from Norvig's corpora, as we believe that they are already hand-selected to some extent.

As error models, we have selected the following combinations of basic models: the basic edit distance consisting of homogeneously weighted errors of the Levenshtein-Damerau type, the same model limited to the non-first positions of the word, and the Hunspell version of the edit distance errors (i.e. swaps only apply to adjacent keys, and deletions and additions are only tried for a selected alphabet).

4   The Speed and Quality of Different Finite-State Models and Weighting Schemes

To evaluate the systems, we have used a modified version of the HFST spell-checking tool hfst-ospell-survey 0.2.4[10] otherwise using the default options, but for the speed

---

[7] http://divvun.no
[8] http://cldr.unicode.org
[9] https://github.com/flammie/purplemonkeydishwasher/
     tree/master/fst-spell-journal/
[10] http://sf.net/p/hfst/

measurements we have used the --profile argument. The evaluation of speed and memory usage has been performed by averaging over five test runs on a dedicated test server: an Intel Xeon E5450 at 3 GHz, with 64 GB of RAM memory. The rest of the section is organized as follows: in Subsection 4.1, we show naïve coverage baselines. In Subsection 4.2, we measure the quality of spell-checking with real-world spelling corrections found in Wikipedia logs. Finally in Subsections 4.3 and 4.4, we provide the speed and memory efficiency figures for these experiments, respectively.

4.1   Coverage Evaluation

To show the starting point for spell-checking, we measure the coverage of the language models. That is, we measure how much of the test data can be recognized using only the language models, and how many of the word-forms are beyond the reach of the models. The measurements in Table 2 are measured over word-forms in running text that can be measured in reasonable time, i.e. no more than the first 1,000,000 word-forms of each test corpus. As can be seen in Table 2, the task is very different for languages like English compared with morphologically more complex languages.

Table 2. The word-form coverage of the language models on test data (in %)

English aspell	22.7
English full automaton	80.1
Finnish full automaton	64.8
North Sámi Hunspell	34.4
North Sámi full automaton	48.5
Greenlandic full automaton	25.3

4.2   Quality Evaluation

To measure the quality of spell-checking, we have run the list of misspelled words through the language and error models of our spelling correctors, extracting all the suggestions. The quality, in Table 3, is measured by the proportion of correct suggestions appearing at a given position 1-5 and finally the proportion appearing in any remaining positions.

On the rows indicated with *error*, in Table 3, we present the baselines for using language and error models allowing one edit in any position of the word. The rows with "non-first error" show the same error models with the restriction that the first letter of the word may not be changed.

Finally, in Table 3, we also compare the results of our spell-checkers with the actual systems in everyday use, i.e. the Hunspell and aspell in practice. When looking at this comparison, we can see that for English data, we actually provide an overall improvement already by allowing only one edit per word. This is mainly due to the weighted language model which works very nicely for languages like English.

The data on North Sámi on the other hand shows no meaningful improvement neither with the change from Hunspell to our weighted language models nor with the restriction of the error models.

Table 3. The effect of different language and error models on correction quality (precision in % at a given suggestion position)

Rank:	1st	2nd	3rd	4th	5th	rest
Language and error models						
English aspell	55.7	5.7	8.0	2.2	0.0	0.0
English Hunspell	59.3	5.8	3.5	2.3	0.0	0.0
English w/ 1 *error*	66.7	7.0	5.2	1.8	1.8	1.8
English w/ 1 non-first error	66.7	8.8	7.0	0.0	0.0	1.8
Finnish aspell	21.1	5.8	3.8	1.9	0.0	0.0
Finnish w/ 1 *error*	54.8	19.0	7.1	0.0	0.0	0.0
Finnish w/ 1 non-first error	54.8	21.4	4.8	0.0	0.0	0.0
North Sámi Hunspell	9.4	3.1	0.0	3.1	0.0	0.0
North Sámi w/ 1 *error*	3.5	3.5	0.0	6.9	0.0	0.0
North Sámi w/ 1 non-first error	3.5	3.5	0.0	6.9	0.0	0.0
Greenlandic w/ 1 *error*	13.3	2.2	6.7	2.2	0.0	8.9
Greenlandic w/ 1 non-first error	13.3	2.2	6.7	2.2	0.0	8.9

Some of the trade-offs are efficiency versus quality. In Table 3, we measure among other things the quality effect of limiting the search space in the error model. It is important to contrast these results with the speed or memory gains shown in the corresponding Tables 4 and 5. As we can see, the optimizations that limit the search space will generally not have a big effect on the results. Only the results that get cut out of the search space are moved. A few of the results disappear or move to worse positions.

4.3   Speed Evaluation

For practical spell-checking systems, there are multiple levels of speed requirements, so we measure the effects of our different models on speed to see if the optimal models can actually be used in interactive systems, off-line corrections, or just batch processing. In Table 4, we show the speed of different model combinations for spell-checking—for a more thorough evaluation of the speed of the finite-state language and the error models we refer to Pirinen et al. [23]. We perform three different test sets: startup time tests to see how much time is spent on startup alone; a running corpus processing test to see how well the system fares when processing running text; and a non-word correcting test, to see how fast the system is when producing corrections for words. For each test, the results are averaged over at least 5 runs.

Table 4. The effect of different language and error models on speed of spelling correction (startup time in seconds, correction rate in words per second)

Input:	1st word	all words	non-words
Language and error models			
English Hunspell	0.5	174	40
English w/ 1 *error*	0.06	5,721	6,559
English w/ 1 non-first error	0.20	16,474	17,911
Finnish aspell	<0.1	781	686
Finnish w/ 1 *error*	1.0	166	357
Finnish w/ 1 non-first error	1.0	303	1,886
North Sámi Hunspell	4.51	3	2
North Sámi w/ 1 *error*	0.28	2,304	2,839
North Sámi w/ 1 non-first error	0.27	5,025	7,898
Greenlandic w/ 1 *error*	1.27	49	142
Greenlandic w/ 1 non-first error	1.25	85	416

Table 5. The peak memory usage of processes checking and correcting word-forms with various language and error model combinations (memory in base 10 megabytes)

Measurement:	Peak memory usage
Language and error models	
English Hunspell	7.5 MB
English w/ 1 *error*	7.0 MB
English w/ 1 non-first error	7.0 MB
Finnish aspell	186 kB
Finnish w/ 1 *error*	79.3 MB
Finnish w/ 1 non-first error	79.3 MB
North Sámi Hunspell	151.0 MB
North Sámi w/ 1 *error*	31.4 MB
North Sámi w/ 1 non-first error	31.4 MB
Greenlandic w/ 1 *error*	300.0 MB[11]
Greenlandic w/ 1 non-first error	300.7 MB

In Table 4, we already notice an important aspect of finite-state spelling correction: the speed is very predictable, and in the same ballpark regardless of input data. Furthermore, we can readily see that the speed of a finite-state system in general outperforms Hunspell with both of the language models we compare. Furthermore, we show the speed gains achieved by cutting the search space to disallow errors in the first character of a word. This is the speed-equivalent of Table 3 of the previous section, which clearly shows the trade-off between speed and quality.

---

[11] Drobac & al. [30] report on how to hyper-minimize finite-state lexicons keeping the Greenlandic lexicon at less than 20 MB at runtime which gives a considerable speed-up of loading with only a small reduction of runtime speed.

4.4   Memory Usage Evaluation

Depending on the use case of the spell-checker, memory usage may also be a limiting factor. To give an idea of the memory-speed trade-offs that different finite-state models entail, in Table 5, we provide the memory usage values when performing the evaluation tasks above. The measurements are performed with the Valgrind utility and represent the peak memory usage. It needs to be emphasized that this method, like all of the methods of measuring memory usage of a program, has its flaws, and the figures can at best be considered rough estimates.

5   Discussion

The improvement of quality by using simple probabilistic features for spell-checking is well-studied, e.g. by Church and Gale [7]. In our work, we describe introducing probabilistic features into a finite-state spell-checking system giving a similar increase in the quality of the spell-checking suggestions as seen in previous approaches. The methods are usable for a morphologically varied set of languages.

The speed to quality trade-off is a well-known feature in spell-checking systems, and several aspects of it have been investigated in previous research. The concept of cutting away string initial modifications from the search space has often been suggested [4, 11], but only rarely quantified extensively. In this paper we have investigated its effects on finite-state systems and complex languages. We noted that it gives speed improvements in line with previous solutions, and we also verified that the quality deterioration on real-world data is minimal.

In this paper, we have reviewed basic finite-state language and error models for spelling correction. The obvious future improvements that need to be researched are extensions, e.g. to errors at edit distance 2 or more, as well as more elaborate models for both model types. The combination of adaptive technologies based on user feedback at runtime and finite-state models has not been researched in spelling correction, but it has shown good results in practical spelling correction applications.

6   Conclusion

We have demonstrated that finite-state spell-checking is a feasible alternative to traditional string algorithm-driven versions by verifying three claims. The language support has been demonstrated by the fact that there is a working implementation of Greenlandic that could not have been successfully implemented without finite-state models, and by giving finite-state versions of the North Sámi and English language models that cover more Wikipedia word forms than the non-finite-state equivalents. In addition, the suggestion mechanism using a weighted finite-state implementation is able to provide better quality suggestions for English and Finnish than corresponding non-finite-state implementations. The efficiency of the finite-state approach is verified by showing a reasonable or greater speed when compared with Hunspell.

Acknowledgements. We thank our fellow researchers in the HFST research group at the University of Helsinki for ideas and discussions.

References

1. Aho, A.V., Lam, M.S., Sethi, R., Ullman, J.D.: Compilers: principles, techniques, and tools, vol. 1009. Pearson/Addison Wesley (2007)
2. Beesley, K.R.: Morphological analysis and generation: A first step in natural language processing. In: First Steps in Language Documentation for Minority Languages: Computational Linguistic Tools for Morphology, Lexicon and Corpus Compilation, Proceedings of the SALTMIL Workshop at LREC, pp. 1–8 (2004)
3. Beesley, K.R., Karttunen, L.: Finite State Morphology. CSLI Publications (2003)
4. Bhagat, M.: Spelling Error Pattern Analysis of Punjabi Typed Text. Master's thesis, Thapar University (2007)
5. Brants, T., Popat, A.C., Xu, P., Och, F.J., Dean, J.: Large language models in machine translation. In: EMNLP (2007)
6. Brill, E., Moore, R.C.: An improved error model for noisy channel spelling correction. In: ACL 2000: Proceedings of the 38th Annual Meeting on Association for Computational Linguistics, pp. 286–293. Association for Computational Linguistics, Morristown (2000)
7. Church, K.W., Gale, W.A.: Probability scoring for spelling correction. Statistics and Computing 1, 93–103 (1991)
8. Damerau, F.J.: A technique for computer detection and correction of spelling errors. Communications of the ACM 7(3), 171–176 (1964)
9. Deorowicz, S., Ciura, M.G.: Correcting spelling errors by modelling their causes. International Journal of Applied Mathematics and Computer Science 15(2), 275 (2005)
10. Huldén, M.: Foma: a finite-state compiler and library. In: Proceedings of the 12th Conference of the European Chapter of the Association for Computational Linguistics: Demonstrations Session, pp. 29–32. Association for Computational Linguistics (2009)
11. Kukich, K.: Techniques for automatically correcting words in text. ACM Comput. Surv. 24(4), 377–439 (1992)
12. Levenshtein, V.I.: Binary codes capable of correcting deletions, insertions, and reversals. Soviet Physics—Doklady 10, 707–710 (1966); Translated from Doklady Akademii Nauk SSSR, 845–848
13. Max, A., Wisniewski, G.: Mining naturally-occurring corrections and paraphrases from wikipedia's revision history. In: Proceedings of LREC (2010)
14. Mays, E., Damerau, F.J., Mercer, R.L.: Context based spelling correction. Inf. Process. Manage. 27(5), 517–522 (1991)
15. Mohri, M.: Weighted automata algorithms. In: Handbook of Weighted Automata, pp. 213–254 (2009)
16. Németh, L.: Hunspell manual. Electronic Software Manual (manpage) (2011)
17. Norvig, P.: How to write a spelling corrector (2010), http://norvig.com/spell-correct.html (referred January 11, 2011)
18. Oflazer, K.: Error-tolerant finite-state recognition with applications to morphological analysis and spelling correction. Comput. Linguist. 22(1), 73–89 (1996)
19. Otero, J., Graña, J., Vilares, M.: Contextual spelling correction. In: Moreno Díaz, R., Pichler, F., Quesada Arencibia, A. (eds.) EUROCAST 2007. LNCS, vol. 4739, pp. 290–296. Springer, Heidelberg (2007)
20. Pirinen, T.A., Hardwick, S.: Effects of weighted finite-state language and error models on speed and efficiency of finite-state spell-checking. In: FSMNLP 2012, pp. 6–14. University of the Basque Country (2012)

21. Pirinen, T.A., Lindén, K.: Finite-state spell-checking with weighted language and error models. In: Proceedings of the Seventh SaLTMiL Workshop on Creation and Use of Basic Lexical Resources for Less-Resourced Languages, Valletta, Malta, pp. 13–18 (2010)
22. Pirinen, T.A., Lindén, K.: Creating and weighting Hunspell dictionaries as finite-state automata. Investigationes Linguisticae 21 (2010)
23. Pirinen, T.A., Silfverberg, M., Lindén, K.: Improving finite-state spellchecker suggestions with part of speech n-grams. In: CICLING (2012)
24. Pitkänen, H.: Hunspell-in kesäkoodi 2006: Final report. Technical report (2006), http://www.puimula.org/htp/archive/kesakoodi2006-report.pdf (referred on September 16)
25. Raviv, J.: Decision making in Markov chains applied to the problem of pattern recognition. IEEE Transactions on Information Theory 13(4), 536–551 (1967)
26. Savary, A.: Typographical nearest-neighbor search in a finite-state lexicon and its application to spelling correction. In: Watson, B.W., Wood, D. (eds.) CIAA 2001. LNCS, vol. 2494, pp. 251–260. Springer, Heidelberg (2003)
27. Schulz, K., Mihov, S.: Fast string correction with Levenshtein-automata. International Journal of Document Analysis and Recognition 5, 67–85 (2002)
28. Shannon, C.E.: A mathematical theory of communications, i and ii. Bell Syst. Tech. J. 27, 379–423 (1948)
29. Wilcox-O'Hearn, A., Hirst, G., Budanitsky, A.: Real-word spelling correction with trigrams: A reconsideration of the Mays, Damerau, and Mercer model. In: Gelbukh, A. (ed.) CICLing 2008. LNCS, vol. 4919, pp. 605–616. Springer, Heidelberg (2008)
30. Drobac, S., Lindén, K., Pirinen, T., Silfverberg, M.: Heuristic Hyperminimization of Finite-State Lexicons. In: The Proceedings of LREC, Reykavik, Iceland (2014)

Spelling Correction for Kazakh

Aibek Makazhanov, Olzhas Makhambetov,
Islam Sabyrgaliyev, and Zhandos Yessenbayev

Nazarbayev University
Research and Innovation System
Astana, Kazakhstan
{aibek.makazhanov,omakhambetov,
islam.sabyrgaliyev,zhyessenbayev}@nu.edu.kz

Abstract. Being an agglutinative language Kazakh imposes certain difficulties
on both recognition of correct words and generation of candidate corrections
for misspelled words. In this paper we describe a spelling correction method for
Kazakh that takes advantage of both morphological analysis and noisy channel-
based model. Our method outperforms both open source and commercial analogues in terms of the overall accuracy. We performed a comparative analysis of
the spelling correction tools and pointed out some problems of spelling correction
for agglutinative languages in general and for Kazakh in particular.

1 Introduction

Kazakh is an agglutinative language that belongs to the Turkic group. It has a complex
and productive derivational and inflectional morphology. While extensive research has
been conducted into the morphological analysis of agglutinative languages [1–3], not
too much work has been accomplished in building tools for the analysis of Kazakh.
Being one of the oldest problems in NLP with arguably the highest demand for a practical solution, automatic spelling correction is one of the basic steps in the analysis
of any language. In this paper we describe a spelling correction method for Kazakh
language that takes advantage of both statistical and rule based approaches. Note that
fixing punctuation, structure, or stylistics is out of the scope of the present work.

The spelling correction can be divided into two tasks: word recognition and error
correction. For languages with a fairly straightforward morphology recognition may be
reduced to a trivial dictionary look up: if a given word is absent from the dictionary,
then most likely it had been misspelled. Correction is done through generating a list of
possible suggestions: usually words within some minimal edit distance to a misspelled
word. For agglutinative languages, such as Kazakh, even recognizing misspelled words
becomes challenging as a single root may produce hundreds of word forms[1]. It is practically infeasible to construct a dictionary with all possible word forms included: apart
from being gigantic such a dictionary would be all but verifiable. For the same reason
the correction task becomes challenging as well.

---

[1] In Kazakh, for example, nominals (nouns, pronouns, participles) produce up to 157 forms, and
verbs - up to 840 [4]. Not to mention derivational suffixes that transform POS of a word upon
inflection.

A. Gelbukh (Ed.): CICLing 2014, Part II, LNCS 8404, pp. 533–541, 2014.

One possible solution that has been successfully applied to agglutinative languages in the past [3, 5] is to use a mixture of lexicon-based and generative approaches by keeping a lexicon of roots and generating word forms from that lexicon on the fly. This method requires a generator of word forms. Following the approach presented by Oflazer and Güzey [3], we developed a word forms generator (referred to as the generator hereinafter) and extended it to implement a tool for isolated-word (i.e. context insensitive) error correction for Kazakh language. We implemented the generator as an FSA whose states correspond to morphemes and transitions correspond to morphological rules. If a given word cannot be generated it is considered a misspelling, and for such a misspelled word our method generates a list of possible corrections. To rank such a list we use a Bayesian argument that combines error and source models. For our error model we employ a noisy channel-based approach proposed by Church and Gale [6]. Our source model is built upon the theoretical aspects that were used for morphological disambiguation in [2].

The developed tool is evaluated in terms of general accuracy, top-k precision, and false positive rate. For the purpose of comparison we also experiment with Kazakh spelling dictionary (KSD) [7], an open source Kazakh spelling corrector and Microsoft Office 2010 (MSO) Kazakh language pack [8]. We show that although our method is more accurate than the open source and commercial analogues, the process of generation of candidate corrections still needs improvement in terms of pruning and better ranking of suggestion lists.

The rest of this paper is organized as follows. Subsection 1.1 briefly outlines our contribution. Section 2 reviews related work. Section 3 thoroughly describes our methodology. Section 4 describes the experimental set up and analyzes the results. Finally, we draw conclusions and discuss future work in Sect. 5.

1.1 Our Contribution

Our contribution can be summarized in two following statements: (i) we have built one of the first morphological disambiguators for Kazakh language that can be used to generate and segment word forms; (ii) based on the disambiguator a spelling correction tool was implemented.

2 Related Work

A large number of studies have been performed on spelling correction problem. Some early approaches were based on comparing a misspelled word to words in a lexicon and suggesting as possible corrections the ones with the minimal edit distance [9, 10]. Another popular approach used in more recent works [6, 11] is based on applying a noisy channel model [12], which consists of a source model and a channel model. These works differ in the way how authors weigh the edit operations and in context-awareness of the source models. While Church and Gale [6] utilize word trigram model, Mays et al. [11] do not consider context. Later Brill and Moore [13] proposed an improved method with more sophisticated error model, where instead of using single insertions, deletions, substitutions and transpositions, the authors model substitutions of up to 5-letter sequences that also depend on the position in the word. An interesting method based on neural networks were proposed by Hodge and Austin [14]. The authors use modular neural system AURA [15], where for checking/correction they employ two correlation matrix memories: one trained on patterns derived from handling typing errors by binary Hamming distance and n-grams shifting, and another trained on patterns derived from handling phonetic spelling errors. Ranking suggested corrections is accomplished by choosing the maximum score obtained from the addition of the scores for Hamming distance and n-grams shifting with the score for phonetic modules.

A classical approach to spelling correction for agglutinative languages is to use FSAs [3, 16, 17]. One of the pioneering works that uses finite state automata for spell checking were presented by Oflazer and Güzey [3]. In the proposed method candidate words are generated using two-level transducers. To optimize the recognizer the authors prune the paths that generate the substrings of the candidate words which do not pass some editing distance threshold. Ordering of suggested corrections is accomplished by employing ranking techniques based on the statistics of the types of typing errors. In a more recent work presented by Pirinen et al. [17], the authors use two weighted FSAs one for language model and second for error model, where the authors reorder corrections by using POS n-gram probabilities for a given word.

One of the tools we compare our method to, Kazakh spelling dictionary (KSD) was developed by Mussayeva [7] and freely available in the form of add-ons to various Mozilla products [18] and OpenOffice extension [19]. KSD is based on Hunspell [5], an open source spelling corrector originally developed for Hungarian. To work with any language Hunspell needs a dictionary and an affix table designed for that language. These two are the essence of KSD. The author reports 51 suffix types included to the affix table [7]. These types represent large grammatical groups like noun case, person suffixes, and verb tense suffixes, etc. Given a word Hunspell searches its dictionary to determine if it is correct. If a given word is not found, Hunspell derives possible word forms by appending suitable suffixes. Each word form consists of a root and only one (at least with default settings) appended suffix. In Kazakh, however, suffixes can be, and usually are, appended into longer chains. In the KSD affix table this issue has been partially overcome by collapsing shorter suffix chains into a single suffix.

3   Methodology

Given an input word, the fundamental task of spelling correction is to determine if it is correct, and if not, to offer a list of corrections. Intuitively to solve the problem one could try to identify the root of a given word and then generate a list of all possible word forms that can be transformed into the target word using no more than some maximum number of edit operations (usually two). If there is a generated word form that has a zero edit distance to the target, then the given word is considered correct. Typically word forms are generated by appending morphemes (Kazakh has only suffixes) to roots (root-first fashion) and to each other using some sort of automaton. This is exactly how Oflazer and Güzey [3] solve the problem for Turkish. However, there is a potential drawback. There are usually more than one candidate roots, especially given that a word can be misspelled. Thus, the same morpheme chain can be generated for several candidate roots involving extra computation. In contrast, if morpheme chains

Algorithm 1. The procedure of generating correction suggestions

---

Require: WRD, MAX_ED {The procedure takes as input a target word and maximum edit distance threshold}
  $correct \Leftarrow$ **false**
  if $WRD \in Lexicon$ **then**
    $correct \Leftarrow$ **true** {Already correct}
  end if
  $suggestions \Leftarrow []$
  $chains \Leftarrow ['']$ {We start from a single null-morpheme chain}
  while $correct =$ **false** and $chains \neq \emptyset$ **do**
    $current\_states \Leftarrow chains$
    $chains \Leftarrow []$ {Empty morpheme chains list}
    for all $c \in$ **fsaGetChains**($current\_states$) **do**
      if withinDistance(c, WRD, MAX_ED) $=$ **true then**
        appendToList($chains, c$) {Append current chain to the list, if there is such a suffix SFX of WRD that editDistance(c, SFX) $\leq MAX\_ED$}
        for all $r \in$ **lexGetRoots**(c) **do**
          $s \Leftarrow$ stringConcat(r, c) {Append the current morpheme chain to a candidate root to get a suggestion}
          if editDistance(s, WRD) $= 0$ **then**
            $correct \Leftarrow$ **true**
          end if
          if editDistance(s, WRD) $\leq MAX\_ED$ **then**
            appendToList($suggestions, s$)
          end if
        end for
      end if
    end for
  end while
  return $correct, suggestions$

---

are generated first (morpheme-first fashion) and then the root lexicon is searched for acceptable[2] entries any given chain is generated only once. Let us consider an example for English. Suppose the following list of corrections was generated: *merci + ful + ly*, *peace + ful + ly*, *beauti + ful + ly*. In both root- and morpheme-first approaches the root lexicon would have been searched once to get three candidate roots. However, in the root-first fashion the morpheme chain *ful + ly* would have been generated three times (once for each candidate root), and in the morpheme-first fashion - only once.

Following Oflazer and Güzey [3] we build the generator FSA, with the exception of using morpheme-first approach. We have considered all inflectional suffixes of nominals and verbs and their appending order, i.e. transitions, described in the classic Kazakh grammar. We have also considered some of the frequent derivational suffixes found in the annotated sub-corpus of the Kazakh Language Corpus (KLC) [4]. As a result, the generator consists of 298 states (allomorphs and sub-categories of 55 distinct suffix

---

[2] Vowel harmony must be accounted for. Also certain morphemes are appended to roots with certain POS. This fact, actually, narrows down root search.

types) and 329 transitions (not counting transitions between allomorphs). Our POS-labeled root lexicon comprising 18230 entries was also derived from KLC.

The developed FSA produces morpheme chains which in turn can be appended to roots from the lexicon to produce inflected word forms. In the setting of spelling correction, however, we need to account for misspellings, and prune less probable corrections. Algorithm 1 describes the process of generating correction suggestion for a given input word and maximum edit distance threshold. Provided that a given word is not in a root lexicon (otherwise the word is considered correct), the process starts with an empty list of suggestions and a list of morpheme chains containing a null-morpheme (an empty string) that can be appended to any other morpheme. The algorithm repeatedly invokes the procedure **fsaGetChains**(*current_states*), which, using FSA, provides a list of chains reachable from a given list of morphemes. The process stops in either of two cases: (i) a word form identical to an input word is generated (the word is correct); (ii) **fsaGetChains**(*current_states*) returns no morpheme chains, because the list provided to the FSA consists of final states only. The pruning is done with the help of **withinDistance**(c, WRD, MAX_ED) procedure, that returns *True* if for a given morpheme chain c target word WRD has at least one suffix (in a sense of a sequence of trailing letters) for which an edit distance between itself and c is no larger than MAX_ED. Indeed, if the distance between a morpheme chain and a word exceeds the threshold there is no need in either developing that chain or searching for acceptable root. To get the suggestions, we search for roots acceptable by the eligible chains. This is done by the procedure **lexGetRoots**(c). Roots are concatenated with morpheme chains, and resulting strings that pass the edit distance threshold are added to the suggestion list.

Once we have a list of candidate corrections produced by the generator, we need to rank it. To this end we use a Bayesian argument that combines error and source models. For our error model we employ a noisy channel-based approach proposed by Church and Gale [6]. Our source model is built upon the theoretical aspects that were used for morphological disambiguation in [2]. Thus, the ranker that for each suggested correction computes a conditional probability of a suggestion being correct given a word:

$$P(s|w) = \frac{P(s)P(w|s)}{P(w)} \tag{1}$$

The denominator $P(w)$ can be dropped as it is common for all suggestions. Here $P(s)$ is a source model that denotes the probability of a suggestion having a given surface form, and $P(w|s)$ is an error model which denotes a likelihood of w being transformed into s. Let us start from describing the computation of the error model probability. As in [6] we compute $P(w|s)$ with respect to the types of possible errors in the following manner. In case of missing a letter (deletion):

$$P(w|s) \approx \frac{deletion(s_{i-1}, s_i) + \alpha}{pattern(s_{i-1}, s_i) + \alpha|V|} \tag{2}$$

where $deletion(s_{i-1}, s_i)$ is a number of times a pair of consecutive letters (s_{i-1}, s_i) was written as s_{i-1} and $pattern(s_{i-1}, s_i)$ is the count of this pair in the training set.

In case of reversing consecutive letters:

$$P(w|s) \approx \frac{reversing(s_i, s_{i+1}) + \alpha}{pattern(s_i, s_{i+1}) + \alpha|V|} \tag{3}$$

where $reversing(s_i, s_{i+1})$ is a number of times a pair (s_i, s_{i+1}) was written in reverse order and $pattern(s_{i-1}, s_i)$ is the count of this pair in the training set.

In case of inserting a letter:

$$P(w|s) \approx \frac{insertion(s_{i-1}, w_i) + \alpha}{N(s_i) + \alpha|W|} \tag{4}$$

where $insertion(s_{i-1}, w_i)$ is the number of times s_{i-1} was followed immediately by w_i and $N(s_i)$ is the count of s_i in the training set.

In case of typing a wrong letter (substitution):

$$P(w|s) \approx \frac{substitution(w_i, s_i) + \alpha}{N(s_i) + \alpha|W|} \tag{5}$$

where $substitution(w_i, s_i)$ is the number of times s_i was written as w_i and $N(s_i)$ is the count of s_i in the training set.

In all cases we use Laplace smoothing where $|V|$ is the cardinality of a set of letter bigrams, and $|W|$ is the cardinality of a set of unigram letters found in the training set of our misspelling-correction pairs dataset. The smoothing factor α was empirically set to $\alpha = 0.7$.

Let us now describe the computation of the source model probability $P(s)$. We compute $P(s)$ following the ideas discussed in [2] and taking advantage of morphological disambiguation. Recall that each suggestion is produced in a segmented form, i.e. a root and (possibly) a chain of morphemes with corresponding POS information. Thus, using a training set we can compute the probability of a morpheme chain using the chain rule. Assuming that a morpheme chain is independent of a root (it indeed depends only on the POS of a root), we consider the probability $P(s)$ to be proportional to the product of the probabilities of a root and a morpheme chain:

$$P(s) \propto P(r\_s) \prod_{i=0}^{n} P(m\_s_i|m\_s_{i-1}) \tag{6}$$

where r_s is a root of suggestion s with its POS information, m_s_i is the i_{th} morpheme with its inflectional/derivational information, and n is the number of morphemes in s. All counts are derived from a training set gathered from the annotated data of KLC. Both $P(r\_s)$ and $P(m\_s_i|m\_s_{i-1})$ are smoothed using Laplace smoothing, with λ empirically set to $\lambda = 0.3$. The rank of a suggestion s given word w is calculated as:

$$R(s) = P(r\_s) \prod_{i=0}^{n} P(m\_s_i|m\_s_{i-1})P(w|s) \tag{7}$$

Finally, we would like to note that the developed FSA can be also used for morphological segmentation. A segmentator can be implemented on the basis of Algirithm 1.

Table 1. Overall accuracy of spelling correction

Tool	Acc.,%	1 err.,% corr.	2 err.,% corr.
Ours	83	85	69
MSO	79	85	31
KSD	53	55	31

The difference is that instead of collecting candidate suggestions we need to collect all segmentations (a root together with a morpheme chain) whose surface form is identical to a given input word. Then the collected segmentations can be ranked using Eq. 7.

4   Experiments and Evaluation

For the experiments we gathered data from KLC [4]. During the annotation, annotators had fixed spelling errors in some of the documents. In KLC each annotated document has its earlier unlabeled version saved. Thus, by a simple comparison of the edited and the original versions of the documents we have collected more than 1800 error-correction pairs. Removing words with more than two errors left us with 1776 pairs. Recall that our ranker requires a training set of error types. Given a relatively small dataset of errors, we resorted to a 10-fold cross-validation leaving out 90% of data for training at each fold. This way we have trained and tested our method on the entire dataset and reported average performance. We compare our method to a Hunspell-based [5] open source Kazakh spelling correction tool KSD [7] and Microsoft Office 2010 (MSO) Kazakh language pack [8]. As these tools do not require training, we run them on the entire dataset without using cross-validation.

We begin with comparing the overall accuracy of the tools. Table 1 shows the overall accuracies broken down by the number of errors. The accuracy is calculated as a per-cent of words for which a correct fix was suggested regardless on which position in a suggestion list it appeared. Our method outperforms the remaining two in both overall accuracy and percent of 2-error words fixed. It is interesting that KSD, which like our method generates word forms on the fly, performs much worse. This can be explained by a small affix table and noisy root lexicon that the tool uses. When we skimmed through the words that our method missed, we found out that the most common reason was the absence of some derivational suffixes from our generator FSA. Thus, the accuracy of our method can be improved by incorporating new transitions into the generator and adding new roots to the lexicon.

Next, in Tab. 2 we compare the tools by another important metric, precision-at-k, which is calculated as a percentage of all correct suggestions that appeared at the first k positions of ranked suggestion lists. As we can see, KSD outperforms both MSO and our method. The latter two perform almost in par for k in the range of 2-7, with MSO leading 5% at $k = 1$ and $k = 10$. For further analysis of the results we have measured the average length of suggestion lists and the lowest rank at which a correct suggestion appeared. The average lengths of suggestion lists for KSD and MSO were 4.74 and 4.89 respectively. For our method it was 99.49. Similarly, the lowest ranks for

Table 2. Comparison by the precision @k,%

k	KSD	MSO	Ours
1	54	42	37
2	73	57	55
3	85	67	67
4	90	75	76
5	94	79	80
6	95	84	83
7	97	87	85
8	97	90	87
9	98	93	88
10	99	94	89

KSD, MSO, and our method were 18, 20, and 535 respectfully. Low ranking of some corrections is explainable: a root or a morpheme chain, or both may be rare, hence our ranker assigns them low probability. In fact, we tried to modify our ranker by removing either source or error model from Eq. 1. Still, the best results were the ones reported and they were achieved when both terms were kept. Long suggestion lists are explainable too: insufficient pruning while generating word forms.

Basically, the overall accuracy can be regarded as the recall of the method, i.e. the coverage of misspelled word forms. Whereas precision at-k is the precision of the method, i.e. a percentage of the fixed words for which a correct suggestion appeared at a reasonable position in the list. It is clear that our method trades recall over precision. However, upon analysis of words for which the most precise tool, KSD, could not find corrections, we found that those were frequent enough word forms and our method ranked many of them at the top. We think that precision-recall trade off is a crucial issue in spelling correction for agglutinative languages, and it definitely needs to be studied further.

Finally, we compare the tools in terms of the false positive rate, i.e. percentage of incorrect words recognized as correct ones. For our tool the ratio is 4%, whereas MSO and KSD have 8% and 13% false positive rates respectively. In terms of this metric our tool turned out to be twice and thrice as accurate as MSO and KSD respectively.

5 Conclusion and Future Work

We have developed a spelling correction tool for Kazakh language based on a morphological disambiguator. Our tool outperformed both open source and commercial analogues, achieving the overall accuracy of 83% in generating correct suggestions. The advantage of our method is that it can be iteratively improved by adding new rules/transitions to the disambiguator and new entries to the root lexicon. Moreover the generator FSA, which is the core of our method, can be also used for morphological segmentation. In this paper we have discussed this possibility.

We also report on some existing weaknesses of our method. In particular, relatively poor candidate correction ranking and pruning of candidate correction lists. Our future work will be directed towards solving these problems, as well as incorporating context sensitivity into our method.

References

1. Koskenniemi, K.: A general computational model for word-form recognition and production. In: Proceedings of the 10th International Conference on Computational Linguistics, pp. 178–181. Association for Computational Linguistics (1984)
2. Hakkani-Tur, D.Z., Oflazer, K., Tur, G.: Statistical morphological disambiguation for agglutinative languages. Computers and the Humanities 36(4), 381–410 (2002)
3. Oflazer, K., Güzey, C.: Spelling correction in agglutinative languages. In: ANLP, pp. 194–195 (1994)
4. Makhambetov, O., Makazhanov, A., Yessenbayev, Z., Matkarimov, B., Sabyrgaliyev, I., Sharafudinov, A.: Assembling the kazakh language corpus. In: Proceedings of the 2013 Conference on Empirical Methods in Natural Language Processing, Seattle, Washington, USA, pp. 1022–1031. Association for Computational Linguistics (October 2013)
5. Németh, L.: Hunspell open source spell checker (2011)
6. Church, K., Gale, W.: Probability scoring for spelling correction. Statistics and Computing 1(2), 93–103 (1991)
7. Mussayeva, A.: Kazakh language spelling with hunspell in openoffice.org. Technical report, The University of Nottingham (2008)
8. Microsoft: Microsoft Office 2010, kazakh language pack (2010)
9. Damerau, F.J.: A technique for computer detection and correction of spelling errors. Commun. ACM 7(3), 171–176 (1964)
10. Levenshtein, V.I.: Binary codes capable of correcting deletions, insertions and reversals. Soviet Physics Doklady 10(8), 707–710 (1966)
11. Mays, E., Damerau, F., Mercer, R.: Context based spelling correction. Information Processing & Management 27(5), 517–522 (1991)
12. Shannon, C.E.: A mathematical theory of communication. The Bell System Technical Journal 27, 379–423 (1948)
13. Brill, E., Moore, R.: An improved error model for noisy channel spelling correction. In: Proceedings of the 38th Annual Meeting of the Association for Computational Linguistics, Hong Kong (2000)
14. Hodge, V.J., Austin, J.: A comparison of a novel neural spell checker and standard spell checking algorithms. Pattern Recognition 35(11), 2571–2580 (2002)
15. Austin, J., Kennedy, J., Lees, K.: The advanced uncertain reasoning architecture, aura. Technical report, University of Canterbury (1995)
16. Alegria, I., Ceberio, K., Ezeiza, N., Soroa, A., Hernández, G.: Spelling correction: from two-level morphology to open source. In: LREC. European Language Resources Association (2008)
17. Pirinen, T.A., Silfverberg, M., Lindén, K.: Improving finite-state spell- checker suggestions with part of speech n-grams (2012)
18. Mussayeva, A.: Mozilla add-ons, kazakh spelling dictionary 1.1 (2009)
19. Mussayeva, A.: OpenOffice, kazakh spelling dictionary (2008)

A Preliminary Study on the VOT Patterns of the Assamese Language and Its Nalbaria Variety

Sanghamitra Nath, Himangshu Sarma, and Utpal Sharma

Department of Computer Science and Engineering,
Tezpur University, Assam, India
{s.nath,utpal}@tezu.ernet.in, himangshu.tezu@gmail.com

Abstract. The speech signal contains various analytical features and one such feature is the VOT or voice onset time which has proved to be a very important feature for classifying stops into different phonetic categories with respect to voicing. Furthermore in order to identify the features of digital speech and language for automatic recognition, synthesis and processing, it is important that the languages phoneme set is analyzed and VOT proves to be very useful in such an analysis. The stops in Assamese, the language spoken by the people of the state of Assam in North-East India, may be classified into three groups according to the place of articulation. They are labials, alveolars and velars. Historically the dental and retroflex stops have both merged into alveolar stops. Also for each group there are two different types based on the manner of voiced/ unvoiced distinction, i.e., aspirated and murmured. This paper focuses on computing and analyzing the VOT values for the stops of the Assamese language and its dialectal variants to provide a better understanding of the phonological differences that exist among the different dialectal variants of the language which may prove to be useful for dialect recognition, translation and synthesis.

Keywords: Voice Onset Time, Stop consonants, Assamese, Nalbaria, dialect.

1 Introduction

Voice Onset Time or Phonation Onset, commonly known as VOT is generally defined as the difference in time (in milliseconds) between the instant of stop consonant closure release and the start of vocal cord vibration. Negative values of VOT mean that the vocal cords begin vibrating before time of end of vocal tract closure while positive VOT indicates that the vocal cords start vibrating after the beginning of vocal tract closure.VOT may also have a zero or near zero value when the vocal cords start vibrating at the time of closure release. VOT plays an important role in perceptual discrimination of phonemes of the same place of articulation. VOT has come to be regarded as one of the most important methods for examining the timing of voicing in stops (especially in word-initial position) and has been applied in the studies of many languages.

A. Gelbukh (Ed.): CICLing 2014, Part II, LNCS 8404, pp. 542–552, 2014.
© Springer-Verlag Berlin Heidelberg 2014

VOT has been studied by researchers from various fields like language typology, phonetics, second language impact on the first language, speaker identification and also dialect and accent detection and identification. VOT is also used as a parameter for speech synthesis.

Most existing studies concentrate on the English language. However no known attempt has been made to examine the VOT patterns in Assamese and its dialectal variants so far. Our aim therefore is to make a comparative study of the VOT patterns of the standard variety of Assamese and one of its dialectal variants and see if any significant difference exist which can be used for dialect identification or for synthesis. For our initial study we concentrate on the stops in word initial positions.

The language we have chosen for our study, Assamese, is the principal language of the state of Assam in North East India. Assamese is regarded as the lingua-franca of the whole of northeast India. The Assamese language is the easternmost member of the Indo- European family and is spoken by most natives of the state of Assam. As reported by RCILTS, IITG over 15.3 million people speak Assamese as the first language and including those who speak it as a second language, a total of 20 million speak Assamese primarily in the northeastern state of Assam and in some parts of the neighboring states of West Bengal, Meghalaya and Arunachal Pradesh and other northeast Indian states[1]. The Assamese language grew out of Sanskrit, however, its vocabulary, phonology and grammar have substantially been influenced by the original inhabitants of Assam, such as the Bodos and the Kacharis. Assamese and the cognate languages, Maithili, Bengali and Oriya, developed from Magadhi Prakrit, the eastern branch of the Apabhramsa that followed Prakrit[2].

During the middle of the 19th century, Bengali was considered the official language of Assam, but British colonizers decreed Eastern Assamese to be the standard Assamese dialect. Presently, however, Central Assamese is accepted as the principal dialect. Several regional dialects are typically recognized. These dialects vary primarily with respect to phonology and morphology. Dr. Banikanta Kakati, an eminent linguist, has divided the Assamese dialects into two major groups, Eastern Assamese and Western Assamese. However, recent studies have shown that there are four major dialect groups, listed below from east to west[3]:

1. Eastern group spoken in and other districts around Sibsagar district.
2. Central group spoken in present Nagaon district and adjoining areas.
3. Kamrupi group spoken in undivided Kamrup, Nalbari, Barpeta, Darrang, Kokrajhar and Bongaigaon.
4. Goalparia group spoken in Goalpara, Dhubri, Kokrajhar and Bongaigaon districts

The Assamese script presently has a total of 11 vowel letters, used to represent the eight main vowel sounds of Assamese, along with a number of vowel diphthongs. The names of the consonant letters in Assamese are typically just the

---

[1] http://www.iitg.ac.in/rcilts/assamese.html
[2] http://en.wikipedia.org/wiki/Assamese_language
[3] http://www.iitg.ac.in/rcilts/assamese.html

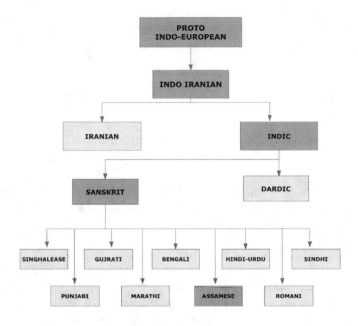

Fig. 1. Linguistic Affiliation of the Assamese language

		Labial			Alveolar			Velar		
		IPA	ROM	Script	IPA	ROM	Script	IPA	ROM	Script
Stop	voiceless	p	p	প	t	t	ত/ট	k	k	ক
	aspirated	p^h	ph	ফ	t^h	th	থ/ঠ	k^h	kh	খ
	voiced	b	b	ব	d	d	দ/ড	g	g	গ
	murmured	b^h	bh	ভ	d^h	dh	ধ/ঢ	g^h	gh	ঘ

Fig. 2. Classification of Assamese Stop Consonants

consonant's main pronunciation plus the inherent vowel . The stop consonants
are classified as shown in Fig.2.

In our study we consider the All India Radio (AIR) variety of Assamese as
the standard form of Assamese and the Nalbaria variety (spoken by the people
in and around the district of Nalbari in Assam) which fall under the Kamrupi
group as its dialectal variant. The AIR variety is the form of Assamese generally
spoken by the readers of Assamese news of All India Radio. Though a number
of dialects of Assamese exist, we have chosen Nalbaria for our preliminary study
specially because it greatly differs from the standard form in terms of accent,

vocabulary and tempo. The paper is organized as follows. Section 2 presents a brief literature review of related works, Section 3 describes the building of the speech corpus, the methodology used, VOT measurements and results and finally some conclusions are drawn from the findings and based on them plans for future work are proposed in Section 4.

2   Literature Review

Lisker and Abramson was the first to describe VOT in their well known cross language study of voicing in initial stops and they defined VOT as the time interval between the burst that marks release of the stop closure and the onset of quasi-periodicity that reflects laryngeal vibration. [1] examined 11 languages and classified them into three groups according to the number of stop categories each language contains. They also suggest that each stop category falls into one of three ranges, 125 to 75 ms (lead), 0 to +25 ms (short lag), and +60 to +100 ms (long lag), respectively. Following Lisker and Abramsons categorization, both the standard variety and the dialectal variety of Assamese fall into the four-category group of languages, however they can be distributed in the above mentioned ranges with slight modifications of the lower and upper end values.

Measurements of VOT before the release of stop closure are stated as negative numbers and called voicing lead, while after the release are stated as positive numbers and called voicing lag [2]. If release and voicing are simultaneous, VOT is considered to be zero.

Depending on VOT, [1] divided languages into two groups: group A languages which have long VOT, over 50 milliseconds, for a voiceless stop but short VOT for voiced: and group B languages which have short VOT, less than 30 milliseconds, for voiceless, but negative VOT for voiced stops.

VOT values for stops are found to vary in relation to the place of articulation [3]. The fact that VOT values get longer when the place of articulation moves from an anterior to a posterior position is confirmed in most languages, nevertheless exceptions exist. Lisker and Abramson (1964) demonstrated that, for both unaspirated and aspirated stops, velar stops have longer mean VOT values than alveolar and bilabial stops.

Lisker and Abramson in 1967 reported that the influence of vowels on the VOT of stop consonants is not significant, however later research prove that the quality of the vowel do effect the VOT of stops. Moreover researchers also claim that in rapid speech, the speaking rate also might influence the stop VOTs [4].

Different tones have different pitch levels, which are determined by the vibrating frequency of the vocal cord. As speculated by many researchers, VOT durations are in fact affected by tone as different tones have different fundamental frequencies and pitch levels. Therefore VOT values may vary when they occur in different lexical tones [5].

3   Experimental Framework

3.1   Speech Corpus

A corpus is developed having two parts, one part for the standard variety of Assamese and the other part for one dialectal variety which for the present study we consider it to be Nalbaria spoken by people in around the Nalbari district of Assam. A list of words having the voiced and voiceless plosives in word initial position is prepared. Since the quality of the vowel following the stop influences the VOT of the stop, for each stop consonant, five words are selected where the stop is followed by one of the five vowel sounds a,e,i,o and u. For e.g., for the plosive /p/ we have five words starting with pa,pe,pi,po and pu. 4 speakers, all male, are chosen, 2 speaking the standard variety and 2 speaking the dialectal variety. Each word is spoken by the speaker 5 times. The corpus is thus prepared by recording the word list in the voices of the 4 speakers. The recording for the dialectal variety was carried out at the Jyoti Chitraban Film and Television Institute, Guwahati, in a sound proof recording room at a sampling rate of 44.1 kHz and bit resolution of 16. The recording for the standard variety was carried out in a sound proof room at Tezpur University, Dept. of Computer Science and Engineering with a Sony recorder at the same sampling rate and resolution.

Table 1. List of stop consonants of Assamese

Sl No.	Stop Consonant	Word
1	/p/	/pani/ water
2	/pʰ/	/pʰatek/ jail
3	/b/	/bamun/ priest
4	/bʰ/	/bʰagɔɹ, bʰagaɹ/ tiredness
5	/t/	/taɹikʰ/ date
6	/tʰ·/	/tʰali/ plate
7	/d/	/dapun/ mirror
8	/dʰ/	/dʰaɹɔna/ idea
9	/k/	/kali/ yesterday
10	/kʰ/	/kʰam/ envelope
11	/g/	/gahɔɹi/ pig
12	/gʰ/	/gʰam/ sweat

3.2   Measuring VOT

The voice onset time of a plosive is defined as the duration between the release of a plosive and the beginning of vocal cord vibration. VOT can be positive, negative or 0.

1. If the onset of voicing follows the release, measure the interval in milliseconds between the release of the plosive until the onset of voicing. This is positive VOT or voicing lag.

. 2. If the onset of voicing coincides with the release, this is 0 VOT. The measurement would be 0 milliseconds.

3. If the onset of vocal cord vibration precedes the plosive release, then measure the voicing duration from the onset of voicing (or the onset of closure if there is voicing throughout), again in milliseconds. This is negative VOT or voicing lead.

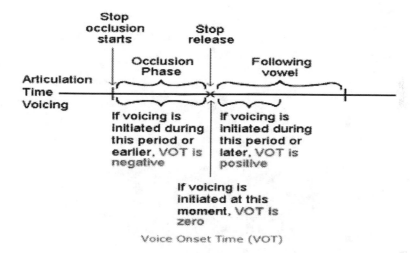

Fig. 3. Voice Onset Time (courtesy: [6])

3.3   Methodology/Acoustic Measurements

The experiment carried out for our research depends mainly on extracting VOT values of the voiced and unvoiced stops of standard Assamese and Nalbaria Assamese. Our analysis was carried out with the help of PRAAT speech analysis software. PRAAT was used to generate the waveform and spectrogram for each word utterance containing the plosive in word-initial position. On each waveform 2 points in time were located: the onset of burst release marked by the onset of low amplitude, aperiodic noice and the onset of voicing marked by the onset of high amplitude periodic energy. Onset of voicing, i.e., starting of vocal cord vibration, can be observed by noticing low frequency periodicity in the wide band spectrogram.VOT was calculated as the latency between release burst and voicing onset. We used signal energy and vocal cord vibration information to locate the beginning of stop release, closure and voicing. Audio monitoring of the signal was also made to check the sound of each segment. The closure release is marked at the beginning of abrupt increase in the energy level and can be easily identified in the signal waveform in PRAAT. VOT values were measured to a

precision that permitted rounding to the nearest 0.5ms. The starting mark was ·
set at the sharp increase of signal energy which signaled the release of /b, d,
g, p, t, k/. The end mark was set at the first upward going zero-crossing which
signaled voicing onset. If a positive VOT or voicing lead is produced, the end
mark is set at the first burst of /b, d, g/ [7]. Furthermore, the VOT values were
measured by two investigators for more reliability.

3.4   Results

The VOT range and mean of Nalbaria dialect and that of AIR variety is shown
in Table 2 and compared with different snapshots.

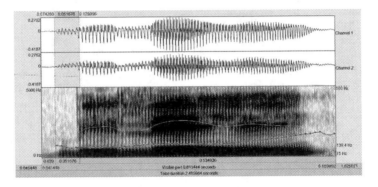

Fig. 4. Waveform and Spectrogram for the word /bilehi/ in Nalbaria having the stop
/b/ in the initial position. (VOT= -52ms)

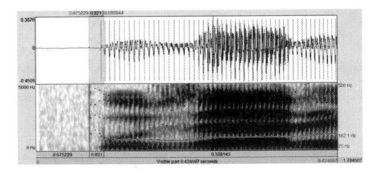

Fig. 5. Waveform and Spectrogram for the word /pujari/ in Nalbaria having the stop
/p/ in the initial position. (VOT= 21ms)

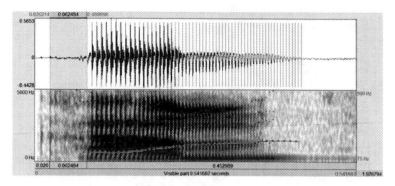

Fig. 6. Waveform and Spectrogram for the word /g^ham/ in Nalbaria having the stop (aspirated) /g^h/ in the initial position. (VOT= 62ms)

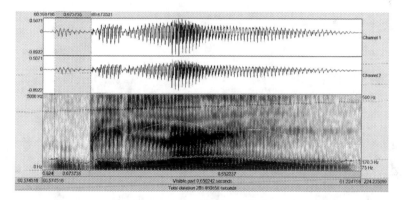

Fig. 7. Waveform and Spectrogram for the word /bilahi/ in AIR having the stop /b/ in the initial position. (VOT= -73ms)

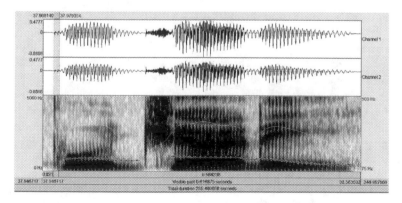

Fig. 8. Waveform and Spectrogram for the word /pujari/ in AIR having the stop /p/ in the initial position. (VOT= 12ms)

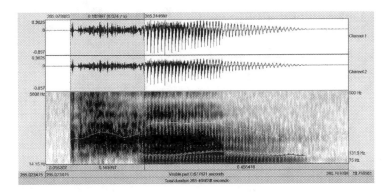

Fig. 9. Waveform and Spectrogram for the word /g^ham/ in AIR having the stop (aspirated) /g^h/ in the initial position. (VOT= 160ms)

Table 2. Range and mean of VOTs for N—Nalbaria variety and AIR—All India Radio variety

Phones	Range(N)	Mean(N)	Range(AIR)	Mean(AIR)
/p/	14:42	24	11:14	12
/b/	-98:-21	-59	-120:-56	-88
/t/	17:28	21	10:24	12
/d/	22:42/-56:-42	-46/26	-110:-70	-89
/k/	22:47	34	21:39	30
/g/	17:45/-58:-14	30/-35	-130:-58	-89
/p^h/	43:73	58	100:150	126
/b^h/	28:72	57	76:130	112
/t^h/	58:79	65	67:120	95
/d^h/	38:62	50	90:120	105
/k^h/	56:125	79	80:150	127
/g^h/	62:91	75	100:160	118

The outcomes of our investigation regarding the VOT values in the All India Radio (AIR) variety and the Nalbaria (N) variety of Assamese are shown in the Table 2. After analyzing the Table we find that the VOT values for the voiceless stops /p/, /t/ and /k/ in the Nalbaria variety extend from +14 ms to +47 ms while those for the AIR variety range from +10 ms to +39 ms. The mean VOT values for /p/ and /t/ are almost same, which does not conform to the general agreement that further back the place of articulation, the longer the VOT. The VOT values for the voiced stops /b/, /d/ and /g/ extend from -98ms to -14 ms /17ms to 45ms in the Nalbaria variety and extend from -130ms to -56ms in the AIR variety. While the VOT values for both voiceless aspirated and voiced murmured in both the varieties of Assamese are higher up in the VOT continuum and extend from +45 ms to +160 ms, the stops in the AIR variety are much more aspirated (almost twice) than its Nalbaria counterparts.

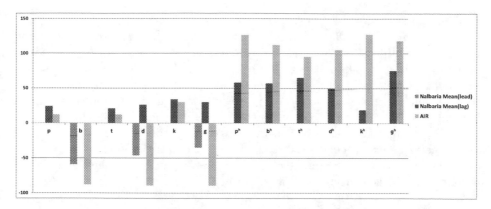

Fig. 10. Mean VOTs of Nalbaria variety and AIR variety

It should be noted that we give two sets of values for /d g/ in case of the Nalbaria variety because otherwise it would have meant lumping both positive and negative values for VOT as members of a single population which would have been misleading. This is similar to the case observed by Lisker and Abramson [1] for the English language. This implies that for voiced stops (Nalbaria) which are alveolar and velar, the onset of vocal cord vibration may follow the plosive release which is not usually the case. One explanation may be that the speakers of the AIR variety, which has been considered as the standard variety of Assamese, follow the common rule of producing the voiced stops in a manner that the vocal cord vibration is preceded by the stop release. While the speakers of the dialectal variant may not adhere to the common rule. Another explanation may be the effect of speaking rate or tempo. Studies show that the speaking rate has an influence on VOT. The speakers of the Nalbaria variety tend to speak fast which may result in the positive values of the voiced stops.

4   Conclusion and Future Work

Firstly, it is found that VOT patterns in the two variants of the Assamese language are similar but not completely identical. As compared to the standard VOT ranges, 125 to 75 ms (lead), 0 to +25 ms (short lag), and +60 to +100 ms (long lag), (Ref:3) the VOT ranges for lead, short lag and long lag for both the varieties of Assamese need to be extended in both directions in the VOT continuum. Secondly, the VOT for the voiced stops in the Nalbaria variety has both positive and negative values and this fact together with the fact that stops(aspirated and murmured) in the AIR variety are much more aspirated (almost twice) than the stops in the Nalbaria variety, can be used for dialect detection.

But it has to be kept in mind that the above conclusions have been drawn based on the corpus which needs to be extended. Currently the corpus we are working on is of a limited size and has been developed from scratch as no relevant corpus for the Assamese language, specially for its dialectal variants, are available.

The comparative study on VOT patterns is to be further extended to all the major variants of the Assamese language. Additional research with more monolingual subjects, as most of our speakers are bilingual, will be needed in order to accurately pinpoint the differences in the VOT patterns of the stops in the different varieties of the Assamese language. It would also be informative to determine whether these dialectal patterns are maintained in unconstrained informal speech rather than limiting the corpus to elicited word lists. Other variables of interest would be the effects of age and gender on the VOT patterns of dialects. Also experiments need to carried out to see if the VOT of one variety of the language can be manipulated and if so how, in order to make a significant difference in the generation of another variety.

Acknowledgments. The authors thank Mr.Pranjul Kashyap, Mr.Samarjit Barman and Mr. Jyoti Prasad Talukdar for their immense help in building the corpus. They are also thankful to DeitY, Govt. of India for financial support.

References

1. Lisker, L., Abramson, A.S.: A cross-language study of voicing in initial stops: Acoustical measurements. In: Proceedings of the 6th International Congress of Phonetic Sciences, vol. 20, pp. 384–422 (1964)
2. MacKay, I.R.: Phonetics: The science of speech production. Little, Brown (1987)
3. Cho, T., Ladefoged, P.: Variation and universals in vot: evidence from 18 languages. Journal of Phonetics 27(2), 207–229 (1999)
4. Miller, J.L., Volaitis, L.E.: Effect of speaking rate on the perceptual structure of a phonetic category. Perception & Psychophysics 46(6), 505–512 (1989)
5. Peng, J.F., Chen, L.M., Lee, C.C.: Tonal effects on voice onset time 14(4), 341–361 (2009)
6. Yang, B.: A voice onset time comparison of english and korean stop consonants 20, 41–59 (1993)
7. Nordhoff, S.: A grammar of Upcountry Sri Lanka Malay. LOT (2009)

Evaluation of Sentence Compression Techniques against Human Performance

Prasad Perera and Leila Kosseim

Dept. of Computer Science & Software Engineering
Concordia University
Montreal, Canada
{p_perer,kosseim}@encs.concordia.ca

Abstract. This paper presents a comparison of various sentence compression techniques with human compressed sentences in the context of text summarization. Sentence compression is useful in text summarization as it allows to remove redundant and irrelevant information hence preserve space for more relevant information. In this paper, we evaluate recent state-of-the-art sentence compression techniques that are based on syntax alone, a mixture of relevancy and syntax, part of speech feature based machine learning, keywords alone and a naïve random word removal baseline. Results show that syntactic based techniques complemented by relevancy measures outperform all other techniques to preserve content in the task of text summarization. However, further analysis of human compressed sentences also shows that human compression techniques rely on world knowledge which is not captured by any automatic technique.

1 Introduction

The goal of sentence compression is to generate a more concise form of a sentence without losing its grammaticality or its relevant content. Sentence compression has been used in several downstream natural language processing applications such as text simplification [1], headline generation [2] and text summarization [3]. In extractive summarization, relevant sentences are extracted from the document collection and re-ordered to create the final summary. However, since these sentences are not processed or modified, they might contain irrelevant content or phrases that do not contribute much to the summary content. As an example, consider the following topic, query and sentence (1)[1]:

Topic: *Microsoft's antitrust problems*

Query: *Summarize Microsoft's antitrust problems, including its alleged illegal behavior and antitrust proceedings against the company*

(1) *Under the schedule set by Jackson in April, the Justice Department and 17 states filed a brief with the court on April 28 asking the judge to break Microsoft into two companies as the remedy for the illegal behavior found in the long antitrust trial.*

---

[1] All examples are taken from the DUC 2007 corpora.

A. Gelbukh (Ed.): CICLing 2014, Part II, LNCS 8404, pp. 553–565, 2014.

The candidate sentence (1) does contain relevant pieces of information, but it may be too long to fit into a summary as it is. In addition, the sentence also contains several pieces of information that are less relevant to the topic and the query given and if it were to be inserted as in, it might lower the relevancy of the overall summary. A few possible shorter or compressed forms of the above sentence include:

(1c1) ~~Under the schedule set by Jackson in April,~~ *the Justice Department and 17 states filed a brief with the court on April 28 asking the judge to break Microsoft into two companies as the remedy for the illegal behavior found in the long antitrust trial.*

(1c2) *Under the schedule set by Jackson in April, the Justice Department and 17 states filed a brief with the court* ~~on April 28~~ *asking the judge to break Microsoft into two companies as the remedy* ~~for the illegal behavior found in the long antitrust trial~~.

(1c3) ~~Under the schedule set by Jackson in April,~~ *the Justice Department and 17 states filed a brief with the court* ~~on April 28~~ *asking the judge to break Microsoft into two companies as the remedy* ~~for the illegal behavior found in the long antitrust trial~~.

To the best of our knowledge, very little previous work has focused on measuring the contribution of specific sentence compression techniques as a means to improve summary content.

2   Sentence Compression Techniques

In this section, we present an overview of previous sentence compression techniques that have been proposed. These techniques are based on syntactic structure based pruning, keyword based pruning, part of speech and syntax related features based machine learning, or a combination of syntax and relevancy based pruning. The following sections present these in detail.

2.1   Syntactic Pruning

Predefined fixed syntactic pruning is the basis of many sentence compression techniques. For example, [4] used complete dependency parses and pruned specific grammatical structures including prepositional complements of verbs, subordinate clauses, noun appositions and interpolated clauses. They achieved a compression rate of 26% while retaining grammaticality and readability of text. In [5], the authors also applied linguistically motivated syntactic filtering. Using the TIPSTER [6] corpus, they identified syntactic patterns which were absent from human-written summaries compared to the original corpus and defined a trimming algorithm consisting of removing sub-trees of specific grammatical phrase structures. They have evaluated their pruning technique on the DUC 2003 summarization task and showed an improvement in ROUGE scores compared to uncompressed length-limited summaries. Finally, [7] describes the sentence compression module of their text summarization system, based on syntactic level sentence pruning. They have implemented a module of compression which filters adverbial modifiers and relative clauses. Their evaluations, performed using the DUC 2007 summarization track, showed an improvement in ROUGE scores after applying their compression technique to their summarization system.

2.2 Machine Learning Techniques

Machine learning techniques differ from fixed syntactic pruning in that the words, phrases or syntactic structures to prune are learned automatically from annotated corpora. [8, 9], for example, remove inessential phrases in extractive summaries based on an analysis of human written abstracts. A syntactic parser is used to identify different types of phrases which are present in the original sentences but not in human written simplified sentences. These phrases along with the main noun or verb they are attached to are used to train a Naïve Bayes Classifier to decide how likely a phrase is to be removed from a sentence. For evaluation, they have used a metric called success rate, which computes the ratio between the number of correct prunings the classifier made that agree with the human annotations over the total number of prunings (the classifier made and the humans made). The authors have achieved a 78.1% overall success rate but have noted a low success rate in removing adjectives, adverbs and verb phrases. However, the effect on overall summary content was not indicated.

On the other hand, [10] introduced semantic features to improve a decision tree based classification. Here, the authors used Charniak's parser [11] to generate syntactic trees and incorporated semantic information using WordNet [12]. The evaluation showed a slight improvement in importance of information preserved in shortened sentences. However again, the effect on summarization was not noted.

[13] points out that text compression could be seen as a problem of finding a global optimum by considering the compression of the whole text/document. The authors used syntactic trees of each pair of long and short sentences to define rules to deduce shorter syntactic trees out of original syntactic trees. They used the Ziff-Davis corpus for their evaluation as well as human judgment. Similarly, [14] describes the use of integer linear programming model to infer globally optimal compressions while adhering to linguistically motivated constraints and show improvement in automatic and human judgment evaluations. [15] have also described a syntactic pruning approach based on transformed dependency trees and a linear integer model.

[16] described another machine learning approach to sentence compression. They first trained a probabilistic model based on Maximum Entropy (ME) to evaluate how likely an edge of a syntactic tree can be removed based on a set of features including the part of speech tags of the surrounding words of the edge, the head of the edge and the modifier of the edge. For the evaluation, summaries were judged by human annotators and results showed that their compression

techniques outperform the baseline algorithm yet underperformed compared to the human annotated compressions.

2.3   Hybrid Methods

The sentence compression technique presented in [17] is based on two main parameters: syntactic pruning heuristics and a relevancy score. Based on these two parameters, three different sentence compression techniques were defined:

1. Syntax-driven sentence pruning: The goal of this technique is to preserve the grammaticality of the sentences while removing predefined syntactic structures that are assumed to always contain secondary information. Hence it is expected that removing these grammatical structures would not remove relevant content significantly. Based on the work of [5] and [4] (adapted to English), six types of syntactic structures are pruned: Relative clauses, Adjective phrases, Adverbial phrases, Conjoined verb phrases, Appositive phrases, Prepositional phrases. This syntax pruning method is equivalent to those presented in Section 2.1.

2. Syntax with relevancy based pruning: This method tries to ensure that the syntactic structures pruned do not contain relevant information. [17] use a relevance score based on tf-idf and cosine similarity with the topic/query. A relevance score for each sub-tree is calculated and predefined syntactic sub-trees are pruned only if relevance score is smaller than a given threshold.

3. Relevancy-driven syntactic pruning: This technique focuses primarily on the relevance score of syntactic structure. Here, no predefined syntactic structures are used. A relevancy score is computed for each syntactic sub-structure and the lowest embedding syntactic structures are removed if their relevance score is below a given threshold. For the sake of preserving the grammaticality of resulting sentences, syntactic sub-structures that are marked as noun or verb phrases are not removed.

2.4   Keyword/Phrase Based Techniques

Keyword based techniques take a more conservative approach and use a predefined list of keywords or phrases to identify less significant parts of the text and remove them from long sentences. [18–20], for example used a keyword list implemented in an ad hoc fashion to omit specific terms. They have evaluated their pruning techniques within their CLASSY [19] summarization system with DUC 2005 [21] and showed an improvement in ROUGE scores. In their participation to the DUC 2006 [22] automatic summarization track, their system scored among the top three based on ROUGE scores. The goal of our work was to apply each type of sentence compression technique for the task of summarization and evaluate them against human compression techniques. To do so, we have performed a series of evaluations based on two factors: the compression rates and content evaluation of compressed summaries. For content evaluation, we have evaluated ROUGE measures [23] on compressed summaries and grammar structure overlap [24, 25, 15] between human compressed summaries and summaries compressed with the automatic sentence pruning techniques. Specifically, we evaluated the following techniques:

Keyword/Phrase Based Approach. We implemented a keyword based sentence compression using the word/phrase patterns described in [18] and [20] plus additional patterns that we learned by analyzing human annotated summaries. The particular keyword/phrase and patterns we used include: meta-data information, temporal words/phrases, attributive words/phrases, keywords and key phrases (e.g. *As a result, In contrast,* etc.) and specific clauses (e.g. clauses that starts with *which, where* or *whom,* etc.).

Machine Learning Approach. As the second approach, we chose the sentence compression system described in [16]. This publicly available sentence compressor[2] uses a set of features based on part of speech (POS) tags, a Maximum Entropy based classifier to prune sentences and a Support Vector Regression Model to select the best candidates of all reduced sentences. The system uses the Edinburgh's Written and Spoken corpus[3] as the training set of original sentences paired with their human written compressed sentences. Additionally, they used a language model created using about 4.5 million sentences taken from TIPSTER corpus [6] in order to rank possible compressed sentences and select the most likely one.

Syntax and Relevancy. Here, we used the work of [17] (see Section 2.3) that used syntactic pruning combined with relevancy measures. The implementation uses the Stanford Parser [26] to generate complete syntactic trees of the sentences. We have evaluated the syntactic pruning based on: syntax alone, syntax with relevancy and relevancy alone.

Baseline Compression Techniques. In order to compare these sentence compression techniques, we implemented a baseline technique that randomly removes words and phrases from summaries to reach a particular compression rate. Using this baseline, we have created compressed summaries with compression rates of 10.5%, 21.5%, 23.4%, 29.8% and 43.1% to correspond to the compression rates of each sentence pruning techniques: keyword based (10.5%), human compression and machine learning based (21.5%), syntax with relevancy based (23.4%), relevancy-driven (29.8%) and syntax-driven (43.1%) (See Section 3.2 below).

2.5   Human Compressed Summaries

To build our human compressed summaries, we have provided a set of summaries to five human annotators and asked them to reduce their length while preserving

---

[2] http://nlp.cs.aueb.gr/software.html
[3] http://jamesclarke.net/research/resources

important content. We provided the evaluators with a set of summaries created using the DUC 2007 summarization task [27] along with the relevant topic and the query used to create these summaries. For this task, we have chosen the summaries created by the best performing system [28] (based on ROUGE measures) at the DUC 2007 summarization track. The human annotators were asked to compress these summaries by removing words or phrases from the sentences that they considered not relevant to the given topic/query. Each sentence was to be considered independently of the others; hence the annotators could not use the context to influence their compression strategies. Human annotators were chosen from a group of undergraduate and graduate students in different science and engineering streams.

2.6 Human Compression Rate

First we evaluated the compression rates of the human compressed summaries and compared these with the compression rates achieved by the automatic pruning techniques. Table 1 shows these word-based compression rates. According to Table 1, the highest compression rate was achieved by the syntax-driven technique. The relevancy-driven technique and the syntax with relevancy based techniques achieved the next highest compressions. Although human compression varies from 18% to 25%, the annotator average (21.5%) seems to be similar to the syntax with relevancy based technique (23.4%) and the machine learning technique (21.5%). In addition, the lowest compression rate was achieved by the keyword based technique (10.5%).

Table 1. Sentence Compression Rates of Different Techniques

Technique	No. of Words	Compression
Original Summary	6237	0.0%
Keyword Based Pruning	5579	10.5%
Annotator 1	5106	18.1%
Annotator 2	5052	19.0%
Machine Learning Technique	4914	21.5%
Annotator 3	4897	21.5%
Annotator 4	4889	21.6%
Syntactic with Relevancy	4779	23.4%
Annotator 5	4657	25.3%
Relevancy-Driven	4381	29.8%
Syntax-Driven	3552	43.1%

2.7 Content Evaluation Using ROUGE

In automatic text summarization, the most standard metric used in measuring summary content is the ROUGE measure [23]. Since we applied sentence compression to text summarization, we measured the effect of the various techniques on content using the ROUGE measure, when sentence compression was applied to text summarization. Therefore our first attempt at content evaluation was based on ROUGE F-measure scores (R-2 and SU4) for the original summaries, the five sets of human compressed summaries and all other automatically compressed summaries. As Table 2 shows, there is a decrease in ROUGE-2 score between the original summaries and the human compressed summaries. On average, the annotators have a ROUGE-2 score of 0.120 and ROUGE-SU4 of 0.172 while the original summaries have a ROUGE-2 score of 0.127 and ROUGE-SU4 of 0.179. The one-tailed t-test shows that for all the annotators (except for annotator 2), the difference between ROUGE scores compared to the original summary score is statistically significant with a confidence level of 95%. The t-test identified four clusters based on ROUGE-2 scores. The first cluster contains the techniques that scored the best ROUGE-2 scores (the original summaries, annotator 2 and the keyword based technique). Compared to the original ROUGE-2 score, the keyword based technique (ROUGE-2: 0.124) does not show a significant decrease in content according the one-tailed t-test with a confidence level of 95%. However, recall from Table 1 that this technique only removes 10.5% of words while other techniques remove up to 20-40% of words. It is therefore not surprising that its ROUGE score be so high.

In the second cluster, we have all other annotators, with a ROUGE-2 score ranging from 0.119 to 0.118. These ROUGE-2 scores are significantly lower than the original summary ROUGE-2 scores. This is somewhat surprising as we would have expected the annotators not to remove too much relevant content; yet considering that on average, they removed 21.5% of the words (see Table 1).

Table 2. Content Evaluation of Compressed Summaries

Technique	R-2	R-SU4
Original Summaries	0.127	0.179
Annotator 2	0.125	0.176
Keyword Based	0.124	0.176
Annotator 1	0.119	0.172
Annotator 3	0.119	0.171
Annotator 4	0.119	0.173
Annotator 5	0.118	0.170
Baseline: Random Compression 10.5%	0.116	0.172
Syntax with Relevancy Based	0.110	0.164
Machine Learning Based	0.107	0.162
Relevancy-Driven	0.106	0.154
Baseline: Random Compression 21.5%	0.102	0.163
Baseline: Random Compression 23.4%	0.100	0.164
Baseline: Random Compression 29.8%	0.085	0.150
Syntax-Driven	0.084	0.134
Baseline: Random Compression 43.1%	0.072	0.137

They were bound to remove some content. Also surprisingly, this cluster contains the random compression 10.5% technique with a ROUGE-2 measure of 0.116.

The third cluster includes: syntax with relevancy based, machine learning, relevancy-driven and two baseline compression techniques (random compression 21.5% & 23.4%). Again, these techniques show significantly lower ROUGE-2 scores than the original summary and the average human ROUGE scores. However, when tested for significance across each other, the ROUGE-2 scores are not significantly different.

The last cluster includes the rest of the techniques: baseline random compression 29.8%, syntax-driven and baseline random compression 43.1% that scored the lowest ROUGE scores. The ranking of the techniques is more or less the same when ROUGE-SU4 scores were used for the task. When words are randomly removed, it hurts the grammaticality and the content of the summaries. However, since ROUGE is only calculated based on bi-gram co-occurrences, it justifies how random removals (10.5%, 21.5%, 23.4% and 29.8%) showed better results than most of the automatic sentence compression techniques.

2.8   Content Evaluation Based on Grammatical Relations

Previous work on sentence compression evaluations have typically focused on two different evaluation methods: ranking of compressed sentences by human judgment or evaluation against human compressed sentences by measuring content overlap. To follow recent trend (e.g. [25, 15]), we have also evaluated the sentence compression techniques and the human annotators based on a metric that takes grammatical relations into account. This metric was first introduced in [24] for automatic summary evaluation with the goal of improving automatic evaluation techniques while taking semantic information into account. The authors argue that it is easy to enhance automatic summary evaluation when a dependency parser is available by counting co-occurrences of dependency grammar structures between the gold standard summaries and automatic summaries as opposed to counting n-gram word co-occurrences as ROUGE does. This technique was used by [25] to evaluate their sentence compression technique. Following them, [15] have also used the same mechanism to evaluate their sentence compression techniques, comparing their results to the work of [25]. Table 3 shows the F-measure calculated for all techniques and all five annotators. The dependency grammar structure F-measure seems to show some interesting results. Here, we observe four clusters of grammar structure F-measure (created based on the one-tailed t-test). The first cluster clearly shows better content evaluation results for all annotators. The second cluster includes: syntax with relevancy and keyword based techniques. Here, unlike the ROUGE measure (in Table 2), the dependency grammar metric has penalized the keyword based technique compared to the human summaries. The third cluster includes machine learning based, relevancy-driven, syntax-driven and random compression (10.5%) techniques. Finally, cluster four includes, the rest of the random removals techniques.

Using this grammar-based content metric, all baseline techniques, where words are removed randomly, have been penalized as expected. In addition,

Table 3. Content Evaluation of Compressed Summaries Against Human Annotations Using Dependency Structure Based F-Measure

Technique	F-Measure
Annotator 2	0.829
Annotator 1	0.819
Annotator 4	0.817
Annotator 3	0.808
Annotator 5	0.806
Syntactic with Relevancy	0.759
Keyword Based	0.748
Machine Learning Based	0.707
Relevancy-Driven	0.706
Syntax-Driven	0.664
Baseline: Random Compression 10.5%	0.661
Baseline: Random Compression 21.5%	0.514
Baseline: Random Compression 23.4%	0.514
Baseline: Random Compression 29.8%	0.400
Baseline: Random Compression 43.1%	0.278

all the automatic sentence pruning techniques have performed significantly better than the random word removal baselines (see Table 3). Finally, the syntax complemented with relevancy while removing 23.4% of the words (see Table 1) outperformed all other automatic pruning techniques. This seems to show that it is useful to have predetermined classes of syntactic structures to remove, but they cannot be removed systematically without first verifying their content. The keyword based technique is comparable to the syntax with relevancy in terms of grammatical F-measure but only removed 10.5% of the words (see Table 1).

3   Discussion

The results of Section 3.4 clearly show that a mixture of syntax and relevancy give the best grammatical F-measure given its compression rate. In order to investigate the precision of this approach further, we measured which types of words and phrase structures the annotators removed. Table 4 shows the syntactic structures removed by the annotators along with the compression rate achieved by removing only this structure and the percentage of such structures removed. For example, by removing only prepositional phrases (PPs), the annotators achieve a compression rate of 34.7% on average; but they only removed 12.4% of all prepositional phrases (87.6% of all PPs were left in the compressed summaries). Apart from individual words, noun and verb phrases, the human annotators removed the same syntactic structures as the syntactic sentence pruning techniques, but with a much more subtle selection. In other words, although the removal of PPs account for a great reduction of words (34.7%), only 12.4% of the PPs were actually removed. In effect, this is what relevance score of the

Table 4. Sentence Compression Rates of Different Techniques

Syntactic Structure	Compression %	Removed %
Adverbial Phrases	2.5%	22.0%
Individual Words	2.7%	0.6%
Verb Words	2.8%	1.7%
Adjective Phrases	4.0%	9.4%
Conjoined Clauses	6.4%	9.4%
Appositive Phrases	14.7%	35.2%
Noun Phrases	15.6%	4.1%
Relative Clauses	17.0%	17.2%
Prepositional Phrases	34.7%	12.4%

syntax with relevancy based approach attempted to do; be more subtle about which structure to remove.

It is interesting to note that none of the techniques we have evaluated remove noun and verb phrases. However, as shown in Table 4, humans do remove some. We therefore analyzed these cases to see why they were removed.

In human compressed summaries, a lower percentage of compression was achieved by removing verb phrases compared to noun phrases (2.8% as opposed to 15.6%). After analyzing the noun phrases that were removed, we noted that human annotators seem to remove proper and compound nouns based on their knowledge level. This seems to be subjective for each individual and reflects the annotator's knowledge and perception of the world. As an example, consider the following sentence, pruned by three different annotators:

(4) Annotator 1: *Myanmar's military government has detained another 187 members of pro-democracy leader Aung San Suu Kyi's party, bringing the total to 702 arrested since a crackdown began in May.*

(5) Annotator 2: *Myanmar's military government has detained another 187 members of pro-democracy leader Aung San Suu Kyi's party, bringing the total to 702 arrested since a crackdown began in May.*

(6) Annotator 3: *Myanmar's military government has detained another 187 members of pro-democracy leader Aung San Suu Kyi's party, bringing the total to 702 arrested since a crackdown began in May.*

Here, Annotator 1 has only removed the adjectival phrase *pro-democracy*; while, Annotator 2 has gone a bit further and removed *pro-democracy leader*. Finally, Annotator 3 attempted to remove the entire phrase *pro-democracy leader Aung San* leaving the remaining phrase, *Suu Kyi*. This choice seems to be rather subjective and more influenced by the individuals and is difficult to capture through syntactic pruning rules or relevancy measures or even learned by classifiers.

4   Conclusion and Future Work

In this paper, we have described the evaluation of various sentence pruning approaches and compared them against human compressed summaries. For this task, we used a set of 25 summaries with a word limit of 250, created from the best performing system [28] based on ROUGE scores in the DUC 2007 summarization track. We have used five sets of human compressed summaries, created using the DUC 2007 summaries to evaluate various sentence compression techniques.

First we have evaluated the compression rate of each technique and compared the results against human sentence compression rates. Human sentence compression had an average compression rate of 21.5% which was similar to the compression rate of the syntax with relevancy based technique and machine learning based technique.

We have performed content evaluations using two metrics: ROUGE [23] and a dependency grammar structure based F-measure [24]. The content evaluation using ROUGE showed that even human compressed summaries tend to lose content and the higher the compression rate, the greater the decrease in content compared to the original summaries. This was clearly visible with the baseline systems where we used different compression rates (10.5%, 21.5%, 23.4%, 29.8% and 43.1%). Further, it also showed the weakness of the word-based n-gram ROUGE measure to capture and evaluate attributes such as grammaticality and relevancy of content when it comes automatic summarization.

In our second series of content evaluation, we calculated an F-measure metric based on dependency grammar structures, introduced by [24, 25, 15]. The results were interesting as they showed that this grammar based metric could discriminate the loss of grammaticality of the naïve random removal baselines. The overall results showed that the highest F-measure was not surprisingly achieved by the human annotators with an F-measure of 0.81 and out of all automatic techniques, the syntax with relevancy based sentence compression technique showed the best result with an F-measure of 0.760. Considering that this technique has a similar compression rate to humans and obtained the best grammar structure F-measure, we conclude that the syntax with relevancy based pruning technique seems to model better what humans do. This seems to show that it is useful to have predetermined classes of syntactic structures to remove, but they cannot be removed systematically without first verifying their content.

By analyzing the human compressed summaries, we found that annotators tend to remove syntactic structures more than removing individual words. In addition, these syntactic structures are similar to the syntactic structures typically used in automatic approaches. However, annotators do not remove these syntactic structures systematically but only in certain circumstances. As future work, it would be interesting to further the investigation of which specific structures humans remove and which are kept. Finding discriminating features, other than syntax or relevancy, would be worth looking into.

Acknowledgement. The authors would like to thank the anonymous reviewers for their comments on an earlier version of the paper. This work was financially supported by an NSERC grant.

References

1. Chandrasekar, R., Doran, C., Srinivas, B.: Motivations and Methods for Text Simplification. In: Proceedings of COLING 1996, Copenhagen, pp. 1041–1044 (1996)
2. Dorr, B., Zajic, D., Schwartz, R.: Hedge Trimmer: A Parse-and-Trim Approach to Headline Generation. In: Proceedings of the HLT-NAACL Workshop on Text Summarization, pp. 1–8 (2003)
3. Knight, K., Marcu, D.: Summarization beyond sentence extraction: A probabilistic approach to sentence compression. Artificial Intelligence 139(1), 91–107 (2002)
4. Gagnon, M., Da Sylva, L.: Text Compression by Syntactic Pruning. In: Lamontagne, L., Marchand, M. (eds.) Canadian AI 2006. LNCS (LNAI), vol. 4013, pp. 312–323. Springer, Heidelberg (2006)
5. Zajic, D., Dorr, B.J., Lin, J., Schwartz, R.: Multi-Candidate Reduction: Sentence Compression as a Tool for Document Summarization Tasks. Information Processing and Management 43, 1549–1570 (2007)
6. Harman, D., Liberman, M.: TIPSTER Complete. Linguistic Data Consortium (LDC), Philadelphia (1993)
7. Jaoua, M., Jaoua, F., Belguith, L.H., Hamadou, A.B.: Évaluation de l'impact de l'intégration des étapes de filtrage et de compression dans le processus d'automatisation du résumé. In: Résumé Automatique de Documents. Document numérique, Lavoisier, vol. 15, pp. 67–90 (2012)
8. Jing, H., McKeown, K.R.: Cut and Paste Based Text Summarization. In: Proceedings of NAACL 2000, Seattle, pp. 178–185 (2000)
9. Jing, H.: Sentence Reduction for Automatic Text Summarization. In: Proceedings of the Sixth Conference on Applied Natural Language Processing, Seattle, pp. 310–315 (April 2000)
10. Nguyen, M.L., Phan, X.H., Horiguchi, S., Shimazu, A.: A New Sentence Reduction Technique Based on a Decision Tree Model. International Journal on Artificial Intelligence Tools 16(1), 129–138 (2007)
11. McClosky, D., Charniak, E., Johnson, M.: Effective Self-Training for Parsing. In: Proceedings of HLT-NAACL 2006, New York, pp. 152–159 (2006)
12. Fellbaum, C.: WordNet: An Electronic Lexical Database. The MIT Press (May 1998)
13. Le Nguyen, M., Shimazu, A., Horiguchi, S., Ho, B.T., Fukushi, M.: Probabilistic Sentence Reduction Using Support Vector Machines. In: Proceedings of COLING 2004, Geneva, pp. 743–749 (August 2004)
14. Clarke, J., Lapata, M.: Global Inference for Sentence Compression an Integer Linear Programming Approach. Journal of Artificial Intelligence Research (JAIR) 31(1), 399–429 (2008)
15. Filippova, K., Strube, M.: Dependency Tree Based Sentence Compression. In: Proceedings of the Fifth International Natural Language Generation Conference, INLG 2008, Stroudsburg, PA, USA, pp. 25–32 (2008)
16. Galanis, D., Androutsopoulos, I.: An Extractive Supervised Two-Stage Method for Sentence Compression. In: Human Language Technologies: The 2010 Annual Conference of the North American Chapter of the Association for Computational Linguistics, HLT 2010, Los Angeles, California, pp. 885–893 (2010)
17. Perera, P., Kosseim, L.: Evaluating Syntactic Sentence Compression for Text Summarisation. In: Métais, E., Meziane, F., Saraee, M., Sugumaran, V., Vadera, S. (eds.) NLDB 2013. LNCS, vol. 7934, pp. 126–139. Springer, Heidelberg (2013)

18. Conroy, J.M., Schlesinger, J.D., O'Leary, D.P., Goldstein, J.: Back to Basics: CLASSY 2006. In: Proceedings of the HLT-NAACL 2006 Document Understanding Workshop, New York City (2006)
19. Schlesinger, J.D., O'Leary, D.P., Conroy, J.M.: Arabic/English Multi-document Summarization with CLASSY—The Past and the Future. In: Gelbukh, A. (ed.) CICLing 2008. LNCS, vol. 4919, pp. 568–581. Springer, Heidelberg (2008)
20. Dunlavy, D.M., Conroy, J.M., Schlesinger, J.D., Goodman, S.A., Okurowski, M.E., O'Leary, D.P., van Halteren, H.: Performance of a Three-Stage System for Multi-Document Summarization. In: Proceedings of the HLT-NAACL 2003 Document Understanding Workshop, Edmonton, Canada, pp. 153–159 (2003)
21. Dang, H.T.: DUC 2005: Evaluation of Question-focused Summarization Systems. In: Proceedings of the Workshop on Task-Focused Summarization and Question Answering, Sydney, pp. 48–55 (2006)
22. Dang, H.T.: Overview of DUC 2006. In: Proceedings of the HLT-NAACL 2006 Document Understanding Workshop (2006)
23. Lin, C.Y.: ROUGE: A Package for Automatic Evaluation of Summaries. In: Moens, M.F., Szpakowicz, S. (eds.) Text Summarization Branches Out: Proceedings of the ACL 2004 Workshop, Barcelona, Spain, pp. 74–81 (July 2004)
24. Riezler, S., King, T.H., Crouch, R., Zaenen, A.: Statistical Sentence Condensation Using Ambiguity Packing and Stochastic Disambiguation Methods for Lexical-Functional Grammar. In: Proceedings of the 2003 Conference of the North American Chapter of the Association for Computational Linguistics on Human Language Technology, NAACL 2003, Edmonton, Canada, vol. 1, pp. 118–125 (2003)
25. Clarke, J., Lapata, M.: Models for Sentence Compression: A Comparison Across Domains, Training Requirements and Evaluation Measures. In: Proceedings of the 21st International Conference on Computational Linguistics and the 44th annual meeting of the Association for Computational Linguistics, ACL-44, Sydney, Australia, pp. 377–384 (2006)
26. Marneffe, M.C.D., Manning, C.D.: The Stanford Typed Dependencies Representation. In: Proceedings of the Workshop on Cross-Framework and Cross-Domain Parser Evaluation, CrossParser 2008, Manchester, pp. 1–8 (2008)
27. Copeck, T., Inkpen, D., Kazantseva, A., Kennedy, A., Kipp, D., Szpakowicz, S.: Catch What You Can. In: Proceedings of Document Understanding Conference (DUC 2007), Rochester, New York, USA (2007)
28. Pingali, P.: K, R., Varma, V.: IIIT Hyderabad at DUC 2007. In: Proceedings of the HLT-NAACL 2007 Document Understanding Workshop, Rochester, New York (2007)

Automatically Assessing Children's Writing Skills Based on Age-Supervised Datasets

Nelly Moreno, Sergio Jimenez, and Julia Baquero

Universidad Nacional de Colombia, Bogotá
{nemorenoc,sgjimenezv,jmbaquerov}@unal.edu.co

Abstract. In this paper, we propose an approach for predicting the age of the authors of narrative texts written by children between 6 and 13 years old. The features of the proposed model, which are lexical and syntactical (part of speech), were normalized to avoid that the model uses the length of the text as a predictor. In addition, the initial features were extended using n-grams representations and combined using machine learning techniques for regression (i.e. SMOreg). The proposed model was tested with collections of texts retrieved from Internet in Spanish, French and English, obtaining mean-absolute-error rates in the age-prediction task of 1.40, 1.20 and 1.72 years-old, respectively. Finally, we discuss the usefulness of this model to generate rankings of documents by written proficiency for each age.

1 Introduction

Children's writing skills increase as they progress in their process of language learning and education. It is commonly accepted that older children write richer and more complex texts than their younger counterparts. However, for any age (e.g. age 8) it is common to observe individuals with big differences in their writing skills. This paper explores the use of an age-prediction model, on the basis of features extracted from texts written by children, to make inferences about their written proficiency. In particular, deviations from this model were used to identify differences in the level of proficiency of individuals.

The development of children's writing skills has been studied considering different levels of analysis. For instance, the evolution of the superstructure of narrative texts has been analyzed [5,12,23] as other approaches consider macrostructure and coherence [23]. Similarly, other features related to the text microstructure, cohesion and syntactic complexity have also been considered [1,5,10].

The manual process of assessing the written production of students is a costly and time-consuming task. Moreover, human cognitive limitations and other psychological factors affect assessments compromising quality and objectivity. Only recently, the need to assist this process with automated tools has been recognized [8]. Current approaches for automatic evaluation of text employ a variety of techniques from natural language processing and machine learning fields. Common methods use features including bag-of-words representations, stylistic features

A. Gelbukh (Ed.): CICLing 2014, Part II, LNCS 8404, pp. 566–577, 2014.

[4], part-of-speech taggers [20,3], syntactic parsers and latent semantic analysis [14,13], among other approaches. Some practical applications of these methods are automatic scoring of essays and written short answers [13].

One of the causes that have limited the widespread use of these tools is their cost. For example, rule-based systems require careful tuning and might frequently need adjustment. Similarly, approaches that use supervised machine learning techniques require a considerable amount of good quality training data to obtain suitable models. These issues have been addressed in other fields by using unlabeled data, cheap sources of supervision or even noisy data. Our approach uses the age of the children as a source of low-cost supervision to build a model for written proficiency assessment.

The proposed method is inspired in a recommender system proposed by Becerra et al. [2], which is able to identify products that offer to users high value for a low price. The approach consists of learning with a regression model to predict the product's price, and deviations of particular products are used to identify interesting outliers. Thus, a product with a large (positive) difference between its predicted and actual price is considered a "good deal" and vice versa. That system, rather than relying on feedback from users, uses the price of the product (which is an inexpensive class variable) to provide predictions of value for the users. That model exploits the average trend according to which more expensive products offer more value to the users. Analogously, we used the age of the texts' authors for assessing written proficiency, as Becerra et al. used products' price for assessing product value.

Recently, the task of predicting author's age from text was proposed, at a relatively large scale, in the past PAN workshop [18]. Most of the proposed approaches used combinations of various lexical, syntactic and stylistic features [7,15]. Similarity, our model followed the approach of Nguyen et al. [16] by combining these features using a regression model.

For experimental evaluation, we gathered from the Internet three collections of documents in Spanish, French and English, which contain narrative texts written by children of different ages. Results obtained using several combinations of text representations, feature sets and regression algorithms were reported and discussed. Although there is considerable variation in the writing skills among children of the same age in our data, the average trend in the proposed model obtained a fairly good correlation between predictions and actual ages.

In addition to the age-prediction experiments, we conducted a preliminary study with volunteers which we called "toy-study". There, we asked participants to predict the age of the authors of a set of narrative texts written by children in order to assess the difficulty of the task for humans. Finally, we reviewed texts that had large differences between their predictions and actual author's age, concluding that these differences provide promising clues for assessing written proficiency of their respective authors. For that, a few illustrative examples are provided and discussed.

2   Predicting Children's Age from Text

2.1   A "Toy-Study"

As an initial approach to the problem of predicting the author's age from text, the following brief study to assess the difficulty of the task for humans was performed. First, we retrieved 58 tales, written in Spanish by children, from a website[1] each one labeled with its author's age. Figure 1 shows the distribution of the texts by age and length. In addition, black squares in that figure represent the average text length for each age showing that the data follows the expected pattern of longer texts for older ages. Second, a subset of 8 texts (1 for each age) was selected trying to avoid that the length of the text could be used as a hint to determine the age of the author (see the dashed line in Figure 1). We prepared a form with those texts removing the author's age and presenting the texts in the following age order: 10, 9, 7, 6, 11, 13, 8 and 12. Finally, 9 volunteers were asked to read the texts and assign the author's age for each one being informed that each text correspond to an (exact) age between 6 and 13 years-old and that there are not repeated ages. All individuals were native Spanish speakers, 6 of them have background in linguistics and 3 in computer science. The obtained answers are shown in Table 1.

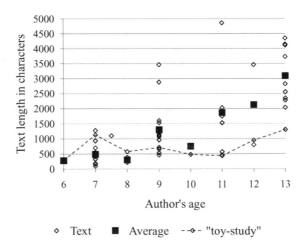

Fig. 1. Length of texts in the "toy-study"

For comparison, we randomly generated 100 sets of responses obtaining a mean absolute error (MAE) of 2.53 years-old on average (baseline), with a standard deviation of 0.68. Note that the value obtained with the same measure by the 9 participants was 2.68 (0.52), which is higher than the error rate of the random baseline. As it can be seen in Table 1, only 4 out of 9 individuals got a

---

[1] From http://www.leemeuncuento.com.ar retrieved in September 2010.

Table 1. "Toy-study" answers sheet (units for all data are 'years old')

	text 1	text 2	text 3	text 4	text 5	text 6	text 7	text 8	†MAE
individual #1	6	9	11	13	10	12	7	8	2.75
individual #2	6	12	13	10	8	11	7	9	3.25
individual #3	9	10	12	8	6	13	7	11	2.00
individual #4	6	7	9	12	11	8	10	13	2.75
individual #5	6	7	13	10	9	12	8	11	2.50
individual #6	6	7	11	13	8	9	10	12	3.25
individual #7	8	10	13	9	6	12	7	11	2.50
individual #8	7	6	12	13	8	10	9	11	3.25
individual #9	8	7	12	6	8	13	9	10	1.88
average	6.89	8.33	11.78	10.44	8.22	11.11	8.22	10.67	2.68
std. deviation	1.17	2.00	1.30	2.51	1.64	1.76	1.30	1.50	0.52
real author's age	**10**	**9**	**7**	**6**	**11**	**13**	**8**	**12**	**0.00**

† Henceforth, MAE stands for mean-absolute error

lower error rate than the baseline. These results suggest that the prediction of authors age based on the text seems to be a difficult task for humans.

2.2   Regression for Age Prediction

In the past, the age prediction task based on text has been addressed mainly using classification models [17]. However, the use of regression models for this task has been only recently proposed by Nguyen et al. [16]. In this approach, age is considered as a continuous and ordinal variable. Our approach is similar to the method proposed by them, in which texts are represented in a vector space model indexed by words, POS tags and bigrams of POS tags.

Generalization is a desirable factor for any prediction model, and in particular, for a regression model for the age prediction task. An alternative to achieve a good degree of generalization is to use simple linear regression models. Although, these minimum-error models provides good interpretability, other discriminative models such as support vector machines [6] have shown to provide better generalization conditioned on a good regularization and low model complexity. For our age prediction model we used a simple linear model and SMO reg. [22,21], which is an adaptation of the well-known SVM model for regression.

In order to measure the amount of overfitting we should expect when training the model with and unseen dataset, the generalization property of the model was assessed by doing random divisions into train and test datasets. A model that learns an effective regularized function from the entire dataset should have a lower error rate than one that only uses a portion for training. In addition, several random splits have to be made to assess the statistical significance of the difference in error rate among the two regression models. In our experiments, data was divided into 66.6% for training and 33.3% for testing in 30 different random splits.

Another issue that concerns generalization is the fact that representations based on words usually involve high-dimensional sparse feature spaces. Although, texts written by children usually contain less than 500 words (see Figure 2), the vocabulary size reaches thousands of words. Using the vector space model, each text is represented as a vector with zeroes in the majority of dimensions. If the number of texts is considerably smaller than the size of the vocabulary, this high-dimensional space leads to overfitted regressors. For instance, in a short text with a rich vocabulary, which isn't common to the rest of the texts, the regression model fits the age prediction due to the occurrence of low-frequency words. This problem can be alleviated by dimensionality reduction techniques such as latent semantic analysis (LSA) [9] or mapping the words to a smaller set of features such as POS tags. In our experiments, we have trained regressors in the vector space model [19] indexed by words or POS tags and in reduced latent semantic spaces as well.

Regarding the distribution of texts per age in the dataset, a uniform distribution is preferable. In our experiments, we provided one balanced dataset (English) and two unbalanced datasets (Spanish and French). Particularly, in the French dataset, 70% of the texts belong to ages between 7 and 8. Thus, a very low error rate can be achieved predicting 7.5 years-old for all instances. Finally, another important factor that must be considered is the fact that the length of the texts is highly correlated with the author age (see Figures 1 and 2). For the task at hand, it is mandatory to normalize all instances in order to avoid misleading predictions regarding this issue. Therefore, the vectors associated to all the used feature sets were scaled in a way that all of them have a Euclidean norm of 1.

3   Experimental Evaluation

The aim of the proposed experiments is to determine to what extend the age of a child can be determined from his (or her) written production. To do this, we tested various configurations of document collections, feature sets and regression algorithms

3.1   Datasets

The data consist of texts labeled with the age of their authors, which were retrieved from different Internet sources (see Table 2). The number of texts for each language is: 484 texts for Spanish, 662 for French and 1,800 for English. The distributions of the number of texts for each age in the datasets in Spanish and French are shown in Figure 2. The English dataset is balanced because it contains 200 texts for each age between 7 and 15 years old. The average length of the texts and their standard deviations are also depicted in Figure 2. Regarding the genre of the texts, the Spanish dataset consists of narrative texts, while the French and English datasets also include descriptive, expository and argumentative texts. Some automatic and manual preprocessing tasks were performed in order to clean up the texts by removing html tags, website names, author's age and other irrelevant information.

Table 2. Languages and source URLs of the collected texts

Language and source URL	Retrieved in
Spanish http://www.pekegifs.com	Jan 2011
Spanish http://www.pequelandia.org	Jan 2011
Spanish http://www.leemeuncuento.com.ar	Nov 2010
Spanish http://www.puroscuentos.com.mx	Jan 2011
Spanish http://www.escritoresninyosyjovenes.es.tl	Feb 2011
Spanish http://www.elhuevodechocolate.com	Jan 2011
Spanish http://www.morellajimenez.com.do	May 2011
Spanish http://www.encuentos.com	May 2011
French http://www.kidadoweb.com	Feb 2011
English http://www.edbydesign.com/	Jun 2011

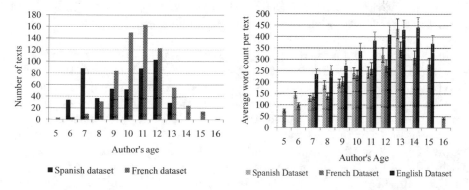

Fig. 2. Number of texts and text length histograms for all datasets

3.2   Features

Lexical and syntactic features were extracted from the texts to produce different representations of the texts using words or POS tags provided by automatic tools such as TreeTagger (Spanish, English and French) [20] and Freeling (Spanish and French) [3]. Additional representations were obtained generating bigrams of words or POS tags. Subsequently, these representations were processed by the filter *StringToWordVector*, provided by WEKA [11], using vector normalization and converting all characters to lowercase. Also, alternative sets of vectors were obtained either by selecting the options 'occurrences' or 'frequencies' in the *StringToWordVector* filter, which means that the occurrence of a feature is represented by a binary indicator variable or the number of times that this feature occurs. Another text representations were provided by LSA [9] and a set of stylistic features (see De-Arteaga et al. [7]).

The naming convention to identify each feature set is as follows. The prefixes 'word', 'TT' or 'FL' identify representations of words and POS tags provided by TreeTagger or Freeling respectively. If these representations are transformed by bigrams or LSA, then the name includes '2g' or 'lsa' respectively. To identify the usage of occurrences or frequencies, an 'o' or 'f' are added to the name. Finally, if the feature set includes the set of stylistic features, then the suffix 's' is added. For instance, 'TT.2g.o.s' represents a feature set of bigrams of POS tags provided by TreeTagger using occurrences and including the stylistic feature set.

3.3   Regression Models

The different sets of vectors, presented in the previous subsection, were used to build regression models to predict the age of the author for an unseen text. Three regression models were used in our experiments. First, decision stump, a "weak learner", which is a single-level decision tree adapted for regression. Second, the simple linear regression, which is the classic least squares estimator for a linear model. Finally, the SMO.Reg model, which is an adaptation of the popular SVM algorithm for regression. For the latter, we use a Gaussian kernel with $\gamma = 0.01$ and the complexity parameter $C = 1.0$.

The measures employed to evaluate the performance of the regression model were the Pearson correlation coefficient (r) and mean-absolute error (MAE). Two baselines are provided using these measures (see Table 3). The first, labeled "age average", is a model that always predicts the average age of the on each dataset. In the second, we produced 100 sets of predictions by generating random predictions of age with the same distribution as the data for each language. The average from these 100 runs and standard deviation is reported. Given that the English dataset has a uniform age distribution, only the first baseline is provided for this dataset. As expected, both baselines obtained a value for r very close to 0.

Table 3. Baselines for experiments

	Age average		Random 100 runs	
	r	MAE	r	MAE
Spanish	0.00	1.91	0.00	2.45(0.07)
French	0.00	1.34	0.00	1.88(0.05)
English	0.00	2.22	na	na

3.4   Results

Table 4 summarizes the results of all the experiments that were carried out using the data mining framework WEKA [11]. Columns "Feat. set" and "#Feat." provide the name of the feature set and the number of these. For the three regression algorithms (decision stump, simple linear regression and SMO.Reg), the average results of 30 runs corresponding to random training-test divisions (66.6%-33.3%) of the datasets are reported. Results marked with * are significantly better than the result obtained by decision stump, using the paired T-test with $p < 0.05$.

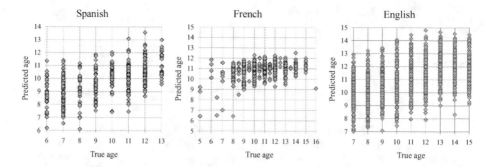

Fig. 3. Scatterplots for all datasets

It can be seen that, in general terms, the SMO.Reg regressor performs better than the other algorithms. Regarding the choice of occurrence/frequency, Table 4 shows that the best results (in bold) were obtained using binary occurrences in all languages. Moreover, there are not major differences between the results obtained by using TreeTagger and Freeling taggers. It is important to note that the Spanish dataset was the only one that obtained the best results using POS tags instead of words. We hypothesize that this result is related to the number of texts on each dataset (Spanish is the smallest). It can also be noted that in all datasets the LSA representation performed poorly in comparison with the simple word representation. Finally, the transformations using bigrams of words and POS tags do not obtain consistent improvements in comparison with using unigrams. Comparing the baselines in Table 3 versus the results in Table 4, it is clear that the results obtained by SMO.Reg with the best performing feature sets ('word.o' and 'TT.o') overcame, with a wide margin, the proposed baselines.

Figure 3 shows scatter plots for each language contrasting the observed and predicted ages obtained with the 'TT.o' feature set and SMO.reg model. The Spanish and English models show a clear upward trend while the French model does not reflect this pattern.

3.5   Some Examples

In this section we present two Spanish examples with large differences between real and predicted age. The first example correspond to a text with a predicted age lower than the actual author age. This text shows less complex syntax, punctuation and co-reference mechanisms. Those features reflect less developed written skills that impede to produce a cohesive and coherent text. The opposite situation happens in the second example when predicted age is higher than the true age.

Example #1: "EL HOMBRE BOMBILLA: Había una vez **un hombre bombilla**, **él** era gordo y con mucho cabello, **él** vivía a las orillas de *un gran río*, el hombre bombilla todos los días iba a bañarse al *río el río* tenia las aguas tan frescas que **el hombre bombilla** se sentía atraído

Table 4. Age prediction results for all datasets. Results marked with * were significantly better in comparison to *decision stump*.

†Feat. set	#Feat.	Decision Stump		Simple Linear Reg.		SMO Reg.	
		r	MAE	r	MAE	r	MAE
Spanish							
word.o	5,892	0.35(0.06)	1.73(0.07)	0.42(0.05)	1.69(0.06)	0.48(0.05)*	1.59(0.06)*
word.lsa	386	0.35(0.06)	1.74(0.06)	0.46(0.05)*	1.65(0.07)	0.41(0.05)	1.70(0.07
word.2g.o	10,691	0.30(0.06)	1.77(0.07)	0.24(0.07)	1.83(0.06)	0.55(0.04)*	1.54(0.06)*
TT.o	63	0.30(0.06)	1.79(0.06)	0.26(0.05)	1.81(0.06)	**0.61(0.04)***	**1.45(0.07)***
TT.2g.o	1,433	0.33(0.06)	1.75(0.07)	0.35(0.06)	1.74(0.07)	0.49(0.03)*	1.62(0.06)*
TT.o.s	82	0.37(0.05)	1.72(0.07)	0.43(0.04)	1.69(0.05)	0.50(0.04)	1.56(0.06) *
FL.o	3,845	0.32(0.06)	1.75(0.08)	0.45(0.04)*	1.67(0.06)*	0.48(0.04)*	1.58(0.06)*
FL.2g.o	4,082	0.35(0.05)	2.16(0.07)	0.44(0.05)*	2.06(0.07)*	0.48(0.04)*	1.94(0.07)*
French							
word.o	6,894	0.10(0.07)	1.35(0.07)	0.03(0.05)	1.36(0.07)	**0.36(0.04)***	**1.23(0.06)***
word.lsa	507	0.06(0.06)	1.35(0.07)	0.09(0.07)	1.34(0.06)	0.19(0.04)*	1.36(0.06)
TT.o	46	0.08(0.07)	1.33(0.07)	0.10(0.07)	1.32(0.06)	0.27(0.06)*	1.29(0.06)*
TT.f	46	0.12(0.05)	1.33(0.07)	0.11(0.06)	1.33(0.07)	0.28(0.04)*	1.27(0.06)*
TT.2g.o	818	0.13(0.06)	1.35(0.06)	0.09(0.05)	1.35(0.06)	0.29(0.05)*	1.34(0.06)*
TT.2g.f	818	0.13(0.06)	1.35(0.06)	0.07(0.06)	1.35(0.07)	0.30(0.05)*	1.28(0.05)*
TT.o.s	55	0.17(0.07)	1.34(0.06)	0.23(0.06)	1.31(0.06)	0.24(0.05)	1.32(0.05)
English							
word.o	6,478	0.57(0.02)	1.72(0.04)	0.50(0.04)*	1.85(0.04)	**0.77(0.01)***	**1.32(0.03)***
word.lsa	1,098	0.34(0.03)	2.04(0.05)	0.37(0.03)	2.01(0.04)	0.60(0.02)*	1.70(0.04)*
TT.o	84	0.42(0.03)	1.96(0.04)	0.35(0.05)	2.06(0.04)	0.49(0.03)	1.84(0.04)*
TT.2g.o	2146	0.56(0.03)	1.74(0.04)	0.48(0.04)*	1.90(0.03)	0.66(0.02)	1.54(0.4)*
TT.o.s	72	0.57(0.02)	1.71(0.04)	0.35(0.06) *	2.02(0.04)	0.52(0.02) *	1.80(0.04)

† o: occurrence, f: frequency, TT: TreeTagger, FL: Freeling, 2g: bigrams, s: stylistic features

con las aguas del gran río. Un día amaneció sucio y contaminado, **el hombre bombilla** se echó *al río* sin darse cuenta que estaba sucio, para el otro día **el hombre bombilla** amaneció enfermo que nunca **le** dieron ganas de ir a bañarse, de miedo de encontrar la muerte en el *rio sucio* y contaminado." [true age 11, predicted age 7.39]

Example #2: "**El niño** que no obedecía: Erase una vez **un niño** que no obedecía a sus padres y siempre se perdía. Sus padres **le** advirtieron que no fuera al bosque pero **él** como siempre no les hizo caso y se perdió. Cuando se había divertido mucho se dio cuenta que se había perdido y entonces grito, !PAPA! !MAMA!. y se sentó en *un árbol* y lloro y lloro. Admitió que se portó mal. Entonces *alguien* **le** contestó. **Le** dijo !Por fin lo has admitido!. *¿Quién* ha dicho eso dijo **el niño**?. He sido *yo*, he sido *yo*. y quien eres *tú*? Soy *yooooooo* el árbol. Y tu cómo te llamas *señor árbol*? Sooooy *el árboool de laaa verdaaad*. *Árbol de la verdad que nunca mientes*, ¿**me** puedes decir donde están mis padres? No **te** acuerdas que hoy **os** ibais a mudar? AH! sí. Pero qué podré hacer para encontrarlos?

Pues supongo que tendrás que viajar. Bien veamos.... La mudanza era a Francia. Paso un día, ando y ando hasta que cayó en las arenas de una playa y dijo **el niño**: ¿Qué es eso que brilla en el suelo?. Resulto que eran diamantes. Por muy joven que sea ya era rico y por los carteles que enseñaban de *él*, al final se dieron cuenta **sus** padres y gracias a los carteles pudo regresar con **sus** padres y les dijo: !PAPA! !MAMA! os prometo que de ahora en adelante os obedeceré y fue todo gracias *al árbol de la verdad*." [true age 7, predicted age 11.91]

In these examples, we can see that the author of the second text uses, in a greater extent, references mechanisms, punctuation marks, and different inflected forms of verbs. With regard to reference mechanisms, in the first example the author uses, frequently, noun phrases for reference and co-reference. Note how the author refers to entities "el hombre bombilla" (in bold) and "el rio" (in italics): mainly referred through noun phrases. In the second example, in addition to noun phrases, the author uses another reference mechanisms to refer to the entities "el niño" (in bold) and "el árbol" (in italics) such as pronominal phrases with personal pronouns (yo, él and tu); object pronouns (le, te and me); indefinite pronouns (alguien) and interrogative pronouns (quién). On the other hand, in the second example we can see a wider range of punctuation marks. For instance, in addition to comma [,] and full stop [,] used in the first text, the second author uses question and exclamation marks [¿?¡!], colon [:] and ellipsis [...]. Finally, let us consider the verbs used in the two texts. In the second one, unlike the first, based on TreeTagger and Freeling tags, it is possible to observe a wider morphological paradigm which includes non-finite verbs (participle and infinitive), and finite forms (principal, auxiliary and modal verbs) in different tense (past, present, future) and mood, and verbal periphrasis (haber+participle). In conclusion, the written proficiency differs importantly and the divergences against the age predicted by the model clearly reveal this situation.

4   Discussion

The results shown in the Table 4, the scatter plots in Figure 3 and the examples in subsection 3.5 indicate that our method can identify patterns which make possible to predict the authors' age. Scatter plots show, qualitatively, that the best results were obtained for the Spanish dataset in which it can observe a clear correlation between true and predicted age. Interestingly, for the three datasets, the linear pattern increasing between true and predicted age does not hold for ages above 12. These graphs also show the rankings obtained for each age. In the French dataset, regression models have difficulty finding appropriate patterns, probably, due to the distribution of the ages of the authors, which is highly concentrated in the range of 9 to 12 years (see Figure 2). This highlights the importance of a balanced dataset for obtaining meaningful models and results.

The examples analyzed in subsection 3.5 show that the deviations of the true age and the proposed model manage to reveal either good or poor writing skills according to the signs of these deviations. Although it is not possible to make a general statement based on the observation of these data issues, this result is a promising research direction. Clearly, in order to probe the hypothesis that the deviations from a prediction model of age are predictors of written proficiency, these deviations should be proved and correlated with a gold standard built on proficiency assessments.

5 Conclusion

An age prediction model for texts written by children between 6 and 13 years old was proposed using words and POS tags as features (unigrams and bigrams) and a regression model for combination. The proposed method obtained considerably better correlations and lower error rates than the plausible baselines. In addition, we showed that humans perform poorly (close to a random baseline) in the age-prediction task from text in a much simpler scenario compared with the experimental setup used for the evaluation of our model. Finally, we showed that deviations of the true age against the predictions provided by our model are useful for assessing the level of written proficiency for children.

References

1. Aguilar, C.: Análisis de frecuencias de construcciones anafóricas en narraciones infantiles. Estudios de Lingüística Aplicada 22(38), 33–43 (2003)
2. Becerra, C., Gonzalez, F., Gelbukh, A.: Visualizable and explicable recommendations obtained from price estimation functions. In: Proceedings of the RecSys 2011 Workshop on Human Decision Making in Recommender Systems (Decisions@RecSys 2011), Chicago, IL, pp. 27–34 (2011)
3. Carreras, X., Chao, I., Lluis, P.: FreeLing: an open-source suite of language analyzers. In: Proceedings of the 4th International Conference on Language Resources and Evaluation (LREC 2004), Barcelona, Spain (2004)
4. Cheng, N., Chandramouli, R., Subbalakshmi, K.: Author gender identification from text. Digital Investigation 8(1), 78–88 (2011)
5. Colletta, J.-M., Pellenq, C., Guidetti, M.: Age-related changes in co-speech gesture and narrative: Evidence from french children and adults. Speech Communication 52(6), 565–576 (2010)
6. Cortes, C., Vapnik, V.N.: Support-vector networks. Machine Learning 20(3), 273–297 (1995)
7. De-Arteaga, M., Jimenez, S., Dueñas, G., Mancera, S., Baquero, J.: Author profiling using corpus statistics, lexicons and stylistic features. In: Online Working Notes of the 10th PAN Evaluation Lab on Uncovering Plagiarism, Authorship. and Social Misuse, CLEF 2013, Valencia, Spain (September 2013)
8. Dikli, S.: An overview of automated scoring of essays. Journal of Technology, Learning and Assessment 5(1) (August 2006)
9. Dumais, S.: Latent semantic analysis. Annual Review of Information Science and Technology 38(1), 188–230 (2004)
10. Furman, R., Özyürek, A.: Development of interactional discourse markers: Insights from turkish children's and adults' oral narratives. Journal of Pragmatics 39(10), 1742–1757 (2007)

11. Hall, M., Eibe, F., Holmes, G., Pfahringer, B.: The WEKA data mining software: An update. SIGKDD Explorations 11(1), 10–18 (2009)
12. Ilgaz, H., Aksu-Koç, A.: Episodic development in preschool children's play-prompted and direct-elicited narratives. Cognitive Development 20(4), 526–544 (2005)
13. Kakkonen, T., Myller, N., Timonen, J., Sutinen, E.: Automatic essay grading with probabilistic latent semantic analysis. In: Proceedings of the Second Workshop on Building Educational Applications Using NLP, EdAppsNLP 2005, pp. 29–36. Association for Computational Linguistics, Stroudsburg (2005)
14. Landauer, T.: Pasteur's quadrant: Computational linguistics, LSA, and education. In: Proceedings of the HLT-NAACL 2003 Workshop on Building Educational Applications Using Natural Language Processing, Edmonton, Canada (2003)
15. López-Monroy, A.P., Montes-y Gómez, M., Escalante, H.J., Villaseñor-Pineda, L., Villatoro-Tello, E.: INAOE's participation at PAN'13: author profiling task notebook for PAN at CLEF 2013. In: Online Working Notes of the 10th PAN Evaluation Lab on Uncovering Plagiarism, Authorship. and Social Misuse, CLEF 2013, Valencia, Spain (September 2013)
16. Nguyen, D., Smith, N.A., Rosé, C.P.: Author age prediction from text using linear regression. In: Proceedings of the 5th ACL-HLT Workshop on Language Technology for Cultural Heritage, Social Sciences, and Humanities, LaTeCH 2011, pp. 115–123. Association for Computational Linguistics, Stroudsburg (2011)
17. Pennbaker, J.W., Stone, L.D.: Words of wisdom: Language use over the life span. Journal of Personality and Social Psychology 85(2), 291–301 (2003)
18. Rangel, F., Rosso, P., Koppel, M., Stamatatos, E., Inches, G.: Overview of the author profiling task at PAN 2013. In: Online Working Notes of the 10th PAN Evaluation Lab on Uncovering Plagiarism, Authorship. and Social Misuse, CLEF 2013, Valencia, Spain (September 2013)
19. Salton, G., Wong, A.K.C., Yang, C.-S.: A vector space model for automatic indexing. Commun. ACM 18(11), 613–620 (1975)
20. Schmid, H.: Probabilistic part-of-speech tagging using decision trees. In: Proceedings of International Conference on New Methods in Language Processing, Manchester, UK (1994)
21. Shevade, S., Keerthi, S., Bhattacharyya, C., Murthy, K.: Improvements to the SMO algorithm for SVM regression. IEEE-NN 11(5), 1188–1193 (2000)
22. Smola, A.J.: Learning with Kernels. GMD Forschungszentrum Informationstechnik, Sankt Augustin (1998)
23. Stadler, M.A., Ward, G.C.: Supporting the narrative development of young children. Early Childhood Education Journal 33(2), 73–80 (2005)

Author Index